The Macmillan Encyclopedia of
Architecture and
Technological Change

The Macmillan Encyclopedia of
Architecture and
Technological Change

Pedro Guedes

M

© 1979 Reference International
Publishers Ltd.

This book was created, designed, and produced by
Reference International, 21 Soho Square, London W1.
The managing editor for this book was Sandra Shaw.
Advisory editor (Building Types) – Peter Allison.
Picture researcher – Andrew Higgott. Special
consultant – Gontran Goulden. The designer was
Julian Holland and the Production Supervisor was
John Cleary.
This book was set in 9 on 10 Times.

First published in Great Britain
1979 by
THE MACMILLAN PRESS LTD
London and Basingstoke
Associated companies in *Delhi,
Dublin, Hong Kong, Johannesburg,
Lagos, Melbourne, New York,
Singapore and Tokyo*

British Library Cataloguing in
Publication Data

Guedes, Pedro
  The Macmillan encyclopedia of
  architecture and technological
  change.
  1. Architecture   2. Building
  I. Title   II. Encyclopedia of
  arch and technological change
  720   NA200

ISBN 0-333-26766-4

*Printed and bound at
William Clowes & Sons Limited
Beccles and London*

Contributors

Introduction by Adolf K. Placzek

Paul Ahm
Alan J. Berman
Michael Brawne
Theo Crosby
Ian Davis
Jolyon V.P. Drury
Kit Evans
Robin Evans
Clare Frankl
Gontran Goulden
Pedro Guedes
Roderick Ham
Cecil C. Handisyde
Dr. Dean Hawkes
Kenneth Hudson

Susan Jellicoe
Dr. Ronald B. Lewcock
Dr. Rowland J. Mainstone
Edward D. Mills
Andrew Rabeneck
Timothy Ronalds
Ivor Samuels
Dennis Sharp
William J. R. Smyth
Peter Stone
John Weeks
Dr. Trevor I. Williams
John Winter
John Worthington
Anthony J. Wylson

The editor would like to acknowledge the
contribution of the late Duccio A. Turin,
Professor of Building at University
College, London, who saw the need for a
book of this kind and guided the early stages
of its development.

Contents

Introduction   8

SECTION 1
Stylistic Periods and Geographical Adaptations

Architecture Adapted to
Climate   12
Primitive Architecture   14
Mesopotamian and Iranian
Architecture   21
Egyptian Architecture   23
Pre-Columbian Architecture   24
Ancient Greek Architecture   26
Roman Architecture   28
Early Christian and
Byzantine Architecture   30
Romanesque Architecture   32
Gothic Architecture   34
Renaissance Architecture   35
Baroque Architecture   37
Rococo Architecture   38
Neo-Classical Architecture   39
Romantic Architecture   40
Islamic Architecture   41
Japanese Architecture   43
Chinese Architecture   44
Indian Architecture   45
Modern Isms   46

SECTION 2
Built Forms and Building Types

LANDSCAPES
Private Gardens and Parks   56
Urban Open Space   59
Specialized Landscapes   62
TRANSPORTATION
Bridges   64
Railroad Stations   68
Airports   70
RESIDENTIAL
Houses   76
Apartments   86
Palaces   88
Hotels   91
INDUSTRIAL
Factories   94
Warehouses   100
COMMERCIAL AND
ADMINISTRATIVE
Offices   106
Skyscrapers   109
Shops, Stores, and
Shopping Centers   111
Banks   114
Exhibition Buildings   116
GOVERNMENT
Civic Buildings   120
EDUCATIONAL AND
RESEARCH
Schools   124
Universities   126
Museums   129
Laboratories   132
ENTERTAINMENT
AND RECREATION
Theaters   136
Movie Theaters   138
Sports Buildings   141
INSTITUTIONAL
Hospitals   146
Prisons   150
DEFENSE, EMERGENCY,
AND PORTABLE
BUILDINGS
Fortifications   154
Emergency Buildings   161

SECTION 3
Structures – Ideas, Elements, Structural Systems, and the Processes of Erecting Buildings

STRUCTURAL DESIGN
AND ITS BASES
Structural Theory   166
Statics and Dynamics   166
Deformation and Strength   167
Design Criteria   168
Structural Design   169
STRUCTURAL
ELEMENTS
Arches   171
Beams and Slabs   172
Columns, Piers, and Walls   173
Domes and Related
Elements   174
Floor Systems   175
Foundations   176
Rigid Frames   177
Shells   177
Trusses and Space Frames   178
Suspension Elements,
Tensile Membranes, and
Cable Nets   180
Vaults   180
STRUCTURAL SYSTEMS
Early Forms   182
Later Wide-Span Buildings:
Timber-, Concrete-, and
Masonry-Roofed Systems   184
Iron-, Steel-, and Reinforced-
Concrete-Roofed
Systems   186
Later Multistorey
Buildings:
Bearing-Wall Systems   188
Hybrid Systems with
Partial Timber or Iron
Framing   189
Fully Framed Systems of
Iron, Steel, and
Reinforced Concrete   190
Recent Hybrid Systems
for Tall Buildings   191
Protection Against Fire,
Earthquake, and Other
Hazards   192
MEANS OF BUILDING
AND CONSTRUCTION
PROCESSES
Construction Plant   193
Measuring Equipment   194
Scaffolding, Centering,
and Formwork   194
Construction Processes:
Problems Associated
with the Incompleteness
of the Structure during
Construction   196
Prestressing   197
From Prefabrication to
Industrialization and
Systems Building   197

SECTION 4
Services, Mechanical and
Environmental Systems

BEFORE THE INDUSTRIAL
REVOLUTION
(Minimal Servicing — The   200
  Age of Wood and Water)
THE 19th CENTURY   202
(The Age of Steam and Gas)
Assembly Buildings   202
Hospitals   204
Offices   206
Schools   208
Domestic Buildings   209
THE 20th CENTURY   212
(The Electric Age)
Assembly Buildings   213
Hospitals   215
Offices   217
Schools   220
Domestic Buildings   222
THE RESPONSIBLE
AGE   224

SECTION 5
Building Materials

Timber   228
Plywood and other
Wood-Based Sheet Material   234
Stone   236
Earth in various forms
(cob, pisé, adobe, wattle
and daub)   241
Brickwork   143
Terra-cotta   246
Roofing Materials and   247
Tiles
Plaster   251
Mortar and Cement   252
Concrete   253
Asbestos   263
Iron   263
Steel   272
Copper   277
Brass   278
Zinc   279
Bronze   279
Lead   280
Aluminum   281
Glass   282
Paint   285
Bitumen and Asphalt   286
Plastics   286

SECTION 6
Tools, Techniques, and Fixings

TOOLS   290
Carpentry   290
The Hammer   291
Axes and Adzes   291
The Plane   292
The Saw   293
The Drill   293
Masonry   294
Glazing and Sealing   295
Finishing and Ancillary
Trades   295
Plastering   295
Painting   295
The Blacksmith   296
Powered Tools   296
Surveying Instruments   297
TECHNIQUES   298
Casting and Molding   299
Extrusion   300
Carving   301
Cutting   301
Drilling   302
Grinding   302
Milling   303
Turning   303
Rolling   304
Forging   305
Bending   305
Stamping   305
JOINTS AND FIXINGS   306
Lashings   306
Homogeneous Joints for
Wood   306
Masonry Joints   307
Nails   308
Screws and Bolts   309
Riveting and Welding   309
Adhesives and Sealing
Compounds   312
Miscellaneous Fixings and
Fastenings   313

Index   314

Introduction

Architecture—not a survey of styles, but of its living, material reality—is the subject of this book: architecture as it is now, and how it has become what it is now. Architectural technology, architectural sociology, and architectural development are the subject.

Enormous forces have changed the face of the earth and its man-made structures. Mass production, electric power, population explosion, wealth, sophisticated machinery—all this is ushering in a new architecture the total power of which we have barely begun to understand. The eternal question of what comes first arises again. Is it the need for a mass architecture which comes first and the vast technology which follows? Or is it the new technology which produces this architecture? Or is it the great increase in human numbers which brings about both the technology and its built results? Does form really follow function, or does it make the function possible? A form that is made possible by new matter? And so, is not architecture the result of a complex and profound reciprocal effect between needs and manifold powers?

The following pages will again, for our time, try to answer these fundamental questions. Encyclopedic surveys of architecture are nothing new; it seems that every age wants its own. In this sense we all go back to Vitruvius, that practical and practicing Roman architect of the 1st century BC who wrote the encompassing great survey of architecture—at least the first such survey preserved in its entirety. The Greeks produced theirs before him, but we have only Vitruvius' word for it; the books are lost. It is, however, clear that the great buildings of Greece were not built without books initiating or interpreting them. From Vitruvius onward, many architects and their works can be mentioned. There is a big jump to Leone Battista Alberti's treatise in the 15th century, a straight line from this to the works of Sebastiano Serlio and Andrea Palladio in the 16th century, then on to Jacques François Blondel's in the 18th century, to Julien Guadet's *Elements et théorie de l'architecture* in the 19th century, and then, in the U.S., to Talbot Hamlin's *Forms and functions of 20th-century architecture*. All these works have one thing in common;

they combine the practical, even the technical, with the conceptual and historical. They look at architecture as a whole, with all its ramifications, causes, and effects. It is in the intellectual tradition of these great works that this book aims to continue.

It is, however, very much written for our time: Post-Victorian, Post-Ruskinian. It will have no use for John Ruskin's famous—and for a while pernicious—distinction, in his *Seven Lamps of Architecture,* between building and architecture, for building *is* architecture. And when Ruskin, in an often omitted footnote to his first chapter—"The Lamp of Sacrifice"—speaks of the "mental arche" which separates architecture from "a wasp's nest, a rat hole, or a railroad station," the answer of this book will be: railroad stations *are* architecture, some of the most exciting and expressive architecture built; architecture is factories, hotels, banks—all the gathering places of life. Wasps' nests are not architecture, not because of a lack of structural subtlety or form, but because they are not designed by humans for human use. It is human design and human use on which the definition of architecture must be based. Ruskin speaks—in the same passage—of architecture as sight; no, it is design, it is use.

This book is basically a story of change: architecture in change, architecture of change. Materials, methods, and structure have been thoroughly treated in engineering and technical literature. Here they are pulled together into the orbit of an architectural encyclopedia. Obviously, such a work has to be organized as an encyclopedia rather than as a dictionary; it is dealing with conceptual areas, not with individual technical terms or isolated details, definitions, or entries. It is to be more than a handbook, a textbook—although it can of course be put to such use, just as Vitruvius' was. The book is arranged in a topical sequence of subject matter. But the perspective is a unified one, because this is *one* book, not a collection of individual articles.

The text begins with a *stylistic survey*. It is not a new architectural history; rather a concise and clear overview of architectural development presented with a global view; from the earliest architecture through Egypt,

Greece, Rome, India, China, Japan, and pre-Columbian America, the development is traced—both geographically and stylistically—through Byzantine, Romanesque, Gothic, Islamic architecture on to the Modern. Special aspects—primitive architecture, architecture adapted to climate—are given consideration in separate entries.

An extensive survey of *building types* follows. The development, forms, and uses of buildings have been a primary interest of architects and architectural writers from the very beginning of architectural literature. What is offered here—anew and with a very conscious aim at updating and inclusiveness—is both the great variety and the incredible range of contemporary building types. Where Vitruvius spoke of forum and basilica, treasury, senate house, theater, farmhouse, and bath and harbor, this book will speak of bridges, railroad stations, airports, factories, skyscrapers, schools, hospitals, movie theaters, and laboratories, etc. The emphasis, again, is on technology and use: how such divergent building types evolved, what technical innovations, social needs, and new materials shaped them and indeed made them possible; how changing organizations and new requirements have affected them and affect them now. How, for instance, does air conditioning relate to architectural design? How will photocopying machines alter the lay-out in office buildings?

Turning to *structure,* an important section of this book, the reader will find a four-step arrangement: from structural design and its bases, through structural elements (arches, beams, floor systems, roof trusses, columns, space frames, cable nets etc.—again, the most ancient elements standing next to the most recent) and the structural systems which combine them—the multistorey building, the wide-span, the load-bearing—and finally, to the construction means and processes that make them possible—centering, scaffolding, formwork, and mass production itself (sequence control, rationalization, etc.).

The mechanical and environmental *services* within a wide range of building types are then discussed: how do steam, gas, electricity affect architectural forms and systems? The answers are historic, or rather, genetic in presentation: before and after the Industrial Revolution, from wood and water to steam and gas in the 19th century, and on to the electrical revolution and into the future.

The guided tour (if we may call it that) then turns to the *material* basis of architecture—bricks, earth, mortar, timber—and on to our present reinforced concrete, glass, iron, and steel and the fabulous new crop of plastics, laminates, and other revolutionary products.

Tools, techniques, and fastenings complete the survey: the mason's, the carpenter's, the metalworker's tools; the joints and fixings, bolts, screws, nails, rivets; how these tools are used and, of course, have been used to erect buildings and to make them last through the centuries and into the future.

While a picture is not in every case worth the thousand words given to it by the Chinese saying, certainly architectural illustrations are worth a great many words. Here we have over 800 carefully selected illustrations to enhance the wide and complex subject matter. Many of them come from the unique slide library of the Architectural Association in London and will, in themselves, form an original contribution. Engravings from ancient sources have in many instances been used. Architect, art historian, and engineer alike should profit from this fine collaborative effort, not only in information, but also in inspiration. But it is also very much intended for the same interested, curious, and concerned layman for whom Vitruvius and Alberti wrote their treatises. The text has been kept as non-technical as a thorough presentation of technical matter in a condensed form will allow. The book will be read as well as used, read for the inexhaustible and fundamental pleasure of looking at and understanding architecture, man's original and most powerful tool in changing his planet and his existence; and used for a new perspective, a stock-taking and an anticipation of future progress.

Adolf K. Placzek
Avery Librarian
Columbia University

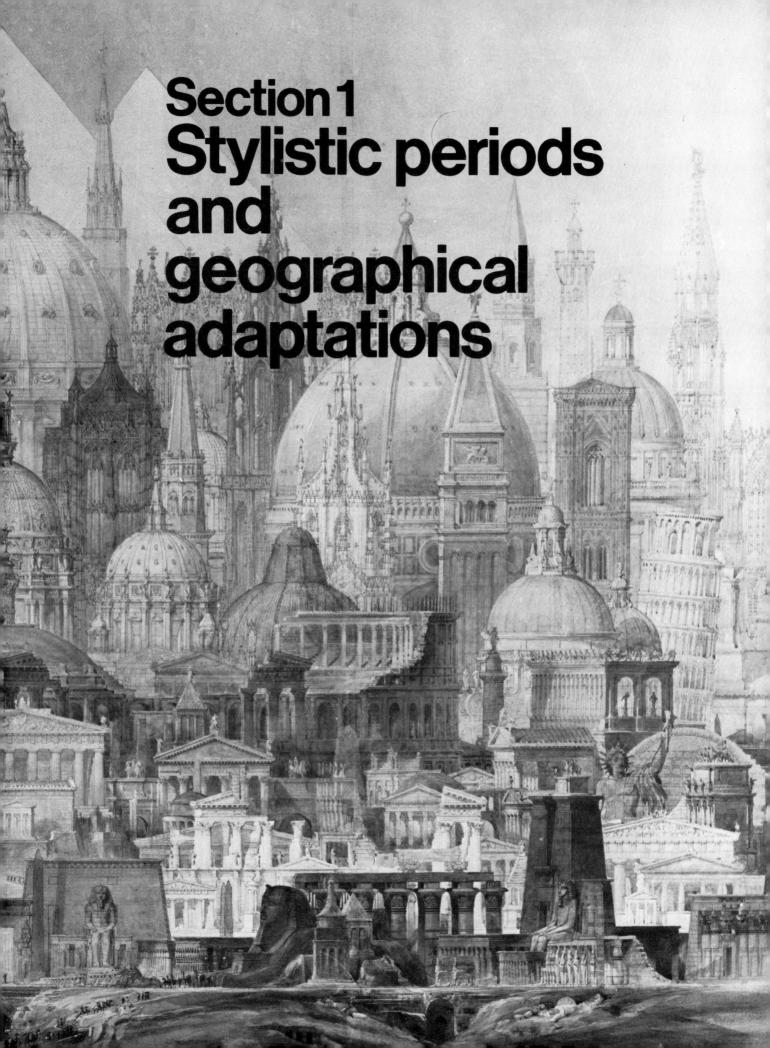

Section 1
Stylistic periods and geographical adaptations

Stylistic periods and geographical adaptations

Architecture adapted to climate

In many regions of the world in which the natural outdoor climate is severe, methods of adapting architecture simply and economically to provide the necessary comfort standards internally have been introduced into indigenous building techniques. These adaptive measures vary depending on the problems presented by the particular local climate.

Hot, dry, tropical conditions

Such conditions can be found in any part of the world within 15° latitude of the equator. This climatic type presents the highest temperatures, with hot, dry air and dry ground conditions, producing dust storms. There is little or no rain and little vegetation. The sunlight at midday is strong, and one of the main problems in building is not so much dealing with the air temperature as with the surface temperature reached by the buildings because of solar radiation. Another problem is that the nights are cool, and the clear sky allows rapid radiation of heat from the buildings and the ground so that temperatures can drop to as low as freezing point at certain times of the year.

The natural means of dealing with this climate is to provide a thick, well-insulated enclosure for living, with only small openings. The walls and roofs are thick and are built of masonry and plaster or, if of timber framing, are packed with either thick thatching of grass and leaves or with clay stones, to provide adequate insulation. It is normal in such architecture to provide high windows relatively small in area to avoid glare from the reflecting sunlight; as humidity is low, cross ventilation is not necessary. In order to reduce the intense radiation to a minimum, such buildings, if they have courtyards at all, have them of very small dimensions so that the sunlight cannot reach the pavement. For similar reasons, buildings are placed close together with only narrow alleyways between them. Such a type of construction provides considerable thermal storage capacity; the walls and roof take some hours to heat up. In a well-constructed building the time delay may be as much as 8 to 10 hours, with the result that the building is reradiating heat into the interior during the cool nights; by sunrise the building will have cooled down and the interior will then preserve its coolness for a large part of the following day.

A fine example of this kind of traditional solution can be seen in the houses of Benin in Nigeria. Each of these has as many as six tiny courtyards, but otherwise presents no openings to the outside. They are built of mud brick, with a thick layer of mud over the timber beams of the roof construction.

Other traditional architectures in hot, dry climates utilize larger courtyards but have found other means to restrict solar radiation. The buildings of the Arabian deserts used galleries or verandas around three or four sides of the courtyards to shade the walls and the floor surfaces outside windows and doors. These were supplemented with reed awnings hanging from the walls. The houses of the central desert region of Persia and the surrounding areas used deep recesses in the walls of the courtyards (called *iwans* or *liwans*) which were covered with masonry vaults. The great depth of the *iwans* ensured that very little radiant energy from the penetration ot the sun into the courtyard could reach the rear of these shaded living areas. Persian houses, in common with those of the Gulf and parts of the Indian subcontinent, also used rooftop wind catchers which faced in the direction of the prevailing wind and led the breeze down ducts in the rear walls of the main living rooms on the sides opposite the courtyard. Another variation used these wind towers for extract, in which case they faced away from the direction of the prevailing winds. A third type incorporated both intake and extract ducts in a single tower, with a vertical wall separating the two shafts. A final type of wind catcher was built halfway up the height of walls of rooms at second-floor level; this led moving air down only a short distance, releasing it just above floor level inside the roof room, which was used for entertaining guests.

In an endeavor to provide cool spaces during the midday heat of the hottest months of the year, Persian architects traditionally constructed underground rooms on the shaded side of the courtyard into which cool air could be led from wind catchers on rooftops above. Sometimes this air was further cooled by being passed over underground streams or canals before being released into the underground rooms known as *serdabs*. Water was likewise used in the courtyards of houses and public buildings, either in the form of still pools or as fountains, to provide cool air and relieve extreme dryness. A final refinement was the planting of shrubs or trees around the pools which both shaded the courtyard and aided in the provision of cool air. Many of these devices are also known to have been used in ancient architectures in Mesopotamia and Egypt.

Modern techniques of dealing with hot dry climates include the provision of a double roof—an upper roof usually made of a non-waterproof material shading the permanent roof—and a double wall in which the outer wall is merely a shading screen, the real support and enclosure being provided by the inner wall. A variant on the double roof is the provision of a ventilated roofspace in which convection naturally removes the heat through openings at the top of the roof pitch, thus reducing the need for substantial thermal insulation in the ceiling below. Finally, the

Middle Kingdom Egyptian house with wind catchers and screened window openings.

House with wind catchers, in Sind, Pakistan.

Farmhouse in Morocco with thick external walls and roof, and a small courtyard providing a shaded area which reduces solar radiation entering the house.

provision of sun-breaker, or brise-soleil elements as shading devices, has become part of the vocabulary of modern architecture. Sun visors, either fixed or movable, can be arranged so that their geometric configuration excludes direct sunlight while allowing relatively large areas of glazing behind them. These devices are used in a wide variety of forms in many climates when direct sunlight and solar heat-gain are a problem.

Warm, humid conditions

These conditions occur in tropical countries in areas which border oceans and inland seas. Here, the most difficult architectural problem is that of dealing with the high humidity for this can only be alleviated by providing substantial air movement across the skin of the inhabitants who will thus obtain relief from the evaporation of perspiration. In order to achieve this, traditional architects raised the living rooms high in the air and opened the whole height and width of walls with latticed screens to allow the free passage of air. Only in situations in which there were no surrounding buildings was it normally possible to achieve this degree of forced ventilation on ground level. Hence in urban conditions buildings tended to rise vertically and to be faced in the upper storeys with latticed wooden screens which often cantilevered out over the streets like balconies. The rooftops of such buildings were likewise surrounded by latticed screens of masonry or wood which provided air movement to sleeping spaces on the roofs, used on the most humid evenings. During the heat of the day, however, substantial thickness of the roof was necessary to prevent the absorption of the heat from the sun. But rooms below such roofs would have been intolerable at night as the reradiation would raise the temperature even higher: hence the need to provide sleeping space in the open air on the roofs. Traditional architecture of this type was characteristic of Cairo, the cities bordering the Red Sea, Turkey, Morocco, and other parts of the Mediterranean, as well as parts of India and the Persian Gulf.

Wind catchers were also used to encourage the movement of air in hot, humid climates, notably in the Persian Gulf, northwest India, and in Pakistan. In particular, the provision of a row of mid-wall wind catchers to direct air just above the floor level across the full width of a room was utilized in Bahrain in reception rooms which were characteristically built on the roofs of houses.

Screened verandas around buildings provide flexible living space with the maximum possibility of the movement of air; they are thus often favored in hot humid climates. Sometimes, as in Sri Lanka, the veranda roof is extended forward to create a double veranda, the outer height of the eaves then being

so low that sky radiation is effectively prevented from entering.

Composite climates

Many climates present serious problems because they combine at different times of the year the characteristics of hot, dry climates with warm, humid climates.

Monsoon climates. Normally hot and dry, this climate changes some two to six weeks before the arrival of the monsoon rains and during this period it becomes warm and humid. Two compromise solutions are possible. The first is designed for the hot, dry condition with the acceptance of the discomfort of the night humidity. The more practical compromise is the construction of buildings in two parts, each of which provides an area suitable for use during one of the two periods. An example of the latter is a building with a first floor with walls and ceilings of heavy construction (and with small window openings) which provides insulation during the long dry periods, and another storey of light materials with pierced or louvered walls that provides cross ventilation and little retention of heat during the short, warm, and humid rainy periods. Alternatively, these two blocks can be arranged on either side of the courtyard. In either case, the family has to move from one part of the building to another as the climate changes. However, this type is characteristic of certain latitudes in the Indian Ocean, Muscat and Oman, southern Asia, and Iraq, and, although they not normally monsoonals, the West Indies and parts of South America.

Small-island climates. Small islands do not retain cloud cover and so nights are cooler and not so humid. The problems of coping with humidity are reduced and therefore cross ventilation is not as important. On the other hand, islands are frequently exposed to sudden storms and cyclones, necessitating effective shuttering of all pierced and louvered openings to prevent the entry of wind and rain.

Maritime desert climates. These climates are normally hot and dry but are subject to sudden changes when winds from the ocean bring humidity across the area. Buildings need to have both thick walls and roofs and also pierced walls to allow air movement. As breezes blow inland from the sea with some regularity at night, it is normal to provide pierced openings facing in that direction.

Savanna, or interior continental climates. These are characterized by dry winters, which can be relatively hot, followed by long rainy seasons which are warm and humid. The same type of solution is usually provided here as for monsoon climates, with the construction of two types of buildings juxtaposed so that comfortable conditions for living may be found through flexibility of use. In the

Open veranda in a Dyak "long house", Sarawak, Malaysia, producing a well-shaded and ventilated living area.

House in Jaipur, India, with protected openings and screened open-roof area.

Thatched buildings in a dispersed village near Rusape, Zimbabwe; a region with a savanna climate.

Cameroons and part of West Africa, for example, buildings include both light, thatched, round shelters with open sides, and square or round clay structures with thick, mud roofs.

In savanna climates where there is a wide daily and seasonal fluctuation of temperature, thick heavy walls are useful in the house interior. Their high thermal capacity effect is then exploited to store heat and reradiate it during cold evenings. The outside walls can remain as thin screens to allow cross ventilation during warm humid periods.

Tropical upland climates. Although the tropics are usually associated with heat and humidity, a number of high mountain masses occur within the tropical latitudes. Here, inhabitants experience extremes of temperature from day to night due to the clear skies which prevail for a large part of the year. There is strong solar radiation, even when the air temperatures are low. There is low humidity and constant air movement. On the other hand, there are distinct rainy seasons, during which the humidity may rise. Such climate demands an architectural solution providing thick walls accompanied by shading, and adequate cross ventilation which is closable with shutters at night. The towerhouses of southern Arabia, the southern High Atlas in Morocco, and parts of India and Pakistan, are good examples of the application of this solution.

Alpine and Arctic climates

Insulation is of major importance, so walls are of thick masonry or wooden logs carefully interlocked at the corners to avoid air gaps. The buildings are raised on stilts or platforms to lift the floors above the snow level. Traditionally, any interstices in the walls or around windows or doors were filled with moss as a wind and waterproofing seal. The roofs are normally made of thick wooden planks or tiles to ensure insulation, and on top of these, especially at the level of the ridge, weights are placed to prevent the roofs from blowing away. A traditional device used in Finland and northern Japan is to fix a ridge pole in the scissor joint between projecting rafters, which is then tied down firmly at the ends to secure the roof against wind. In these countries it is also common to fix an outer frame across the roof which is weighted down with stones suspended from it at the eaves. In some areas roofs are made with a steep pitch to shed snow and reduce the possibility of moisture penetration. In others, roofs are arranged in such a way that snow is encouraged to settle on them to increase their thermal insulation.

In such houses it is usual to keep the

fireplace or stove and the chimney in the center of the structure away from the outside walls. This ensures that no heat is lost except to warm the surrounding rooms. Walls facing the direction of the prevailing winds are usually windowless unless the winds come from the south.

Traditional wooden house near Lac Neuchâtel in Switzerland.

Primitive architecture

The term primitive has a number of different meanings, and confusion between them is a frequent cause of misunderstanding.

First, within the context of any culture, art historians have used the term "primitive" of early phases in the historical development of architecture, in the belief, now thought to be debatable, that the early builders would have erected buildings differently had their knowledge of technology been more advanced.

Second, "primitive" is a late 19th-century term applied to cultures which were not part of the "evolutionary" development of civilization in any of the major centers. It was assumed that their social patterns closely paralleled early phases in a development of the great civilizations. In modern times it has become apparent that these cultures were not necessarily in a formative stage of development but frequently represented mature systems of social organizations and technology in their own right. There are no longer any implications of technical or social inferiority or suggestions that they belong to an earlier stage of development which would necessarily lead toward a great cultural maturity. Nor is there any implication that architecture in this context is merely a distant and imperfectly understood version of a more sophisticated architecture elsewhere.

Houses in the High Atlas region of Morocco, with thick walls and small closable openings.

Nevertheless, these two uses of the term are here combined for convenience, since the innovations produced in the early phases of architecture in western Europe, and other areas in which great civilizations developed, frequently parallel in type and technology those of modern primitive societies in the rest of the world.

The Old Stone Age

The earliest evidence we have of the construction of shelters dates from the Old Stone Age. The simplest were crude stone walls converting rock overhang shelters into caves. Although evidence of what was built elsewhere in open ground is scanty, traces of surface structures have been claimed for two or three sites in western Europe. Tent-shaped structures have also been interpreted from the dating of the Magdalenian culture on the walls of the Font-de-Gaume cave, and elsewhere in the Dordogne in France. It seems likely that Old Stone Age man built crude artificial shelters like the summer huts of spaced saplings still built by the Paiute Indians of Nevada, or those of the Alacaluf of Tierra del Fuego.

More substantial evidence for Old Stone Age dwellings of the most remote antiquity comes from south Russia where, because caves were lacking, small groups of hunters built themselves winter quarters. Archaeological evidence shows them to have been roughly oval dwellings, apparently built of mammoth tusks covered with skins weighed down by slabs of stone. An important aspect of these houses, as of those of other Old Stone Age cultures and of many primitive dwellings to the present day, is that they were partly excavated into the ground. This made the construction of the roof simpler, and ensured better insulation against cold, since in winter the ground retains warmth for a long period. One difficulty in these pit dwellings was to prevent the entry of rainwater or damp from melting snow. For this reason a trench was frequently excavated around the edge of the pit to serve as a drain.

Other semi-subterranean dwellings of the Early Stone Age have been found in southern Russia. A rectangular hollow 13 ft. (4 m) wide and 40 ft. (12 m) long at Pushkari had three hearths in a row down its length and contained mammoth bones. There were some 30 holes in the floor which may have held thin uprights. Even longer rectangular hollows have been found at other places, notably at Kostenki. One measured 18 x 115 ft. (5.5 x 34 m) and contained nine hearths. It is possible that it was not a single dwelling but the site for a row of tents or for a multiple dwelling covered by a roof of leather or skin.

Later Stone Age sites in Russia provide evidence for similar dwellings or groups of dwellings. One at Timonovka has six subterranean dwelling pits, 10–12 ft. (3–3.5 m) wide and 38–40 ft. (11.5–12 m) long; two of them contained hearths, one with a conical chimney of clay-covered bark. It seems likely that these dwellings were almost completely sunken underground as the pits are between 8.5 and 10 ft. (2.5 and 3 m) deep, with timber linings to their sides. The entrances were apparently down narrow sloping ramps about 3.5 ft. (1 m) wide. It is reasonable to assume that the roofing was of logs laid horizontally and heaped over with earth, like the earthhouses of the semi-polar Eskimo in Siberia. This group of houses was associated with ancillary storage pits, open-air hearths, and working places for flint knappers. A sunken dwelling at Mezhirich in the Ukraine, excavated with great care, was 10 ft. (3 m) wide and 25 ft. (7.5 m) long, and was built from the skeletal remains of no less than 95 mammoths. The walls of this structure were made of skulls and long bones and the roof was supported on a frame of mammoth tusks, and probably of tree branches as well; it was covered with animal skins kept in place by the weight of tusks and reindeer antlers, with mammoth shoulder blades holding down the edge of skins on the skull foundation walls.

At the end of the Old Stone Age the advanced reindeer hunters of the North European Plain used portable tent structures supported on wooden poles and held down by glacial boulders. Again, they resemble the

Completing a roof on the bamboo framework of a large ceremonial house in a Wingel village in the Sepik area of New Guinea.

tents of some of the modern Eskimos. It is likely that even when the hunters occupied territory where caves and rock shelters were available, they supplemented those drafty shelters with artificial screens of skins.

The Middle Stone Age

About 8000 BC the end of the last Ice Age brought the evolution of a new culture—the Middle Stone Age—which was marked by enormous advances in man's hunting and food-gathering techniques. Dwellings were adapted to a seasonal mode of life. Archaeological evidence suggests that in summer only small temporary shelters covered with skins, branches, and twigs were used. To withstand the cold of the winters, however, the small groups of hunters needed stouter dwellings. In general, they built semi-sunken houses, generally close to a river or stream, although in some sites the floors were now only slightly below the ground level. Low walls were probably made of banked earth taken from the original excavation, and pitched roofs, covered with branches and possibly turfs, were brought down over these low walls. There is no evidence in the form of postholes to suggest that Middle Stone Age people understood the principle of framed construction, although an occasional post seems to have been erected to give head clearance at an entrance.

Most of the Middle Stone Age houses excavated in Europe have slightly sunken floors with a floor surface made of brushwood or bark. They have a framework of branches, sometimes raised on posts of substantial size (14 in./350 mm thick), which were chopped through with crude stone adzes. A framework was made of close-set saplings (2–4 in./50–100 mm diameter) either brought together at the top, or supporting a separate roof frame. The covering of walls and roofs was usually thatch, made from reeds or grass. Wattle and daub construction was found to have been used in houses excavated in Belgium, while one Middle Stone Age hut at Bockum, Hanover, stood entirely above ground. Like most of these houses, it contained a hearth made of pebbles. Most of the Middle Stone Age sites in the Middle East seem to have been open encampments that already had village characteristics. In Palestine, villages of round wattle and daub huts, 23 ft. (7 m) in diameter, with sunken floors, have been found. Near them were circular storage pits with white plastered sides. The pits may have been used for storing grain. The inhabitants of these villages buried their dead within the dwelling places.

The New Stone Age

With the transition from Middle Stone Age food collecting to New Stone Age plant

cultivation and animal domestication, houses evolved which were more substantial and commodious than any occupied by the earlier hunters and fishermen. Occurring first in the old world of the Fertile Crescent, somewhere between 8000 and 6000 BC, and subsequently spreading west into Europe and east further into Asia, the husbandry of the New Stone Age brought a concept of further special-ization of space, so that houses containing several rooms were soon being built. The old grouping of a few shelters clustered together grew into a new social invention, the village: a cluster of houses within walking distance of field and pasture, but usually not built on arable ground.

The earliest New Stone Age town that has so far been excavated is that of Jericho, which dates from the 8th millennium BC. It has mud-brick houses of circular or elongated oval plan, some consisting of a single room and others of as many as three. The walls of these rooms were sunk below the level of the ground and wooden steps led down into them. Walls and floors were plastered with mud; it is thought that the form of the buildings was domed in wattle and daub, with doorways lined with a wooden door frame. Burials took place within the houses in graves about 3.5 ft. (1 m) under the floor. Some of the round houses were built on stone foundations.

The earliest evidence of an effective village farming community is that excavated at Jarmo in Iraq, c. 7000 BC. Jarmo was a permanent village of about 20 rectangular huts, each with two or more rooms, with a population of more than 150 people. The walls were of layered clay and there were clay floors covered by reed mats. Each of the houses had an open alley or a small court on two of its sides. A clay oven and a base for a silo were built into each house.

Early New Stone Age villages seldom had a

Reconstruction of a house with a structure of lashed and woven saplings, Singleton Open Air Museum, Sussex, England.

proved area of more than 1,650 x 1,650 ft. (500 x 500 m). Their population was rarely more than about 500 people. The architecture appears to have been entirely domestic in nature, although certain archaeologists have proposed dubious interpretations of some buildings as "shrines." The almost bewildering variety of regionalized and successively varied decorative styles of pottery suggests that the houses were also quite likely to have been decorated by painting, although there is very little evidence since few houses have survived much above foundation level. The development of agriculture brought increasing complexity into the picture of housing types and styles as the principle of settled life based on husbandry spread into regions with differing climates and building materials. It is probably not very fruitful to ask exactly where any particular building element was "invented" or first discovered. Wherever timber was available, houses in wood planks hewn with stone adzes began to be erected during the New Stone Age. Even in Egypt and Mesopotamia there is evidence for such buildings. In Egypt, predynastic houses with walls of vertical wooden planks lashed together with hide passed through slots cut in the corners of the faces of the boards were used. By removing some of the ties, the sides could be opened to allow the free circulation of air. Such a house had the advantage of being easily dismantled and reerected in the desert during the annual inundation. Elsewhere in Egypt and Mesopotamia the development of elaborate reed huts, built on reed platforms, took place. Sometimes they were mud plastered, while in other cases, the reeds were left exposed.

In New Stone Age Europe, wood was the most abundant and convenient material. With rain or snow in winter, a sloping roof was necessary, usually in a tent shape with a central ridge pole supported on a line of posts. Other posts supported the walls, which were made of wattle or split saplings, plastered with dung or clay. In the forested Lower Danube region, the substantial rectangular houses were walled with split tree trunks filled in with wattle and daub, and each had a central hearth as well as a separate front entrance porch attached to one end. In south Russia, long timber houses had similar porches, and there is evidence of roof finials with elaborate spiral-shaped moldings of clay. In the Rhineland and further north in Scandinavia, "long houses" up to 106 ft. (32 m) in length were built. Their form, and the effort of erecting them, suggests an organized community cooperation derived from a close-knit clan or tribal structure. The "long houses" were divided into two parts; one apparently with a raised floor. The plan is determined by posts set at intervals in three parallel rows, the central one supporting the ridge pole. Be-

tween the poles were split timbers set upright in the ground; these lined one end and wattle and daub lined the other. The existence of raised floors is partly argued because of the lack of any evidence of hearths or ovens. On the other hand there is evidence of numerous small granaries raised on posts.

Excavations at Trelleborg in Denmark revealed two "long houses" each approximately 100 ft. (30 m) long; one end of each house was used communally by a number of families, the other as a stable for beasts. The gable roof, approximately 10 ft. (3 m) high, sloped down to the ground on one side while the other side rested on a wall approximately 6.5 ft. (2 m) high. Other similar "long houses" in Denmark are estimated to have held 50–60 families. They resemble the huge communal houses of British Columbia, where a stone-based culture survived until the 19th century.

A second type of European New Stone Age house was the lake dwelling of the Alpine region. Houses of this type were either built on piles driven into the water, or placed on a framework of beams laid over soft marshy ground on the edge of the lakes. They were rectangular with gable roofs, and many had two rooms: an anteroom and an inner chamber. The walls were of split saplings and wattle and daub. There were open hearths in the anterooms, with drying frames before them. The inner rooms also contained hearths, and supports for raised couches or benches. It is thought that food was prepared in the outer room, where there was normally a clay oven for breadmaking. In these houses the positions of clay hearths and baking ovens had evidently become standardized. In front of each house there was a planked forecourt, presumably as a place for working and sitting.

The Late New Stone Age in Europe was a period of change when houses became much simpler. In the Third Danubian phase in central Europe, huts were one roomed, and only 15 ft. (4.5 m) square with a large central pit, apparently serving as a storage silo for

Reconstructed view of a lake village near Glastonbury, England.

Model of a Pile dwelling typical of the late Stone Age and Bronze Age. From Riedsschachen, Württemberg, West Germany.

grain, and an adjacent hearth in the sunken floor. The walls were made of thin saplings, which apparently were bent together to form a ridge. Some of the timber houses had apse-shaped ends partitioned off as separate rooms.

There is evidence that New Stone Age houses in Europe were quite well furnished; in the Orkney Islands, where there were no trees, articles which elsewhere would have been made in wood and have perished without trace, had to be translated into durable stone. We find here fixed beds, dressers of at least two tiers of shelves, and various wall cupboards or niches for keeping objects. Near-Eastern tombs contain model stools and couches in clay which bear similar testimony to the evolution of furniture in this period. Light came from the central fireplace in each room; there was unlikely to have been a chimney, merely a hole in the roof or a gap under the eaves.

Dating from the 1,000 years between the middle of the 7th and 6th millennia BC, a number of sites have been found which indicate the growth of cities with populations of up to 3,000 people. In Jericho, the walls were built of long cone-shaped bricks on stone foundations. They were covered with fine gypsum plaster, stained with red ocher and highly burnished. The houses consisted of several rectangular rooms linked by wide, open doorways with rounded jambs. The houses communicated with each other through courtyards and open spaces; there were no narrow streets. One or two of the buildings may have been used for cult purposes.

At Khirokitia in Cyprus (c. 5500 BC), nearly 1,000 round domed huts were grouped to form a small city. The high conical hut was built of mud brick covered with mud plaster. Entering from the cobbled street through a wooden-framed doorway, the visitor passed down a short flight of plastered steps to the sunken main floor, which was finished with beaten mud. In the center of the house was a hearth of baked clay. Part of the plan was covered by an upper gallery raised on two square limestone pillars, the floor of the gallery being constructed of wooden beams covered with brushwood and beaten mud. Wooden-framed niches or closets were set into the stone pillars. On the lower level a round stone table served for eating. The gallery used for sleeping or storage was approached up a ladder-like flight of stairs. One or two small square openings in the dome served as windows, while a narrow central hole allowed smoke to escape from the fire below. Adjoining flat-roofed shelters, open on one side, served as living spaces in hot weather and for the stabling of animals. Modern descendants of this kind of construction are to be found in eastern Syria; variants in stone occur in southern Italy (the ''trulli'') and in other parts

Remains of a group of stone houses at Skara Brae, Orkney, Scotland.

of the Mediterranean. At Catal Huyuk in Anatolia there was a large New Stone Age town of rectangular houses built in mud brick on stone foundations without doors; the houses were apparently entered down ladders through openings in the roofs. This consolidation of the early pit dwellings into an urban complex occurred also in the pueblo cultures of North America, in northern China, and in places along the northern fringe of the Sahara, where they survive to the present day.

At the end of the New Stone Age, metals came into use, mainly copper. At this period the growth of the Halaf culture in northern Mesopotamia witnesses a combination in one building of the circular dome construction with a rectangular antechamber covered by a thatched pitched roof.

The Bronze Age

The production of bronze, based on the addition of tin alloy to the copper which had already been in use for several millennia, began in about 3000 BC in the Middle East. It was followed by a concentration of power and wealth which produced vast, fortified, walled cities incorporating great civic buildings and palaces. But the use of bronze spread into more primitive societies in Europe and Asia, where the Bronze Age witnessed architectural developments which were not based on urban culture but which were, nevertheless, more sophisticated than earlier primitive architectures.

Evidence of Bronze Age carpentry techniques can be seen in the use of mortised beams found in submerged structures off the Essex coast in England, and in the invention of pile shoes. The latter were small transverse timbers under the uprights of the Swiss lake

Trulli houses in Puglia, Italy. These small circular houses built of dry stone have internal spaces covered in corbeled domes. Similar forms are found in Sardinia, Malta, France, Ireland, and elsewhere.

General view of the remains of Catal Huyuk in Anatolia.

settlements, introduced to minimize sinking. The large funeral mounds or "barrows" of Britain contained wooden chambers with upright timbers along the sides, and horizontally laid tree trunks to form the ceiling. On the European continent there are similar timber structures within barrows, some of them with inverted V-shaped structures of thick oak beams. Many of the barrows, both in Britain and on the continent, were surrounded by palisades and ditches lined with poles. Frequently, entrance was possibly only through a gateway between two substantial oak posts.

Two "woodhenge" ritual sites have been found in Britain. At Arminghall, near Norwich, eight greak oak uprights were set in a horseshoe plan surrounded by two concentric ditches. At Woodhenge in south Wiltshire, six concentric and roughly circular rows of large uprights may have formed the substructure of a circular roofed building; a similar interpretation has been given to the concentric timber and stone circles near Avebury. Alternatively these buildings may have been open-air temples akin to Stonehenge, possibly associated with astronomical observations.

The use of large trunks for uprights was made possible by the availability of sharp bronze-headed axes. Soon afterward we find the introduction of true log houses. At first we find foundation platforms laid on logs in marsh settlements (for example in Yorkshire, England, in the late Bronze Age). In continental Europe, and later Britain, large round timber huts 20 ft. (6 m) and more across were built with substantial uprights close together and were covered with conical roofs. A typical farm group had a main circular hut, a second sub-rectangular, and a third smaller shelter, possibly a kitchen, together with several small structures raised on posts suggesting storage sheds. The walls of the rectangular buildings were apparently constructed of horizontal logs. The whole group was enclosed in a low bank.

The increase in prosperity in the late Bronze Age is well attested in archaeology from waterlogged sites in central Europe. At Wasserburg Buchau there was an early occupation, (c. 1100 BC), of 38 small rectangular one-roomed log cabins; a second occupation, (c. 900 BC), contained evidence that bronze tools were plentiful and that there were nine large farmhouses each built around three sides of a yard and constructed in the log-cabin method with interlocking timbers. The whole was enclosed with a palisade of pine stakes with defensive platforms at several places.

The Iron Age

During the 1st millennium BC the use of iron spread out from the great centers of civil-ization at the eastern end of the Mediterranean and India into many surrounding primitive societies. We know most about the effect of the harder, sharper, iron tools on primitive buildings from archaeological excavations in Europe. One development was a far greater variety in buildings than had hitherto existed. Communities became larger and crafts more sophisticated. Construction was still largely in timber, but it was now frequently adzed to a rectangular shape and jointed with great sophistication. The use of grooving to join horizontal and vertical planks for walling was introduced, as was floor construction of planks on wooden beams.

In Iron Age Britain the circular patterns of house plans continued to predominate. At Little Woodbury, Wiltshire, a large circular house had timber walls contained within two rings of posts, with a projecting, solidly built timber porch which appears to have had some ornamental superstructure. There were four massive central uprights, apparently to take the ends of crossbeams from the other walls in order to carry a heavy thatched roof. The use of four central posts suggests an opening, probably to allow the escape of smoke and to permit natural lighting in the center of the plan. Another circular house at Woodbury included numerous pairs of postholes, probably for drying frames for corn and hay, comparable with those used by the Maori. Groups of four postholes, set in a square of a little more than 6 ft. (2 m) dimension, probably supported small granaries similar to those used today in many parts of Asia and British Columbia for storing food. There were also deep pits, perhaps with wooden lids, for storing roasted grains, shallower pits for water containers, and small domed ovens of plaster for roasting grain to prevent germination. The Woodbury group was surrounded by a palisade which at the entrance had a number of postholes suggesting a gate tower.

Fragments of timber construction from the Iron Age period in England have been excavated in many places and demonstrate the continuity between prehistoric and recent vernacular construction. Details of regular mortised holes in rectangular oak beams to take wattle hurdles for plastering and grooved oak planks resemble details used until recently in rural England.

In places where forests were distant, clay, wattle and daub, and stone construction continued to be used following practices which had been established earlier.

Summary

It can be seen that primitive architecture can be loosely categorized into five types of building: open wind breaks; pole constructions of spaced saplings filled in with a light open-work lattice of twigs, covered with

Part of the stone circle at Stonehenge, Wiltshire, England.

Model reconstruction of an Irish Iron Age hut.

Reconstructed interior view of an Iron Age dwelling, Britain.

A house under construction in Dombo Shawa, Zimbabwe. Saplings are bound with bark strips to form a structural framework which is enclosed with a covering of grass sheaves.

grass or reeds; pole constructions of spaced saplings, with a light open-work lattice and covered with plaster, e.g. wattle and daub; frame constructions of close-set shaped posts, with an infilling of timber, wattle and daub, or masonry; and solid constructions of logs, planks, turf, mud, adobe, mud brick, baked brick, or stone. Other special types were necessitated by shortages of materials, or encouraged by particular qualities of flexibility and utility in easily available materials; the Eskimo house of snow and ice is an example of one, and Central African palm-thatched huts of the other.

In general, houses are more substantial where the climate is rigorous, but the view that climate is the most important determinate of building type, dominating other factors such as inherited cultural and social patterns, is now no longer regarded as valid. The Indians of Nevada built an open shelter of brushwood for use in sub-zero winters, and a similar phenomenon existed in northern central Canada. Similarly the Chinese in the Peking area built courtyard houses with paper-thin walls in a climate with a long and freezing winter, so that adaptation to the climate had to be accomplished at personal level by wearing thick padded clothing.

Another view—that the available building materials were generally more important as determinants of building form than living patterns or social cultural traditions—can also be disproved. Only Eskimos on the central Arctic shore make igloos of snow; other Eskimos, east and west, where snow is just as

abundant, build houses of driftwood timbers and animal skins in the same domed shape and with the same internal disposition of uses. Frequently, completely different structural systems for the same living patterns exist side by side—as in the mud-walled houses and thatched dwellings in northern Nigeria, the western Sudan, and many other parts of Africa.

Techniques of building and, of course, social structures and cultural patterns, can migrate with the movement of people from one region to another. This explains both the appearance of house forms in regions with climates for which they are not strictly suitable, and the existence of quite different structural techniques side by side. Once established however, both construction techniques and patterns of living tend to be very resistant to change and may persist for thousands of years. Certain areas of the world have developed structural techniques or house forms which generally predominate over all others; e.g. the flat-roofed, plastered courtyard buildings of the Mediterranean, Iraq, Iran, and northern India; the gable-roofed framed buildings in brick or stone of northern Europe; and the raised thatched framed structures of southeast Asia and the Pacific.

During the last century, anthropologists studying modern primitive societies have added new dimensions to our understanding of architecture which have important consequences. Firstly, in revealing the existence of many forms of social structure very different from those hitherto known, which are reflected directly in their buildings and settlement patterns: examples are the "long houses" of North America and parts of Africa and Asia; the pueblo cultures of Arizona and New Mexico; and the boat-structured societies of the Indonesian archipelago. Secondly, in leading to the realization that built forms may not reflect the true concept of the architecture held by the people who built it; i.e. that the shape of the living pattern can on occasion be more complicated and even quite distinct from the built form. Thirdly, in pointing to the conclusion that among settled primitive societies, dwellings are not thought of only as structures or bases in which to eat and sleep, but have values and even something approaching a spirit of their own. The orientation, the position of the hearth or door, a particular relationship of the parts may offer religious or magical protection, express some belief, or denote rank in society. A whole world of symbolism may parallel the utilitarian world of material objects and buildings; the visual aspects of architecture denote meanings to the primitive man which do not lie very far below his consciousness and may even be part of a conscious rational system of delineating his own relationship to the real world.

Mesopotamian and Iranian architecture

Sophisticated Mesopotamian architecture first developed in a relatively small area in the lower plain of the Euphrates-Tigris valley, although a related culture, probably colonial Mesopotamian, appeared soon afterward on islands in the Persian Gulf. Under the Assyrians, Mesopotamian architecture covered a large area extending northward to the sources of the rivers and into Syria. The conquest of Mesopotamia by the Persians, which encouraged the Persians themselves to erect fine buildings, extended the region of developed architecture far to the east into Afghanistan and to the north into central Asia; their sway extended down the Persian Gulf to Oman and eventually into Asia Minor and Egypt. The last great Persian empire was the Sassanian which, for half a century, extended its realm to encompass southern Arabia and Egypt.

The main creative phase of Mesopotamian architecture was a period of about 400 years (3100–2700 BC), which predates written history. An ancient practice of erecting mud houses in clustered settlements had already led to the formation of raised mounds or *tells*, created by the alternate destruction and reconstruction of buildings. On such *tells* religious buildings began to be erected, the platforms growing increasingly higher as with successive generations earlier temples were razed and new structures built on their ruins. The temples were rectangular buildings diverging about 45° from the direction of the cardinal points, and containing T-shaped courts (possibly covered with roofs). Entering on one of the long sides, the worshiper had to turn at right angles to face the altar or hearth, which often had a podium opposite it. Small cells or rooms surrounded the court and were entered from it. A characteristic feature of the construction was the lining of the walls with rows of niches, or alternatively with staggered buttresses; these devices suggest that the origin of the construction lay in half timbering—however, by the period of the earliest discovered monuments, all the construction was in baked brick or mud brick due to a shortage of timber (except for short lengths of date palm trunk). Brick building techniques included the use of the arch, the dome, and the vault (although the latter appears to have been used at first only for underground burial chambers). Buildings at Waraka and Uruk of the late 4th millennium BC (the Uruk period) have wall surfaces and pillar columns decorated with a mosaic of red, black, and brightly colored flat terra-cotta cones arranged in geometric patterns.

During the first period at Ur, painted and relief decorations were used not merely as ornaments, but to emphasize structural forms (temple of Al 'Ubaid near Ur, c. 2600 BC). By

One of the staircases leading up the ziggurat at Ur in Sumeria. Built c. 2350 BC, it led to a temple which no longer exists: the staircase's baked brick is excellently preserved.

this time some temples were beginning to be reduced to a single rectangular *cella*, entered through a door near the end of one of the long sides and approached to its site on the top of a high rectangular platform by a long straight flight of stairs.

The two centuries of Akkadian rule (c. 2350–2150 BC) were characterized by changes marking the steady infiltration of Semitic immigrants from the steppes. Large flat bricks, approximately 6 in. (150 mm) square, replaced the smaller fired bricks of earlier times. At Tell Brak a palace was built over the sanctified grounds of an earlier temple, an indication of the elevated status afforded to themselves by the god-kings. The palace was nearly square, with walls 33 ft. (10 m) thick, a huge cutting containing the entrance gates to the axially related main court behind. There were three smaller court complexes surrounded by relatively narrow, roofed rooms.

Almost contemporary with it was the erection of a great temple at Ur in the 22nd century BC. This, the most splendid example of Neo-Sumerian architecture, was built on a massive platform, the ziggurat, which was approached up long ramped staircases. Little is known of the temple building on the top, as it did not survive, but there is evidence to suggest that it had a central sacred chamber, with a raised platform where the ritual marriage ceremony of a virgin to the Sumerian god was performed.

There were also Neo-Sumerian "low temples," without a raised platform or ziggurat, with an axial approach to a broad *cella* and an equally large *antecella*. The focus of the axial sequence was a throned niche at the back of the *cella*. Palaces followed closely similar plans; it seems likely that two processes were at work, the humanization of the gods— access to them being regulated by court ceremonial — and the deification of the king,

E-Num Mah Temple at Ur (c. 3000 BC).

who was surrounded by semireligious ceremonial in his relationship with his subjects.

Mesopotamia was then swept by a new wave of western Semitic peoples who eventually came to power and produced a new series of great rulers, culminating in Hammurabi of Babylon. By far the most important architectural monument from this period of the Western Dynasties is the palace at Mari, which dates from the 19th and 18th centuries BC and is extremely well preserved. This colossal complex (560 x 383 ft./200 x 120 m), orientated by its sides to the cardinal points, contained more than 260 rooms focusing on two large courtyards used for royal ceremonial purposes. Around these courtyards there were the main throne rooms and a large number of smaller courtyards with rooms grouped around them. The rooms had no windows; light penetrated through the very tall doorways. The walls were plastered and richly adorned with paintings.

The Assyrians rose to power in the 13th century BC. Their chief buildings were huge palaces built in two linked complexes, one grouped around the entrance court, from which the throne room and administrative rooms were reached, and the other grouped around the residential court. Other smaller courtyards surrounded them in diminishing stages. Arches, but not columns, were used in these buildings. The chief palaces were those at Ashur (13th century BC), Nimrud (9th century BC), Khorsabad (722–705 BC), and Nineveh (7th century BC). The older palaces, culminating in Khorsabad, were conceived as symbols of the cosmos. Squares were favored in proportioning, particularly in the outlines of buildings and as a shape for courtyards. The basic planning unit contained a court and a shallow transverse hall of the same width; this would be placed on the south side in the hotter plains, and on the north side in the northern hill regions. The longitudinal room frequently had a high door in one long side and contained a hearth. The throne room, the heart of the palace, was of this plan. The great gates and doorways were flanked by giant winged bulls and fantastic animals, calculated to inspire awe in the beholder who passed between them. The rooms were lined with monumental wall reliefs in stone, and decorated above with wall paintings and sometimes with glazed tile decorations.

In 1200 BC the Chaldeans settled and ruled in Babylon, establishing a number of other city states along the Euphrates; rivaling the Assyrians for control of Mesopotamia. The Chaldean or Neo-Babylonian period was characterized by palaces containing a series of parallel complexes, each focusing on a large square courtyard, with a single or double transverse room on the southern (i.e. the shaded) side. The palaces, and even the city walls and gates, were richly embellished with

glazed brick reliefs, such as those of the famed Ishtar gate.

With the fall of Assyria and the extinction of the Chaldean Neo-Babylonian Empire by the Persians, the historically integrated, cultural, and artistic complex we normally call "Mesopotamian" came to an end. The Archaemenid Empire, which succeeded the Neo-Babylonian Empire in its sway over Mesopotamia, had its center on the Iranian plateau. A new religion came to the fore, that of fire worship, resulting in the building of the first open-air altars, behind which small sacred buildings containing the eternal fire were located. The strongest evidence of their architecture lies in excavations of the great palaces, which are contemporary with the archaic and classical period of Greek architecture: the mainly ruined palaces of Cyrus at Pasargadae, of Darius and Xerxes at Persepolis, and of Artaxerxes at Susa. The style of these buildings was clearly derived from the architecture of the mountainous, timber-rich regions, having flat roofs carried on timber beams. These beams were supported on tall slender stone columns with bracket capitals. The latter were often carved to represent pairs of winged griffins or sacred bulls. The walls and gateways were embellished with fine stone carvings in low relief. Tombs were often constructed by hollowing out the living rock and providing full-scaled facades carved from the faces of cliffs.

The heirs of Alexander did not hold Iran and southern Mesopotamia for very long; the Parthians (c. 250 BC) conquered them and continued in power until they were replaced by the Sassanians in the 3rd century AD. Parthian architecture introduced remarkable innovations derived from a combination of the axial symmetrical planning of the classical world, with the deep-shaded recessed spaces of the *liwans*, the origins of which are possibly

Reconstruction of the Ishtar Gate rebuilt by Nebuchadnezzar (604–561 BC). The gate is decorated with animals in glazed brick relief.

Monumental winged bull guardian figure from the entrance to the throne room in the Palace of Sargon II, Khorsabad.

Capitals and bases similar to those of the hypostyle Hall of Xerxes at Persepolis, from a rock-cut tomb.

related to the nomad tent, which is also open on one side. A characteristic plan, such as that of the palace of Nysa, has four *liwans* surrounding a central courtyard. A number of great palaces in this style were built with superimposed colonnades or arcades flanking the central *liwan,* notably those at Hakra, Ashur, and possibly Ctesiphon, which some authorities ascribe to the Parthian period. The succeeding Sassanian style continued these traditions, evolving, however, a complicated plan form for a palace in which a great audience forecourt was followed by a huge *liwan* with a central domed audience room behind it, fronting onto a spare courtyard surrounded by residential accommodation. In some plans, such as that at Firuzabad, vaulted banqueting halls paralleled the main volumes of the central axis. At its greatest extent the Sassanian Empire stretched from Syria to northwest India; its influence was felt far beyond these limits. Sassanian religious buildings focused on domed fire temples for the Zoroastrian religion, the dome frequently being supported on four piers with arches spanning between them, creating the effect of an open pavilion, at the center of which stood the fire altar.

Parthian and Sassanian city planning is characterized by the development of a circular city plan (Merv, Hahra, Firuzabad), with the royal palace at its center and a defensive wall at the periphery; it is a type of plan which owes something to a cosmological analogy, and something to the practical problem of defense in a fragmented, feudal, and often discordant society.

Egyptian architecture

The earliest buildings were constructed of reeds and rushes (often reinforced or finished with mud) and thick planks of wood. Brick was first used for important buildings in late predynastic times. In the 1st dynasty, mud-brick buildings were occasionally finished with thin slabs of marble or stone, but only in the 3rd dynasty was stone employed for an entire building. Ordinary dwellings were always built simply of timber, reeds, and brick, and these materials were often used in palace buildings. Most columns were designed on the theme of plants and flowers; the shafts representing clusters of reeds, and the capitals lotus buds or open lotus flowers; alternatively, capitals resembled papyrus or palms. From the massive architecture of some early funerary temples developed the use of the square pier, which was afterward refined into eight- and sixteen-sided pillars, occasionally fluted.

The earliest temple so far discovered, which was not a funerary temple, is a 5th-dynasty sun temple with a court, bounded by passages and containing in the center a huge obelisk—a square tapered shaft crowned by a pyramid top, which acted as a symbol of the sun-god, and in which he could be invoked to take up his residence. The earliest tombs were pits covered with sand or rubble, and from these evolved the *mastaba* tombs of the first two dynasties. They were usually built of brick and reproduced, in bas-reliefs and wall paintings, all the amenities of life which the departed needed to take with him for greater happiness in the next world. From the *mastaba* evolved the step pyramid of the 3rd dynasty, the earliest being constructed as the focus of the funeral complex of King Zoser at Saqqara, which is known to have been built by the royal adviser and architect Imhotep (c. 2780–2680 BC). The vast step pyramid was almost 200 ft. (60 m) high and had attached to it a funeral temple, an audience hall, and ancillary buildings to provide a suitable setting for the life after death of the king and his attendants, all enclosed by a niched limestone wall. From the step pyramid there developed the bent pyramid and eventually the true pyramid; the sides of which were perfect isosceles triangles. The largest of these were built at Giza for the kings of the 4th dynasty (c. 2600–2480 BC). They were constructed of enormous blocks of stone faced on the outside with a smooth limestone surface, which in most cases has disappeared. There were

Reconstruction of the hypostyle Hall of Xerxes at Persepolis (485 BC).

Court of the Temple of Khons, Karnak (1200 BC).

Egyptian column capitals from a number of temples. Stylized plant forms appear in some of the capitals. The shafts were often carved to represent clusters of reeds.

several false tomb chambers to disguise the position of the real one, evidencing the danger of tomb robbery which became a more serious problem with each succeeding dynasty. By the Middle Kingdom (2100–1650 BC) rock-cut tombs were predominant, incorporating the forms of mortuary temples with forecourt, hall, chapel, vertical shaft, and burial chamber. By the time of the New Kingdom (1580–1075 BC) both royal and private tombs were normally cut deeply into the rock, and great care was taken to conceal their precise location. To this end mortuary temples were eventually erected separately, at a distance from the tomb itself.

Most of the surviving temples date from the time of the New Kingdom, although even these were frequently modified by subsequent additions (Great Temple of Ammon at Karnak). Characteristically, a temple was aligned at right angles to the river, its monumental pylon facade being approached from a landing stage down an avenue of ram-headed sphinxes. The facade of the temple was embellished with bas-relief figures of the gods and of god-kings, some giant-sized so that they could be appreciated at a distance, and others smaller to be read at close quarters. Set into the pylon were four or six masts, themselves embodiments of the presence of gods, and, in front, one or two obelisks to invoke the presence of the great sun-god. A gateway gave passage through the pylon into a colonnaded courtyard which acted as an assembly point for entry into a great transverse columned hall (the hypostyle hall), in which an elevated central passageway with the largest columns supporting the roof allowed clerestorey light to enter the ceremonial way, the sides of the hall being seen as a forest of columns lost in gloom. Behind the first hypostyle hall the worshiper often passed through a second hall to the dark sanctuary with surrounding treasure rooms and sacristies. The floor level stepped up slowly as one passed through the temple, while the ceiling height diminished as the rooms became darker, until the final confined mystery of the sanctuary itself was reached.

A remarkably different monument was that of the Mortuary Temple of Queen Hatsheput at Deir-el-Bahari (c. 1480 BC), which had a series of pillared colonnades on three sides of three superimposed terraces linked by gigantic ramps. The terraces were planted with aromatic trees and looked across the city of Thebes to the facade of the Temple of Ammon at Karnak on the opposite bank.

Ancient Egyptian architecture was revived under the Ptolemies, the successors of Alexander the Great, who built numerous temples of traditional style of which the finest examples that survive are the Temple of Horus at Etfu and the temples on the island of Philae.

(See also HOUSES—EGYPTIAN)

Pyramids of Giza. These vast structures, which contained the tombs of the pharaohs of the 4th dynasty were originally faced in smooth limestone.

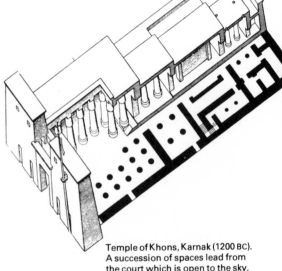

Temple of Khons, Karnak (1200 BC). A succession of spaces lead from the court which is open to the sky, through the hypostyle hall lit by a clerestory to the sanctuary.

Pre-Columbian architecture

There are two cultural areas in pre-Columbian architecture. One is the Middle American, focused on Mexico, Honduras, and Guatemala, and the other is the Andean in South America, focused on Colombia, Bolivia, Ecuador, and Peru. Essentially the two are distinct cultures, with different social structures and religious and architectural traditions.

Middle American

Architecture in this area begins about 1000 BC (the pre-Classic period), when the Olmecs built the first large-scale ceremonial centers in their homeland along the coast of the Mexican Gulf. Such a ceremonial center, which became characteristic of Middle American architecture throughout its history, consisted mainly or entirely of religious and government buildings that were used only at the prescribed times when the populations of the surrounding agricultural district congregated to watch religious ceremonies. Although these buildings had pyramidal shapes, the tops were always truncated to allow a platform for the erection of a temple shrine, which for many centuries was patterned on a domestic house and made of wood and thatch. The pyramid was therefore essentially an elevated platform conceived in the shape of a geometrically ordered mountain or hill; the stairway to the temple led up the middle of one side. Subsidiary structures included platforms and courtyards, often spread out over immense areas. The greatest of the pre-Classic centers was La Venta.

The Classic period began around AD 100 in Central Mexico and represented a development from the pre-Classic period of the same kind of culture. In this period, pre-Columbian architecture achieved its highest levels, especially in scientific learning and craftsmanship. Teotihuacán, a small farming community in the valley of Mexico, became a great urban center with far-flung political and cultural influence. The focal point of the community was a ceremonial complex that was laid out in a regular grid pattern of temples, squares, compounds, and houses, many of them decorated with polychrome frescoes. The most important of the ceremonial buildings were built during the 2nd century AD. These included the Pyramids of the Sun and Moon and a number of shrines along the central Avenue of the Dead, which ran from the Citadel and the Temple of Quetzalcoatl to the Pyramid of the Sun. Teotihuacán was destroyed in about AD 750.

The Maya culture in the Yucatan peninsular of Guatemala and in Honduras also developed during the Classic period. The oldest and largest city of the Mayas was Tikal, began c. 200 BC and fully evolved by c. AD 300. Tikal had a ceremonial center which covered 1 sq. mi. (2.6 sq. km) and included the highest pyramid of the Mayas—about 57 ft. (17 m) high—together with government buildings, the whole surrounded by habitations in the manner of suburban developments. This city, and the other city centers of the Mayas, had identifiable differing styles for the various types of architecture: public buildings, temples, sanctuaries, palaces, monasteries, ball courts, observatories, and dance platforms. Buildings had complex ground plans, and some had corbel-arched roofs, executed in rubble and dressed stone. They often imitated, or carried representations of, pitched thatched hut roofs, which were prototypes for the corbel-arched stone buildings. The door lintels were of wood, which often carried fine carvings. The grouping, proportioning, and designs of facades of Mayan buildings reached a high level of sophistication.

The late Classic period (c. AD 600–900) shows more flamboyance, aesthetic development, and growth of regional styles, particularly those of northern Yucatan, where cities such as Chichén Itzá and Uxmal gained especially fine buildings. Soon after this period the great centers were abandoned for reasons which have never been fully understood. The post-Classic period (c. AD 900 to the Spanish Conquest in the 16th century) saw the growth of the Toltec state in central Mexico, a militaristic society with a high degree of civic and social organization focused on ceremonialism. One branch of the Toltecs spread to Yucatan, where they revived the Mayan culture with a strongly Toltec architectural style. Here they built pyramid temples that had columns with a feathered-serpent figuration and which were dedicated to the Feathered Serpent, the deity imported from central Mexico. A more pure Mayan renaissance took place at Uxmal, where the most magnificent single building was the Governor's Palace in which the upper wall surfaces were decorated with small pieces of cut stone resembling those of geometrical textile patterns.

In the central Mexican plateau the Aztecs were the heirs of the Toltecs, and developed a very well-built religious architecture, in which an illusion of great height in the temple pyramids was combined with a love of large, open, ceremonial spaces.

Andean

Culture in South America, once begun, developed independently of that in Central America. Although, in the early centuries, there were separate civilizations in different regions, by the time of the European conquest in the 16th century the whole area of the Andes was completely unified under a single political-administrative organization, known as the Inca Empire.

The earliest architecture is that of the ceremonial center at Chavín, which was reconstructed several times, and consisted of various temple platforms containing a series of interlinked galleries and chambers on different levels. It is thought to have been built during the period 1200–400 BC. Andean culture rapidly developed between 400 BC and AD 400 and was characterized by a development of technological skills in building and weav-

Temple of the Warriors, Chichén Itzá, Yucatan, Mexico (12th century AD). The columns of the hypostyle hall precede the pyramid which was surmounted by a sanctuary.

"El Castillo," Chichén Itzá, Yucatan, Mexico (12th–13th century). The long and steep staircase up the center of the pyramid leads to a small and elevated temple.

Inca sanctuary, Machu Picchu, Peru (12th–15th century). The accurately shaped large blocks of granite are fitted together without mortar.

General view of remains of buildings and terraces at Machu Picchu, Peru.

Entrance to the Treasury of Atreus at Mycenae (c. 1350 BC). The best preserved of the Mycenaean "Tholos" tombs. The entrance passage leads into an underground chamber of corbeled stonework circular in plan and covered externally by earth.

ing. Peruvian architecture reached its highest development from AD 400–1000. The construction of huge pyramids in stone, the largest being the Huaca del Sol near Trujillo, testifies to the high degree of organization of the society. Houses and temples were built of large stone slabs, and the lintels were carved in high relief with catlike motifs.

The Tiahuanaco style, which was prominent c. AD 1000, is typified by the stepped pyramid of Acapala in Bolivia, which is faced with dressed stone. At the same site a monolithic stone gateway has exceptionally fine stone jointing cut in basalt and sandstone, and fitted together without mortar.

The period AD 1000–1300 saw the development of circular and rectangular tower tombs in finely dressed stonework. They are fundamentally of two types, the *kulpi* or house-tomb, and the *chullpa* or funerary monument. Both are usually crowned by an oversailing cap of stonework, and contain rudimentary corbeled domes and vaults finished to perfectly smooth shapes.

The Inca Empire, founded in the period between AD 1000 and 1300 focused on Cuzco in Peru and was characterized by a grid-iron system of city planning with streets converging on a central plaza and secondary plazas distributed according to the contours of the sites. There were wide main avenues and narrower secondary streets. The major public buildings were located in the central plaza or on the main avenues; they included a city temple, a governor's residence, a large public granary, a public meeting hall, and an inn for travelers and guests. On a neighboring hill there was usually a strongly walled fortress. Towns were connected by a system of roads with bridges crossing streams and aqueducts to transport water.

Ancient Greek architecture

Architecture developed to a high level of sophistication in the Aegean Islands around 2000 BC, in the palaces of Knossos and Phaestos in Crete. These and other palaces, of which only fragmentary remains have been found, were rebuilt a number of times after damage by fire and earthquake. The last restoration took place after the enormous volcanic eruption of Thera (c. 1475 BC) which destroyed and damaged many buildings and cities in the islands.

Soon afterward (about 1400 BC) Mycenae, previously a strong and prosperous fortified base on the mainland, became powerful enough to attack the palaces and cities of Crete and emerged as the dominant power in the Aegean, a position it retained until the Dorians invaded Greece from the north in about 1100–1000 BC. Although they have many similarities, particularly in points of stylistic detail, the buildings of Crete and Mycenae nevertheless reveal important differences of architectural character. The palaces of Knossos and Phaestos had elaborate clusters of rooms, ceremonial and utilitarian, around vertical light shafts. These were connected by labyrinthine passages (hence the legendary origin of the labyrinth in the Theseus legend). At the center of the palaces was a single giant courtyard surrounded by tiered, columned loggias. At the level of the court, closely related to the throne room, was a temple sanctuary in which the high priest or king was given a focal throne. The large size of the courtyard is believed to have been due to its function in some ceremony, presumably the bull leaping depicted in frescoes in the palaces. The only exception to the somewhat conglomerate and irregular character of the palaces was an axial ceremonial approach up a flight of stairs; this rose within the palace through a full storey to the throne room. There were underground channels providing excellent sewerage and drainage throughout the buildings. Cretan architecture is characterized by its use of inverted columns, which taper from a broad-capped capital down to a narrower pad base; they were of turned wood painted bright red. The remainder of the construction consisted of areas of stonework interspersed with timber reinforcing frames, so that the structural system could be said to be essentially half-timbered. The walls were plastered and decorated in bright colors, sometimes with frescoes, which were occasionally modeled in low relief.

On the Greek mainland, the fortified palaces of Mycenae, Tiryns, and Pylos exhibit the common characteristic of a circuitous defensive entrance through the castle walls leading to a formal colonnaded gateway. There follow one or two courtyards, more or less axially arranged, before the facade of the main palace structure, the king's throne room or *megaron*, was reached. This low-gabled, colonnaded facade was constructed largely of timber with infilling panels of brick or stone covered with plaster; at the center of the building was the royal hearth flanked by four columns which supported a gabled roof with clerestory lighting. The surface treatments included alabaster panels and even decorative pieces of lapis lazuli. The outside walls of the castles were built of huge blocks of cyclopean masonry, polygonal in shape with very fine jointing. Those of Mycenae contain the so-called Lion Gate (c.1450 BC) which contained in the heraldic emblem above the lintel an early example of stone sculpture. Outside the walls of Mycenae was one of a number of subterranean domed tombs approximately 50 ft. (15 m) in diameter, popularly known as the "Treasury of Atreus" (c.1350 BC); the corbeled dome is beehive in shape using finely shaped and coursed ashlar.

The Mycenaeans were conquered, and their buildings fell into ruin, after invasions from the north in the 12th century BC. There followed a hiatus of five centuries before sophisticated architecture began to be revived about 700 BC. The newly emergent architecture owed some important aspects to the Mycenaeans, particularly the use of the *megaron* plan for their sacred buildings, the homes of the gods, and their positioning on an *acropolis*, i.e. a naturally fortified outcrop of rock, resembling those used by the ancient Mycenaeans. The facade of the sacred building was likewise developed from that of the *megaron*, as a gabled colonnaded portico. Unlike the *megaron* throne rooms, these sacred buildings were freestanding and could be approached from any side. In acknowledgment of this, the portico was extended during the 7th century BC as a colonnade of wooden posts surrounding the building (a *peristyle*). In an endeavor to make the building permanent, the materials of its construction were changed during the same century to stone, always retaining, however, the details of its timber ancestors. Hence, in the Doric order the *triglyphs* represented the ends of the cross beams, the *guttae* the pegs used for fastening them, and the *metopes* the spaces between them. A further development was the concept of monolithic construction in which the joints were so fine that they were indistinguishable—an expression of the desire for a unique, timeless quality in the houses of the gods. At the same time, the simple Doric order of fluted cylindrical shafts, now tapering up from a broad base, was developed. The columns emerged without a base from the floor of the building and had a simple "pad" capital related to those of ancient Crete. The floor was elevated on a few steps above the natural ground; the roof was constructed

above massive timber trusses with stone tiles.

This form of temple remained essentially the same for hundreds of years, but was continuously refined in form and proportions. The Ionic style, developed in the 6th century BC, was particularly favored east of the Aegean Sea, along the coast of Asia Minor, while the Doric style was favored in mainland Greece, and also in the Greek colonies of Sicily and southern Italy, where many of the best early Doric temples may still be seen. This early period was also notable for the construction of huge temples measuring as much as 195 x 345 ft. (60 x 110 m). The Ionic temples of Samos, Ephesus, and Didyma are all of this scale; the Doric temples at Selinus and Agrigento in Sicily are also huge. There are other fine early Doric temples at Corinth, Paestum, and Syracuse.

The character of the temple went through many changes as Greek concepts of religion, society, and philosophy developed. While conservatively retaining all the features derived from timber construction, the temples first evolved a noble balance and symmetry of parts expressing strength and power, with forms of refined grace (Temple of Zeus at Olympia, c. 470–457 BC). The growth of importance of Apollo as the embodiment of perfect youth led to a lessening of the emphasis on power, and an increase in the expression of vitality and perfect beauty. It was probably as much a desire to express lifelike muscular vigor as a desire to achieve optically correct proportions that led the Greeks of the Periclean age to give greater importance to optical refinements, so that almost every line in the Propylaea and Parthenon (447–438 BC) in Athens is curved. Particularly important in adding a sense of anthropomorphic life to these buildings was the use of *entasis* on columns. A restrained use of sculpture, carefully ordered into composite designs between the structural elements or into friezes on the outer or inside walls, drew attention to the link between the shapes of the human body and those of the architecture. Much of the stonework was left in its natural color, but certain areas of moldings and sculptures were painted, mainly in vermilion red, indigo blue, and gold. Occasional touches of viridian green and silver were added. Each temple housed a single statue of a god or goddess, usually executed in marble, ivory, bronze, or gold.

It is noticeable that the interiors of the temples grow in importance and sophistication of architectural design during the period of the Periclean age; the altar, however, and hence the assemblage of the congregation, was outside the entrance on the eastern axis. The temple was never approached axially, but always at an angle so that its three-dimensional form could be observed. It was situated on a *temenos*, or sacred enclosure

Doric order. From the portico of the Parthenon in Athens.

Ionic order. Athenian example from the 5th century BC.

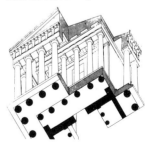

Schematic axonometric of part of the Parthenon at Athens (447–438 BC) designed by Ictinus and Callicrates.

Theater at Epidaurus (c. 350 BC), the best preserved example of its kind. Greek theaters were sited in natural amphitheaters.

Caryatid figure from the north porch of the Erechtheum on the Acropolis at Athens (420–393 BC).

they had earlier. The palace at Pelae, and houses which in Delos remain well preserved up to upper storey level, had elegant internal colonnaded peristyles around which the rooms focused, with walls of paneled stone or stucco imitating marble, and mosaic or pattern floors in rich materials. The ceilings were coffered, as the earlier temples had been. Cities began to be planned on regular grids, and axial approaches to public buildings were introduced. It was an age in which architects thought of designing groups rather than individual buildings. Many cities had large council halls with wide-spanning roofs (Miletus, 175 BC) and double-storeyed *stoas* (Stoa of Attalus, Athens, c. 150 BC—recently restored). Theaters began to be larger and more impressive (Epidaurus, c. 300 BC, Pergamun, c. 220 BC). They were superbly designed from the point of view of both acoustics and theatrical effect.

Little is known of Greek architectural theory, but there is considerable evidence to suggest that it was closely allied to the work of the philosophers, showing an understanding of geometry and proportioning, and deriving the latter from the mathematical progressions; arithmetical, geometrical, and harmonic.

The failure of modern research to produce by analysis a single coherent system of design which convincingly fits the buildings of the Periclean age suggests a more complex and flexible theoretical approach to design than the apparent simplicity of the buildings has hitherto led us to expect.

At its greatest extent the Alexandrian Empire introduced Greek architecture in the late 4th century BC to all countries between Italy, North Africa, and Egypt in the west, and the Indus valley and Bactria in the east.

surrounded by walls, with the gateway, the *propylaea*, often given a form second in importance only to that of the temple itself; it was usually some type of double colonnade open on both sides. Flanking the *temenos* were other ancillary buildings, notably one or more *stoas*, simple colonnaded buildings open to the *temenos*, which provided shelter from the sun in hot weather.

Periclean buildings are also remarkable for their combination of Doric and Ionic elements; in addition, several purely Ionic temples were built in Athens, the temple of Athena Nike (c. 425 BC) and the Erechtheion (c. 420–405 BC). In the latter, anthropomorphic columns, known as *caryatids*, were used to carry a heavy stone entablature, thus completing the fusion between architectural elements and the human form. At the temple of Apollo at Bassae (c. 430–400 BC), designed by Ictinus, one of the architects of the Parthenon, the Doric order is used on the exterior while Ionic and Corinthian columns (the latter making its first known appearance) on the interior. In the 4th century a number of highly refined and delicate late temples were built (Tegea, Priene, and particularly Ephesus, c. 350 BC).

The Hellenistic period (c. 323–331 BC), following the conquests of Alexander and affected by the sudden growth of prosperity and the new contact with oriental luxury, was marked by a shift of emphasis toward secular building. Religious buildings continued to be erected (Didyma c. 300 BC onward, the Olympeion at Athens, 174 BC onward), but the essential problems had already been solved, and development could only take the direction of overrefinement or experimentation with the Corinthian style. City walls, gates, castles, palaces and houses, and public buildings now received more attention than

Roman architecture

Although relatively little is known of Republican architecture, it seems to have evolved by the 3rd century BC into an individual style, part Etruscan, part derived from the late styles of the Greek colonies in Italy. The Hellenizing tendencies grew stronger with the conquests of Greece and Asia Minor and, by the time of Caesar and Augustus, Roman architecture was beginning to adopt some of the oriental qualities of eastern Hellenism, in particular an extensive use of vaults and domes. At first employed in the public baths, themselves probably oriental in origin, these structural features made possible the highly original complex buildings of Hadrian's time, such as the Pantheon in Rome (c. AD 100–125 in its present form) with a dome of 141 ft. (43 m) diameter. This essentially Roman architecture flourished until it reached its apogee in the great public bath ensembles, such as the Baths of Caracalla (c. AD 215) and the Baths of Diocletian (c. AD 306). Thereafter, Roman buildings of size and splendor continued to be built but less frequently; the Empire was converted to Christianity c. AD 323 (and after this date the late Roman style is dealt with under the heading BYZANTINE).

Early Republican architecture before c. 140 BC was limited to Italy. Following the conquests of the wars of Carthage it spread to Sicily, Spain, North Africa, and Greece, and eventually embraced the Mediterranean basin, extending north to Britain and Germany and east to Mesopotamia. Under the Empire, the Romanization of all regional architectures led to a marked uniformity of style throughout this huge territory. The impetus was largely the essentially military organization of the government, resulting in uniform policies and attitudes and the regular transfer of personnel from one region to another.

Roman architecture was always characterized by the use of arcuated systems of construction, derived from the Etruscans and subsequently the eastern Mediterranean, clothed by colonnades of Hellenic type used to articulate and lend scale. Buildings of complex planning evolved and were of an increasing size and majesty, made possible by a developing technology and science of building. The imaginative variations and growth in scale of the architecture were paralleled by the increasing size of the metropoles throughout the Augustan age, a trend which was only clearly brought to an end with the difficulties which preceded the early Byzantine period.

The buildings were organized using a rational system of planning based on symmetrical distribution of parts on either side of axial vistas. The vistas were marked by gateways, columns, stairs, triumphal arches, and obelisks, and flanked by pairs of sculptures or fountains. Axes were related by intersections at a visual feature, or at right angles in an open space, and the proliferation of axes to cover the landscape could proceed indefinitely. Within this ordered system variety could be obtained by changing building forms, materials, and the scale and treatment of openings.

Roman planning systems are well exemplified in the design of palaces: Nero's Domus Aurea, the "Golden House" in Rome, was a palace laid out on one axis in a straight line and following in the center the shape of the Oppian Hill, in the open plan of a colonnaded villa. While most of the main rooms look out across colonnades to views of the city, there was a trapezoidal courtyard in the center which separated a wing arranged around a large rectangular court on the west from the east wing, which centered on an octagonal domed hall from which radiated five entertaining rooms.

In the unsettled state of the later Empire an imperial palace assumed a military character,

Corinthian order, from the Temple of Castor and Pollux, Rome (AD 6).

Tuscan order, based on the Doric order, with no flutings but incorporating a base.

Pont du Gard, Nîmes, France (19 BC). This well-preserved three-tiered aqueduct is constructed of masonry without mortar.

ABOVE: Colosseum, Rome, (AD 70-80). The upper storey was added in AD 222–24. The three tiers of regular arcading are framed by a series of three superimposed orders; Tuscan on the lower storey, followed by Ionic and Corinthian.

BELOW: Pantheon, Rome, (AD 118–128). The large circular space with a diameter of 141 ft. (43 m) is covered with a hemispherical dome and lit by a single round opening at its crown.

Colosseum, Rome, (AD 70–80): interior view by G.B. Piranesi (c. 1750).

Arch of Constantine, Rome (AD 315). Originally surmounted by a quadriga, this arch commemorates the victory of Constantine over Maxentius.

Composite order. Invented by the Romans, combining characteristics of the Ionic and Corinthian orders, it was often used in triumphal arches.

as for example the Palace of Diocletian at Split, Yugoslavia. On a square plan, dissected by two straight axes lined with arcades, the palace complex was enclosed by a strong defensive wall which opened into arcaded galleries on the southern side facing the sea. There were three main gates on the land side, on the north, east, and west at the ends of the main axes. From the wider north-south axis, which corresponded in the palace complex to the peristyle of a Roman house plan, access was obtained to the four major subdivisions of the plan in the corners of its square shape, and to the area along the sea which contained the public and private rooms of the imperial residence; the former included the military barracks, the mausoleum of the emperor, a temple, and residences for officials and attendants. Many later palaces and castles followed models of this type.

The Romans often employed a system of proportioning based on circles and squares, semicircles and double squares, owing, perhaps, to the widespread use of semicircular arches, vaults, and domes. For example, in the Pantheon, the height of the walls is equal to the radius of the dome. In the same building the skillful use of a convex floor surface, which emphasizes the central vertical axis of the building, and its radiating surface pattern in circles and squares of alternating tones, testify to the subtlety of architectural design in the Augustan age. Light enters through a single opening, a wide circular hole in the center of the dome, which increases the feeling of enclosure within a huge globe-shaped cavern, symbolic of the remote universe of the gods. In contrast to the complete integrity of this internal conception, the external facades are disappointing: the Greek temple-type portico failing to achieve adequate preparation for the majesty of space contained within. In this sense of an introverted plastic architecture, Roman buildings repeatedly achieve qualities undreamed of by the Greeks, while externally not surpassing their tactile and sculptural achievements.

One Roman functional innovation, the public hall, in which courts of law were held, evolved from a Hellenistic structure of columns and lintels, with a trussed roof (the ancestor of the Early Christian basilica), to the vast concrete vaulted structure of the Basilica Nova of Maxentius (AD 310–313) in Rome, which served as an inspiration for later medieval and Renaissance churches.

Early Christian and Byzantine architecture

The transference of the capital from Rome to Byzantium in AD 330 is usually taken to mark the end of Roman architecture. Once Constantine had finally officially recognized Christianity by the Edict of Milan in 313 a number of basilican churches were erected—mainly in Rome—between c. 330 and 340, and very little other building took place in the city. It is for that reason that the name Early Christian is given to the period in Italy from c. 330 to the establishment of the Lombard Kingdom in 568.

Before c. 330, Christians had to worship in secret for fear of persecution. Their meetings were held either in private houses or in catacombs outside cities. The basilican churches founded in and around Rome during and shortly after Constantine's time include some of the most famous shrines of Western Christendom. Several have been entirely rebuilt, such as the great original basilica of St Peter's (330). Most of the others have been much altered; a typical example of Early Christian architecture in Rome is St Paul's-without-the-Walls, destroyed by fire in 1823 but faithfully reconstructed in its original form.

Early churches usually had a high central nave, together with a low flanking aisle on either side, separated by rows of columns, which were generally Corinthian and often rifled from older pagan buildings; there were round-headed windows in the wall above them, to light the interior; an apse at the east end; and a narthex or vestibule at the west end. Thus, except for the general plan, the churches present no new structural or architectural features; but the plan is important, for it was the ancestor of the great medieval churches of western Europe. Dating from the same period as these basilican churches, some round churches were also erected, usually over the tombs of martyrs and hence called *martyrions* but sometimes merely serving as baptisteries: in Rome, S. Costanza (330), the Baptistery of Constantine (430–440), S. Stefano Rotondo (c. 470); and, at Nocera, the Baptistery (350).

Basilicas were also built as places of worship by Constantine in his new capital at Byzantium (Constantinople). One of these was S. Sophia, destined to be rebuilt to an entirely different design two centuries later; also at this time the Church of the Nativity at Bethlehem (330) was constructed. Circular churches were also erected; over the sepulcher of Christ in Jerusalem a famous *martyrion* was built, much of which still survives; while it is clear that at Bethlehem a polygonal east end was built to surround the underground manger in which Christ was born. A large basilica adjoined the Holy Sepulcher in Jerusalem.

There are two interesting later basilicas in Ravenna: S. Apollinare Nuovo (493–525), and S. Apollinare in Classe (534–539); both contain fine mosaics and were erected by Theodoric, the Gothic King of Italy.

The art characteristic of the developed Byzantine Empire can be traced back to the period just before the reign of Justinian, c. AD 500. The style had enormous influence on both the East and the West. Early Byzantine art may to some extent be regarded as Roman art transformed under influences of the East. It reached a high point in the 6th century, rose again for a short time to new heights during the 11th and 12th centuries and still survives among Greek Orthodox communities.

Byzantine art was a development of Roman art influenced largely by Persian architecture and Greek culture, all of which found a common meeting ground in Constantinople. The dominant Byzantine art form was architecture. As in Early Christian times, the two chief types of church were the basilican and the circular or centralized. Of the latter type, the chief examples are SS Sergius and Bacchus (526, Constantinople), and San Vitale (526–547, Ravenna). The outstanding example of a building which combined the longitudinal qualities of the basilica with the centralized volume of the *martyrion* was the church of Holy Wisdom (Hagia Sophia) at Constantinople, which was built 532–537; it was designed by Anthemius of Tralles and Isidorus of Miletus.

Byzantine architecture in all its phases is characterized by the prominence given to the dome in contrast to the simple sloping wooden roof of the Early Christian basilica. At Hagia Sophia, the great central dome is buttressed by semidomes. The dome had been used in Roman architecture on circular buildings; but Byzantine domes generally rise from square bases, the transition from square to circle being effected by means of "pendentives." The vast dome of Hagia Sophia, 107 ft. (33 m) across, has its pier supports largely hidden from view in the central space by screened aisles and galleries so that it appears to float effortlessly, an effect which is enhanced by a ring of windows which pierce the dome at its lower edge. The contrast of void and solid, light and dark, produces an atmosphere of mystery in this and other Byzantine buildings which is wholly unclassical. The reflected light shimmering from the myriad facets of the flat surfaces of mosaic decoration heightens this unique effect.

Brick was the main material used for the construction of Byzantine churches; it was covered externally with plaster, and internally with thin marble dadoes and mosaics above. Byzantine decoration was flat and incised, in contrast to the bold modeling of Western surfaces. The Roman Corinthian capital was modified into a convex shape, and the foliage carved on it was either "windblown" in character or was decorated with new forms of ornament. Above it was placed a new feature, the "dosseret," a block from which rose arches, which, as in Roman work, were always semicircular. Columns, usually made of marble, were strengthened against earthquakes by bronze annulets.

Byzantine architecture of the period of Hagia Sophia was markedly concerned with mathematics and theory. The historian Procopius wrote of the great church, "through the harmony of its measurements it is distinguished

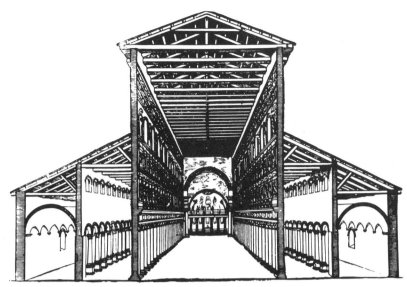

ABOVE: Reconstructed view of the original basilica of St. Peters, Rome (330).

ABOVE: Byzantine capitals from Constantinople and Venice. Capitals with deep dosseret blocks were popular and provided useful enlarged springing platforms for arcading.

RIGHT: Hagia Sophia, Constantinople (AD 532–537). The minarets were added when the church became a mosque in 1453.

ABOVE: Exterior apse of S. Vitale, Ravenna, Italy (526-547). The domes, built of interlocking earthenware pots, are covered by tiled roofs.

by indescribable beauty!'' Elsewhere, he observed that the design of ''a spherical shaped Tholos standing upon a circle makes it exceedingly beautiful!'' The architect, Anthemius, was the outstanding mathematician of his day, and is reported to have described architecture as ''the application of geometry to solid matter.''

By the 9th century, the Byzantine style was widespread throughout the countries of the Near East and eastern Europe, where the Greek Orthodox religion was followed, and was beginning to appear in Russia. Symbolism had now begun to dominate church architecture, each building being conceived as a microcosm of all earth and sky, as a setting of Christ's life on earth, and as a record in visual images of the liturgical year. Even the colors used in the internal decoration assumed significance in this triple symbolism.

These Byzantine churches followed the plan of a Greek cross, that is, a central, domed space with four short square arms (evolved c. 7th century). This form of church eventually became almost universal, focusing in the brilliantly lit central space, which dissolved mystically into the dark screens and galleries in the arms of the cross. Examples are to be seen in the small Metropolitan Cathedral in Athens, and at churches in Daphni, Mistra, Salonica, and Stiris.

Romanesque architecture

The generic term Romanesque is sometimes applied to embrace all the styles of architecture which, in most European countries, followed the Early Christian style and preceded the introduction of the Gothic style, c. 1200. It is often subdivided into pre-Romanesque, which includes the Lombardic, Carolingian, and Ottonian or Rhenish styles as well as Saxon and Romanesque proper, which is taken to have begun c. AD 1000. Romanesque architecture is not identifiable with Romanized Europe, for one finds a few Romanesque buildings in Scandinavia and Poland, which were never Roman colonies; whereas in southeast Europe, which was once Roman, the Byzantine style was followed (see EARLY CHRISTIAN AND BYZANTINE ARCHITECTURE). Romanesque is the term used for architecture in the countries of Europe where the Roman Catholic Church prevailed, while Byzantine is adopted in countries where the Orthodox or Greek Church was supreme.

All Romanesque architecture is, to some extent, a natural development from ancient Roman or classical architecture, but there was a period, commonly called the Dark Ages— between the collapse of the Roman Empire in the 6th century and the reign of Charlemagne in the 9th—when a reduced form of Roman architecture still persisted. In Italy, this is called late Early Christian, in England and France there are a few churches which could be similarly classified; in Spain there was the Visigothic style.

From the ancient Roman tradition, the pre-Romanesque architects adopted characteristic features: the semicircular arch, the groined cross vault, and a modified and simplified form of the Corinthian column with its capital of acanthus leaves. Occasionally, at an early period, they used carved fragments of antique buildings. They made important advances upon Roman structural methods in balancing the thrust of heavy vaults and domes by means of buttresses, and in substituting thinner webs supported on the curved stone ribs for the thick vaults used by the Romans.

The developed Romanesque style utilized easily comprehensible plans, volumes, and forms derived from the spanning of openings with semicircular arches, vaults, and, to a lesser extent, domes. Parts of the building were clearly demarcated from each other by articulating elements, such as moldings, string courses, vertical shafts, and archivolts; these were often emphasized by carved or painted decorations. A whole building could frequently be seen to be composed of a series of structural bays, each more or less self-contained, structurally and visually static, which could be removed or added to the plan at will. This sense of completeness and repose, Romanesque owes to its classical ancestry. In other respects, however, there is a striving for mystical and dynamic effects, in the contrasts of light and shade produced by rows of relatively small openings in large interiors, and in the upward thrust of the nave and aisles beyond classically balanced proportions.

The growth of saint worship in the medieval church led to systems of staggered and radiating apses at the east ends of religious buildings. The desire to educate illiterate people about the Bible encouraged the spread of painted scenes from the Old and New Testaments across the wall surfaces, and into the glass of the windows, creating a rich interior effect which must, at times, have paralleled that of Byzantine architecture.

The oldest buildings that can properly be called pre-Romanesque are found in Italy, such as S. Pietro, Toscanella; S. Maria in Cosmedin and S. Giorgio in Velabro, both in Rome; parts of S. Ambrogio, Milan, and Torcello Cathedral. All these are of the 8th–9th centuries and have been called Lombardic because Lombard kings ruled Italy from 568 to 774, their capital being at Milan. Charlemagne became Holy Roman Emperor in 800, and his cathedral at Aachen in Germany (796–804) is the first important Carolingian building. This is usually taken to be the beginning of pre-Romanesque in northern Europe.

The Romanesque period lasted two centuries, and was the great age of European monasticism. The Benedictine Order was

Cathedral Church of Vassili, Moscow. The bulbous domes are derived from Tartar sources.

Ste. Madeleine, Vézelay (1100). The nave employed some of the earliest pointed cross vaults in France.

Reconstructed cross section of the 3rd abbey church at Cluny, Burgundy, France (1088–c. 1121).

founded in Italy in the 6th century, but its chief buildings were of later date and of Romanesque character. During the period 900–1100, the Benedictines were followed in due course by the Cluniac, Cistercian, Augustinian, Pre-Monstratensian, and Carthusian Orders. The religious monastic orders were the chief vehicles for the dissemination of new architectural ideas during the Romanesque period. It is estimated that, at their height, they controlled half of the wealth of Europe, and almost all were wholly international in distribution and organization.

The 11th and 12th centuries also saw great activity in castle building, and the science of military architecture developed rapidly as the result of warlike contacts with Moslems during the Crusades. The architectural work of the Romanesque period therefore consists almost exclusively of monasteries, cathedrals, parish churches, and castles. Very few domestic buildings have survived, for there was hardly any middle class below the feudal lord in his castle. In England, the so-called Jews' Houses at Lincoln are a rare exception.

Romanesque churches generally followed the basilican plan, with aisles and an apse; but transepts were often added, and sometimes the aisles were continued around the apse to form an ambulatory. Towers were now popular, the campanile having been invented centuries earlier; Italy contains several graceful examples, while the churches of the Rhineland have gabled towers, and in England (e.g. Sompting in Sussex, 11th century) there are small towers of the same type. Numerous centralized churches, too, were built during this period, based upon the precedent of the Early Christian baptistery, and thus ultimately derived from ancient Roman temples and tombs. The Knights Templars built four circular churches in England, taking as their model the Church of the Holy Sepulcher in Jerusalem, originally founded by Constantine. Arcading (rows of semicircular arches, sometimes interlacing) was freely used as decoration on the towers, the walls, and even the interiors of churches. Windows were comparatively narrow and round headed. Doors also had round arches over them, the tympanum (the space between the lintel of the doorway and the arch) being usually carved. The jambs or sides of the doorway were decorated with crude and even grotesque carved moldings and small shafts were recessed in them. The roofs of Romanesque churches were steep in northern Europe, whether the building was vaulted or not, but of a much lower pitch in Italy, where the simple tiled roof of the Early Christian basilica was followed. Stained glass was introduced toward the end of the Romanesque period. Some of the finest examples of the use of stained glass can be seen at Canterbury Cathedral, England, Poitiers Cathedral, France, and at Augsburg Cathedral, Germany.

RIGHT: Abbey church at Laach, Germany (1093–1156).

ABOVE: Romanesque capitals. The cubiform capital (lower figure) is the basis of richly shaped capitals of the later Romanesque period as shown (upper figure) from the church of St. Sebald, Nuremberg, Germany.

RIGHT: S. Ambrogio, Milan, Italy. The plan dates from the 9th century and was completed in 1140. It includes the only surviving atrium among Lombard churches.

Benedictine abbey church at Jumièges, Normandy (1037–66); predating the use of vaulting in Norman France.

Cloister of S. Paolo fuori le Mura, Rome, Italy (1241). The delicately carved twisted twin columns are inlaid with glass mosaic.

Amiens Cathedral, France (1220–88), by Robert de Luzarches. The flying buttresses, a characteristic of French Gothic architecture, are lightened by delicate tracery carving.

Bourges Cathedral, France (1190–1275). The principal portal has a double doorway surmounted by a tympanum containing scenes from the Last Judgment.

Sens Cathedral, France, begun in 1143. View of nave looking toward the east end.

Gothic architecture

The architecture of the central Middle Ages was termed Gothic during the Renaissance because of its association with the barbarian north. Having lost its derogatory overtones, the term is now used to describe the important international style in most countries of Europe from the end of the Romanesque period in the 12th century to the advent of the Renaissance movement in the 15th century in Italy; and in the 16th century elsewhere. It was essentially the style of the Catholic countries of Europe including Hungary and Poland; and it was also carried to Cyprus, Malta, Syria, and Palestine by the Crusaders and their successors in the Mediterranean. The forms that were developed within the style on a regional basis were often of great beauty and complexity. They were used for all secular buildings, as well as for cathedrals, churches, and monasteries.

The inception of Gothic architecture dates to about 1140, when Abbot Suger initiated the redesigning of the chevet of the Abbey of Saint Denis, outside Paris. It attained its highest excellence in France and England. In Germany, Belgium, Holland, Spain, and Switzerland it was largely borrowed from one of those two countries, especially from France. In Italy, because of the strength of the classical Roman tradition, it developed in a different, and more restrained, direction; and, for the same reason, it quickly succumbed to the Renaissance, which itself echoed that tradition.

The characteristic feature of Gothic architecture at all its stages and in all the countries mentioned is the pointed arch. Up to comparatively recent times, some scholars attributed the origin of this feature to the shape produced by interlacing of semicircular arches in an arcade; but now the invention of the pointed arch is universally ascribed to the Middle East, where it was used on a large scale in the great mosque of Samarra in Iraq (mid-9th century) and Ibn Tulun at Cairo (876–879). The idea may well have been brought to France and England by the Crusaders.

The pointed arch is, however, more significant in Europe because its introduction enabled Gothic builders to solve many problems, especially those surrounding the use of ribbed vaulting (in Durham, 1109, and the abbeys of Caen, 1115–20), and thus to develop the elaborate and highly rational system of vaulting and buttressing which is the real basis of Gothic architecture, differentiating it from the heavier Romanesque. One result of the improved system of vaulting and, eventually, flying buttressing, was an increase of window area in the walls between buttresses, because these walls no longer had to carry the main weight of the roof; therefore they could be thinner and pierced with ease. The pointed arch was now used over windows, and these lancet windows were grouped in twos or threes under an enclosing arch, the remaining enclosed space being pierced with small circular openings. Later the stonework between all the various windows and openings in the group was reduced to slender stone bars, and the whole enclosed group of openings became a single window; the upper portion within the arch being filled with tracery, consisting at first of geometric patterns, then of flowing patterns, and finally of quasi-rectangular openings forming a grid. The effect of a grid (giving the name Perpendicular to Late Gothic in England) was mainly due to the introduction of horizontal transoms in the larger windows. The increased glass areas became a field for magnificent displays of stained glass which eventually dominated the interior character of the churches, such as the cathedral at Chartres.

The Gothic style developed rapidly in the region around Paris from the mid-12th to mid-13th centuries. Many great cathedrals were built or refashioned in the new style, of which Paris (begun 1163), Chartres (1194), Bourges (1190), Reims (1211), and Amiens (1220) were the most significant. With the collapse of the tower (550 ft./168 m) and the choir vaulting (157 ft./48 m) at Beauvais in 1284, the upward striving of the continental cathedrals was brought to an end. Although Cologne (1248) and Ulm (1377) were intended to be bigger in various ways, they had to wait until the 19th century for completion.

Meanwhile, English Gothic had taken another direction, concentrating on developing Romanesque complexity of forms in relatively long, low, diffuse buildings with extremely rich patterning and decoration. English Gothic came to an end with the final flowering of the Perpendicular style, seen at its best in King's College Chapel, Cambridge (1444), the naves of Winchester (c. 1480), and Canterbury (c. 1400). Flying buttresses, perhaps the most dramatic features of French Gothic cathedrals, are only occasionally found in England, e.g. at Westminster Abbey, which is the most French in character of the greater English Gothic churches. Fan vaulting, on the other hand, is a distinctly English invention. Open timber roofs are seen at their best in English Gothic churches and secular halls, the enormous oak roof of Westminster Hall (1397–99) being the finest example. This is a hammer-beam roof, a type peculiar to England. The so-called Tudor arch is also distinctively English, though a similar form had been used in Islamic buildings.

The development of Gothic in other European countries differed in many respects from that in England. Thus France had no late Perpendicular phase, but produced a Flamboyant phase unknown in England, though it is occasionally found in Scotland. Italian Gothic is again different from English and French, concentrating on great spans with simple arches on basilican plans, and some Italian cathedrals have striped exteriors of black and white marble. Many great German churches were built of brick.

Stained glass had begun to be used during the Romanesque period, but its full development occurred in Gothic times, and all other branches of craftsmanship made great advances. Color was freely used in the interior of the greater churches, not only in the form of stained glass but also by painting moldings and in wall paintings of religious subjects.

The theory of architecture during Gothic times derived from Scholasticism, the great medieval philosophical system which attempted to resolve all comprehensible conflicts in the real world by its special technique of "argument" then "counter-argument" (anti-thesis), followed by "resolution." Art historians have recently claimed (Panovsky) that this process of thinking could be applied to determine the direction of architectural development in the Gothic style, as well as to solve detailed problems of design.

Renaissance architecture

Beginning in central and northern Italy, the renaissance of the 15th century in architecture spread in the early 16th century to northern Europe, Spain, and Portugal, and soon after to the European colonies in America, Africa, and Asia.

The cultural center of Florence was the birthplace of Renaissance architecture; Rome, which contained many of the classical ruins from which the Renaissance architects first drew their inspiration, was a small city by comparison.

Filippo Brunelleschi (1377–1446), a young goldsmith of Florence, having been unsuccessful in the competition for bronze doors to the baptistery there in 1401, decided to turn to architecture, and visited Rome in order to make a firsthand study of the ancient buildings. His designs for completing the unfinished cupola of Florence cathedral were successful in 1420. He later built the Pazzi Chapel adjoining S. Croce in Florence; and this work may be regarded as the first Renaissance church attempting to wholly employ the principles of Roman architecture.

The fundamental tenet of the new Renaissance theory of architecture was the idealization of architectural forms in accordance with the concept of universals expounded in Platonic philosophy. The essential tool to be used to achieve this perfection was mathematics, which, in the shape of Euclidean geometry, could be observed to be present in forms of natural life and was, indeed, believed to underlie nature and the proportions of the human body. Brunelleschi himself discovered that geometrical mathematics could be used to establish the laws of visual perception in perspective. Architectural design therefore

RIGHT: Diagrams of nave arcading from different periods of English Gothic architecture.

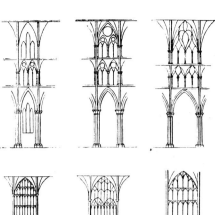

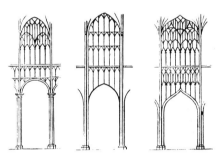

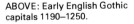

ABOVE: Early English Gothic capitals 1190–1250.

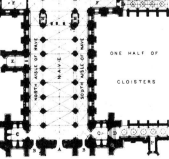

RIGHT: Plan of Wells Cathedral, England, built in stages (1180–1425). Only one-half of the cloister is shown.

Ospedale degli Innocenti in Florence, Italy (1419), by Filippo Brunelleschi. The columnar arcade with semicircular arches became a popular Renaissance theme.

Detail of the facade of the Palazzo Rucellai in Florence, Italy (1445–50), by Leone Battista Alberti. This was the first Renaissance palace to use a series of superimposed pilasters and entablatures.

Tempietto in the cloisters of S. Pietro in Montorio, Rome, (1502–10), by Donato Bramante. Based on Roman peripteral temples, it displays an early use of the Doric order.

became primarily concerned with the use of pure geometric forms, with the effects of careful visual proportioning, and with the mathematical relationship of objects in space.

During the next generation, Renaissance architecture was confined to Florence and its neighborhood, the main exponents being Brunelleschi himself and Michelozzo. By the middle of the 15th century, the movement had spread owing to the influence of Leone Battista Alberti (1402–72), a characteristic Renaissance man who obtained a doctorate in law and was talented as a poet and musician. Turning to architecture in middle age, he designed his first building in 1446; and wrote a book on architecture, *De Re Aedificatoria*, published in 1485, after his death. The subsequent importance of Alberti was partly due to the fact that he was the first architect since ancient Roman times to write on the principles of architectural design, and that he based these largely upon his observation of Roman architecture. In due course, he was followed by other Italian writers who adopted the same procedure. They included Sebastiano Serlio (1475–1554), Giacomo Barozzi da Vignola (1507–73), Andrea Palladio (1508–80), and Vicenzo Scamozzi (1552–1616). All of them were able to quote from the work on architecture written by the ancient Roman architect Vitruvius, the manuscript of which seems to have been discovered in 1414 and was published in Rome in 1486. Thus, architectural design in Italy, and ultimately the rest of Europe, became to some extent a matter of Roman and Greek precedent.

In Italy, the impatient genius of Michelangelo (1475–1564), a sculptor-painter turned architect late in life, rebelled against so much pedantic dictation; and his deviations from orthodoxy were partly responsible for the growth of Mannerism and led ultimately to the Baroque style. His masterpieces were the design of St Peter's in Rome (1546–64) and the Campidoglio (1539). Raphael (1483–1520) similarly rejected narrow canons of architecture, observing that the ancient Romans themselves had produced buildings rich in diversity. His disciples, Giulio Romano (1492–1546) and Baldassare Peruzzi (1481–1536), were among the leaders of the subsequent Mannerist phase. A generation later, Palladio led a return to classicist design in a prolific career of villa, palace, and church design focused on Vicenza (Villa Rotunda, c. 1550) and Venice (S. Giorgio Maggiore, 1566 and Il Redentore, 1576).

The stages by which the new Renaissance doctrines of architectural design gradually overcame the prevailing Gothic tradition varied from country to country. Traces of medieval influence remained in the northern Renaissance, such as the sinuous outline of gables, high-pitched roofs and, in many facades, the predominance of openings over areas of solid wall. The three channels through which the new gospel was communicated were: Italian crafts-

Palazzo Vendramini Calergi, Venice, Italy (1481), by Pietro Lombardo. The facade, which faces onto the Grand Canal, does not return around the sides of the building, and is composed to accommodate the traditional Venetian fenestration.

S. Giorgio Maggiore, Venice, Italy (1565), by Andrea Palladio. The composition is based on the idea of superimposing two temple facades on one another — a tall narrow one for the nave and a wide low one to accommodate the church aisles.

Bramshill House, Hampshire, England (1605–12). The design of the Jacobean porch by Gerald Christmas is based on plates of the Dutch engraver Dietterlin.

men and architects working in other countries; Italian books on architecture, such as those mentioned above; and visits of architects and wealthy aristocrats to Italy in order to study the Roman ruins. In the first category are crafts-men such as Pietro Torrigiano (1472–1528), who executed the monument of Henry VII in Westminster Abbey in London, England (1512), and the group of Italians who worked in France at Fontainebleau, Amboise, etc, for François I (1515–47), among them the architect Sebastiano Serlio. Italian books on architecture were translated and sold throughout Europe; but other, less authentic works on the orders of architecture and on classical design began to appear from the printing presses of Antwerp and Amsterdam. Some of them were produced by engravers with little knowledge of architec-ture and resulted in mere caricatures of classical buildings. Their effect on architectural design was naturally unorthodox, which explains the often curious architecture of the late 16th and early 17th centuries in such buildings as the châteaux of the Loire, Elizabethan and Jaco-bean houses and colleges in Britain, and the gabled town halls of German, Dutch, and Flemish cities. Some French architects of the late 16th century—among them Philibert Delorme (1500–70) and Jacques Androuet du Cerceau (c. 1515–c. 1590)—had studied ancient Roman architecture firsthand in Italy, as had the English architect John Shute (d. 1563) who published *The Chief Groundes of Architecture* in 1563. In 1624, Sir Henry Wotton (1568–1639), a cultured diplomat, published *The Elements of Architecture*, a small book based almost entirely on Vitruvius, but reinforced by per-sonal study in Italy.

Inigo Jones (1573–1652), although a poor man, managed to spend years of study in Italy, concentrating mainly on the works of Palladio; and, after returning to England, he was commissioned to design the Queen's House at Greenwich (1617–35) and the Banqueting House at Whitehall (1619–22). These two remarkable buildings introduced to England the genuine Renaissance architecture of Italy. Sir Christopher Wren (1632–1723) did not visit Italy himself, and his architecture was never so wholeheartedly Italian as that of Jones. He, and still more some of his followers, Thomas Archer (1668–1743), Nicholas Hawksmoor (1661–1736), and Sir John Vanbrugh (1664–1726), showed some sympathy with the Ba-roque. In the early 18th century, however, there was another swing of the pendulum in England back to the orthodox scholarship of Palladio and his advocate Inigo Jones. Hence its practitioners were called the Palladian School of architects. They included such famous names as Lord Burlington (1694–1753), James Gibbs (1682–1754), William Kent (1685–1748), and Colen Campbell (1673–1729) (see NEO-CLASSICISM AND ROMAN-TICISM).

Baroque architecture

Baroque was originally a term of abuse, implying misshapen, given by later classical critics to the architecture of the whole of the 17th century, and including in some areas the early part of the 18th century. In fact, the style was born as a result of the confluence of a number of tendencies in the architecture of the late Renaissance in the 16th century. At the beginning of the 17th century various works were executed in Rome which signaled the advent of the new style by displaying an interest in dramatic, almost melodramatic effects, and favoring a giant scale, a rather congested modeling of surface—with the plac-ing of moldings and columns close together—and a new emphasis on the importance of unity. All the elements were, in fact, sub-servient to this overriding idea of oneness. Hitherto, in the Renaissance and Mannerist styles, individual parts of the building had formed perfect compositions in themselves, so that the whole was built up of balanced parts; the effect was static and measured. The character of Baroque is a dynamic clustering of forms, with the individual elements being subordinated, to culminate in a vital and invigorating whole. Baroque buildings cannot therefore be understood in terms of an intel-lectual system—instead they must be experi-enced emotionally, the mind and feelings reacting to the orchestration of forms and spaces.

The idea that forms could be manipulated in defiance of classical canons goes back at least to Michelangelo (1475–1564). It was he who experimented with changes of scale, intro-ducing, in his designs for the external facades of St Peter's and the Campidoglio in Rome, the use of the colossal order which runs through two or three storeys; the scale is emphasized, as it had been in the ancient Roman baths and palaces, by the jux-taposition with smaller columns related to sculpture at eye level. The clustering of columns and moldings to suggest lines of force in buildings, often in conflict with one another, had also been developed in Michelangelo's earlier experimental designs, such as the Laurentian Library in Florence. Now these effects were further emphasized by the use of curves of contrasing shape. From the volute there developed the idea of the concavo-convex line, which linked late Renaissance gable facades, such as that of the Gésu in Rome with Leone Battista Alberti's Santa Maria Novella in Florence (1456-70) and hence with late Gothic gabled facades. All these tendencies are best exemplified in the work of two great Baroque architects in 17th-century Rome: Francesco Borromini (1599–1667) and Giovanni Lorenzo Bernini (1598–1680) (Palazzo Barberini, 1628, S. Carlo alle Quattro Fontane, 1633, St Peter's Square,

S. Carlino alle Quattro Fontane, Rome (1633), by Francesco Borromini. The undulating wall and a free use of the classical vocabulary accentuate the presence of this small facade in a narrow street.

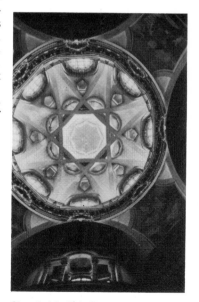

Chapel of the Holy Shroud, Turin Cathedral, Italy (1694), by Guarino Guarini. The dome has 16 intersecting binding arches supporting a large lantern allowing light to flood through some of the open panels of the structure.

Facade of the Cathedral of Santiago de Compostela, Spain (1738), by Fernando Casas Novoa. The mass of detail on the facade is disciplined by the strong lines of its form.

Elaborate gilding and embellishment were hallmarks of the Rococo style. Here, at the Amalienburg Pavilion in Munich, Germany (1734–39), the sequence of rooms opening upon each other creates a marvelous sense of lightness.

1656, the Scala Regia in the Vatican, 1633). The illusionist and theatrical aspects of Baroque made possible the fusion of painting, sculpture, and architecture in a way that had never previously been attained. In Bernini's Cornaro Chapel in Santa Maria della Vittoria (1646), the visitation of an angel to St Teresa is accomplished in a swirl of marble drapery and fluttering wings which is given a shimmering other-wordly intensity by the natural light flooding down from hidden windows behind the broken pediment of the architectural frame; radiating gilded rays emphasize the direction of the light and add a dynamic drama to the scene. In a similar way, Pietro da Cortona's (1596–1669) great ceiling mural in the vaulted salon of the Palazzo Barberini transforms the shallow curving surface into an extraordinary construction of colossal stone shapes surmounted by giants and crowds of floating figures. Da Cortona's own works in architecture are more subdued but are, nevertheless, splendid examples of the early Baroque style (SS Martina e Luca, 1635). Although Baroque facades are frequently independent compositions, distinct from the architecture of the interior, both convey an impression of compressed forms bulging and swaying with energy and force, and are expressive of a powerful unity.

Baroque architects felt emboldened to experiment with unusual and somewhat bizarre forms; the oval, the octagon, and star shapes were used, often interpenetrating with one another. Guarino Guarini (1624–83), one of the principal architects of the High Baroque, built a number of churches in Turin, Lisbon, and Paris. His Palazzo Carignano in Turin further developed the original and bizarre characteristics of the Baroque, while retaining a strict mathematical control, as Borromini had done. His complex interpenetrations of convex and concave forms, and his breaking through of vaults and domes with windows to provide unexpected shafts of light, had considerable influence on the Baroque styles of Austria, southern Germany, eastern Europe, Spain, and Portugal. In all of these countries, and in the new colonial world overseas, Baroque architecture flowered; in some areas it lasted another hundred years.

The Baroque desire to impress led to innovations such as the contrasting effects of construction and release, produced by leading the spectator into a small antechamber from which he emerged into a vast reception hall or cathedral nave, and to such effects as the infinite vista, made possible by continuing the architectural lines of a building into the surrounding landscape, where they could be led off to the horizon. The most influential example of the latter concept was executed at Versailles, where the king could be "seen" to be visually dominating France (see PALACES). Baroque architecture is, therefore, often iden-

tified as an expression of autocratic power, and this was true to the extent that great expense was often incurred in erecting its monumental grandeur. Nevertheless, Baroque architecture often achieved its effects by economical illusionist means, which might more justly be criticized for the rapidity with which they decay if neglected. Plaster, gilt, and imitation marble were common ingredients. The aim was not so much to express autocratic control as to achieve a magnificent effect and for a time this aspect of the Baroque even found a role in northern European countries such as England (Blenheim Palace, 1705).

By retaining the qualities of grandeur, large scale, and magnificence, but reducing those which introduced congestion and bizarreness, Baroque could also assume a Classicist guise. This form was especially popular in France and the Low Countries, (the east front of the Louvre, 1665, and the Royal Palace in Amsterdam, 1648–55).

Rococo architecture

Rococo architecture is a natural development from Baroque architecture. Like Baroque, the term originally began as one of derision, implying decorative rockwork and shellwork such as was found in grottoes and gardens. Rococo, again like Baroque, was less an intellectual, logical style, evolved from a rational use of architectural elements, than a purely visual style designed to create an emotional response.

In Rococo architecture the emphasis is very much on interior effect, facades being relegated to an almost utilitarian role, although retaining some of the lightness and delicacy that was so highly prized in the interiors. Whereas lines and forms had been clearly articulated in the Baroque, following Renaissance principles, the breaking up into separate distinct parts was abandoned in the Rococo, and forms were now allowed to flow freely into one another. An example can be seen in the transition from Baroque concavo-convex curved moldings, clearly made up of arcs of circles, into the continuous undulating curved profiles of the Rococo, in which oval and parabolic forms abound. The chief characteristics of Rococo expression are lightness, gaiety, playfulness, and wit—the emphasis is on delighting the eye. A spirit of freedom is implied, in contrast with the majesty and grandeur of the preceding Baroque.

Rococo was born when the nobility moved away from Versailles and returned to Paris with the rest of society on the death of Louis XIV in 1715. The new town houses were built in a free graceful style without the columns and heavy moldings of the old palace. Suites of rooms now opened freely into one another, and french windows gave easy access to the surrounding gardens. High windows and doors allowed light to penetrate everywhere, so that only the shallowest of moldings were necessary to decorate the surfaces; the effect of continuity of space and of brilliance of lighting was enhanced by reflections from countless mirrors. Corners were curved to eliminate sharp angles, and delicate gilt stucco work traced lines across the pastel walls to unite all the elements into an overall pattern. Where Baroque forms and lighting were variable and dramatic, Rococo forms and lighting were unified and restrained.

The period after 1730 saw the introduction of asymmetrical decoration into Rococo. This, again, seems to have begun in Paris and was taken up and popularized by the decorative schemes of the painter François Boucher (1703–70).

In their concern with the provision of an impressive sequence of spaces in a palace, Baroque architects had introduced the concept of the grand stairway; this gradually increased in size until it became the main feature in many of the larger palaces in Germany and Austria. The removal of walls dividing the stair hall from the entrance permitted an unbroken movement from the entrance to the main reception salon on the second floor which, in such palaces as Brüchsal (1730) and Würzburg (c. 1735) in Germany, was accompanied by exquisite transitions of character from the lower to the upper levels. At Würzburg the columns supporting the oval vaults under the stairs, which bridge the carriage drive through the building, seem to dissolve in multiple spatial and light effects. This illusion is heightened by the great ceiling painting by Giovanni Battista Tiepolo (1696–1770) spanning the upper stairs across a volume which seems to grow ever wider as one moves upward. Here the painting finally releases the volume of the upper stair into the infinite illusion of a painted blue sky.

The greatest of all Rococo churches is that of Vierzehnheiligen (1744) by Johann Balthasar Neumann (1687–1753) who started the building on existing foundations. The relatively geometric forms of the exterior are internally transformed by light curved screens of superimposed galleries into a series of five interpenetrating oval vaults; these focus on a delicate shrine in the center of the floor representing a vision of 14 saints encircling the infant Christ. Again, the integration of sculpture, stucco work, wood carving, gilt, and illusionist paintings with the architectural forms and materials is so skillfully handled that the building seems more an illusion than reality, an effect enhanced by the light streaming from numerous windows hidden behind screens and sculptures. No form or architectural space is separate and comprehensible—all are interwoven and interacting. Walls are no longer wholly enclosing, they are

"dematerialized." Yet any individual detail is tangible and can be admired for the exquisiteness of its shape and execution.

Rococo was eventually replaced as a style by Neo-Classicism, following a reaction against it which began in the 1740s.

Palace of the Belvedere, Vienna, Austria (1721–23), by Lucas von Hildebrandt. It demonstrates a playfulness typical of the 18th century in its white stucco and decorated end pavilions.

Neo-Classical architecture

This style spread through Europe in the second half of the 18th century as a reaction from Rococo and the excesses of the late Baroque. The term "Classicist" is used for any architecture or art which revives the principles of Greek and Roman art; Neo-Classicist is specifically used for the revival which occurred in the 18th century. Its particular quality is its faithfulness to the new science of archaeology which grew directly with the enthusiasm created by excavations at Herculaneum, Paestum, and Pompeii (1736 to 1756). A number of finely illustrated volumes were published on the archaeological finds, and the German archaeologist and art historian Johann Joachim Winckelmann (1717–68) was the first to interpret the art of the classical world in terms of its qualities of directness, balance, and correct proportion. The theorists of Classical architecture, Abbé Laugier (1713–69) and Carlo Lodoli (1690–1761), demanded that simplicity and rationality, derived from first principles rather than from mere imitation of Greek and Roman splendor, should dominate architectural design. The movement held strong ethical implications, rejecting the frivolity of Rococo social life and expressing a desire to restore the "ancient Roman" virtues to public and private life. At the same time, the engravings of Giovanni Battista Piranesi (1720–78) inspired a vision of Roman architecture which emphasized its formal and spatial qualities.

At first the Neo-Classicist style was largely

Lansdown Crescent, Bath, England (1792), by John Palmer. As befitted a spa dating back to Roman times, its streets and squares were designed in an austere Classical style.

High above the Danube, the Walhalla at Regensburg, Germany (1830–42), by Von Klenze, appears as a Greek temple transported to northern Europe.

Plate from Thomas Hope's *Household Furniture and Interior Decoration* (1807). The interior decoration is inspired by an interpretation of classical mythology.

Cannon Foundry – an engraving from C.N. Ledoux's collection of schemes for institutional and other buildings for an ideal city (c. 1804).

rievitch Zakharov (1761–1811) in Russia. Decoration, including classical enrichments, was restrained and sometimes dispensed with altogether.

Although in the 18th century few architects took Neo-Classical principles to their logical conclusion, in the best work the orders were used structurally rather than ornamentally, columns supported real entablatures and were not merely applied to the walls. The principle of the Baroque and the Rococo, whereby unity in a facade was achieved by subjugating the various elements to the central focus, was rejected in favor of a new clarity of parts which was also reflected in the assembly of the volumes of the buildings, which were sometimes almost brutally juxtaposed.

The morality implicit in the Neo-Classicist movement found an echo in late 18th-century projects for visionary social schemes, such as ideal cities and Ledoux's government Salt Works. Although these ideas were bold in conception, they always proved too expensive or too impractical to be realized.

Romantic architecture

Although Romanticism is often considered a movement in opposition to Classicism, in architectural terms both Classicism and Romanticism in the 19th century incorporated ideals which were beyond the ordinary everyday world. Both embraced concepts of nobility, grandeur, virtue, and superiority and looked to other ages for their inspiration. Classicism, like Gothic, claimed to be a "natural" style on the grounds that it had evolved as a direct response to materials and techniques in building—in fact, both styles were used poetically and romantically.

The roots of Romanticism go back far into the 18th century. Its earliest manifestation is usually taken to be the English landscape garden. Here, a deliberate reconstruction of the chance effects of nature evoked the Arcadian idyllic landscapes of Nicolas Poussin (1594–1665) and Claude Lorrain (1600–82). The introduction of a ruined temple or chapel into a landscape garden gave it a literary connotation. A ruined temple had more poetic significance than a new one because it was more evocative of the passage of time and the decay of human endeavors; the transfer of attention from the perfection and rationality of the complete building to the broken and partial qualities of a ruin provokes meditation on its philosophical, spiritual, and emotional qualities. The view that finds a ruined temple beautiful is a Romantic view, though the temple may once have been classically perfect.

inspired by Roman buildings. Its chief exponent in England was Robert Adam (1728–92), who had spent several years in Italy and Yugoslavia studying ancient Roman buildings and interior decoration. After 1758 Adam created a new style in domestic decoration based on Roman interior design; he rejected the prevailing Palladian style, since it applied inside the house the heavy columns and cornices of Roman external facade architecture. He replaced it with a style in which all massive details were omitted, and walls and ceilings were divided into geometric shapes by the lightest of moldings, delicately enriched by classical Roman ornaments such as swags of wheat husks, fluted fans, and winged sphinxes. This was all executed in molded plaster, painted in simple pastel shades. Delicate classical motifs, with panels of bas-relief sculpture, were also incorporated in the design of fireplaces and door frames.

After the publication in 1762 of James Stuart's and Nicholas Revett's *The Antiquities of Athens*, the severity of Greek Doric architecture began to be influential. It was accompanied by a belief in the inherent nobility of early architectural styles—a conviction that architecture and society had been at their purest and best in their simplest and most primitive forms. The emphasis on simplified and even severe elements culminated in the creation of an architecture of pure geometric forms—the cube, pyramid, cylinder, and sphere—in the work of Etienne Louis Boullée (1728–99) and Claude Nicolas Ledoux (1736–1806) in France, Sir John Soane (1753–1837) in England, Friedrich Gilly (1772–1800) in Germany, Benjamin Henry Latrobe (1764–1820) in the U.S., and Adrian Dmit-

With the growth of the Industrial Revolution, the expansion of the middle classes in England and France together with the isolation of educated and cultivated groups, led to a nostalgia for what was past in preference to the new material values. Eventually the Romantic attitude became an indirect protest against the inadequacy of contemporary reality, its social conditions, and its corrupt political tenets. Taken up by the newly rich, so that it finally became middle class in attitude, the Romantic movement, nevertheless, permeated the whole of society.

The Gothic Revival began in the mid-18th century with architectural conceits of the literati, such as Horace Walpole's (1717–97) Neo-Gothic villa of Strawberry Hill, near Twickenham, England (c. 1755), and William Beckford's Fonthill Abbey, also in England (1796–1807). The latter was a large country house in the form of a cathedral with an immense central octagonal tower. In the early 19th century, Gothic came to be seriously considered as the correct style for building new churches; it became closely identified with various religious revivals and was advocated by architects like Augustus Welby Pugin (1812–52) as the only suitable style, provided it was based on a correct understanding and careful imitation of the original Gothic of the Middle Ages. By 1830 the style most imitated was that of the 13th century, but the architect Sir Charles Barry (1795–1860), a Classicist by preference, used the English Perpendicular style when, with Pugin, he executed the British Houses of Parliament. Ruskin, however, in the mid-19th century, advocated the richer Venetian Gothic. In the latter part of that century scholarly imitations gave way to more original adaptations of Gothic and Romanesque principles by British architects such as William Butterfield (1814–1900) and George Edmund Street (1824–81) and, in America, Frank Furness (1839–1912) and Henry Hobson Richardson (1838–86).

Closely aligned to the Gothic revival was the taste for the "picturesque." Also originating in landscape architecture, but soon applied to rural building, the taste for the picturesque led to the cultivation of asymmetry, irregular building outlines in both plan and elevation, and an excess of variety, ornament, and whimsy.

The Gothic style dominated 19th-century architecture in many countries, notably England and America, even for civic and commercial buildings. Under the influence of Romanticism, severe Neo-Classical ideals were abandoned in favor of styles richer in decoration, more picturesque in composition, and more literary in their allusions to the past. The Neo-Classical style was replaced by a form of Revivalism, either Greek Revival, Roman Revival, or Renaissance Revival. Architects in the 19th century often resolved the confusion of allegiances demanded by the exponents of these various Romantic styles by resorting to a form of eclecticism, favoring Gothic for religious and educational buildings,

and sometimes town halls and law courts, and Classical Revival styles for houses, palaces, and commercial buildings. There were, however, curious anomalies, such as the preference for Gothic styles by many designers of railroad stations. In North America, the Greek Revival came closest to assuming the proportions of a national style. The style was also that most favored in Germany, Austria, and Russia. Under the various French empires of the 19th century the Imperial Roman styles were revived to give authority to the new regimes; the same course was followed in Italy.

Eclecticism and the "battle of the styles" came to an end under the impact of industrialization in building, the use of new materials, and the eventual triumph of common sense. The unsuitability of the Revival styles for many of the new building forms, such as the skyscraper, was eventually recognized—although one of the highest buildings, the 52-storeyed Woolworth Building, New York, was executed in the Gothic style in the early 20th century by Cass Gilbert (1859–1934).

Islamic architecture

Islamic architecture focuses on a religious architecture deriving ultimately from the teaching of Muhammad, the prophet who lived in Arabia at the beginning of the 7th century AD. It contains elements taken from the creative traditions of many conquered countries. One of the outstanding achievements of Islam was the successful fusion of many diverse cultures into a new whole, which eventually asserted its universality throughout the Islamic world.

Royal Pavilion, Brighton, England. Designed by John Nash in 1815–21 in an Islamic Indian style, it is one of the most theatrical of buildings, creating an impression of fantasy and opulence.

All Saints, Margaret Street, London, England (1850–59), by William Butterfield. The brick structure of the building is richly patterned in black and red — "constructional polychromy."

St Pancras Station Hotel, London (1867) designed in the Gothic style by Sir George Gilbert Scott, built as an independent building in front of the large single arch train shed.

Courtyard view of the Great Mosque in Damascus, Syria (706–716).

Portal of The Royal Mosque, Isfahan, Iran (1612–30), built by Shah Abbas.

The first surviving monuments are those of the Ommayad dynasty of caliphs, who transferred the capital from Arabia to Damascus in Syria in AD 661. With this move the Romano-Byzantine character of Ommayad art was firmly established. The Great Mosque in Damascus (706–715) was built within the walled precinct of a Roman temple, and its Roman corner towers were used as platforms for the call to prayer. Both the mosque and the octagonal shrine, The Dome of the Rock, erected a few years earlier in Jersualem, used a late Roman structural system and mosaic decoration of Byzantine type. Other decoration included window grilles in geometric patterns derived from late Roman floor and ceiling patterns.

A major function of the leading mosque in each city was, besides congregational prayer, its use for the delivery of the Friday sermon by the caliph or his representative. The mosque plan was early crystallized to include a columned or arcaded prayer hall, an open-air courtyard, and a high surrounding wall. The wall on the side nearest Mecca, the *qibla*, indicated the direction of prayer and sometimes included a niche or *mihrab*. The sermon was delivered from a flight of steps, the *minbar*. Sometimes one or more minarets were built on the entrance side of the mosque from which the call to prayer was made.

In the mid-8th century the Ommayad dynasty was replaced by the Abbasids, with their capital in Baghdad instead of Damascus. From then onward Persian and Asiatic influences began to figure strongly in Islamic architecture; brick was used more often as a building material, and ornament of oriental type appeared over the interior and exterior surfaces.

In Spain, where the last survivor of the Ommayad dynasty ruled after fleeing from Syria, the Syrian tradition continued to flourish and develop. In the late 8th century the Great Mosque of Cordoba contained many features in plan and decoration which derived from the East, but it also contained innovations from the Iberian Peninsula. One of these was the introduction of two rows of superimposed arches above the columns to increase the height without weakening the structure. The result was an effect of complicated spatial geometry, with striped horseshoe arcades stretching upward and outward—an effect which inspired architects in the next two centuries to add complicated superimposed scalloped arches, interpenetrating each other, in the richly decorated *mihrab* areas of the mosque.

The rise of the Turkish tribes to positions of power in the Islamic world saw the introduction of new architectural forms and types of expression. One of these was the domed mausoleum, such as the Mausoleum of Isma'il the Samarid at Bukhara (c. 940). Here the ancient flat brick of the Sassanians is used to create architectural features and a wide variety of decorative patterns over the interior and exterior surfaces.

The extent of Turkish power steadily increased until by 1100 it embraced almost all of Asia Minor, which from then onward remained thoroughly Turkish. The earlier prohibition against the building of monumental mausoleums by Moslems was to a large extent broken, although the presence of a tomb for a secular ruler could often be justified by including it in a teaching building, a *madrasah*.

In Persia a traditional royal audience hall, the open *iwan*, of the kind preserved from Sassanian times at Ctesiphon, was incorporated into the mosque. Like the royal audience hall, the *iwan* had behind it on the side nearest Mecca a domed chamber, which became in the mosque the domed space in front of the *mihrab*. The earliest mosques of this kind had probably only one *iwan* on the *qibla* side. But by the time of the rebuilding of the Great Mosque in Isfahan in the 12th century a large *iwan* faced into the court on each of the four sides, the additional ones apparently serving as *madrasah* platforms for the teaching and study of the Koran. From then onward this Persian type of mosque became increasingly influential; not only was it copied in mosques from Cairo to India, but it was the model for *madrasahs*, *caravanserais*, and many other Islamic buildings. The type reached its culmination in the Royal Mosque built by Shah Abbas at the southern end of his great Maidan in Isfahan (1612–30).

The entrance to many medieval mosques was accentuated by placing a pair of minarets over the gateway. In the Royal Mosque at Isfahan a twin pair flank the great *iwan* on the *qibla* side. But the chief glory of the Royal Mosque gateway is its "stalactite" vault. This epitomized a particular Islamic tendency to transmute and develop structural features for decorative effect. For the stalactite vault began as a series of superimposed squinch arches bridging the corner between a dome and the square plan below it. Quite early it was realized that the squinch arches could be interlocked by placing the second row so that it rose from the points of the first row, and so on. By also projecting the center of each arch outward, the upper row cantilevered over the first, to create a sloping surface bridging the corner. Experiments revealed that such vaults could be evolved to fit perfect geometric patterns in plan; by varying these, wonderfully complex and diverse effects could be produced in the vault. The final step was to achieve the quality of pendants, like stalactites, which seemed to drop down to provide a magical support for each tiny vault. Built in the time of Shah Abbas these exquisite vaults are, like surfaces in his buildings, entirely clothed in ceramic tiles of breathtaking perfection in design and execution.

Turkish architecture was characterized by its extensive use of domes, a tendency which was reinforced by its contact with the architecture of its great enemy, Byzantium. In particular, a type of mosque was evolved with a large central dome over what had formerly been the courtyard; a development which doubtless resulted from the vicissitudes of the northern climate. After the conquest of Constantinople in 1453, and the conversion of Hagia Sophia into a Friday mosque, Turkish architects became preoccupied with the attempt to surpass this masterpiece of Byzantine design. Eventually the great architect Koca Sinan (1489–1578) did so, in the Sultan Selim Mosque at Edirne, which not only had a slightly larger dome (102 ft. (31 m) in diameter), but avoided the multitude of supports in the earlier building by supporting the dome on eight massive columns. The latter are placed close to the walls, and the opening of the interior through a number of large windows created the effect of one large, airy, unencumbered space. The sense of unity is likewise emphasized on the exterior by the placing of four high minarets close to the rising mass of the central dome. In the 16th century, after the Turkish conquest of Syria, Egypt, and North Africa, Turkish taste predominated in much of the Islamic world. But by the 19th century European influence was making increasing inroads into the traditional architectural and decorative values of Islam.

(See also HOUSES—ISLAMIC.)

Japanese architecture

Traditional Japanese buildings were all of timber-framed construction except for city walls and the outer walls of forts. The roof span was achieved without trusses, so that spans were limited to the length of timber available; quite large roofs were, however, made possible by employing rows of internal columns.

The earliest pre-Buddhist Shinto shrines, Ise and Naiku, are raised on piles and have thatched roofs, showing in these and other ways an affinity with southeast Asian examples. The forms of the shrines were originally similar to those of palaces and dwelling houses, and when eminent people died, the early Japanese consecrated their relics as objects of worship. As the building material was wood, with a frail thatched roof, the life of a Shinto shrine in its original materials was on an average about half a century; since any major alteration, such as reroofing, involved a new dedication of the building, the practice grew of rebuilding Shinto shrines completely on adjoining sites every 25 years or so. But in rebuilding these structures the original forms

of the ancient shrines were always faithfully preserved. Shinto shrines were associated with sacred elements in the landscape, and consequently were sometimes placed in remote sites necessitating and encouraging the careful relationship of buildings to landscape with a skillful use of stairways and ropes linking one temple building to another (Kurongdanji Temple, Kyoto).

Buddhism reached Japan by way of Korea (c. 575), bringing with it the highly developed architecture of China. Two types of Buddhist temple were built. The first, imported from Korea, had all its elements arranged on a central axis in a symmetrically disposed enclosure. Entering through a gateway with columned porticoes front and back, the visitor was confronted first with a high pagoda and then directly in line with it, and behind, a transverse prayer hall (*Kondo*) in which the statues of Bodhisattvas were placed; finally in the rear and on the same axis, another transverse hall (*Kodo*) contained further treasures and statues of Bodhisattvas. This type is exemplified in the Shitennoji Temple, Osaka. The second arrangement has the pagoda and *Kondo* standing side by side to the right and left of the central axis, which continues through to the *Kodo* in the rear. This is the classic arrangement seen at the early temple of Horyuji, near Nara, begun in AD 607 and reconstructed in AD 670. The latter was built by monks from China, and is the best evidence we have for the temple styles of the great Tang period in China. The framed constructions in Japanese cypress represent the oldest timber buildings still standing in the world. The pagoda is five storeys high, of square plan with separate roofs over each floor projecting far out on corbeled carved brackets of Chinese type. The pagoda is crowned with a tall finial of metal rings and bells, marking the upper limit of a tall central

Kiyomizo Temple, Kyoto, Japan (1633). The timber structure is constructed out of interlocking members without the use of nails.

North gate of the Imperial Palace, Kyoto, Japan.

Imperial Palace, Kyoto, Japan, last rebuilt 1855. Interior view, characteristic of the medieval form of Japanese palaces.

post; about 100 ft. (30 m) high, which runs down to the ground in the center of the structure. The essential features of Buddhist temples remained the same throughout the succeeding centuries, although with many variations in size and layout. The simplicity of the earliest buildings, with low-pitched roofs exhibiting subtle curves, was later replaced by steeper roofs with pronounced curving lines; this was accompanied by greater enrichment and sometimes heaviness of detailing.

The later Shinto shrines retained their links to the earliest shrines and therefore often had thatched roofs and natural wood rather than painted structures. In the finest of these buildings, proportion, simplicity, and the eloquent use of natural materials are the dominant characteristics. Later Shinto shrines exhibit influences from the Buddhist temples, and from later Chinese mainland styles.

(See also HOUSES—JAPANESE.)

Chinese architecture

Very little Chinese architecture survives from before the 6th century AD because it was built almost entirely of wood. Raised on a stone or earth platform, a characteristic Chinese building had a framework composed of a number of heavy, tree-trunk columns carrying cross members on which the posts supporting the sloping rafters were placed. The roofs were of thatch or of thick clay tiles, usually projecting beyond the line of the walls with a wide overhang carried on a system of interlocking brackets. Owing to the weight of the roof, and the fact that the roof construction was not trussed in wood but simply depended on the strength of the beams spanning between the posts, the width of the building was strictly limited, unless there were internal posts; on the other hand, the length of the building could in theory be any number of bays.

Such a system of construction lends itself to the "pavilion layout," in which each building is conceived as a single rectangular unit, arranged in relation to other similar buildings which could be increased in number according to the scale of the accommodation; the pavilions were arranged around courtyards, or connected by open galleries. The most characteristic arrangement of pavilions was to place them transversely across a continuous axis which passed through their centers; such a system was used for early Taoist temples, in which the visitor passed through first the entrance pavilion, then a courtyard leading to the reception pavilion, with behind it another courtyard, then the main ceremonial pavilion. Palaces and large houses followed the same plan, the number of courtyards and pavilions on the main axis proliferating with the increasing importance of the building. Secondary axes sometimes allowed transverse buildings

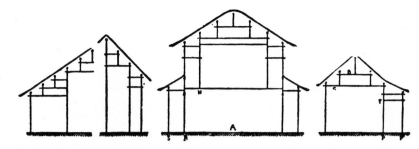

Diagram of Chinese roof framing composed of posts and beams. The triangulated truss was not part of the Chinese structural vocabulary.

to move away from the central axis at right angles; alternatively, and more commonly, there were parallel minor axes on either side of the central axis.

Shang dynasty buildings (c. 1766–1122 BC) were always roofed in thatch, and the walls between the timber columns made of wattle and daub. In the capital An-yang a timber hall some 98 ft. (30 m) long has been uncovered.

During the Chou dynasty (1122–221 BC) there is evidence of the common use of lookout towers, one of which is described as being nine storeys high, apparently the prototype of the later pagoda. Clay roof tiles were in use before 770 BC, and baked bricks soon afterward as infilling materials for the spaces between the columns.

The Han dynasty (206 BC–AD 220) was a period of great palace buildings. One is described as being 492 ft. (150 m) long and 328 ft. (100 m) high. Such palaces were flanked by tall timber towers and brick or stone towers. Buddhism entered China at the end of the Han period, and in the subsequent centuries a great number of Buddhist temples and monasteries were built. Each had a pagoda as a shrine or memorial, which signaled its position, as well as a series of transverse halls on a central axis following the pattern of the Taoist temples already described. The pagoda was the Chinese equivalent of the Indian Buddhist shrine, the *stupa*, which contained a relic of a holy man to serve as a model for the virtuous life. The Chinese pagodas served as relic houses in the same way, but took their pattern from the earlier Chinese timber towers of which they were an enlargement and refinement, now reaching up to a height of 12 storeys. Through the middle of each pagoda ran a mast which projected at the top and was ringed with metal disks reminiscent of the sacred umbrellas which in India served to mark the importance of the center of the *stupa*. More permanent pagodas were erected in brick and stone, the multistorey character of the timber pagodas being represented by projecting string courses with architectural features borrowed from timber work indicated in relief. The earliest surviving pagoda is that at Sung Shan, a 12-sided stone building of Indian shape and detailing (c. 520).

Chinese culture reached a high point under the Tang dynasty (AD 618–907). Apart from a few masonry pagodas, however, very few

Chih-hien-Tien (the Temple of Heaven), Peking, was built in the Ming Dynasty, work beginning in 1420. Set on a three-tier marble terrace and approached along axial paths.

Tang buildings survive. We can obtain an impression of Tang wooden walls, such as were used for both temples and palaces, from the surviving Chinese-influenced *Kondo* of a number of temples at Nara in Japan. The earliest similar structure to survive in China itself is the relatively small main hall of the Temple of Nan-Ch'an Ssu in northern Shansi province (782). Like this hall, almost all later Chinese buildings were of one storey, though a purely decorative attic might be added to increase the impression of size. During the Tang period the system of clustering corbeled brackets at the top of each column to reduce the span of the rafters and beams began to be developed in an elaborate yet logical way.

The Sung style of architecture (AD 960–1279) saw the widespread adoption of the distinctive curve of the roof, the line of eaves now curving up at the corners and the ridge having a sagging silhouette. The origin of this curved roof is not fully understood; it is relatively easy to achieve in the Chinese system of construction, since longitudinal curves can be produced by varying the heights of the columns and transverse curves by varying the lengths of the cross beams which are supported on the vertical posts above the main cross beam—there are no trusses, which would necessitate the straight lines of triangulation. It is now believed by most scholars that this curvature of the roof and the ridge was borrowed, because of its attractive appearance, from southeast Asia and Indonesia, where curved roofs seem to be associated with boat symbolism. Accompanying these sweeping roofs, the bracket clusters of the Sung period become more complex and varied. The projecting cantilever arm (*ang*) which hitherto had been no more than a kind of second rafter, anchored at the inner end to a cross beam, now becomes a free lever arm, balanced on the bracket cluster and supporting, at each of its ends, main purlins which run longitudinally to support the roof. The resulting system has something of the dynamic balance of Gothic architecture. In the best examples the buildings using this principle achieved wide spans without any cross beams.

Sung architecture is often considered to be the high point of Chinese building design; architects of this period were much more adventurous in experimenting with interlocking roofs and with buildings at different levels than their successors in later centuries. The interiors were also much more elaborate, employing a language derived from miniature buildings, and crowning the whole with ceilings with rounded vaulted cornices, and even cupolas executed in wood over the principal images.

Under the Ming dynasty (AD 1368–1644) China once more became a great power. In 1421 Peking was made the capital, and the Forbidden City was established on the site which it still occupies today, though it has been considerably enlarged and entirely rebuilt. The Chih Hua-ssu Temple, however, is a standard architectural group, completed in 1444, which is still well preserved. The architecture of the Ming and the subsequent Ching dynasties is simpler than that of the earlier styles, except in the richness of its decoration; the interlocking brackets, for instance, having degenerated into an intricate frieze. Its attraction lies in the pleasant grouping of halls and pavilions, and in the sweeping expanses of tiled roof which dominate and order the intricate and brightly painted decoration.

(See also HOUSES—CHINESE.)

Indian architecture

The Indian subcontinent was the focus of one of the world's major cultures, which extended its direct influence to the Indonesian archipelago and Indochina in the east, and to East Africa in the west.

The earliest civilization was that of the Indus Valley, which had an urban architecture that was sophisticated and highly developed. The baked brick cities of Harappa and Mohenjo-Daro were laid in a regular grid plan; the houses were on two or more levels surrounding paved courtyards, containing drains and garbage disposal chutes.

The Aryan architecture which succeeded produced many large cities in the Ganges valley dating back before 600 BC. There have been limited archaeological excavations, but these have revealed such buildings as the fine Mauryan palace at Pataliputra, erected by Chandragupta. Built of brick, with superstructures of wood, its splendor is probably reflected in the Buddhist friezes at Sanchi (2nd century BC).

The Buddhist stupa is said to have been instituted by the Buddha himself, but the earliest datable examples are from the reign of Asoka (3rd century BC). Its hemispherical form symbolized the universe; the incorporation, from the 2nd century BC onward, of four richly decorated gates reflecting ancient solar cults may be seen at Sanchi and Badami.

The earliest Vedic temples were of wood and were thatched; only their foundation fragments survive. Some indication of their circular, square or cruciform character may be glimpsed from rock-cut caves of the 5th century BC. Early Buddhist shrines were similar in form and likewise mainly of wood with thatched roofs. The temples of a later phase (Mahayana) were built by the Andhras for Buddhist communities in the Deccan and were hewn out of rock, imitating timber structures; a characteristic temple had three aisles, the central nave flanked by rows of stone pillars leading to an apse

Yi-he-Yuan (the Summer Palace), Peking (late 19th century). The richly decorated timber structures are surmounted by the distinctively curved lines of tiled roofs.

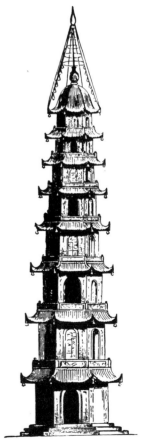
19th-century drawing of a Chinese pagoda — the porcelain tower of Nankin.

The Great Stupa at Sanchi (c. 150 BC), a prominent example of ancient Bhuddist architecture in India.

The Great Temple, Madura (1623). One of the greatest of all Hindu temples in the south, with massive gate towers (gopura) appearing beyond.

Taj Mahal, Agra. Approach view built by Shah Jehan in 1630.

containing the *chaitya* (shrine). Among the finest examples of these sanctuaries are those at Karli, Ajanta, Ellora, and Bhaja. Similarly, groups of monasteries (*viharas*) hewn out of natural rock cliffs were attached to these temples in imitation of earlier freestanding monasteries of which only foundations remain. In them, a number of small cells are grouped around a central space representing the courtyard. Buddhist art reached a zenith in the late Andhra period, in the Great Stupa at Amaravati.

The conquests of Alexander introduced Western art and architecture to India. Foreign influence is apparent in Kushan art: Corinthian capitals and Greek-style statues were found at Gandhara. The Guptas, however, saw a golden age in the middle of the 1st millennium AD, when foreign influence subsided and Hinduism was revived, evidenced by the freestanding Durga temple at Aihole. The Pataliputra and Mathura monasteries and the Buddhist university of Nalanda are among the last Buddhist buildings to have been built in India until modern times. The towered sanctuary, or *sikhara*, appeared for the first time at the Durga Hindu temple. The richness of its reliefs contrasts with the surrounding unadorned surfaces, which retain a simplicity of form.

In Hindu eyes, religious architecture is seen as the production of magic replicas of sacred beings, the temple (*vihama*) being the house and body of the deity, the plan a square of smaller squares dedicated to deities around the central Brahma, the Mandala. The shrine consists of a perfect cube, the *sikhara*, being twice its height. Building was based on the *sastras*, craftsmen's books. At Khajuraho (c. 12th–13th centuries) the *sikharas* were parabolic rather than pyramidal, rising in a crescendo of curves. Many Indian temples in the ornate Gujerati style, covered in deep relief carvings, were subsequently destroyed by earthquakes and by Mahmud of Ghazna. The style had reached its culmination in the Jain temples of Mount Abu, such as the 10th-century Dilwara shrine and the 13th-century Tejpal temple; built of white marble, its delicate lacework ornamentation veils the architectural construction. On the eastern side of the continent, elaborate sculptural ornamentation merged with the architecture in the Seven Pagodas, rock-cut temples of the Pallavas at Mahabalipuram, 8th–9th century AD. The Shore Temple is particularly fine; its door faces the sun rising from the sea, and it is the only survivor from a number of such temples.

The first Islamic mosques in India were built using columns from destroyed Hindu and Jain temples, such as the Quwwat al-Islam Mosque in Delhi (begun late 12th century), where the Qutb Minar, the largest of all minarets, was erected c. 1232. The Bahmanis built the Great Mosque at Gulbarga, with an immense prayer hall covered by a multitude of small domes (c. 1367). Persian influence introduced by the Moguls modified the austerity of the buildings of the early sultans, bringing with it the bulbous dome, cupolas at corners over slender pillars, and lofty, vaulted gateways. The finest architecture dates from the reigns of the Emperor Akbar (who built the royal city of Fatehpur Sikri and tombs for himself and his father, Humayun, as well as the Red Fort of Agra) and his grandson Shah Jahan. The latter continued to embellish the palace-fort of Agra, and undertook the construction of a larger and more splendid palace in the Red Fort of Delhi. His other works include the Jami Masjid at Delhi, the city of Shahjahanabad, and the tomb for his queen, the Taj Mahal.

The Portuguese discovered the possibility of sea trading to India at the end of the 15th century. They conquered and reestablished enormously rich merchant cities. Their colonial architecture reached its apogee in the great churches and palaces of Cochin, Goa, Diu, and Daman, on the west coast, and São Thomé south of Madras. It influenced the Dutch and French, and subsequently the British, until the mid-18th century, producing a polyglot style which reached its climax in the prolific building activities of the Nawabs of Oudt (c. 1770–1856). In the last 80 years of British rule, an Empire style predominated, culminating in the construction of New Delhi as the Imperial capital, designed in 1911 by Sir Edwin Lutyens (1869–1944) and Sir Herbert Baker (1862–1946).

After the partition of India, the influence of Le Corbusier was profound, deriving from his design for the new capital of Punjab, Chandigarh (based on a fusion of ancient Indian principles with his own theories), its government buildings (1951–55), and various buildings in Ahmadabad.

In Pakistan, Louis Kahn (1901–74) was commissioned to design the new capital of Islamabad (c. 1962).

Modern Isms

The chief characteristic of 20th-century architecture is its plurality. Some critics have erroneously suggested that there has been a single evolutionary Modern Movement in modern architecture as such. Indeed there have been many modern movements. The main revolution in architecture began with the new master problems that emerged as long ago as the 1780s when a vast amount of monumental symbolistic building began and when new problems of a specifically public architectural character were met by the architects of the period. For example, the museum buildings of that period were considered as temples built to emphasize the holiness of art, but the desire for monumentality which spread through new building types such as prisons, hospitals, and educational establishments, soon evolved into a pluralism of styles. This became an obsession of the Revivalists of the 19th century.

It was not until the 1880s that a desire for a truly modern style emerged and even then it was by no means articulate, although in some ways it prefaced the whole of the work of the early 20th century. By the turn of the century, architects, sensible to the changes that were going on in society, science, technology, and psychology, were struggling with the problems of identification, of architectural ideals (ornamentation and structure), and the increasingly important notion of providing an architecture appropriate to its time. The Art Nouveau did not successfully produce the necessary transition from the stylistic Revivalism of the 19th century into the new world of the 20th. It did however provide a bridge—via Expressionism—between the individualism of the Art Nouveau designers and the collective work of the architects who were associated with the International Modernism movement of the late 1920s.

As early as 1925 Eliezer Lissitzky (1890–1941) and Jean Arp (1887–1966) had despaired of the multiplicity of approaches to architecture and art and had set out in their little book *Die Kunstismen* ("The Isms of Art") a disparaging list of the various tendencies to be found at that time. They hated Expressionism and its many manifestations and they sought to define the new art in terms of the Constructivist aesthetic and abstractionism. The art of the 20th century in their and many other people's definition was to be based on nonobjective ideas, on technical possibilities, and on the symbolism of the machine. In the following analysis an attempt has been made to isolate the major phases that architecture went through in the 20th century.

Eclecticism

This overworked term covers a multitude of interpretations. It is usually applied to any building that incorporates a mixture of the historical styles. It has been argued by some critics that it refers to the whole output of European and American 19th-century architects who were so adept at changing their stylistic stance in midstream; others insist it is a label to describe stylistic illiteracy. Correctly, the term means "discriminate borrowing," and in the work of the best 19th-century architects it was an effective means of design.

Among the best-known Eclectic architects of the 19th century and early 20th century were Henry Richardson (1838–86) and Louis H. Sullivan (1856–1924) in the U.S., the British architects Auguste Pugin (1812–52), Richard Shaw (1831–1912), Sir George Scott (1811–78), and Alfred Waterhouse (1830–1905), the French architect-encyclopedist Eugène Viollet-le-Duc (1814–79), and the

Swiss/German theorist Gottfried Semper (1803–79). In their work there was a concern for constructional logic although in almost every case the structure was hidden by an appropriate cloak of stylistic ornament.

Structuralism

Joseph Paxton (1803–1865), with his design for the Crystal Palace for the "Great Exhibition of the Industry of all Nations" held in Hyde Park in London in 1851, began a new era in architecture. At the time his building was described as a "cruciform cathedral with double aisles." It consisted of a nave 1,848 × 72 × 64 ft. (563 × 22 × 19.5 m) crossed by a semicircular roofed transept of the same width and having a total length of 408 ft. (124 m). Galleries above the aisles ran around the sides of both nave and transept. It was constructed entirely in prefabricated glass, iron, and wood sections and was erected in less than 20 weeks.

The Crystal Palace made a new kind of building design possible. John Ruskin (1819–1900), the architectural profession's mentor, deplored it, but it proved to stimulate Revivalism. Even allowing for the growing interest that had been seen in iron construction in Britain in the previous 40 years before its erection the architectural possibilities of combining the new material with large areas of glass was left to Paxton. Numerous exhibition halls, locomotive sheds, and other large-scale "engineering" types of structure followed.

Monumentalism

In architecture, one aspect of individualism stands out: the idea of building monuments. It was to pervade 20th-century architecture and is still with us today, being based on a general notion that (to quote the Czech architect Adolf Loos from 1908) "the form of an object should last" and that implicitly there are some forms which have eternal validity. Loos himself made this abundantly clear when in 1922 he submitted an entry for the Chicago Tribune Tower competition in the form of a huge Doric column.

It was, however, among the German pioneers of modern architecture that Monumentalism really took hold. With an enormous admiration for Karl Friedrich Schinkel (1871–1941) they were able, on the one hand, to argue for the abolition of architectural eclecticism and, on the other, for design structures which were patently derived from a fondness for Neo-Classic order. Examples include the work of Peter Behrens (1868–1940) for AEG (particularly the Turbine Shop, 1909), and of Hans Poelzig (1869–1936) in his designs for the Posen Tower (1910) and for the Breslau Centennial Exhibition (1913). Mies van der Rohe (1886–1969) never gave up

Crane Memorial Library, Quincy, Massachusetts (1880–83), by H.H. Richardson.

Swan House, Chelsea, London (1875), designed by Richard Norman Shaw.

Turbine Shop, AEG Factory, Berlin (1909), by Peter Behrens.

South view of Hill House, Helensburgh, Scotland (1902–03), by C.R. Mackintosh.

Helsinki Railroad Station entrance, Finland (1906), by Eliel Saarinen.

Casa Vicens, Barcelona, Spain (1878–80), by Antoni Gaudí.

his interest in monumental buildings; Le Corbusier (1887–1966) rooted part of his own theory of a "new architecture" upon it; and in France, Tony Garnier (1869–1948) and Auguste Perret (1874–1954) used a vocabulary of Neo-Classicism.

National Romanticism

The National Romantic period in architecture extended from the 1860s well into the 20th century. Bolstered by ideas of national aggrandizement, this self-emulating style fed on particular local historical motifs and devices as well as the associative aspects of the great historical periods in architecture so beloved by the eclectics. In some cases it parallels the work of those architects normally referred to as Art Nouveau designers, but its aspirations were much wider than those of the international "proto modernists." For example, the work of Charles Rennie Mackintosh (1868–1928) in Glasgow can be seen to be an important part of the international Art Nouveau, although it strove to evoke the Celtic mysteries as well as Scottish baronial architecture. In Finland, the work of Eliel Saarinen (1873–1950), Lindgren, Gesellius, Lars Sonck (1870–1956), and Gallen-Kallela (1865–1931) was a powerful vindication of nationalistic interests. In Britain, Richard Norman Shaw (1831–1912), and later Charles Voysey (1857–1941), were particularly interested in creating a new English vernacular style and the whole of the English Garden City Movement, which began during the last decade of the 19th century, was a typical homegrown example of National Romanticism.

The peculiarly idosyncratic work of the Catalan architect Antoni Gaudí (1852–1926), although individualistic to the extreme, is very much related to the nationalistic tendencies of the Catalonians at the end of the 19th century, as is the work of Gaudí's contemporary Domenech i Montaner. In Germany, the railroad station at Stuttgart designed in the 1910s by Scholar and Paul Bonatz (1877–1951), and influenced by the success of the Helsinki terminal, is also a monument to a specific strain of German nationalism—echoed later in the hideous buildings of Albert Speer (b. 1905) for the National Socialists. The whole of this work is characterized by buildings having a heavy appearance (with the exception of the British houses) and a constructional basis in heavy masonry.

Fin de Sièclism (Art Nouveau)

Around the turn of the century the European art world was shocked into realizing the presence of the new art, or L'Art Nouveau as it is now generally called. This artistic movement reached its height at the Paris Exhibition

House in the Avenue Foch, Nancy, France (1902), by Emile André.

Detail of porch, Hôtel de St Cyr, Brussels, Belgium (1900), by Gustave Strauven.

Model factory pavilion, Machinery Hall, Werkbund Exhibition, Cologne, Germany (1914), by Gropius and Meyer.

of 1900 and it faded from popularity just before World War I. At first it was seen as an outrageous and decadent phase but the flashy, twisting, undulating, naturalistic ornament of the style led to a breakthrough in creative innovation. The then still prevalent 19th-century Eclecticism was an inspirational source, but the Art Nouveau, in its many national expressions, soon produced a form language of its own. It covers many separate developments in art and architecture during the period 1880–1910. It attempted to define a progressive and modern style at a time when significant changes were occurring in society in modes and manners, and when new ideas were apparent in science and technology, both of which required new means of expression.

Essentially the new arts indicated the new-found freedom of the individual. In addition there was a desire to comply with the requirements of a Neo-Classical or Gothicist viewpoint. Many designers returned to nature for inspiration and to an expression of natural forms through the use of wood (the designers of the Nancy, Glasgow, Paris, and Viennese schools) and glass and the wonder of building in iron (both wrought and cast). Architects such as the Belgian Victor Horta (1861–1947) explored the possibilities of spanning large spaces (cf. the work of the earlier engineers) in decorative cast iron e.g. L'Innovation Store, Brussels (1900). In the work of Charles Rennie Mackintosh (1868–1928) in Glasgow, and Joseph Maria Olbrich (1867–1908) and Josef Hoffmann (1870–1956) in Vienna, an equally inventive but much more restrained interest in geometric forms was shown.

The Art Nouveau provided a breeding ground for a new approach to individualism and design, and a timely synthesis of metal, glass, and applied ornament and construction.

Radicalism

It was largely the individualists who demanded a radical shift in emphasis from the buildings of the past to the design of those which met the demands of modern life. "There must be," wrote the architect Hans Poelzig in 1906, a "correct use of materials and construction consciously adapted to give advantages over the old use of decorative embellishments without losing what can be learned from the mastery of tectonic problems in the past."

In the ensuing years the Deutscher Werkbund in Germany (founded in 1907) took on the role of radicalizing art, architecture, and industry through "form." "Without form," Hermann Muthesius (1861–1927), the Werkbund inspirator, wrote in 1911, "we should still be living in a crude and brutal world." Form came to be seen as synonymous with "good design"—freedom from visual pain! Muthesius elaborated a theory of

types (or standards) which he presented in Cologne at the Werkbund Exhibition of 1914. Individualists such as Henry van de Velde (1863–1957) saw typecasting as an offense to the artist's own freedom to create by will and intuition. The model factory at the Werkbund Exhibition designed by Walter Gropius (1883–1969) can be viewed as a typical example of a building erected to set a standard of good design and to be copied and applied universally. Later the effects of Muthesius' and Gropius' views on type and standardization (and by inference, prefabrication) can be seen in the mass-housing schemes in Germany in the period after 1924 and, of course, in the work of the Bauhaus.

Constructivism

The term Constructivism was first used by Vladimir Tatlin (1885–1953) in 1914 when he was involved in the pre-revolutionary Cubo-Futurist movement. The so-called Constructivist Movement in the USSR was however initiated by the two sculptor brothers Naum Gabo (b. 1890) and Antoine Pevsner (1886–1962) soon after the 1917 Revolution. The aims of the movement were compounded in their *Manifesto on Realism* issued in 1920. They, together with a group of artists closely associated with them and dedicated to objectivism in art—Kasimir Malevitch (1878–1935), the Suprematist painter, and the architects Vladimir Tatlin (b. 1885) and El Lissitzky (1890–1941) — acknowledged debts to the French Cubists, the German Expressionists, and the Italian Futurists. The new movement was vastly different from its predecessors although in terms of technique and preoccupation it followed many of the innovations of these slightly earlier groups. Gabo later wrote, "We all called ourselves constructors . . . instead of carving or molding a sculpture of one piece we built it up into space . . ." Constructivism was, like the elementalism of *De Stijl*, with which it became closely associated in the mid-1920s, a passionate pleading for ideas on form and space in architecture as well as in the other arts.

The most significant early example of Constructivist architecture was Tatlin's design for a "Monument to the Third International" made in 1920 and based on a spiral which clearly had associations with engineering structures going up in the USSR at that time, as well as buildings such as the Eiffel Tower. Most of the earlier Constructivist architecture schemes remained as unbuilt projects, but as late as 1957 Gabo built his enormous "Construction in Space" placed outside the Bijenkorf store in Rotterdam. The Constructivist influence on architecture can be seen in many projects of the so-called pioneers of the Modern Movement, in particular projects by Mart Stam and Marcel Breuer (b. 1902), and

Reconstruction of Vladimir Tatlin's model for a monument to the Third International of 1920.

Unbuilt project (1924) designed by El Lissitzky and Mart Stam.

Rusakov Club, Moscow (1927) by Konstantin Melnikov: general view.

Sketch for Einstein Tower, Potsdam (1920), designed by Erich Mendelsohn.

Glass skyscraper project for Berlin (1920) by Mies van der Rohe.

Project for an electric power station (1913) by Antonio Sant'Elia.

even the work of James Stirling (b. 1926) indicates close affinities with the earlier work in his use of glass and steel.

Expressionism

There was no Expressionist school of architects as such, although a number of architects and designers were closely associated with the Expressionist movement in Germany between 1910–23. The term, however, was definitively applied to a group of Dutch architects in the magazine *Wendingen* in 1918. Since that time it has become a thoroughly accepted term used to describe the work of those architects who prefigured the International and Functionalist period of the Modern Movement. Thus it covers individualist German architects of the pre-World War I period such as Hans Poelzig (1869–1936) and Bruno Taut (1880–1938), the Taut Brothers in their postwar phase, and such visionaries and utopians as Finsterlin, Erich Mendelsohn (1887–1953), Hans Scharoun (b. 1893), the Luckhart Brothers, Krayl, and for a short time Walter Gropius.

The postwar German phase of Expressionism was dominated by the activities of the "Glass Chain" group whose leader was Bruno Taut. His inspirational source was the work of Expressionist poet and lexicographer Paul Scheerbart. Although it is clear that Mies van der Rohe (1886–1969) had no time for the group's visionary and romantic work, he did however design a number of projects in the early 1920s in an Expressionist idiom including a skyscraper project sheathed in glass with reinforced-concrete cantilevered floors and a concrete parking lot. The theme of the Glass Chain group was architecture in glass and concrete (colored as well as transparent) and this led to many novel designs. In contrast, the work of the so-called Dutch Expressionists (or Phantasts) appeared traditional. Their sources were to be found in vernacular forms and through their buildings they expressed their affection for warm materials such as stone, brick, and thatch; unlike the Germans they had no utopian programs.

The strange concrete structures built in the 1910s at Dornach, Switzerland, by the anthroposophist Rudolf Steiner (1861–1925), are also categorized as Expressionist in intent.

Futurism

The first of the anti-art (and later anti-architecture) avant-garde groups was the Futurists. Filippo Marinetti (1876–1944), a poet, issued the first Futurist manifesto in Paris in 1909 in the pages of *Le Figaro*. Although it was concerned specifically with Futurist poetry, it extolled the beauty of speed and movement and condemned academic art, libraries, and museums. Later, the Futurist movement became concerned with the expression of their ideas through painting and sculpture and theater.

Short of an architectural futurist, Marinetti co-opted Antonio Sant'Elia (1888–1916) into the group, having seen his drawings for a multilevel concrete towered "New City" in a student exhibition in Milan in 1914. Sant'Elia was killed in action in 1916. However, Marinetti had published the *Futurist Architecture Manifesto* in July 1914. It proclaimed that Futurist (Marinetti added this word where required although it had not been used by Sant'Elia) architecture "is the architecture of calculation, of audacity and simplicity; the architecture of reinforced concrete, of iron, of glass . . . and all those substitutes for wood, stone, and brick which make possible maximum elasticity and lightness!" The document continued, in its highly rhetorical way: "Let us throw away monuments, sidewalks, arcades, steps. Let us sink squares into the ground, raise the level of the city."

The manifesto had a limited influence at the time (possibly because of the nullifying effects of the war) but it was rediscovered and reactivated as the inspirational source of architect-planners in the 1950s.

Neoplasticism

In its precise meaning the term Neoplasticism relates to the theory of pure plastic art that the Dutch painter Piet Mondrian (1872–1944) derived from Cubism. It consisted in the exclusive use of the right angle in a horizontal position, and the use of the three primary colors contrasted with or incorporating in various canvases the three non-colors: white, black, and gray. This theory, developed between 1912-17 (the year of the foundation of the De Stijl group at Leyden in Holland), had a pronounced influence on Dutch architects. The first fully integrated neoplastic house can still be seen today in Utrecht and was built to the design of Gerrit Rietveld (1888–1964) for Mrs Schröeder-Schradar. It consisted of a number of vertical and horizontal planes built largely out of concrete and a series of three-dimensional facades similar to the individual paintings of Piet Mondrian. The De Stijl sculptor Vanton Gerloo also produced a number of studies which can be seen as essays in architectonics. The founder of the De Stijl group, Theo van Doesburg (1883–1931) and Cor van Eesteren (b. 1897) produced projects both for individual buildings and for planning schemes which incorporated Neoplasticist ideas.

The Neoplastic aesthetic was revived in the postwar period by the artist-critic Baljeu and displays a close affinity to some of the earlier constructivist ideas of the Russians and Germans as well as to the De Stijl group's ideas.

Bauhaus style

The Bauhaus was a school of art and design

Study for buildings for a modern metropolis (1914) by Mario Chiattone.

Architectural project (1923) by Theo van Doesburg and Cor van Eestern.

founded in 1919 by the architect Walter Gropius (1883–1969) which was successively housed at Dessau and finally, for a short period in 1933, in Berlin. It commenced as a school based on Expressionist principles and its staff included the world-famous painters Lyonel Feininger (1871–1956), Paul Klee (1879–1940), Wassily Kandinsky (1866–1944), and Oskar Schlemmer (1888–1943). As a design school it revolutionized industrial design, graphics, theater, photography, and film. In 1923, when Laszlo Moholy-Nagy (1895–1946) joined the Bauhaus, the school became obsessed with Constructivist design and the geometry of primary forms and colors. By this time the Expressionist phase was over. Although "building" was at the center of the curriculum at the Bauhaus, an active architecture department was not created until 1927 when the Swiss architect Hannes Meyer (1889–1954) took over from Gropius as head of school. When he left, the school was finally headed by the German architect Mies van der Rohe.

The Bauhaus was the creative center of artistic experiment during the 1920s and it became internationally known through its publications and exhibitions and also, most importantly, through the work of its architect heads who were in the front line of the European avant-garde. The influence of the Bauhaus design methods can be seen in numerous consumer products from bent metal furniture and hanging globe lamps to the black, block, lowercase lettering to be found on exhibition posters the world over.

CIAM and International Modernism

The Congrès Internationaux d'Architecture Moderne (CIAM) was an organization set up by Le Corbusier (1887–1966) and Sigfried Giedion (1893–1968) at Hélène Mandrot's château at La Sarraz, Switzerland, in 1928. It was the organization that consolidated the common form among all those architects concerned with the new architecture. It was international in scope and eventually spawned a worldwide series of conferences and splinter groups. It devolved through a bitter dispute with the younger generation of architects at Otterlo, Holland, in 1959.

CIAM was the major organization through which the ideas of modern architecture and urbanism became known to the world. Its founder, leader, and champion Le Corbusier was the vital force that held the organization together. His pronouncements took on almost biblical significance, the repercussions of which are still felt today. By providing the organization with an effective voice, Le Corbusier and his close associates influenced housing policy of governments, local authorities, and private individuals throughout Europe. In 1929 the second CIAM Congrès was held at Frankfurt under the auspices of

Schröder House, Utrecht, Holland (1924), by Gerrit Rietveld.

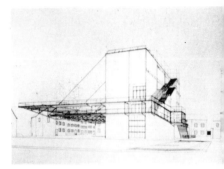

Peterschule Project for Basel, Switzerland (1926), by Hannes Meyer and Hans Wittwer.

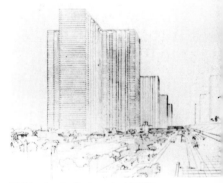

Voisin plan for Paris project (1923) by Le Corbusier.

Bauhaus Building, Dessau, Germany (1926), by Walter Gropius.

Housing scheme for Frankfurt (1925), by Ernst May.

Farm buildings, Gutes Garkau, Germany (1923), by Hugo Häring.

Falling Water, Bear Run, Pennsylvania (1936), by Frank Lloyd Wright.

Secondary Modern School: teaching block and gymnasium, Hunstanton, Norfolk, England (1950–54), by Alison and Peter Smithson.

the then city architect Ernst May (b. 1886), and out of this congress came the recommendations for existence minimum dwellings. The CIAM style of architecture was characterized by cubic, white-surfaced, flat-roofed architecture, usually set in an arid landscape. In 1933 the CIAM 4 congress took as its theme "The Functional City" and issued the *Athens Charter*, a report which was to affect the planning and reinstatement of cities throughout Europe (e.g. the MARS plan for London in 1938).

Organicism

Organicism, or organic architecture, is another vague term that covers a number of interpretations. Its modern usage seems to stem from the American mid-19th century sculptor Horatio Greenough (1805–52), but its more particular use as a description of architecture that sympathizes with its environment comes from the early work of Frank Lloyd Wright (1867–1959) and the Prairie School. It is the very antithesis of the geometrical organized facadism of those architects who believe that architecture should intrude on the environment in the Classic, Neo-Classic, and Gothic sense. A whole line of architects and designers can be seen to relate their work—however vaguely—to Wright's ideas, and there is a stream of organic architecture running through American, English, German, and Dutch design. Apart from Frank Lloyd Wright, the principal apologists for American organic architecture are Claude Bragdon and Henry-Russell Hitchcock. In Germany, the two main figures concerned with "organic" ideas were Hugo Häring (1882–1958) and Hans Scharoun (1893–1972). Currently with a new interest in organic architecture, the work of Bruce Goff (b. 1904) stands out for its inventiveness.

Paolo Soleri (b. 1920), and Herb Greene are seen to be part of the main stream of organic architectural development.

Utilitarianism

The individual work of British housing architects after the building of Letchworth in 1903 has been described as a kind of utilitarian architecture. Numerous publications were produced on low-cost housing, particularly publications aimed at the private buyer which set out economic solutions for low value sites as well as alternative cheap forms of construction in timber, brick, and metal. However, it was World War II that created a desire for utility buildings in Britain, and it was in the housing area that architects and designers worked for economic, small-scale solutions. A number of organizations were set up in the immediate postwar period in Britain (and similarly in parts of the U.S.), to cope with the problem of the population explosion

and returning soldiers. In Britain the utilitarian house par excellence was known as the "prefab." Originally this was to have a life of 15 years but in some cases examples still remain 35 years later. These houses were prefabricated in factories and brought to the sites ready-built for immediate assembly. The effect of this eventually was to create an atmosphere in which "system building" could take over the role of individually designed dwellings. In the 1950s in Britain a vast program of system-built housing was started, the detrimental effect of which is to be seen in many towns and on many estates. The Greater London Council (GLC) alone has had a repair bill of around 30 million pounds to put this ill-advised housing program to rights. Utility unfortunately was not a qualitative argument.

The New Brutalism

Although the term "The New Brutalism" now has a derogatory tone, it was originally meant to indicate a certain type of architecture of the 1950s. It was introduced by the British architects Peter and Alison Smithson (b. 1928 and 1923), although the Swede Erik Asplund (1885–1940) lays claim to an earlier version, "neo-Brutalism." At first it was applied to describe the buildings of Mies van der Rohe (1886–1969) adherents in Europe, whose interest was predominantly in a display of his precise technology of glass and steel. It reflected a puritanical phase in which the servicing systems of a building were openly on display and not concealed in ducts or by covers (e.g. Hunstanton school, Norfolk, England, by A. and P. Smithson, 1954). Later it was applied to buildings which imitated the exposed concrete finishes (*béton brut*) in Le Corbusier's work and therefore more freely applied to those architects (particularly British, German, and Dutch) who associated with the so-called Team X (i.e., Aldo van Eyck (b. 1918), James Stirling (b. 1926), the Smithsons) as well as other designers such as Denys Lasdun (b. 1914) and Gottfried Bohm. The South Bank Arts Buildings in London (notably Queen Elizabeth Hall and the Hayward Gallery) by Greater London Council (GLC) architects brought the term in disrepute: engineers renamed it the "bunker" style. The whole phase, important though it was inside the European cliques, is still vague: "The New Brutalists," in the words of its chief apologist Reyner Banham, were never sure whether what they championed was ethic or aesthetic.

Metabolism

The term Metabolism was first applied to architecture at the World Design Conference, Tokyo, 1960. It was introduced as a new system of thinking about the problems of cities such as Tokyo and included among its advocates the architects Kiyonori Kikutake, Fumihiko Maki, Masato Otaka, and Kisho Kurokawa, as well as the graphic designer Kiyoshi Awazo. Together they signed the first declaration: "Metabolism 1960—a proposal for a new urbanism." Kenzo Tange (b. 1913) was an early and important convert to the Metabolist cause which regarded "human society as a vital process, a continuous development from atom to nebula." The group concentrated on the new order of relationships between man and the environment.

The early Metabolist terminology was based on organic and cybernetic analogies. However, as their ideas developed they soon came to resemble earlier historical visionary projects, and by the time they came to be built the visionary element was lost in the face of the need to build realistic, earthquake-proof, concrete buildings. The best-known examples in this genre are Tange's radio and press center at Kofiu (1964–66) and Kurokawa's Nagakin Capsule Tower, Tokyo (1972).

Post-Metabolism

This term was first used in a special issue of the *Japan Architect* in 1977. Its use implies an attempt to summarize some of the very divergent currents that characterized the Japanese architectural scene at that moment. It was a reaction to the "meta-architecture" of the earlier Metabolists. Although by no means a group, the Post-Metabolists are interested in exploring such things as the nature of the house in the city, and are concerned with intricate design on small sites and polemical schemes. Mozuna Monta's "anti-dwelling" in Hokkaido, Toyokazu; Watanabe's doctor's house in Ibaragu; and Hiromi Fujii's Miyajame house are important examples of one extreme end of the Post-Metabolist approach. These schemes tend to deal in terms of irony and negation of the environment, while Hiroshi Hara's own house, Group Zo's House for a Magician, and Dam Dan's fantasy villa are at the other extreme, and are celebratory. Isozaki is seen to be the most successful exponent of Post-Metabolism.

Post-Modernism

With the failure of CIAM-type Modernism a new body of theories are emerging as alternatives to Modern Movement ideas. It is by no means coherent. Alternatives suggested range from revivals of pattern book principles of the 19th century, a new interest in vernacular forms adapted to modern needs, a much more strict interpretation of the theatrical element in Modern Movement architecture proper, a distinctly confused revival of Frank Lloyd Wright's organic views, and a

return (under the influence of Schumacher's book *Small is Beautiful*) to the low-rise, high-density developments of the interwar period. The design by Paolo Soleri (b. 1920) for a city on the mesa at Arcosanti; Lucien Kroll's medical faculty at Woluwe, near Brussels, Belgium; the articulated designs of the so-called New York Five; and the books and works of the Venturis, are often cited as examples of this burgeoning new phase.

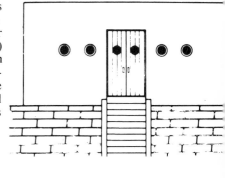
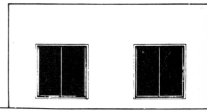

Annihilation House, Mutsuura, Japan (1972) by Takefumi Aida and Associates: front and side elevations.

Nakagin Capsule Tower Building, Tokyo, Japan (1972), by Kisho N. Kurokawa and Associates.

House for Mrs Robert Venturi, Chestnut Hill, Pennsylvania (1962–64), by Robert Venturi.

Municipal Art Museum, Kitakyushu, Japan (1972–74), by Arata Isozaki.

Medical students' residences, University of Louvain, Brussels, Belgium (1974–75), by Lucien Kroll.

Section 2
Built forms and
building types

Landscape and external space

Court of the Lions, the Alhambra, Granada, Spain (1377).

Villa Lante, Bagnaia, Italy (c. 1564) by Vignola.

Private gardens and parks

All designed landscapes have their origins in the domestic garden, which itself was originally inspired either by agricultural patterns or by natural scenery.

The first garden was an idealization of Sumerian irrigation networks in the Tigris-Euphrates basin c. 2000 BC. Then came the first roof gardens—the legendary Hanging Gardens, Babylon, which were made possible by bitumen covering and earth packed into the haunches of arches. The hunting park was a later invention of the Assyrians, who domesticated the horse. The basic form of the original Persian paradise garden (as it was called by the Greeks)—a square or rectangle enclosed against wind and dust, and cooled by intersecting water channels and groves of trees—has remained constant, but transplantation to Spain (AD 750–1492) and India (1526–1707) in the wake of the Moslem conquest brought local variations. In the Court of the Lions at Granada, clusters of slender stone pillars and fan vaulting replaced groves of trees. In India, the Mogul emperors made more spectacular use of water, particularly in Kashmir. Narrow irrigation channels for trees developed gradually into broad canals to cool the air for humans, cascades became more elaborate, water channels flowed through arches beneath buildings.

Gardens were sometimes planned as sequences of traditional squares, and many Islamic gardens were imbued with symbolism: the square represented the world; the octagon, man's struggle toward spiritual regeneration; and in religious gardens, the circle of the dome stood for perfection and eternity. A specifically Mogul development was the tomb garden, combining Persian design with the Tartar nobleman's tradition of building his tomb for use as a pleasure dome during his lifetime.

In China, landscape design began with the hunting park, such as that laid out in 100 BC by a Han emperor. It contained towers for both contemplation and observation, and an artificial lake in which the mystic Islands of the Blest, dwelling place of the immortals, appeared and disappeared in mist. Chinese Taoism regarded all natural landscape as divine. Gardens, however small, were an attempt to create in microcosm the solitary landscape of the mountains. Site and layout were dictated by geomancy; within the gardens the prime attributes of harmony and peace were achieved by careful blending of the basic design elements. Rocks, hills, and mountains represented the *yang*, or active cosmic male force; and still water represented the *yin*, or passive female force. In city gardens, boundaries were planted to exclude everything but the sky wherever possible. Such concepts owed much to the painters and poets, who continued to be the main inspiration of landscape design throughout Chinese history.

Early Japanese gardens were bare rectangles

of raked sand or quartz for court ceremonies and Shinto rites, but growing appreciation of the native landscape of volcanoes and island-dotted seas soon led to the introduction of trees, rocks, hills, and ponds, arranged to simulate a favorite landscape or as simple expressions of Shinto animism. They also represented the male and female cosmic forces. Respect for the personality of each stone or plant has always been a key factor in Japanese garden design.

Chinese culture reached Japan c. AD 600 and was a dominant influence for about 400 years. Large pleasure parks with lakes, islands, and bridges were laid out, culminating in *shinden-zukuri*, a formula combining Chinese architectural symmetry with free-flowing Japanese landscape. Buddhism was a yet more powerful and longer lasting invader (c. AD 552), under whose influence three new garden types emerged. These were: first, Buddhist paradise gardens, which were mandalas of seemingly naturalistic lakes and symbolic pavilions (c.1100–1550); second, Zen Buddhist gardens of contemplation—dry landscapes like Ryoan-ji and Daisen-in in Kyoto (1488–1513), whose rectangles of raked sand and arrangements of rocks attempted to symbolize the universe; and third, tea gardens (c.1550), originally bare except for stepping stones, in order to concentrate attention on the ceremony, but later enriched with trees, rocks, stone lanterns, and water basins. The tea garden led to the secular stroll gardens of the 17th century, such as Katsura.

In Western civilization, the principal basis of design until the 18th century was geometry, as in the highly cultivated enclosures of 8th-dynasty Egypt, with their arched vine pergolas and symmetrically arranged rectangular pools for boating and hunting fowl. In classical Greece, where there was little water, gardens, apart from the sacred groves of the temples, were mainly utilitarian or simple paved courts. The pots of short-lived plants grown for the Festival of Adonis may have been the beginning of gardening in pots.

Unlike Eastern gardens, which were not necessarily associated with dwellings, the Roman garden was essentially an extension of the house. Beginning with the farmhouses around Rome, the country villas of the wealthy had, by the 1st century AD, taken on all the splendor and axial formality of Roman civic planning, with terraces, vine-shaded walks, extensive *allées* for exercise, even arenas. Clipped hedges, topiary, and sculpture were the chief adornments; with either a canal or *nymphaeum* when water was available. Occasional attempts at romanticism, as at Nero's Golden House (c. AD 64) and Hadrian's Villa at Tivoli (AD 118–128) were only tentative. City gardens at Pompeii (1st century AD and inspired by Hellenism) were totally enclosed essays in the creation of imaginative space, partly through the medium of landscape fantasies

painted on the rear walls of the peristyle overlooking the garden court, and partly through axial integration of open and enclosed spaces. Sculpture and small canals, as permanent elements, were supplemented by a growing number of plants from other countries.

Medieval gardens were, for the most part, reworkings of earlier themes, except for the bowling green: from the paradise to the cloister garden or enclosed castle court was only a short step. The sense of protective enclosure that was one of their chief functions declined in importance as nature appeared less menacing and by the 15th century the gardens of the Italian Renaissance began to project into the countryside, their geometric forms contrasting, and yet in harmony, with the natural landscape.

The Italian garden fell into two periods: the Renaissance until about 1550, thereafter Mannerism and Baroque. The Renaissance was based on a revival of imperial Rome and culminated in Giacomo Barozzi da Vignola's (1507–73) Villa Lante at Bagnaia, in which the garden divided up the house into two parts. Up to that point all classical landscape was concerned with finite space; thereafter, the search began for the infinite. Mannerism represented individuality. Its finest example is the Villa Gamberaia at Settignano where many "moods" of the mind are embraced in one composition. The Baroque was, in principle, collective and based on the theater of spectacle. Among its optical innovations was play on illusion, for instance making distances recede or come nearer by use of "false" perspective. The garden theater appeared with its green wings and footlight shades, as at Villa Marlia, near Lucca. Romantic conceits such as rocks, grottoes, secret fountains, and giants joined the permanent elements inherited from Rome—evergreens, stone, and water.

The Boboli Gardens in Florence, home of Marie de Medici (1573–1642), who was to become Queen of France, were the first instance of a great garden appearing to be carved out of woodland. This was one of the three main sources of inspiration for the revolutionary designs of André Le Nôtre (1613–1700) at Versailles and elsewhere, the others being the hunting forest with its straight intersecting rides, and the pseudo-romantic water landscapes of Touraine where Richelieu (built 1627–37) also presented an entirely new geometric concept of a city as part of a domestic layout. In Le Nôtre's designs, the garden ceased to be an extension of the house, which became just one object in a grand composition of solids and voids axially arranged, the scale expanding as it receded from the house. He introduced into landscape the idea of three radiating avenues (the *patte d'oie* or goose foot) that had first appeared in the Piazza del Popolo, Rome. Near the house, blocks of woodlands (*bosquets*) concealing a variety of incidents were outlined by hornbeam *charmilles*, a new

Ryoan-ji, Kyoto, Japan: a garden of contemplation.

Sento Gosho, Kyoto, Japan: a stroll garden.

Courtyard garden of the House of the Vettii, Pompeii.

Garden theater in the Renaissance garden of the Villa Marlia, laid out c. 1690.

Grand Canal and *tapis vert,* Versailles (1667-88), laid out by André le Nôtre.

Blenheim Palace, Oxfordshire, England. Landscape by Capability Brown, early 18th century.

type of clipped hedge about 10 ft. (3 m) high, above which projected free-growing tree branches. Features used to penetrate the woodlands were: the grass *tapis vert*, another novelty; contemplative canals, often of considerable length; and *parterres de broderie*, an earlier French invention (c. 1638), with box hedges and colored sands and flowers arranged in patterns, inspired by the rich fabrics of contemporary costume. Le Nôtre's fountains at Versailles were of an unprecedented scale, involving an exceptionally elaborate pipe system for distributing water around the garden. Unlike Mogul or Italian gardens, which relied on gravity from immediate natural sources, water had to be brought at vast expense from outside the Versailles catchment area; a wheel at Marly to lift water from the Seine was a failure.

Garden designers throughout Europe were, for the most part, under the influence of Versailles, an exception being the glass-fronted vine terraces of Frederick the Great (1712–86) at Potsdam, Berlin, which anticipated modern functionalism.

French classical formality on a much smaller scale was also reflected in North American gardens for many years. Some original features did, however, emerge in Virginia, where the isolation and extent of tobacco estates called for layouts that included "dependencies"—schoolhouses, smokehouses, slave quarters, burial grounds, etc. Where houses were situated on bluffs overlooking rivers, for ease of movement, grass terraces known as "falls" were carved from the sloping ground.

In England, political fears of French domination and a dislike of autocratic classicism in an age of dawning liberalism gave the final impetus to a new approach to landscape, known as the English Landscape School. Having begun as a literary movement, it was based, up to the middle of the 18th century, on interpretations of Ovid, Virgil, and other Augustan poets, and inspired by the heroic period of painting, particularly Claude Lorrain (1600–82) and Nicolas Poussin (1594–1665). Stourhead, in Wiltshire (1740–60), is an allegory of man's passage through the world, based on Virgil's *Aeneid*, and includes the hero's visit to the underworld. The movement's leading practitioner in the early stages was William Kent (1685–1748). His chief contribution lay in transposing the art of painting into landscape. In his humanist landscapes, Palladian mansions stood in well-wooded scenery, often opened up to give views of "eye-catchers" (arches, columns, or temples) in the surrounding countryside. With Charles Bridgeman (d. 1730), Kent invented the "ha-ha," an invisible ditch which separated the inner park from the outer and kept cattle at a distance without interrupting the line of vision. He also initiated the planting of clumps of trees in parkland, possibly an aesthetic rationalization of the needs of

pheasant shooting and fox hunting (which had taken the place of stag hunting, obviating the need for long straight rides).

After 1750, the movement broke away from literary classicism and split in two. One camp was headed by Lancelot (Capability) Brown (1716–83), a professional landscape designer. He evolved a standard design formula based on rolling land, water, tree clumps, and woodland belts to give seclusion. The house, without any formal approach, stood directly on grassland. His technique with water was to create artificial rivers by damming streams and making lakes with sinuous curves whose ends were concealed by planting, land forms, or bridges, as at Blenheim Palace near Oxford. Brown did away with all formal planting and terraces, but this trend was later reversed by another professional, Humphry Repton (1752–1818), who esteemed utility and convenience above beauty. Repton reinstated terraces with flower gardens outside the house, partly as a result of pressure to accommodate the influx of new plants from abroad.

Two amateurs, Richard Payne Knight (1750–1824) and Sir Uvedale Price (1747–1829), rejecting Brown's shaven lawns and pastoral scenes, started a rival movement, the Picturesque. Knight, taking the painter Salvator Rosa as his model, favored wild cascades and savage rocks which would bring out nature's awesome aspect. Price sought beauty in everyday objects like gnarled trees and advocated that roughness and intricacy should be the basis of landscape design.

Another element in the complex makeup of the School was *chinoiserie*, a theme which first appeared c. 1687 as details of Chinese culture percolated to Europe and reached its English climax c. 1760 with the pagoda in Kew Gardens designed by Sir William Chambers (1723–96). Widening travel and the urge to escape into romanticism resulted in eclectic gardens in many countries: the Parc Monceau in Paris (then privately owned) contained Dutch, Chinese, Turkish, Roman, Egyptian, and Tartar artifacts. In the first half of the 19th century, cultivation of imported plants from other continents greatly increased in scale, thanks to the large greenhouses made possible by the invention of cast iron. In England, the chief pioneer, at Chatsworth in Derbyshire, was Sir Joseph Paxton (1803–65), architect of the Crystal Palace of 1851 (see EXHIBITION BUILDINGS). At about the same time, genetic research by Gregor Mendel (1822–84) stimulated hybridization and breeding of new strains, enabling hardier versions of exotic plants to be grown commercially for use outdoors. The new plant vocabulary that became available was put to various uses. Most popular was carpet bedding—close planting of brightly colored, half-hardy annual flowers in geometric patterns. Rock gardens were another product of the mania for plant collecting. Conifers like the wellingtonia brought variety of tone to parklands, but were later accepted as symbols of Victorian gloom.

The Industrial Revolution had two consequences affecting garden design. The first was the rise of the middle class, whose desire to emulate the upper class led John Claudius Loudon (1783–1843) to create a new landscape concept, the suburban villa garden. This incorporated the features of an aristocratic park, but on a reduced scale, in a style known as the Gardenesque. Conservatories were popular and so were croquet lawns after 1856. The second result of the Industrial Revolution was the gradual shrinkage of the labor force, offset initially by Budding's mowing machine (1832), and the advent of piped water.

In 1871 William Robinson (1838–1935) launched a campaign rejecting flower beds, fantastic topiary, and artificiality of any kind. He advocated informal planting of indigenous wild flowers and shrubs, partly on ecological grounds, and a return to the tradition of the English cottage garden. He found an ally in Gertrude Jekyll (1843–1932); she accepted Robinson's ecological approach but used her painter's insight to turn plant relationships into works of art. Her use of color and plant arrangements established the herbaceous border as a major feature of any English garden, and won worldwide acceptance.

Robinson's theories on the use of indigenous plants (though not their arrangement) were taken up in the 1930s in Brazil, where the painter-gardener Roberto Burle Marx (b. 1909), rejecting the outworn forms and unsuitable European plants of current Brazilian garden design, brought in from the jungle native plants which had, until then, been despised as weeds. The luxuriant forms of the larger species contrasted dramatically with the plain concrete of modern buildings. Small plants were used as if they were paint to create abstract designs on the ground—almost a return to carpet bedding but on a permanent basis and without the geometry. Some designs only made use of different types of grass. Clarity of design was preserved by thin metal or timber strips between plant groups just below ground level, to inhibit root spread.

The placing of buildings is still conditioned by site and climate, but modern building techniques have made it possible to use sites which previously could not have been exploited. Nature and architecture are more closely interwoven, as at Frank Lloyd Wright's "Falling Water" at Bear Run in Pennsylvania (1937–39). Plate glass, night lighting, and central heating (air conditioning in hot climates) have brought about an interflow of outside and inside which earlier was rare for climatic reasons. The Kaufmann house built by Richard Neutra (1892–1970) at San Jacinto, California (c. 1942), is an outstanding example of this idea in action.

Kronforth Garden, Theresiopolis, Rio de Janeiro, by Burle Marx.

California was also the scene, in the 1940s, of a breakaway in the design of small gardens, led by Thomas Church (b. 1902), Garrett Eckbo, and Lawrence Halprin (b. 1916). Designed for open-air living in a warm climate, the new gardens focused on the heated swimming pool, previously a random addition like the tennis court (which poses greater problems of integration within a design). Although each garden was tailored to the individual owner and site, Church's designs were homogeneous in their blending of biological form, geometry, and natural scenery. His curved pool at Sonoma had worldwide influence, even in colder countries. His materials—timber decks, sand, stone, and concrete, combined with casual planting—also have the merit of being easily maintained, another reason for their popularity.

Owing to the scarcity and cost of labor, maintenance has become a major factor in garden layout. Designers make widespread use of groundcover plants to reduce weeding, and adjust their plans to take into account the requirements of lawnmowers, pop-up irrigation, electric hedge clippers, and other aids made available by mass production. Garden owners are no longer a small elite: their numbers are so vast that large industries have been set up to cater to their needs.

Garden at Sonoma, California, by Thomas Church.

Urban open space

The first urban open spaces and public parks evolved from the domestic garden or aristocratic private park.

Kubla Khan's Peking (1279–1367), built in a hunting park between two earlier cities, retained the original fishing lakes and dominating tree-clad artificial hill, but superimposed a pattern of rectangles, so that the final layout was a balanced design of geometry and natural landscape—a blend of Confucianism and Taoism. The Ming (c. 1409) added a second hill (Coal Hill) on the central axis, made from material excavated from the canals. Shah Abbas I of Persia's Isfahan (1598), on the other hand, was a sequence of traditional square gardens held together by the strong line of a single long avenue, the (Chahar Bagh)—a layout that has largely dictated the shape of the modern city.

Cities have, on occasion, formed part of a total landscape layout. Louis XIV's Versailles is the most famous example but the idea was quickly copied in Germany where princelings compensated for their lack of resources by eccentricities like the four-mile (6 km) avenue reaching from the center of Kassel to the most spectacular cascade in Europe on the far side of the palace grounds. Even more exaggerated are the 32 avenues radiating out from the Margrave's palace at Karlsruhe, nine of which compose the framework of the new town. Pierre L'Enfant's plan for Washington D.C. (1791) was in a different category. Though based on Versailles, the plan was for the whole city and was so well related to its natural setting beside the Potomac that it lost much of its authoritarian rigidity.

In England, the 18th-century landscape revolution led to the concept of an idealized countryside brought into the city. The pioneer city was Bath, where the architect John Wood (1704–54) applied garden aesthetic techniques to city planning. The Royal Crescent faces what appears to be unlimited parkland, only because trees arranged like a wood in the foreground conceal the city behind. The New Town in Edinburgh consists of a geometric layout, one side of which—Princes Street—faces open parkland with the magnificent backcloth of the medieval city and castle. Apart from the royal parks, London was enriched with countless open squares with hard surfaces for traffic; these were beginning to be enclosed for use by residents as shared but private romanticized property (Russell and Cadogan Squares designed by Humphrey Repton (1752–1835)). The climax was Regent's Park (early 19th century), the northern end of a monumental route from St James's Park to Marylebone Fields, designed by John Nash (1752–1835) and Humphrey Repton in collaboration. The scene around three sides is a sequence of residential terraces that seem to be palaces; the north was intended to be left open to the countryside. The idea of a shared but private romantic park was continued on a smaller scale in the Ladbroke estate, Holland

Park (1846), in which small private gardens opened at the rear onto a strip of parkland wide enough for forest trees.

Under joint pressure from the sociologists, concerned with the urban conditions of the Industrial Revolution, and the businessmen, for whom better health meant better production from employees, the first parks solely for public recreation began to be laid out in the 1830s. Victoria Park in east London (1845) was soon followed by the more adventurous Battersea Park on the swamps south of the Thames. The principles of design were those of the aristocratic park adapted to public use. There was no central focal point (the mansion, from which all views had been planned) and crossviews became paramount. Boundaries were densely planted to conceal ugly building developments on the perimeter and to act as dust filters. Except for a wide open space where crowds could assemble to watch displays, the woodland paths were designed to disperse people into small groups, separating them by ground modeling, trees, and shrubs. Flower beds were (and continue to be) a spectacular feature. Unlike earlier English lakes, the Battersea lake reflects the influence of Japan, for the islands are placed in such a way that the water turns in on itself, adding mystery to boating.

The three great innovators in 19th-century urban park design were Joseph Paxton of England (1801–65), Jean Alphand of France, and Frederick Law Olmsted (1822–1903) of the U.S. Paxton's interests were economic as well as scientific, sociological, and aesthetic. In 1843, he designed Birkenhead Park in the northwest of England, financed by the rise in land values around the perimeter of the park, and the first to be wholly owned by the public itself. For the 1851 Exhibition he evolved a new landscape aesthetic when he placed the Crystal Palace next to the Serpentine and over existing trees, showing that functional architecture could harmonize with romantic landscape. When the exhibition buildings were subsequently reerected in Sydenham, Paxton's lavish Italianate layout of the grounds was enriched by spectacular fountains fed by gravity from water recirculated and pumped by steam engines into reservoirs at the top of the glass towers at either end of the facade (these also contained the flues); 11,788 jets used 120,000 gallons (545,520 liters) of water per minute. Incorporated in the layout was an ultra romantic pool of prehistoric monsters, which still exists.

Alphand was appointed by Baron Georges Haussman (1809–91), in about 1853, to provide a unified system of romantic parks for Paris, to counteract the geometry of his plan for the central area. It was the first example of coordinated urban landscape planning on a grand scale, but unlike the Ringstrasse in Vienna, the Paris boulevard (the word is a corruption of the bulwark it replaced) was never completed as a greenbelt linking the parks. Nevertheless, within the network, there was full scope for the imagination. Perhaps the first aesthetic reuse of industrial waste was at the Parc des Buttes-Chaumont (c. 1863), laid out on a disused quarry whose extraordinary shapes were incorporated in an artificially romantic concept. Alphand introduced cement garden details like seats and balustrades that resembled rough tree branches.

Frederick Law Olmsted studied both Paxton and Alphand and brought park design into the modern age, winning the Central Park competition, New York, in 1857. Although technically not one of his best designs, it introduced new ideas which finally disengaged public parks from the ethos of the aristocratic park. The rectangular park was framed by tall buildings that towered above any boundary tree; hence Olmsted designed for the whole area a small-scale pattern that had been seen in embryo at Battersea—the principle of breaking up crowds and using foreground rather than middle distance as a screen to the outer environment. The crossroads that sliced across European parks, making them dangerous to pedestrians and disruptive of space, were here partly underground. Subsequently, Olmsted's imaginative inventiveness transformed American thought. He made proposals for the first university campus (Berkeley, 1865) to be laid out on romantic rather than classical principles. Housing estates like Riverside were laid out on sinuous curves "to imply leisure, contemplativeness, and happy tranquillity." In his landscape system for Boston, the parks were physically linked. He proposed the first National Park to incorporate natural scenery (where ingenious solutions to the problem of pedestrian access were subsequently evolved, such as the rise and fall walkways in the Everglades, 1935). As the landscape architect for the famous 1893 Columbian Exposition, he comprehended architecture in a unified landscape design.

Modern ecological planning that seeks to unite city dweller with the countryside was initiated by Cadbury Brothers at Bournville, England, in 1898. It was later developed by Sir Ebenezer Howard (1850–1928) as a satellite town in his *Diagram for a Garden City* (1898) and was first realized at Letchworth in 1905. The more mature Welwyn Garden City followed in 1920. The first system of automobile/pedestrian segregation was invented at Radburn, U.S. (1927) by Clarence S. Stein (1882–1975) and Henry Wright. After World War II, the first English new cities were conceived as cities *within* parks. Local neighborhoods (as at Harlow, 1948, Frederick Gibberd), set in open green landscape, revolved around the center. Later new cities were compacted until Milton Keynes (c. 1970), in contrast, proposed a city of 250,000 persons segregated into independent, landscaped areas separated by

Central Park, New York (1858-70), by Olmsted and Vaux.

roads, ground modeling, and trees indigenous to the countryside.

In contrast to the romantic English, Le Corbusier (1887–1966) evolved a revolutionary mathematical concept in his theoretical *Ville Radieuse* (1935), in which city dwellers were to live in widely spaced point blocks with unlimited sun, air, and communal gardens. At Marseilles, he built a fragment of the idea, l'Unité d'Habitation (1947), a self-contained "neighborhood" raised on *pilotis*, or stilts, leaving the land below the building free. Le Corbusier's proposals, although magnificent architecturally, were considered inhuman and proved ecologically unsound owing to abnormally increased wind pressures and down drafts, lack of intimacy, and loss of space and amenity through parking lots. One of the few city designs by Le Corbusier to be realized is Chandigarh, beautifully related to the Himalayas. The ultimate mathematical city is Brasilia, shaped like an airplane poised on the shores of its lake.

After World War I, in sympathy with the expanding liberal society, public parks were planned for active games rather than passive contemplation. Bos Park, Amsterdam (conceived 1928, begun 1934), was designed by a team of botanists, engineers, architects, sociologists, and city planners. It was made from flat land below sea level and was predominantly devoted to physical exercise: a long regatta canal and (from the excavations) a massive hill with toboggan slide; areas for tennis, hockey, cricket, children's soccer, and play; a stadium and an open-air theater. These areas appear to be carved out of a forest in which there were riding and nature trails, nature reserves, and an experimental farm. No park since has been so comprehensive, or so ecologically well balanced. Recently, the idea of a farm as part of an urban park has been taken up elsewhere. Bos Park is also the parent of the modern country park with its emphasis on physical exercise among natural surroundings, and on direct contact with local crafts. The interpenetration of country into existing cities in an overall landscape plan was first realized in Stockholm (c. 1940). The plan allowed for "green fingers" to penetrate to the central area, where streets and public open spaces were enriched (under the Director of Parks, Holger Blom) with movable flower containers that were to be copied throughout Europe, seats among flowers, wild flowers along the Mälarstrand, and climbing sculpture for children. The junk or adventure playground was invented by C. Th. Sørensen at Emdrup, Denmark, in 1943.

In the 1960s many urban squares became pedestrianized. The sculptor Naum Gabo (b. 1890) had issued a famous manifesto in 1920 urging kinetic sculpture to be placed "in the squares and streets." This ideal was realized in 1966 at Lovejoy Plaza, Oregon, U.S.; an abstract rock and water complex in which pedestrians participated. The designer, Lawrence Halprin (b. 1916), stated that it was inspired by the High Sierra of California. Roof gardens appeared as novel attractions to the public in crowded cities. In the 1930s, a walled flower garden complete with small trees and shrubs was constructed on the roof of Derry & Toms' store in Kensington, London. The roof garden above Harveys of Guildford (1953, G.A. Jellicoe) was designed as a sky garden of planted islands set in water that reflected light. The roof garden of the Place Bonaventure Hotel, Montreal (1967, Sasaki, Dawson, Demay Associates), is a continuous romantic water landscape separating outer and inner top-floor rooms.

The automobile has created new problems

Bos Park, Amsterdam, Holland (1934): aerial view.

but also new landscapes. The Westchester Park system (1922, Gilmore D. Clarke) was unlike the traditional boulevard in that it was a concept of a sequence of roads *within* a park. The U.S. and Germany both pioneered the landscaping of multilane, long-distance, divided highways to ensure that they fitted the existing countryside; the immediate new scenery had to be related to fast-moving automobiles. In cities, the automobile has been creative on the roofs above multistorey parking lots, notably in the lakeland scene at the Kaiser Center, Oakland, California (c. 1958, landscape architects Osmundsen and Staley). In the London squares, skill has been shown in relating new underground parking lots to the trees above.

As if in escape from the mathematical planning of cities like New York, modern machinery has been deployed in Florida (Cocoa Isles 1957, Eugène Martini and Associates) and the Bahamas to create spectacular new seascapes. Islands have been created by dredging the seabed to aid navigation and pumping the excavated material inside barricades of piles. Shoreline swamps have been converted into marinas. In a single operation, a walking dragline can excavate water channels on, for example, its left side, swing across to the right and dump the spoil to build up fingers of building land, the dimensions being dictated by the throw of the dragline, usually about 100 ft. (30 m). The extraordinary animal patterns that have emerged are biological and can be interpreted as corresponding to the same urge toward nature that had created the Tennessee Valley Authority (1933) as an economic use of nature's recurring resources, and the subsequent ecological studies in the U.S. summed up by Ian McHarg in his book *Design with Nature* (1971).

Specialized landscapes

Universities

The traditional European universities were introverted and based on the quadrangle evolved from the monastic cloister, hence a distant descendant of the original Persian paradise garden. The U.S. took a more extrovert approach and pioneered the way to the modern university. At the University of Virginia (c. 1817) Thomas Jefferson (1743–1826) created a classical rectangular grass campus with one end open to the countryside, which was thus drawn into the university. Little similar feeling for landscape developed until Olmsted's design for Berkeley in 1865. Here he grouped the buildings along a mall or *tapis vert* sloping down the hill and "shooting at the sunset beyond the Golden Gate." The semiformal central group was set in a romantic landscape that contained the residences, and this has been the basis of many modern campuses, such as Buffalo.

Apart from Berkeley, the American universities developed as great architectural compositions deriving from the Ecole des Beaux Arts in Paris. The first biologic, as opposed to geometric, university in the Western world was Aarhus in Denmark (1932–, architects Kay Fisker, C. F. Moller, and P. Steegman), where the guiding spirit seems to have been the landscape architect C. Th. Sørensen. The buildings are brick, domestic in scale rather than monumental, for the most part ivy covered, and so arranged around the valley that the grouping seems subsidiary to landscape.

In England, the new universities in the 1960s were a total break with tradition. They were asymmetrical, their grouping dictated by the site and tending to be severe in design and unaccommodating to climate, an exception being Stirling University in Scotland, beautifully sited in existing parkland (architects Robert Matthew, Johnson-Marshall & Partners, begun 1967). In Italy, a large modern residential college has been fitted into the Urbino scenery without disruption to history (1970, architect Giancarlo de Carlo), while in Vancouver, Simon Fraser University (1963, architect Arthur Erickson) is poised dramatically on a mountain top, but is accessible from the city by automobile.

Cemeteries

The oldest surviving cemetery is probably that at Giza in Egypt, laid out in the 3rd millennium BC by Cheops in the shadow of his own pyramid tomb. Arranged in streets, the stone mastabas of the royal household, conforming to Egyptian belief in life after death, housed not only the preserved bodies of the dead but material provision for their well-being in the next world.

Roman cemeteries, based on those of the Greeks, with their slab and pillar memorials, had *columbaria* for urns containing ashes when cremation became the normal method of disposal.

The early Christian burial ground of simple grass mounds or inscribed stones evolved almost unchanged into the 17th-century churchyard; its haphazard rows of graves dominated by the church. In the 1830s, soaring populations led to the construction of large nondenominational, commercially owned cemeteries around most big cities. Marble was a favored material for monuments. In England, winding drives and groups of cypress or yew trees eventually made romantic settings for monuments that were increasingly grandiose and sentimental. Elsewhere in Europe layouts followed a grid pattern. Unlike the church, the chapel was subsidiary to the landscape.

After World War I, the need to mechanize maintenance of innumerable rows of graves in military cemeteries, plus a revulsion against monuments to human waste and seas of white marble crosses, resulted in the lawn cemetery; grass, trimmed by lawnmower, sweeps through glades of trees and shrubs planted in memory of the dead, a treatment also adopted in crematoriums (the first in England opened in 1874). Forest Lawn Memorial Park in California begun in 1917, with concert halls, movie theater, and art gallery, represents the ultimate in this genre.

In Scandinavia, greater simplicity was sought. At the Woodland cemetery and crematorium, Stockholm (1915–50), designed by Gunnar Asplund (1885–1940), a tall cross and the bare upward sweep of an artificial hill convey spiritual qualities that are both Christian and universal. Danish cemeteries, after World War II, imposed restrictions on the height and material of headstones, while one section of Mariebjerg crematorium, Gentofte, is reserved for graves marked by boulders.

Simplest of all are the mass graves of hundreds who perished in the siege of Leningrad. At the Piskarevskoe cemetery, row after row of plain grass mounds about 30 ft. (9 m) long and 3 ft. (90 cm) high, flat topped with sloping sides and only a low stone at each end to indicate the year, bear witness to the city's ordeal.

Industry

Although there were many early attempts to ameliorate the environmental effects of industry, the first comprehensive landscape plan on a long-term scale was for the Hope Valley Cement Works in the Peak District National Park, England (1943, landscape architect G. A. Jellicoe). When the quarries are worked out, the area will be returned to recreation, enriched by boating and fishing lakes, and wind-protected camping sites. Because of this innovatory concept all industrial waste is now planned for alternative use. The soil from a road tunnel in north London, for example, was reused to create the "Guinness hills" (c. 1959). Throughout congested England, waste areas of all kinds are being reconstituted as amenity land or for agriculture.

Appreciation of landscape as a business as well as a sociological asset in industry is now almost universal in the Western world. The most spectacular example is probably General Motors in the U.S. (architect Eero Saarinen (1913–61)), but unique as an idea is the Angli shirt factory at Herning, Denmark (1970, architect, C. F. Moller; landscape architect C. Th. Sørensen), where two circles of landscape—one the factory and one a sculpture garden—are encompassed within an abstract work of landscape art. Recognition of human values in landscape is seen in the modesty and restraint of the gigantic underwater generating station in the historic region around the Rance estuary, Brittany. Psychologically, the art has now been medically recognized for its healing capacity, as at Glostrup Hospital, Denmark (c. 1970, landscape architect Sven Hansen), and such formal recognition is spreading throughout the Western world.

Lovejoy Plaza, Portland, Oregon: water landscape by Lawrence Halprin (1966).

Cocoa Islands, Florida (1957): scheme for a marina by Eugène Martini and Associates.

Woodland Crematorium, Stockholm, Sweden (1940), by Gunnar Asplund.

Hope Valley Cement Works in the Peak District National Park, England (1943): model for 50-year plan by G.A. Jellicoe.

Transportation

Bridges

Bridges are technological artifacts and can be classified in various ways: by the material of which they are built, by their structural form, or by their method of construction.

Until the 19th century, the most common bridge materials were timber and stone, or artificial stone. Cast iron appeared toward the end of the 18th century, followed by wrought iron, and then by steel in the middle of the 19th century, although the other materials continued to be used for many years. Reinforced concrete appeared toward the end of the 19th century, and prestressed concrete toward the middle of the 20th century. The principal materials now being used are reinforced concrete, prestressed concrete, and structural steel; timber is used occasionally and aluminum very rarely.

The structural form of a bridge depends on whether its principal action is in compression (typically arches), in tension (typically suspension bridges), or in bending (typically beams and cantilevers). The shape of the structure is designed to resist those actions.

The structural form is, of course, related to the material. Stone can be very strong in compression, but it has little or no tensile strength, therefore it is generally used only for arch bridges or piers. Timber is strong in tension, compression, and bending in the direction of the grain, but it is only available in pieces of limited length. It has therefore found its greatest use in frameworks where a structure can be built up from fairly small pieces. Cast iron is strong in compression and has some tensile strength, but it is unreliable in tension. Cast-iron beams were used through much of the 19th century, with tension flanges which were larger than their compression flanges, but the arch is a form more suited to the material. Wrought iron is strong in tension, but it was much more expensive than cast iron in the 19th century. One expedient which was used to counteract its high cost was to combine cast and wrought iron, with wrought iron used for the main tension members.

Steel is an excellent structural material, strong in tension and compression, but it is expensive and was, at first, regarded with suspicion. For this reason it did not displace the other forms of iron as quickly as might have been expected. As higher strengths of steel became available and new methods of jointing were devised, its structural possibilities expanded. Reinforced concrete can be made strong in tension, but it is heavier than structural steel for the same job. For this reason, long-span beams in reinforced concrete are not feasible. However, reinforced concrete arches are feasible, and because of the material's tensile possibilities, they can have much longer spans and are lighter in

Lord Mayor's procession of 1827 passing under the unfinished arches of "new" London Bridge designed by Sir John Rennie. From an engraving of 1828.

weight than masonry arches. In prestressed concrete it is possible to provide compression even without an arched form, and prestressed concrete beams with spans greater than 650 ft. (198 m) have been constructed.

The material and the structural form are also related to the method of construction. In order to build a stone arch it is necessary to build a centering for it which is virtually a bridge in itself, and similarly most reinforced concrete bridges are cast in forms supported by falsework. Where deep or fast water has to be crossed, it may be physically impossible or too expensive to provide support for a temporary bridge. In that case, the bridge may be built by cantilevering out so that the partly built permanent structure carries the construction loads with perhaps the help of temporary cables or props. Another method is by prefabrication of major elements or parts of the structure before they are placed in their final position. All these methods may be combined in various ways.

Foundations

The Romans, the technological ancestors of modern bridge builders, already used quite sophisticated construction methods for the foundations and superstructures of their bridges. They used two kinds of bridge foundation. One consisted of driven timber piles and the other of spread footings of masonry or timber grillages. For the construction of river piers they used timber cofferdams or boxes formed from driven piles from which the water could be pumped. As far as we know, nothing more advanced than this was used during the Middle Ages and river foundations were fairly shallow; consequently, foundation failures were frequent. A more advanced form of cofferdam had appeared by the 17th century, consisting of double walls with puddled clay in between. Modern cofferdams are most commonly made from steel sheet piling.

The open caisson, a prefabricated box open at the top and bottom that sinks as excavation proceeds, and which may or may not form part of the permanent foundation, was first used during the 17th century for building the Pont Royal (1685) in Paris, designed by Jules Hardouin Mansart (1646–1708).

The pneumatic caisson, a closed-top box using compressed air inside to counterbalance the water pressure, was suggested in the 17th century, but it was first used for a bridge foundation at Rochester, England, in 1850 by Wright, and by Isambard K. Brunel (1806–59) for the Wye Bridge at Chepstow, England, shortly afterward. It was first used in the U.S. for an arch bridge built in 1869 over the Mississippi at St Louis, by James Eads (1820–87), and by John A. Roebling (1806–69) for the Brooklyn Bridge, which was completed in 1870 by his son.

Another technique, used for piers in deep water, is to construct an artificial island through which the foundation can be dug. This technique was first used in 1930 near San Francisco.

Modern bridge foundations still use all these techniques. Piles may be of timber, steel, reinforced concrete, or prestressed concrete, and cofferdams and caissons are made from steel, concrete and, more rarely now, from timber.

Bridge superstructures

Roman bridges were built from masonry or timber. There are, of course, no surviving Roman timber bridges, but they were probably trestle bridges with raking struts. Many Roman stone arch bridges have survived because they were built from carefully shaped blocks which fitted together without mortar. Roman bridges had semicircular arches and piers which were generally wide enough to take the thrust from an unbalanced arch. The spans of a viaduct could therefore be constructed, one after another, with maximum reuse of the centering.

During the early Middle Ages most bridges were timber trestles. From the 12th century onward, stone arches were constructed with varying soffit profiles, many of them with humpbacked roadways suitable only for pedestrians and pack animals. These differed from the Roman bridges which were usually fairly level. Lime mortar was used, and structures were generally made with a rubble core and an outer facing of ashlar, so that they were generally less sound and durable than

The bridge over the Tagus at Alcantara, Spain (c. AD 105); one of the best preserved Roman bridges. It was built without mortar.

Roman bridges. A few bridges were built with ribbed arches but developments in bridge technology in the Middle Ages never paralleled the advances made by the church builders of the same period. Bridge technology showed few advances beyond those of the Romans until well into the Renaissance.

During the 17th and 18th centuries, the stone arch bridge was refined with larger spans, flatter arches, and thinner piers. Waterloo Bridge over the Thames, finished in 1817, was designed by John Rennie (1761–1821). One of his innovations was the prefabrication of the arch centerings, which were constructed onshore then floated out and jacked up into position.

The first all iron bridge, at Ironbridge in Shropshire, England, was built in 1779 by Abraham Darby (1750–91). In America, Thomas Paine (1737–1809) designed a 400 ft. (121 m) span in cast iron which was never built. The castings, which were made in England, were eventually used to construct an arch bridge over the Wear at Sunderland, England, with a span of 236 ft. (72 m) in 1796. The structural concepts upon which these bridges were based were borrowed from timber and masonry building.

British engineers learned how to use cast iron as a material in its own right, and a number of cast-iron arch bridges were built early in the 19th century. Thomas Telford (1757–1834) was responsible for a number, and John Rennie (1761–1821) built several including Southwark Bridge (1819), in Lon-

Medieval masonry bridge at Taggia near S. Remo, Nervia Valley, Italy.

Bayonne Bridge, New York (1931), by Othmar H. Amman. It is the largest steel arch bridge in the world, with a span of 1,675 ft. (511 m).

Concrete arch bridge: the Colorado St Viaduct in Pasadena, California (1938).

don, with a central span of 240 ft. (73 m).

The timber trestle bridge of the Middle Ages developed into the truss. Andrea Palladio (1508–80), in his *Four Books of Architecture* (1570), gave examples of four types of timber truss bridges. Timber bridges which were a combination of truss and arch were built in Germany and Switzerland in the 18th century. The great development of trusses for bridges took place in America, and many different truss configurations were invented there. Many American timber bridges were covered, because timber lasts longer when protected from the weather. The Howe and Pratt trusses were originally of timber with iron rods for their web tension members. Eventually, trusses made entirely of iron were used.

During the 19th century, cast-iron beams were used for railroad bridges of fairly short span, and trusses with various combinations of cast and wrought iron were also used. The first all-iron trusses in America were made with true pinjointed connections, but well before the end of the century riveted connections were universal. The use of wrought-iron sections and plates built up by riveting developed, and engineers like Robert Stephenson (1803–59) and Isambard K. Brunel applied techniques learned from shipbuilding and boilermaking. Robert Stephenson's railroad bridges at Conway and Menai, in Wales, (c. 1850), were hollow box girders of wrought iron with the trains running inside them; they were fabricated on the ground and jacked up into position. The spans at Menai were 460 ft. (140 m).

The first large bridge built of steel, and possibly the first to be built by cantilevering,

was Eads Bridge (1847), consisting of masonry piers carrying trussed arches of steel with spans of 500 ft. (152 m). The arches were built by cantilevering out from the piers using temporary towers and stay cables. A truss is a way of making large structures out of small pieces, and it is particularly well suited to cantilever construction. Eads Bridge was the forerunner of many large trussed arch and trussed beam bridges of steel.

The Forth Bridge, built between 1882 and 1890 by Sir John Fowler (1817–98) and Sir Benjamin Baker (1840–1907), was the first long-span railroad bridge built of steel. It consisted of trussed balanced cantilevers with 680 ft. (207 m) arms connected by 350 ft. (106 m) suspended spans. The main members were enormous tubes made from pieces of plate joined by rivets. A number of large-span cantilever railroad bridges were built during the following 50 years, some of them the most ungainly bridges that have ever been built.

During the first half of the 20th century, steel trusses in various forms were used for most large bridges, and reinforced concrete gradually replaced steel as the material for smaller spans.

The first reinforced-concrete bridges were made in shapes more suited to stone, steel, or timber, and the first bridges which we recognize as modern, using reinforced concrete in a way that makes structural use of its plastic properties, were the beautiful, three-hinged arches built in Switzerland from 1905 onward by Robert Maillart (1872–1940). These led to the deck-stiffened arches which he built after 1923, which separate the functions of arch action and bending action in a very elegant way.

In recent times there have been a number of significant developments. In steel construction, the use of friction grip bolts and the development of welding have largely replaced riveting, and there is far greater use of plate girders and box girders as a result of these technical advances. In concrete construction, prestressed concrete in a variety of forms is more widely used.

Eugène Freyssinet (1879–1962) was the first to use prestressed concrete for substantial bridges. He built five bridges across the Marne with spans of 240 ft. (73 m) in the 1940s. Subsequently, Dyckerhoff and Widmann in Germany developed the technique of in situ cantilever construction in prestressed concrete using adjustable forms which moved outward from the supports as the bridge was constructed. Bendorf Bridge (1962) has a main span of 690 ft. (210 m). The Pine Creek Valley Bridge in California is a more recent American example.

In France, the technique of cantilever construction has been developed by Freyssinet's successors using precast concrete segments stressed together with resin joints of negligible thickness. The first bridge built like this was completed in 1963 over the Seine at Choisy-le-Roi, the segments being handled by floating cranes. The Oléron Bridge, on the west coast of France, was the first precast segmental bridge constructed with a specially built erection gantry. The technique has more recently been adopted in the Americas, where it was used for the approach spans of the Rio-Niteroi Bridge in Brazil, and for several bridges in the U.S.

The Rombas Bridge in France, built in 1974, used temporary towers and cable stays to enable the bridge to be built by cantilevering continuously over the supports rather than by balanced cantilevering. The Byker Railway Viaduct at Newcastle, England, which is still under construction, is being built by a combination of continuous cantilevering using temporary props, and balanced cantilevering, with simple lifting rigs sitting on the ends of the cantilevers.

Another recent technique, which is applicable to long viaducts with moderate spans, is segmental construction, where all the segments are cast at one end of the bridge and jacked forward, producing an extrusion. An example of this technique is Olifant's River Bridge in South Africa which is 3,363 ft. (1,025 m) long and has spans of 148 ft. (45 m).

In steel too, the cantilever construction technique is mostly used for substantial bridges in the form of welded plate girders or box girders constructed from prefabricated segments. Where they are of moderate span they normally have a reinforced-concrete deck which acts together with the steelwork to form a composite section, as on the Sapele Bridges in Nigeria, built in 1968. Larger spans, where weight is more critical, usually have a steel battledeck which also takes part in the main beam action as on the Zoo Bridge at Cologne, which has a main span of about 745 ft. (227 m).

Suspension bridges

The first suspension bridge with a level deck hung below the cables was built by James Finley (c. 1762–1828) in Pennsylvania in 1801. Thomas Telford's suspension bridge across the Menai Straits of 1825, with a span of 570 ft. (173 m) used chains made from wrought-iron I-bars which were prefabricated and lifted into position from a pontoon. None of the early bridges had longitudinal stiffening girders, and the Menai Bridge was later stiffened with substantial parapets. Wires were also used for suspension cables, and in France, in 1829, L. J. Vicat first used the method of constructing a wire cable in place by spinning out the wires.

John Roebling (1806–69) built a suspension aqueduct in Pennsylvania (c. 1850), and a railroad suspension bridge with a span of 820 ft. (250 m) at Niagara Falls, completed in 1855. The cables of the latter were spun from wrought-iron wires, and it had two decks with bracing between them. Roebling added cable stays to prevent dangerous oscillations. Roebling's Brooklyn Bridge of 1883 had a span of 1,600 ft. (488 m), and was the first suspension bridge to have steel wire cables. All the early suspension bridges had masonry towers, but most of those built in the 20th century have had steel towers.

The failure of the Tacoma Narrows Bridge, due to aerodynamic instability, led to the use of stiffening trusses instead of plate girders, and recently in the Severn Bridge, England, (Freeman Fox, 1966) to an aerodynamically shaped, steel box-girder deck used in conjunction with inclined hanging cables. The deck of the Severn Bridge was constructed as a number of slices which were floated out, hoisted up from the catenary cables, and welded together. The Humber Bridge in England, which is still being built, uses the same type of deck and suspension but the towers are made of reinforced concrete and were built by slipforming.

A number of bridges using cable stays were built in the 19th century, as well as bridges which used both cable stays and catenary cables; but the post-1945 cable-stayed bridges have been constructed with decks made from plate girders, from box girders in steel, in reinforced concrete, or in prestressed concrete. A well-known, but not very typical example is the Maracaibo Bridge by Morandi in prestressed concrete. The Theodor Heuss Bridge over the Rhine at Düsseldorf has a conventional tower arrangement—a mast at each side of each end of the main span of 745 ft. (227 m), and three widely spaced, parallel cables. The Kniebrücke at Düsseldorf (main span 1,050 ft. (320 m)) has a single mast and four widely spaced, parallel cables between the two carriageways. Both of these have steel decks.

The tendency now is to use many cables at fairly close spacing. This enables the bridge to be built without temporary cables (bridge over the Rhine at Rees). The Brotonne Bridge in France has a single mast at each end of the main span of 1,040 ft. (317 m) and 21 cables close together. The deck is a box girder of prestressed concrete made from precast webs and flanges cast in situ. A bridge over the Columbia River, Washington, has two masts connected by a portal beam each end of the main span of 975 ft. (297 m), and closely spaced cables at each side of the deck. The deck is constructed from precast concrete segments 79 ft. (24 m) wide, transported to the site on barges and lifted by a traveling rig. (See also: STRUCTURAL ELEMENTS–ARCHES, FOUNDATIONS, TRUSSES; CONSTRUCTION PROCESSES; BUILDING MATERIALS—IRON AND STEEL.)

Clifton Suspension Bridge, Bristol, England (1836-64), by Isambard Kingdom Brunel. The clear span of 702 ft. (214 m) represented a major achievement in 19th-century suspension bridge construction.

Brooklyn Bridge, New York designed by John A. Roebling and completed by his son in 1883. The first suspension bridge to use steel-wire cables and one of the first to use pneumatic caissons for foundations.

Verrazano-Narrows Bridge (1959-64): view from Staten Island. Suspension bridge with stiffened deck.

Railroad stations

The basic planning requirements for railroad stations have not changed very much since the opening of the first passenger railroads; there must be a forecourt where passengers arrive, a ticket hall, covered waiting space, and, preferably, covered access to the trains. Since provision must be made for both arriving and departing passengers, circulation, segregation, and duplication of services have always been familiar problems to designers.

Early designs 1830–50

Railroads were developed in Europe and the U.S. at about the same time, but in very different circumstances. Railroad systems as we know them started in 1830 with the almost simultaneous opening of the Liverpool and Manchester Railway in England, and the Baltimore and Ohio Railroad in the U.S. The first railroad stations date from this time. Liverpool's Crown Street Station had all the elements of a modern station with a train shed open at both ends. Mount Clare Station, Baltimore, still survives—a small brick polygon resembling a tollhouse.

Station buildings were placed on one or both sides of the tracks, across them, or in combination in U or L form. The train shed became the general form, adopted where it could be afforded. The layout of buildings and shed (or sheds) often depended on the site and on whether the station was at the end of the line or on a through line. For administrative purposes it was more convenient to have all the buildings on the same side of the tracks; however, this meant that some passengers had to cross them, and that was considered dangerous even when there was not much traffic. In the double-sided station, passengers departed from one side and arrived at the other so the accommodation had to be duplicated. In a U-shaped station, the sides could be connected, but in a through station this could only be done if the cross buildings was raised over the tracks. In the mid-19th century, the two-sided station was the most common type, both in Europe and the U.S. London's original Euston Station (now demolished) was the first important station of this type. It was constructed between 1835–39, and the architect was Philip Hardwick (1792–1870), with Robert Stephenson (1803–59) as engineer. First- and second-class passengers had separate entrances and accommodation. Its Greek Doric arch conformed to contemporary thinking that the railroad station was a gateway to the city.

At this time there was already a requirement that railroad stations should look important. The French Beaux-Arts-trained architects understood this better than the British, who, although excelling in railroad technology, allowed station architecture to join the battle of the styles without realizing that stations were a fundamentally new building type and should be approached from first principles. When critics spoke out it was not against the planning concept of the building, but the particular style which had been chosen.

In Germany, a type of Romanesque architecture was favored and became a recognized railroad style, but in Britain there was something of everything, and small country stations were usually designed in the vernacular of the area rather than in any recognizable railroad style. In Britain, money was readily available, encouraging good quality work, while the compactness of the country helped quick and economical development. British stations set the pattern for the world.

In the U.S., things were very different. Stations were far apart and most of the money had to be spent on tracks and equipment. Station buildings were a secondary consideration and, often, were inferior to those in Europe. As late as the 1880s American stations were criticized for lacking the normal amenities found overseas. Early stations were small and domestic in scale and appearance, without covered platforms, except for the overhanging eaves of the buildings. Timber construction was employed almost exclusively, and as a result fire was a common hazard. Every type of plan was tried in America, but station amenities developed very slowly, and while the Boston Station in Kneeland Street (1847) was well equipped, the new station of the Hudson River Railroad, opened in 1861, offered little except separate entrances for ladies and gentlemen.

Innovations 1850–60

By 1850, the experimental stage of railroad station building was over in both Europe and the U.S. Wood for fuel gave way to coal, and improved headlamps made possible the operation of night trains. Passenger amenities increased and the larger stations provided shelter all the way from the street to the train. New stations everywhere were built on a larger scale, partly to leave room for future expansion and partly because they began to incorporate hotels and office buildings. One-sided, two-sided, and U-type stations continued to be built. Of the three, the one-sided was the least flexible, but on a restricted site it was sometimes the only feasible solution, as at Newcastle, England (1845), where the building is tangential to the curving tracks.

In their original form, both King's Cross Station (1851–52), designed by Lewis and Joseph Cubitt, and Paddington Station (1852–54), designed by Isambard K. Brunel (1806–59), were two-sided stations. They functioned adequately enough until more tracks had to be laid to cope with increased traffic; arrangements then had to be made for passengers to cross from one platform to another. The best answer was the U-type station, which until then had featured only rarely. The pacesetter here was the Gare de l'Est in Paris (1847–52), designed by François Duquesney (1800–49), which was considered for many years to be the finest station in the world. Its plan owed much of its success to the introduction of a "salle des pas perdus," or concourse, in the head building (an idea first used in the Gare du Nord, Paris (1847) by Leyonce Reynaud). This enabled passengers to move from one platform to another without crossing the tracks. The sides of the U-plan housed departure and arrival accommodation respectively and each had a large court for waiting vehicles.

Up to this time train sheds had been mainly of timber construction, with cast-iron stancheons, and had, by 1839–40, reached spans of 72 ft. (23 m) at Bristol, Temple Meads, by Brunel. Iron, either in truss or arched form then took over, and the Gare de l'Est set the standard for the 1850s. The double arched shed of King's Cross, spanning 105 ft. (32 m), was, however, originally made of timber. It soon deteriorated under the attack of steam and smoke and had to be replaced with iron arches in 1869. Triangular iron trusses were sometimes used, for example in the second Gare du Nord in Paris (1861–65), designed by Jakob Ignaz Hittorf (1792–1867), but they lacked both the grace and excitement of the arched form.

Iron train sheds were now the dominant feature of stations, but they were visible, with some exceptions, from the inside only. Notable exceptions were King's Cross Station and the Gare de l'Est where the train shed was expressed by a great semicircular arch in the center of the main facade. Great arches became a prominent feature of railroad station facades, and their popularity was only exceeded by that of the tower, which became the world symbol of railroad stations. For Chicago Grand Central, for example, S. Beaman designed a tower 247 ft. (75 m) high.

In the U.S., the first important railroad station, Union Station, Providence, Rhode Island, was designed in the German Romanesque style in 1848 by Thomas Tefft. Thirty years later it was considered to be one of the 20 best buildings in the country. Wood was used in many of these structures as it was readily available and often much cheaper than iron. Lattice trusses of various types were introduced. Some, designed by Ithiel Town (1784–1844) and William M. Howe, spanning 150 ft. (46 m), were used in Philadelphia. Howe later developed a lattice truss in metal to span 166 ft (51 m) at the Great Central Station, Chicago (1855).

The era of the great train sheds 1860–1918

The 1860s initiated the period of the great train sheds. St Pancras in London (1868-69), designed by W. H. Barlow and R. M. Ordish, spans 240 ft. (73 m) and springs from platform level in a great curve to meet at a ridge 100 ft. (30 m) above the tracks; the shed is 689 ft. (209 m) long. It was often imitated, particularly in the U.S. In Grand Central Station, New York (1869–71), Isaac C. Buckhout and John B. Shook produced a semicircular arched shed from imported metal. Although it was expressly intended to rival St Pancras, for some reason it fell short in its main dimensions; however, as the largest covered space in the U.S., it was a great tourist attraction. The Pennsylvania Railroad, which specialized in vast train sheds, built the last of the great single-span sheds at Pittsburgh in 1898, with a span of 240 ft. (73 m). The ribs, fixed at one end, were mounted on rollers, so they were free to move. Then, mainly because of high maintenance cost, giant single-span train sheds began to die out, but even so, the day of the impressive shed was not over yet. The popularity of the railroad hotel spread from the U.S. to England. Built across the ends of the tracks and the train shed, it often became the building by which the station was recognized. The grandest hotel of them all was built in High Gothic style at St Pancras (1868–76), designed by Sir George Gilbert Scott (1811–78).

The styles of American stations were nearly as varied as those elsewhere. Henry Hobson Richardson (1838–86), a famous American

Facade of King's Cross Station, London, England (1851–52), by Lewis Cubitt.

Monumental facade of the railroad station at Strasbourg, France (1878-83), by Jakobsthal.

Concourse of Rome Terminal Station (1931-51) by E. Montuori and others.

architect of his time, used rough, massive masonry, and fortress-like detailing in a number of small stations for the Boston and Albany Railroad Company in the 1880s. Railroad companies now set out to be impressive, and concourses inside the buildings (as distinct from within the train shed) became monumental. Union Station, St Louis (1891–94), is a typical example. Some stations were very inconvenient. In the Illinois Central Station in Chicago (1892–93), designed by Bradford Gilbert, the great shedlike waiting room was placed on the first floor over the tracks. Railroad stations had become magnificent, trains were faster, safer, and more comfortable; but passengers sometimes had to walk as much as 900 ft. (273 m) between the street and the trains.

Little was done to find new solutions to the problems of station design although electric traction and lighting encouraged some important innovations. The Gare d'Orsay in Paris (1897–1900), designed by Victor Laloux, had everything under a huge glass roof with flanking arched windows; the entrance was at street level, and various offices and services were grouped around open wells over the platforms. In New York Central Station (1903–13), designed by Reed and Stem, the tracks were put underground on two levels; the train shed disappeared and in compensation the concourse was 125 x 375 ft. (38 x 113 m), with a height of 120 ft. (36 m). In spite of this exaggerated magnificence, the station is one of the most successful ever built.

In Germany, the best examples of great train shed stations were Frankfurt am Main (1879–88) by Eggert and Faust, with a vast eight-span shed, and Leipzig (1907–15) by Lossow and Kuhne, one of the largest and most comprehensive main buildings ever built. Earlier, the station at Hamburg (1903–06), designed by Reinhardt and Sossenguth, had a steel and glass shed with no masonry. The station building flanked by twin towers bridged the through tracks with entrances and exits at its ends. In England, most of the big stations had already been built, but a national romantic wave was passing through Europe, producing Helsinki Central (1901–14) by Eliel Saarinen (1873–1950), a truly great building without a train shed, and Milan Central (1913–30) by Ulisse Stacchini, where the single splendor of the five-arched steel sheds contrasts with the extreme ugliness and inconvenience of the building. In New York's Pennsylvania Station (1906–10) by McKim, Meade, and White (recently demolished), the waiting room and concourse were as big as train sheds. Large multispan train sheds were unpopular because of high maintenance costs. This difficulty was overcome by the American engineer Lincoln Bush (1860–1940) who invented the Bush shed in 1906—a concrete

system using moderate space with a height of only 16 ft. (5 m); smoke and fumes escaped through slotted vents. At first, the vaults were carried on metal stanchions, but later these were replaced by concrete, reducing maintenance.

Post-World War I developments.

Great station building in the old and luxurious tradition stopped with World War I, and concrete and glass replaced exposed metal, notably at Reims (1930–34), designed by Le Marec and Limousin, and at Le Havre Maritime (1936), designed by Urbain Cassan. In Holland, where much of the track is raised above ground level, a type of small station was developed with services at ground level and a light and airy steel and glass train shed containing waiting rooms, restaurants, and other accommodation at track level.

Britain had a brief but bright railroad station revival in and around London in 1932–38, due almost entirely to the work of two men: Frank Pick, head of the London Passenger Transport Board, which controlled the subway system, and the architect Charles Holden (1875–1960). Pick was interested in modern design in all its forms and Holden designed more than 30 surface stations for him. He used concrete, brick, and glass in basic geometric shapes, and his stations were convenient and easy to use.

Generally speaking, European station concourses became more human in scale and better planned and equipped. The most prominent among these is the Rome Terminal (1931–51), designed by E. Montuori and others. Here a great open space is covered with a curved and cranked concrete roof, with glass on three sides and a boldly cantilevered canopy overhanging the vehicle arrival area. The design is generally regarded as the finest recent example of railroad architecture.

Airports

The term airport should properly include the entire range of runways, taxiways, control towers, cargo, maintenance, parking, and administrative buildings which, together with one or more passenger terminals, make up the typical modern airport complex. This account will deal primarily with the design development of the passenger terminal building, only referring to airport infrastructure where this has a specific effect on the design of the terminal itself.

In principle, the airport is no more than a transportation interchange, the interface between different modes of travel. Up to 30 or 40 years ago, the most familiar example of such an interchange was the railroad station, where people traveling by public transportation, private vehicle, or even on foot, assembled to join another type of transportation system consisting of regular train services running on fixed routes, at comparatively high speeds, to specific destinations. It is understandable that in comparison with the, by then, highly developed railroad stations, the earliest airports were very simple, consisting of little more than a waiting room at the side of the grass landing field. This was all that aircraft of the 1920s required. Since aircraft capacities were small, the number of travelers was insignificant in comparison with other means of transportation. Air travel was so much of an adventure in itself that expectations, in terms of amenities on the ground, were limited. Only gradual improvement in the number and standard of facilities available, in at least the larger airports, took place during the years up to the end of World War II; but with the swift development of jet aircraft during the 1950s, major changes had to take place.

Compared with the Douglas DC-3, a typical 1930s-designed twin-propeller aircraft still flying on most world airlines at that time, the first Boeing 707 in service flew 2.5 times as fast and had eight times the range and payload of the prewar machine. Ten years later, the Boeing 747 and other wide-bodied jets could carry nearly five times the payload of the 707 and, if required, could travel twice as far. The effects of this explosion in aircraft productivity on the design of airports were, and still are, profound.

The most obvious effect is one of scale. In 1950, only 14 in every 100 people traveling by public transportation between cities in the U.S. went by air. In 1970, this figure had risen to 77 in every 100, and the number of passenger-miles flown had risen to 13 times the number flown in 1950. At the same time, the number of air passengers flown by all the world's airlines, including the U.S., rose from 30 million to 385 million, and it has been predicted that this will rise to more than twice that number by 1980. Increases in passenger numbers of this magnitude were inconceivable to airport managers and designers during the 1950s and 1960s, and even in the mid-1970s there were many who felt that the energy crisis might at last set a limit to the growth of air travel figures.

Because growth in air travel has dramatically outstripped even informed prediction during the last 30 years, it follows that few airports have been designed to handle efficiently the numbers of passengers and others who actually use them. Because of their size and complexity, most modern airports take many years to build. All too often, by the time they are finished they are found to be too small, inadequately equipped for the intensity of traffic actually encountered, and insufficiently flexible, either in design concept or in built form, to adapt to changed and constantly changing circumstances.

T.W.A. Flight Center, Kennedy Airport, New York (1960-62), by Eero Saarinen. Its winged shape in sculptural concrete suggests the theme of flight.

Passenger terminal functions

Before considering design concepts in detail, it is worth reviewing the various functions that the passenger terminal has to perform.

Passengers departing by air arrive at the terminal by various modes of transportation and must proceed to one or more check-in zones, where their travel documents will be issued or checked and their hold baggage received, weighed, and tagged. Baggage is at this point transferred to a subsystem within the terminal, which must be designed to ensure that baggage and passenger meet again at their destination. From check-in, passengers move to the first general holding area or concourse before proceeding to separate gates for individual aircraft. For international flights, the move from check-in to concourse is by way of passport and immigration controls; once past these controls, passengers have moved from the "landside" to the "airside" of the terminal, through what is, in effect, an international frontier. This separation between the two parts of the terminal must extend through all parts of the building, and throughout the airfield itself.

The concourse contains most of the amenities required by travelers, such as shops, snack bars, and restaurants. At international airports, duty-free sales to travelers are an important and increasing source of revenue to the airport. A substantial area must therefore be allocated to this requirement, with consequent security problems in dealing with bonded goods and large sums of money. From the concourse, passengers may be directed straight to the gate leading to their aircraft, or more frequently to some form of holding area at which a final check on passenger lists can take place. Events in recent years have necessitated careful security measures for all air passengers, and electronic and hand-search procedures for people and carry-on luggage generally take place at this point. From here, passengers either enter the aircraft directly, by way of an adjustable bridge known as an air-jetty or "jetway," or are transferred there by some kind of bus.

Passengers arriving by air follow a somewhat similar procedure in reverse, but with different checks, controls, and priorities. Entering the terminal on the airside, their first wish will be to find their luggage, if any, and then get away by means of public or private transportation. For international passengers there will be immigration and passport controls to negotiate before they can proceed to the baggage collection point, then customs examination once they have located their luggage.

Despite the broad similarity of departure and arrival passenger flows, the vast number of people involved in each case in an airport of any size make it essential to separate these flows almost completely. In the smaller and simpler types of terminal this can be done laterally, with an arrivals section arranged alongside the departures section, each usually at ground level. In very large airports this principle is sometimes adopted on a grander scale by having completely separate arrivals and departures buildings. In most cases, however, separation is achieved vertically, with departing passengers entering the building at an upper level and, ideally, staying at that level until entering the plane, while arriving passengers descend to ground level as soon as possible after leaving the aircraft, and stay on that level for all subsequent procedures. While this approach is simple to establish in principle, it is frequently compromised in practice by conflicts between the passenger circulation and other equally important systems within the terminal. Of these, the baggage distribution system is usually the most critical in its effect both on the design of the building and on the successful operation of the terminal as a whole.

Baggage handling

To cope with the enormous increase in the quantity of passenger baggage handled in airport terminals, various systems and approaches have been developed. It has become a convention, for example, to separate the departing passenger from his heavy luggage as early as possible in the departure process, sometimes even at a drive-in checkpoint before parking. Once it is weighed and tagged with flight number and destination, the

baggage is transferred by some form of mechanical conveyor system to a point on the airside where it can be finally transported to the aircraft hold by motorized carts. On the larger aircraft, baggage is stowed in containers before being put into the plane.

For arriving passengers, baggage is taken from the aircraft on the same system and unloaded, either directly or via a short conveyor, to some form of "carousel" or "racetrack" which distributes bags to the waiting passengers standing around it. Once collected, the baggage movement through the rest of the terminal is the passenger's responsibility.

In a typical two-level terminal departure baggage is conveyed to first-floor level as soon as possible after it is received, and distributed at that level. Arrivals baggage should remain at first-floor level from the aircraft hold right through to passenger collection from the continuous reclaim tracks. This is because passengers enter and leave the aircraft at an upper level conditioned by the height of aircraft floors above ground—over 17 ft. (5 m) in the case of a 747—whereas baggage is stowed in the hold of the plane, which is much nearer the ground, and conveyed to or from this by carts moving on the ground. Various forms of conveyor-belt link between aircraft and terminal have been attempted, particularly where there is a direct passenger link between the two. These have rarely proved satisfactory because of the conflict between the conveyor and the need to retain all-round access to the aircraft at ground level, so that all the other servicing operations can take place during its brief stay at the terminal.

Central well of Charles de Gaulle Airport, Roissy, Paris, completed 1974. Passengers reach the terminal along a sunken road and connect with the satellite buildings by moving sidewalks.

While these principles of baggage handling and movement are as simple to establish as those for passenger movement, there are a number of practical problems that limit their effectiveness in use. Firstly, all mechanical systems are prone to occasional or eventual malfunction. And when breakdown does occur the strain on both passengers and airport staff becomes acute. Secondly, although they are mechanized, all these systems depend on human beings to operate them effectively, and human beings also perform poorly on occasions. Thus, despite the checks built into the system, bags will sometimes be wrongly labeled, or wrongly routed despite having the right label. Once misdirected, for whatever reason, recovery is difficult and time-wasting for both passengers and airline staff. This situation is complicated in many airports by division of responsibility between the airlines, who undertake the care of the passenger at check-in, and the airport management, who deal with baggage until it reaches the aircraft.

Thirdly, baggage handling, particularly between the terminal building and the aircraft, which might be on a remote location a mile or more from the terminal, can be a uniquely unattractive job, with the personnel concerned exposed to extremes of weather, acute noise from aircraft both flying and on the ground, dust from jet blast, and constant exposure to atmospheric pollution from aircraft exhaust and fuel. The volatile labor relations familiar in most docks of the world have, not surprisingly, been carried into the baggage-handling crews of many major airports. When a strike occurs, it tends to cripple the conventional baggage-handling system as effectively as any mechanical breakdown.

These problems are inherent in the system, but there have been a number of attempts to overcome or at least reduce them by modifying the approach to the basic problem. Generally, however, these have tended toward increased mechanization, as, for example, at Charles de Gaulle Airport near Paris, where all baggage movement takes place below airfield level, with baggage inside the main terminal having to descend to, or rise from, a huge basement serving seven "satellites" where the aircraft are located. Despite the technical interest of the system, it has not proved significantly superior to more conventional arrangements, nor has it apparently enabled the airport management to reduce the number of baggage-handling staff, or to avoid reliance on them in case of breakdown.

In other airports, such as Seattle-Tacoma and Zurich, conventional baggage-cart links between aircraft and terminal are retained, but baggage circulation and sorting are guided and controlled by computer. This reduces the unskilled staff but places a great reliance on a small number of technically skilled staff.

Terminal design

A survey of the world's airports in the late 1970s shows that most terminal designs belong to one of four groups. These are: linear, pier, satellite, and mobile; terms which summarize the essential characteristics of each type. Many of the larger airports operate combinations of two or more types.

Linear This is the simplest terminal concept, and most of the airports founded in the early days of air transportation began in this way. The terminal is arranged so that aircraft can park in a line against or around it, with as direct a link as possible between the parking curb, the building, and the plane. It is particularly suited to small airports dealing with internal flights, and is still typical of first-generation airports in small or emerging countries. Examples occur at most smaller U.S. cities, while international versions can be seen at Helsinki and Iraklion in Crete.

The fundamental drawback of the linear approach is that it must be limited in size, and therefore in the number of aircraft and passengers that it can handle in peak conditions. This is because the safety rules governing the movement of planes on the ground and the large wingspans of modern passenger aircraft determine the distance that each must be spaced from the other at the terminal airside. Consequently, a large number of planes would require a very long terminal, and result in long passenger distances from check-in to aircraft. The logical solution to this problem would be to multiply the number of check-ins, baggage conveyors, and other facilities, but this has generally been ruled out for cost and staffing reasons, particularly in international airports where the complications of customs and immigration have to be incorporated.

It is significant, however, that the inherent simplicity and human scale of this configuration has caused designers to seek ways of overcoming its disadvantages when applied to very large airports. One solution that has emerged in several forms is to take a number of compact linear terminal units and arrange these in a larger linear complex along some form of circulation spine. In these cases, each self-contained unit is linked to the others by some form of mechanical transportation system, and the whole complex is very dependent on sophisticated mechanisms and controls. The most impressive current example of this approach is Dallas-Fort Worth where up to 14 semicircular linear terminal units, each accommodating up to 15 aircraft, will be dispersed along both sides of a transportation axis served by an automatic rapid-transit system. A similar but smaller example occurs at Hamburg-Kaltenkirchen in Germany, while in France the proposed Terminal Two at Charles de Gaulle Airport will consist of four pairs of linked curved linear terminals. In these, and similar cases like Toronto and

Kansas City International Airports, distorting the airside into a convex curve enables it to be wrapped around the smaller area containing all central facilities. A compact and well-organized rectangular example is the British Airways terminal at John F. Kennedy International Airport, New York.

Pier The enlargement of early types of linear terminal led to the development of the pier system. In its simplest form, a walkway or pier, double-decked in the case of terminals with vertical passenger separation, is extended from the original passenger building, long enough for several aircraft to be parked against one or both sides. In many cases "gates" and holding lounges are added to the pier so that passengers can be moved quickly through the terminal to lessen the pressure on its central facilities. The concept is simple, and is only limited practically by the surrounding airfield configuration and the travel distances considered tolerable for passengers. Because so many early airports were extended in this manner, in a piecemeal way which showed little apparent concern for the convenience of passengers, the pier system has become largely symbolic of the inadequacy of most airport design.

There are several examples of pier-type airports designed from inception, of which O'Hare International, Chicago, is the most impressive. Still handling the largest annual throughput of passengers in the world, O'Hare consists of one international and two domestic main terminals serving a total of 12 piers. The complex has shown itself flexible in adjusting to increasing aircraft sizes and passenger numbers per plane, but has achieved this at the cost of extremely long walking distances, exceeding a mile between the furthest gates, which might be considered unacceptable in an airport with so much interline activity. The Central Terminal of Frankfurt/Main in Germany is a later example of a new pier design, where four main piers serving 36 aircraft positions are liberally supplied with moving walkways to assist passenger movement to and from curb and plane.

Satellite Awareness of the limitations of even well-designed pier system terminals led to the evolution of the satellite concept. In this, the central terminal with all its major passenger-processing facilities is supplemented by a number of remote buildings at which the aircraft are parked. These buildings, the satellites, contain passenger-holding amenities which can be more economically concentrated than on extended piers. The satellites themselves can be situated in the optimum airfield locations relative to the runways, whereas the central terminal can be located at the point most appropriate to site traffic access. Thus the requirements of land and air vehicles, so contradictory in many respects, can each be met in the best way.

It is clear that the success in practice of any such concept will depend on the effectiveness of the link between the central terminal and its satellites. Ideally, if satellites are to be suitably remote, connections to the terminal must be underground to permit completely free movement of aircraft around them. Charles de Gaulle Terminal One is a dramatic example, where access to and from seven satellites arranged around a circular central building is gained by way of moving walkways which descend below ground to pass under the aircraft taxiways and rise again inside the building. Elsewhere, the satellites are developed more as expanded nodes at the end of long, surface-built transit piers linked back to the central terminal. Tampa International is such a case, where transit is accomplished by automatically controlled electric trains traveling at 1,000 ft./min. (300 m/min.). Seattle-Tacoma International is an example of a brand new airport incorporating all of the concepts described so far, having a central linear terminal with two piers attached and, in addition, two remote linear satellites reached by underground rail transit tunnel loops.

Mobile Lounges Though inherently flexible as a concept, the physical constraints of the link between central terminal and satellites— expensive tunnels or overhead transit links— and the fixed location of the satellites themselves out on the airfield, all serve to limit flexibility in practice. If the facilities provided in the satellite could be linked to the terminal or aircraft transit system, then a new dimension of flexibility could be achieved. This objective led to the development of the mobile lounge or transporter concept, where the passenger-holding lounges are actually mobile, designed to drive between terminal and aircraft wherever the latter is parked on the airfield. This involves acceptance of transit vehicles moving at surface level—only a completely suspended airfield would avoid the need for this—but in practice this has not proved a major problem in view of the high standard of ground control now possible with radio and radar communication. The mobile lounge concept should not be confused with the familiar, and rarely satisfactory, practice of busing passengers between terminal and aircraft, since this usually involves them being exposed to the outside environment at each end of the journey. The mobile lounge is actually a part of the terminal's passenger-holding facility, but designed to move to and from the plane, adjusting its height to any aircraft in service. The first and most famous example of this idea is Dulles International Airport, Washington D.C.

Once again, despite the apparent adaptability of the concept, it has not in practice been found to resolve all the problems of airport terminal design. It has not proved economic to design really advanced mobile

Charles de Gaulle Airport, Paris, France (1974). The ring-shaped central building is surrounded by seven satellites around which the aircraft are parked.

lounges, which will provide all the facilities usually expected by passengers in conventional waiting areas. As a result, passengers tend to stay away from the mobile lounges until the last moment before departure; this means these times are advanced to prevent late aircraft departure, and therefore result in a longer static waiting time, in vehicle and aircraft, than for other systems. Again, the mobile lounges require considerable skill to operate efficiently, particularly when connecting to the aircraft, since even minor damage caused during this maneuver can have substantial financial consequences in terms of delayed flights and disrupted schedules. It is significant that no other pure mobile-lounge serviced airport has been built since Dulles. Despite the flexibility of the principle, most subsequent designs have seen the approach as a supplementary solution of peak conditions of overload, to be used alongside more conventional static solutions of one of the other types described. Used in this way, the flexibility is retained, but the concept is reduced to just another runway bus service.

Traffic management

This assessment of past and current airport terminal types has concentrated almost exclusively on the interface between airport airside and landside—how passengers move between aircraft and terminal processes. Equally important to the functional success of the terminal is the efficiency with which very large numbers of people—inward- and outward-bound passengers, airport and airline staff, and also the considerable quantities of "meeters and greeters" generated by the increasing number of air passengers—can be moved between the terminals and their surrounding cities. Detailed considerations of high-volume traffic management, automobile parking and its proximity to the terminal, and the disposition and flow of people within the terminal—all these have a profound effect on its success or failure, both in broad functional terms and as a personal experience for each traveler passing through it. Empirical methods of calculating these factors and assessing their interaction have now given way to elaborate analysis by computer and the formulation of "mathematical models" of the airport to be designed.

There have, of course, been many attempts to devise radical alternatives to the conventional airport terminal. One solution popular with architectural students is to treat the airport like a vast aircraft carrier with all accommodation below ground, so that runways and taxiways can be freely disposed. This inevitably leads to very costly and technologically demanding systems for dealing with the fundamental problems which

remain—getting passengers in and out of the aircraft, and to and from the airport. Most other radical approaches involve some degree of cooperation from aircraft designers. If, for example, aircraft fuselages could be "containerized," passengers could join the plane at a downtown terminal which is then driven straight to the aircraft on the runway, with no need for an intermediate passenger terminal at the airfield.

Other more immediately realistic solutions require the development of vertical takeoff and landing (VTOL) passenger aircraft, similar in principle to the Hawker Harrier fighter-bomber. These would deal with all short- and some medium-haul flights between large cities, bringing this intense larger volume traffic nearer the city centers and lessening the load on conventional out-of-town airports so that they can continue to deal with a growing international flight load. Such a proposal may clarify the terminal designer's objectives, but it will not obviate the need for some form of city center terminal, with most of the familiar problems to solve, plus a massive extra environmental problem from VTOL aircraft noise and pollution. While there will undoubtedly be some developments in this direction, it seems most probable that airport design, in at least the immediate future, will tend toward refinement and evolution of already familiar types.

Dulles International Airport, Washington D.C. The passenger terminal building, completed in 1963 by Eero Saarinen, is a vast rectangular concourse. Aircraft are reached by mobile lounges which dock in at the terminal air side.

Residential

Thatched house on pilotis fronted by a shaded open area. Near Iquitos, Amazon region, Peru.

Reconstruction of a Roman atrium house at Pompeii. The various spaces of the house face onto the open courtyard.

Royal long house in North Sumatra (c. 1770). An elaborately built, symbolically detailed and decorated structure.

Houses

The characteristic form of dwelling adopted by a particular region is dependent upon a number of primary determining factors—climatic, sociological, or economic. In predominantly hot climates, rooms have tended to be open or grouped around an open courtyard, while in cold climates rooms were often placed together in a compact block to facilitate heating. Where security from attack was a primary concern, or where the seclusion of women was a major determinant, there were no windows facing outward at ground level and the building might even rise into a tower. Often, for economic or sociological reasons, a number of separate habitations might be joined together end to end, or one above the other to produce a multiple unit structure; where an extended family or a tribe was concerned, a multiple unit structure might share common facilities, and even provide one continuous social space serving as a communal focus to all the units, as in the Iroquois "long house" of the northeastern U.S.

In traditional, stratified societies, great importance was placed on distinctive forms of expression for houses of the different ranks of society. Nevertheless, the houses of the richer strata were usually generated from the same basic forms and living patterns as those of the humblest buildings. It is important to note however, the ambiguous relationship which frequently existed in primitive societies between the houses of the gods and those of the rulers. The religious buildings were often elevated forms of domestic architecture given a higher symbolic character. Subsequently, however, a reverse process might take place and the ritualistic and symbolic aspect of the temple might be transferred to the palace of the ruler as he assumed divine authority. Although practical considerations always played a major role in domestic architecture, these seldom dominated over religious, social, or status considerations, and a conception of the dwelling as part of a universal system of order—the observance of which protected the inhabitants from the wrath of the gods or of fate—was as important as the due acknowledgment of the rank and status of the family in the social order.

With the growth of belief in individual freedom in late Gothic times, the house came to be thought of as a vehicle of self-expression, and architects began to be commissioned to develop the facades and the main reception rooms, ultimately transforming the fundamental forms of the house from those which followed traditional, social patterns to an almost endless variety of original concepts.

Prehistoric and primitive

The earliest forms of constructed dwellings

Adobe-built village, Taos, New Mexico.

Traditional low-eaved timber chalet at Château d'Oex in the Alpine region of Switzerland.

were crude open shelters and temporary tents thatched with grass or rushes. During the Neolithic period the tendency of human beings to engage in settled or semi-settled agriculture or pastorialism led to the development of relatively permanent constructions. Bases of the earliest primitive huts showed that they generally consisted of a single room which was submerged partly or wholly underground. A hole in the center of the room functioned as the hearth. At quite an early period, small porches were introduced in front of the main entrance and may have acted both as windbreaks and as tunnels in which an attacker was put at a disadvantage. These huts were the basic forms from which domestic architecture developed. (see PRIMITIVE ARCHITECTURE.)

Although early house constructions were made of frameworks of saplings or bones, covered with hides or grass thatch or leaves, there was an increasing tendency to build in more permanent materials, e.g. stone, layered clay, mud brick. Conical or dome-shaped huts utilized circular or oval plans, and similar shapes were possible in the more permanent kind of materials providing they were roofed with the same forms, i.e. domes or dome-shaped roofs. However, as soon as straight beams were introduced to cover the space, whether it was constructed with a timber framework or a masonry wall, rectangular planning became more efficient.

The elementary form of the multi-room house was one in which a single room was subdivided into several sectors. This principle is used in the tents of the nomadic people of the Near East and North Africa, and was found in many parts of the world in the houses of the late Neolithic peoples. The early examples of the *megaron* on the Greek mainland, for example, had a rectangular hall with a semicircular apse at one end, which was divided from an anteroom by a wall containing a central door.

The earliest forms of the northern European "long house" were of the bay type, the largest individual houses having as many as 8 to 15 bays subdivided into several rooms. The remarkable New Stone Age settlement of Khirokitia in Cyprus had round sleeping cum living spaces flanked by kitchens, workshops, and stores which were clearly defined as separate spaces.

The early Egyptian houses were built of wood or unbaked brick within a walled compound, and took the form of square or rectangular buildings, each containing a central room flanked by smaller rooms on either side. These buildings were usually roofed by a terrace and later came to have a second storey reached by exterior stairways.

The "courtyard house" in antiquity

The formation of cities in many areas bordering on the Fertile Crescent at the eastern end of the Mediterranean, in the period from the 9th to the 7th millennium BC, was accompanied by the development of houses surrounded by high walls with no openings except for the entrance doors, which looked inward for light and ventilation to central courtyards. The earliest firm evidence we have of the evolution of such building types is in the excavation of temples, such as Tell Agrab in Mesopotamia, themselves believed to be derived from undiscovered domestic buildings.

Houses from about 2600 BC have been found in which the principle of grouping rooms around a court is highly developed. The house of the priest attached to the "oval temple" at Khafagae had rooms on all four sides of a court, with the main room running transversely across the southern side of the court. This principle of a transverse room relating to a court on its southern side (in predominantly cold climates) or on its northern side (in predominantly warm climates) was developed in Mesopotamia throughout the succeeding millennia.

By the time of the Assyrian Kingdom, at the beginning of the 2nd millennium BC, houses could be characterized as focusing on two main courts, surrounded by single-storey blocks, containing long, narrow rooms; one court, the "entrance court" (*babanu*) apparently contained the public reception and administrative rooms. The other, the "residential court" (*bitanu*) clearly had smaller units for living purposes. There were other subsidiary courts for more specialized activities, which clustered around the larger units. The great palaces of Nimrud, Khorsabad, and Nineveh, built during the great period of the Assyrian Empire in the 9th to 7th centuries BC, although of immense size and scale, were all of this type.

In Egypt the earliest remains of houses date from the 3rd dynasty, and show the typical house to have been a small rectangular structure about 17 x 13 ft. (5.5 x 4 m), divided into two zones by a longitudinal division, one zone for reception and the other for private quarters. The 4th dynasty, however, introduced L-shaped rooms (presumably containing bed recesses) and the provision of two entrances, one on a main processional street and the other on a service alley; inside the latter there was a granary. Other Old Kingdom houses were known from tomb models. They generally had a front courtyard surrounded by a high enclosure wall entered through a central gateway. There was a columned portico or an awning on columns at the rear of the courtyard, behind which was grouped a small number of enclosed rooms. Larger houses were of two storeys, with a columned portico on each floor and a stairway ascending to a roof terrace. Rooms were roofed with vaults and half vaults, or with timber beams. Half cupolas protuded from the roof terrace, facing north to catch the prevailing wind and direct it down ventilating channels to the rooms below. The courtyard served as a garden in which vegetables, fruit, flowers, and sometimes palms were grown; in the center there was usually a rectangular pool containing fish and reeds.

Urban houses, on the other hand, were entirely built over from one edge of the site to the other, with the exception of courtyards. They were well evidenced at El-Mahun, where both large and small houses dating from the Middle Kingdom (c. 2131–1785 BC) were built on the edge of the desert. The largest type had only one entrance, giving access via a vestibule to two parallel corridors which divided the eastern area from the remainder of the plan on the west. The former contained three courts surrounded by offices with a set of granaries at the rear. The other corridor led to the back of the main reception court, emerging in the colonnade in front of the transverse reception room which faced the cool northern breezes. A central door in this

Wattle and thatch shelter reconstruction at Singleton Open Air Museum, England.

Samoan house with woven palm retracted screen.

A "black house" on Harris, Scotland. Built of stone with a turf thatch roof.

Borres House, Gordes, France (c. 1660) built in dry stone with a corbeled roof on a rectangular plan.

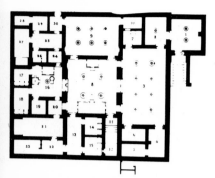

Plan of Egyptian house: the house of Vizier Nekht at El Amarna.

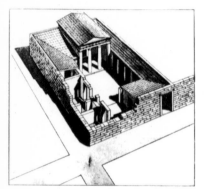

Typical Greek house: a restoration drawing of a house at Priene.

transverse room gave access to a square hall with a raised roof supported on four columns so that clerestory light could enter. This room, the so-called "hypostyle hall" was the kernel of the house, the main living room. On the west of it was the master's bedroom with a separate recessed sleeping bench. A door at the rear of the hypostyle hall led into a second living space surrounded by further bedrooms. The service court, with its own portico facing north, lay behind this. A third longitudinal strip to the west contained the women's apartments with its own court in the center, workmen's quarters on the south and servants' quarters on the north. Animals were kept in the house in stables which were reached from the entrance doorway. Roof ventilators brought cooling air down over the sleeping alcoves; there was also a bathroom near each bedroom. A large house of this kind could accommodate 40–60 people.

In the period of the Great Egyptian Empire during the New Kingdom (1580–1085 BC), houses of exactly the same basic forms continued to be built. The best evidence comes from Akhnaten's capital of Amarna, where a standardized house unit in the artisans' village had a front courtyard in the west, a central hypostyle hall with a masonry bench built onto its walls, and a bedroom and a kitchen at the rear. The staircase rose from the kitchen to the roof terrace, which was probably provided with an awning. The orientation of the courtyard to the west ensured bright sunlight at the end of the day while a covered area on the southern side caught the northern breeze.

Ancient Cretan houses may have owed something in their original planning arrangement to Egyptian houses. Their nucleus seems to have been a central living hall with a reception space in front of it, opening through a loggia to a courtyard or to the outside world. The domestic quarters were situated at the rear, behind the central hall. The facades of city houses were represented on faïence plaques which have been found in excavations (c. 1700 BC). It has been deduced that these houses had lower walls of rubble masonry, with a timber-framed upper structure resting on these, filled with sun-dried mud brick, which was plastered and decorated with colored paints. There was one entrance from the street at ground level, presumably into the vestibule with the main living space behind it, and there were frequently two storeys above ground level; it is assumed that these contained the private living spaces of the family. Many of the houses are represented as having a roof room.

On the Greek mainland the house seems to have developed from the prehistoric circular hut of bent saplings and thatch. A group of seven circular mud-walled houses with conical thatched roofs, arranged around a courtyard with an enclosing rectangular wall and a central door, have been found in a tomb model from the island of Melos. An alternative, without a courtyard, was the apsidal thatched house with a vestibule.

Our knowledge of Greek houses of Hellenic period is limited to those very few small houses which have been unearthed in excavations. At Dystus in Euboea, houses of the 7th century BC were entered down a narrow passage from an entrance door in the street facade. On one side there was a small room, which we know from Vitruvius' description of the classical Greek house to have probably been a porter's room. In a few houses there is evidence of other rooms off the entrance passage; these were stables for animals. The visitor then emerged into a courtyard which, according to Vitruvius, should have been given a portico on three sides, although the examples at Dystus did not usually have this. The main living room (oecus) had some resemblance to the megaron, at least in having a portico of columns between flanking walls in front of it. To the right and left of the portico thus formed were two bedrooms (thalamus and ante-thalamus), while around the courtyard were dining rooms and other rooms for common use, bedrooms, and service rooms. The portico in front of the family room faced south for light and warmth in winter, while one of the flanking colonnades faced north and provided cool shade in summer. The women's apartments were behind the main living room, or on an upper floor around the courtyard, approached up a staircase in one corner of the court.

Over 100 houses of the first half of the 4th century BC have been found at Olynthus near Salonika. The houses had a standard 16 ft. (4.5 m) frontage on a street running east and west. Parallel service alleys behind the houses carried the drains. The houses were all of two storeys; on the first floor the entrance doors, recessed under a portico, led directly into a courtyard, on the opposite side of which a wooden-pillared colonnade in front of a long corridor (pastas) gave access to a number of rooms which included a kitchen with a sunken hearth for a fireplace, a bathroom, lavatory, and stores. A granary and possibly a stable might also be entered from the courtyard. The most important room, however, was the dining room on the north side, placed on a corner so that it could have windows for cross ventilation on at least two sides.

Houses of the second half of the 4th century BC are well evidenced from excavations of the model planned city of Priene. It is particularly noticeable that, in the main reception room with its portico of two columns between flanking walls axially related to the courtyard, the old megaron type of plan, first seen in its developed form in the Mycenaen palaces, still persists. All the main rooms are on ground level, and only the women's· retiring rooms, and possibly a private living room, were on an upper floor.

Hellenistic domestic architecture is best represented in a number of fine villas and houses, preserved in some cases up to second-storey level, on the island of Delos in the Aegean. A characteristic town house focused on a small central court with a slightly sunken, mosaic-lined pool in the center, surrounded on all sides by symmetrical, double-storeyed colonnades with balustrades on the upper level. All the rooms opened into this courtyard, which was entered down a narrow passageway from the street door. The decoration was monumental in style, with inlaid marble or molded plaster representing rich dados, cornice moldings, sculptured friezes containing scenes from mythology and, in the upper part of the walls, exterior-type, fielded-ashlar stonework containing niches for lamps. Doors and ceilings were of paneled, hand-carved woodwork.

In the Western Mediterranean, the earliest civilization, the Etruscan, had unicellular and bicellular dwellings (as evidenced by the copies of them in tombs at Veii), but the most common dwelling had a number of cells placed symmetrically on two or three sides of a larger central room, the center of which was open to the sky in the largest examples (copied in tombs at Vulci, Perugia, Poggio Gaiella, near Chiusi, and Tarquinia).

Evidence of developments in the Etruscan dwelling after the 5th century BC is found mainly from literary sources. An entrance corridor without a roof led to a rectangular central area, now called an *atrium,* which had a symmetrical roof sloping downward toward the *concluvium* (opening in the roof) that directed rainwater into a shallow basin below, an *impluvium.* Opposite the *atrium,* which was flanked by family rooms, was a large reception room (*tablinum*) covered by a pitched roof, which was in turn flanked by two or more rooms or recesses (*alae*), which isolated the *tablinum* from the rooms at the sides of the *atrium.* There was often a garden at the rear of the house. This plan seems to combine the Etruscan tradition with the Greek concept of the reception room, which had reached the center of Italy from the Greek colonies in southern Italy and Sicily. Multistoreyed tenements containing several distinct dwellings were also built in Etruscan times.

The Roman *domus* is well known from extensive literary and archaeological sources. In its town-house form it may be characterized as having a vestibule, often flanked by shops, beyond which one entered the *atrium* with its *impluvium.* Opposite the entrance vestibule and on the long axis of the *atrium,* was the *tablinum* which now opened both into the *atrium* and into a larger planted courtyard surrounded by a peristyle that lay beyond. The rooms, or *alae,* at the sides of the *tablinum* led from the *atrium* into the colonnades of the peristyle living courtyard, off which there were dining rooms (*tricliniums*), reception rooms, and deep open rooms (*exedras*). Dining rooms had permanent masonry couches around three sides of the serving tables. In larger houses both indoor and outdoor dining rooms were provided, and there was a main reception room (the *oecus*), at the end of the long axis running from the *tablinum* through the peristyle court. It was used for celebrations, and in the grandest houses might be lined internally with a colonnade.

By the period of the consolidation of the Empire under Augustus, only the wealthiest citizens could afford private houses. In Rome and its port, Ostia, concentrations of population meant increasingly high ground values and rents, and multiple dwellings rising vertically began to replace the horizontally spreading courtyard houses. These tenement houses, the so-called *insulae*, were no longer designed around courtyards but were, on the whole, opened toward the outside with groups of rooms in a row which were approached along access balconies. Rudimentary collective sanitary facilities were grouped around the stairwells.

There is no evidence of kitchens, chimneys, or heating in the *insulae*. Cooking apparently took place over braziers. The houses were built of brick and concrete, with wooden floors, and frequently rose to great heights. In Rome, after a great fire destroyed a section of these tenement houses during Nero's reign, building regulations were introduced for stricter control of the height limit of 70 ft. (24.5 m) which had been introduced under Augustus; during the reign of Trajan it was lowered again to 60 ft. (18 m). Six- and seven-storey buildings were common at an even earlier period in the Mediterranean world, as is evidenced from the descriptions of the destruction of Carthage in 146 BC. The exterior appearance of such buildings seems to have depended on their fine brickwork, highlighted with string courses, columns, and pillars of stone, with vermilion paint outlining doors and windows.

During the Late Empire, in the 3rd and 4th centuries AD, the pressure of population began to drop and individual houses reappeared at Ostia and other cities in Italy. Unlike the tenements, with their large windows for light and air, these houses tended to look inward to courtyards, and had central heating with wall flues. Their simple plan grouped a series of small rooms in a row facing across a wide loggia or arcade to an internal court with a decorated wall opposite the loggia, possibly containing fountains. It is simpler and less formal than the Republican

Reconstruction of the House of the Tragic Poet at Pompeii. The *impluvium* can be seen in the foreground with a view through the *tablinum* of the peristyle beyond.

Perspective view of a typical Roman courtyard house: House of the Vettii, Pompeii.

Reconstructed section and plan of the House of Pansa, Pompeii.

Wooden house, Trelleborg, Denmark (c. AD 1000), from a reconstruction of a Viking settlement.

House of half-timbered construction in Denmark.

or Early Empire plan, and speaks of graciousness, charm, and comfort, with a certain air of elegance that is found in the last flowering of Roman architecture.

Roman country villas were characterized by the articulation of functionally differentiated building nuclei. The *atrium* was replaced by a courtyard in the form of a peristyle, which also served as a farmyard. Around it were the owner's quarters, quarters for the servants and farm manager, rooms for equipment and facilities, shelters for the animals, and barns and storerooms, and each of these nuclei might have its own subsidiary courtyards. In many cases, the country villa was surrounded by strong defensive walls. Later, the concept of the villa was a center for elegant life and entertainment was emphasized, although it always retained its role as the controlling unit of an efficient farm.

The Northern European house

The Northern European house owes its essential character to its origins in the post-and-beam architecture of prehistoric times. In the Roman Empire north of the Alps, buildings were frequently half-timbered, and a characteristic form of basilican shed was used for farm buildings, barracks, and for the homes of laborers on the large estates. This barn-like type of structure had a high, central, hipped roof supported on two rows of columns, surrounded by a lower lean-to roof on all four sides. It doubtless owed its origin as much to building traditions in the north, such as those of the "long house," as to its antecedents in the Mediterranean. The latter influence would have introduced clerestory lighting between the central and outer roofs; the former would probably have encouraged the fusion of the outer and central roofs into one great hipped roof spreading over the whole plan—the predominant form of later medieval Europe. From pictorial representations such as those on Trajan's Column under the Romans, we know that both types existed. Archaeological excavation has shown that the northern Roman laborers' dwelling was sometimes partitioned, presumably so that in one house both animals and laborers could be housed, a tradition that persisted in some rural communities until modern times.

The main house of the Roman villa in northern Europe usually developed from a simple form of the "colonnaded villa" into a "courtyard house," with the courtyard serving as the farmyard, which was therefore considerably larger than the urban courtyards of the Mediterranean. This courtyard house was often entered through a great arched gateway—the ancestor of a type of farm which has survived to the present day in northern France, the Low Countries, and western Germany. While these houses relate in building technique and living patterns more to the Mediterranean than to northern European traditions, they are nearly always associated with barn dwellings of indigenous northern type. As the villas passed from the hands of private farmers into those of large land magnates or the Imperial government, it was the barn dwellings which survived as the focus of life after the old villa had been converted to mundane farm uses. This kind of house persisted throughout the succeeding centuries, as is evidenced by archaeological remains dating back to the 7th and 8th centuries, which have been found in many parts of Europe.

The desire to emulate the Romans by erecting habitations in more durable material than timber and plaster is evident from the remains of a number of 8th- and 9th-century palaces. But the generally unsettled nature of the times led to a new emphasis on defense, resulting in the double-storeyed dwelling. The hall or common living room was on the first floor and was entered by an outside staircase. The ground floor or undercroft was used only for storage. There is ample evidence that such houses were built in England before the Norman Conquest in 1066, and they are known elsewhere in Europe. In some cases the upper room was supported on an open arcading of stone or of brick. They are closely related to houses of the 10th century onward which were built over a basement wholly or partly below ground level.

The advantages of such a house were not exclusively defensive; in towns and villages a storage basement was particularly useful for merchants, while on farms it could be utilized for the storage of valuable crops and food. In type they persisted through the Middle Ages in many urban contexts and also in manor houses. The undercroft, at least, was of masonry and in England and some parts of Europe the whole house was of stone with dressed ashlar facings around the openings and at the corners; roofs were tiled or shingled. Internally, there was a fire in the center of the early halls, later it was moved into a fireplace on one side. Smoke escaped through an opening in the roof and many accounts tell of sudden fires when the flames rose too high from the hearth.

By the 11th century it was common for English houses of this type to have either an upper sleeping chamber within the roof space at one end or ancillary accommodation beyond the great hall with sleeping space. Garrets above the ancillary wings were used as sleeping spaces for children and servants. Such buildings were common in the 12th century and continued in use throughout the 13th century, although by this time the great halls were beginning to descend once more to ground level. A striking feature of the new cities founded in the 13th century, both in England and France, is that the market places were commonly surrounded by elevated tall houses built over open arcading. There was no glass in such houses, only wooden shutters which were closed at night. From the 12th century onward large houses for the nobility began to have separate kitchens.

An important later development was the concern to construct a large hall without the use of columns or piers to provide intermediate support for the rafters. This gave rise to the development of the "king post" roof in the 11th century, and in the 14th century to the "queen post" roof. Another solution which first appeared c. 1350, was the use of the hammer-beam truss (as in Tiptoft's Manor in Essex, England). In the north and west of England large hall dwellings were constructed by using large, curved tree trunks, "crucks," which were placed on the line of the outer walls and met at the ridge to support the ridge beam onto which the ends of the rafters rested.

The medieval town house was generally conceived as a single building with the upper parts overhanging the street. At ground level were the shop and the kitchen, the former with a basement underneath reached down a straight flight of stairs. A passageway led to a courtyard at the rear where there was usually a well and sometimes a second building which contained straw and firewood, a laundry, and servants' rooms. Sometimes baths and hot rooms were provided near the kitchen or laundry. The main living room was situated on the first floor.

In the 14th and 15th centuries there is evidence of increasing luxury and splendor in the use of larger external windows, more concern with decoration and spatial effect, and a striving after symmetry. In England two new features appeared to add comfort to the feudal manor house. The first was the addition of a porch to the external doorway, and the second was the conversion of the space under the chamber into a parlor—a private room for the family or a quiet place for conversation. The timber beams which supported the plank floor above were carved into an ornamental ceiling, and the wall was paneled. From then onward in England the parlor tended to become the principal first-floor room of the house, with the main bedchamber above it. The hall gradually shrank in size, being retained as the central circulation space between the entrance doorway and the stairway.

Renaissance houses

The Renaissance with its deliberate imitation of classical patterns and conscious return to classical standards of value—although, of course, at that time they were frequently

misunderstood—led to the organization of the Florentine house as a symmetrical composition around a large arcaded courtyard; this contained corridors on all three floors giving access to rooms which were orientated outward. A central doorway in the facade led through an imposing vestibule to the courtyard; there, a broad staircase of straight flights led up from a corner, or, ideally, from the two corners opposite, to the *piano nobile* where a large central hall, placed on the axis of the main facade and often two storeys high, served as the center of all social and official activities. Around it were private reception rooms and often a library and a chapel. On the third floor were the family rooms, dining rooms, and bedrooms. An attic storey below the flat or low-pitched roof contained the servants' quarters. On the first floor were business offices and storerooms. Externally, the facade was divided into three storeys, the lowest being heavily rusticated and the uppermost gently embellished in delicate ashlar work. The whole facade was crowned by a great cornice. This 15th-century pattern, developed in the Medici, Strozzi, and Rucellai palaces in Florence, became the model for almost every Italian Renaissance palace in the 16th century.

The town palace and villa architecture of Andrea Palladio (1508–80) in the mid-16th century was to have a profound effect on the whole of Europe. He used classical temple porticoes, complete with pediments and raised podiums, to prepare the visitor for grand cubic or circular reception rooms, sometimes crowned with domes, which formed the central spaces of his villas. Smaller rooms, ranged on two or three levels, formed the

The form of the medieval town house, with its timber-framed structure projecting forward over the street, is used in this 17th-century house in Oxford, England.

Half-timbered yeoman's house: Wealden Hall House in Kent, England (c. 1500).

Riccardi Palace, Florence, Italy (1444) by Michelozzo.

Courtyard of the Strozzi Palace, Florence, Italy (1489), by B. de Majano.

Villa Capra, La Rotunda, near Vicenza, Italy (1550–57), by Andrea Palladio.

Place des Vosges (Place Royale) in Paris, France. An urban square laid out in 1610 for Henry IV.

corners of the plans, and there were sometimes cross axes which emerged through other porticoes (Villa Capra, La Rotunda, near Vicenza). The axes of the porticoes were extended into the landscape in wide paths and flights of stairs flanked with statues. Ancillary farm buildings were linked to the main house to make majestic forecourts lined with colonnades resembling Greek stoas, sometimes extending far out like arms embracing the landscape (Villa Trissino).

Palladio's style was distinguished by lucidity, balance, and a harmony of proportion derived from the use of simple numerical relationships repeated and correlated throughout the design. His influence was disseminated partly through his own published work, in which many of his buildings were illustrated. Designs based on his ideas continued in Italy until the 18th century and spread in the 17th century to Germany, Holland, and England, as in the Queen's House, in Greenwich, England, designed by Inigo Jones (1573–1652).

From the 16th century onward there was an increasing tendency for all city dwellers except the very richest to live in multi-family buildings, which externally adopted many of the attributes of aristocratic palaces. Rooms were arranged in a number of apartments on each floor, or extended vertically over several floors on a narrow frontage. In the early 17th century even dwellings for the lesser nobility assumed this form in Paris and London (Place Dauphine and Place Royale, Paris, and Covent Garden, London).

The Baroque age inherited from the Italian Renaissance a taste for varied scenographic effects which were now carried to extraordinary visual lengths in house architecture by developing the Roman sequence of spaces into a vestibule, courtyard, and monumental staircase, followed by suites of rooms of varied shape: square, rectangular, circular, elliptical, or cruciform. The importance of the monumental staircase grew until it became the central feature of late Baroque and Rococo houses, achieving immense size in the inverted pyramidal volumes of Austrian and southern German palaces (Brüchsal, Würzburg, and Pommersfelden). Similar scenographic effects were created externally in Baroque houses (Palazzo Carignano in Turin, Italy, and Blenheim in Oxfordshire, England).

The enfilades of rooms opening one upon the other, which was a prediction of the 17th century, was abandoned with the advent of European Rococo. Previously the rooms had been arranged with all their doors opening in line with one another, providing long axial vistas with the rooms disposed symmetrically above them. In the 18th century the preferred arrangement was for small suites of rooms, preserving the illusion of long vistas by the use of mirrors which gave the impression of a succession of volumes. This illustionistic effect was paralleled in a light decorative surface plaster work which replaced the monumental mural decorations of the preceding century.

At the beginning of the 18th century the influence of Palladio was revived by the school of Lord Richard Burlington (1694–1753). The classic purity of the stately symmetrical houses of the Georgian era represents the late fulfillment of the classical Renaissance in northern Europe. Their interiors were more varied than their exteriors. The hall was generally conceived as an imposing entrance, but there were usually projecting galleries at second-floor level, where all the principal rooms were situated. Octagons, ovals, circles, and squares were combined to produce fine effects.

The Neo-Classical house

Palladian country houses in England reveal an essential contrast with their Continental counterparts. Whereas Renaissance and Baroque villas had extended their architectural discipline into the landscape around them in formal terraces and avenues, the English classical houses have an essentially romantic character, standing as they do in landscaped gardens shaped to an ideal vision of nature derived from the Italian and French landscape painters of the late 17th century (Stowe, Pain's Hill). The pictorial effect of these settings, which recall the scenery of the Roman *campagna* as painted by Claude Lorrain (1600–82) or Nicolas Poussin (1594–1665), was at first peculiar to England. In its literary, poetic allusions to a golden Roman age it heralded the Romantic movement.

Late 18th-century domestic architecture in England was dominated by the style of Robert Adam (1728–92), which was based on his studies of the excavations of ancient Roman houses in Pompeii and Herculaneum. The rooms became even more varied in shape and the painted walls and ceilings were divided into geometric shapes by the lightest of moldings, delicately enriched by classical Roman ornaments in which swags and flutes predominated. Rooms were flooded with light by increasingly large windows with knife-edge glazing bars, and the comforts of private bathrooms and toilets adjoining dressing rooms began to be introduced.

The Romantic house

The Romantic movement led to a revival of interest in historic styles, culminating in an interest in Gothic houses. The first of these, Horace Walpole's converted house at Strawberry Hill in England (c. 1750–70), was followed by the great cathedral house of Fonthill

Monticello, Virginia, a Neo-Classical house remodeled by Thomas Jefferson between 1796 and 1808.

Blaise Hamlet, near Bristol, England (1803–11), by John Nash.

Terraced housing in the industrial town of Burton-on-Trent, England (1865).

(1796) and by other Gothic Revival houses in France, Germany, and North America. An admiration for the simplicity of rustic life led to the creation of the Hameau, a charmingly fragile pastoral village at Versailles for Marie Antoinette, and subsequently to a fashion for rustic cottages such as those advocated in Plaw's *Ferme Ornée* (1795) and realized by John Nash (1752–1835) in his thatched cottages at Blaise Hamlet in England (1811). With their irregular planning and variety of novel silhouettes, these buildings exhibit a taste for the Picturesque in architecture which was to profoundly influence 19th-century domestic building in Europe and North America. A house could now appear as sham castle, as Neo-Classic villa, or as Tudor hall. The number and variety of rooms increased in proportion to the increasing affluence of the middle class. In addition to drawing rooms, dining rooms, and bedrooms, there were now smoking rooms, billiard rooms, study and gun rooms, as well as parlors, pantries, sculleries, and kitchens. In a reaction against the terrace house, the middle class accepted the semi-detached house, a pair of houses resembling a detached villa.

Housing of the Industrial Revolution

One effect of the great expansion of industry at the beginning of the 19th century was the introduction of mass housing in the industrial towns. Back-to-back row houses were often three or four storeys high with staircases of a cramped winder type passing through each room. Legislation restricting such housing was introduced after 1832, and a series of movements for reform, such as the Prince Consort's Model Houses of 1850, led to the eventual provision of kitchens and toilets and the elimination of back-to-back planning in the second half of the 19th century.

The evolution of the modern house

After 1860 a taste for simpler architecture, with less insistence on the imitation of rich and elaborate styles from the past, led to concern for the use of simple building materials and good craftsmanship. The English socialist William Morris (1834–96), with his slogan of "Fitness for purpose" encouraged tidier planning and simpler furnishing. By the beginning of the 20th century, English domestic architecture was everywhere beginning to take on its familiar semi-rural character. Garden City development at Turnham Green (1876) by Norman Shaw (1831–1912) and at Letchworth (1903) by Barry Parker (1867–1941) and Sir Raymond Unwin (1863–1940) became the prototype for less successful suburban patterns which were copied throughout the world and led to a decline in traditional city life.

Typical suburban development in England post World War I.

Entrance hall of an early 20th-century English house in the Rural Revival tradition.

Perspective view of Leys Wood in Sussex, England (1868), by Richard Norman Shaw.

Frank Lloyd Wright's Robie House in Chicago (1909).

Project for freehold maisonettes with individual raised gardens by Le Corbusier and P. Jeanneret (1925).

Low-rise houses in the Byker redevelopment area, Newcastle-upon-Tyne, England (1970) by Ralph Erskine.

The "wall" built along the edge of the site at Byker, by Ralph Erskine (1970).

In the U.S. the revolutionary architecture of Frank Lloyd Wright (1869–1959) produced a house in which rooms flowed into one another without separating doors. By projecting the rooms in wings at right angles to each other, Wright was able to introduce light on all sides and provide effective cross ventilation for hot summers. Winter heating was introduced under the masonry floors, and the fusion between house and garden was emphasized by extending the roofs outward in overhanging eaves, car ports, and covered terraces. Some of Wright's innovations were made possible by 19th-century engineering developments in the use of new materials: steel and concrete. This fusion of new structural possibilities with the principle of "Fitness for purpose" of the Arts and Crafts movements might be said to have given rise to the 20th-century house.

In the period between the two World Wars, improved notions of hygiene led to the provision of bathrooms and indoor sanitation as a normal part of domestic design. Suburban living in detached houses was now the ideal, although apartment living was accepted in the largest cities. In the 1920s, Le Corbusier and other avant-garde architects introduced the idea of "functionalism," of the house as an instrument for living. The house is raised on *pilotis* (freestanding columns or piers) which allow the parking of cars and drying of clothes beneath the living area. The second-floor living room rises through two storeys, the kitchen and dining space flowing into it on the lower level and the bedrooms overlooking it on the upper level. Above, the house is enlarged by a flat roof garden, and sometimes a roof room.

Le Corbusier also applied his concept of domestic living to multi-dwelling apartments, which he exemplified in a full-scale model in the Paris Exhibition of 1925. Each double-height living room was to be fronted by a double-height terrace where plants might be grown. The living units would be grouped in immense rectangular blocks raised on *pilotis* so that the landscape would run through undisturbed beneath them. In this way the ground area would be liberated to become a continuous parkland for the use of residents. Cement roads through this parkland would give access for vehicles. This vision of a new type of apartment building was not realized until after World War II, with the construction of the Unité d'Habitation at Marseilles, which was quickly followed by the erection of a number of other such buildings in France and Germany. They were characterized by the provision of a shopping street halfway up the height of the buildings, and of theaters, gymnasiums, and schools on the roof terraces. Now, in the last third of the 20th century, Le Corbusier's vision of high apartment buildings is being replaced by that of

housing which is once more close to the ground, although, where economy is a consideration, the dwellings may be closely packed together in terraces or stacked in four or five floors of apartments and maisonettes, as in the Byker development (begun in 1970) at Newcastle, England, designed by Ralph Erskine (b. 1914).

The Islamic house

Islamic houses were essentially of two types. In inland mountain regions, such as those of Yemen and High Atlas, they related to an ancient Arabian type of tower house in which the harem and cooking areas were at the top of the house, the reception rooms usually in the middle, with storerooms and stalls for animals below. In the remainder of the Islamic world, dwellings were spread out horizontally around a courtyard. Frequently, and perhaps under the influence of the ancient Egyptian model, the courtyards contained pools and fountains and were surrounded by flowering plants and shaded by fruit trees.

The Islamic house was normally divided into two clearly differentiated zones: the reception area for men and the harem or private living area for women and children. In some regions each zone had its own courtyard onto which the rooms faced; in other regions there was only one courtyard, the harem rooms being relegated to a dark zone behind an arcade or reception room at one end of the court. In the northern hemisphere the main rooms were usually on the south side of the courtyard to avoid the penetration of the sun; the other side was sometimes colonnaded or arcaded. In colder regions there might be a second main living room on the north side to catch the winter sun. Roof ventilators, to catch the breeze blowing across the top of the built-up area, were introduced in early Islamic architecture following the precedent of ancient Mesopotamia and Egypt.

From Parthian and Sassanian architecture in Persia, Islamic architecture in the 9th century adopted the use of the *iwan*, a deep, vaulted reception space, completely open on the courtyard side so that people sitting under the shade of its roof felt that they were sitting in the open air of the courtyard. Because of the extremes of climate from summer to winter, Persian houses had often had two such *iwans*, one facing north and the other facing south; this principle was often adopted elsewhere, even in areas without such extremes.

The Persian practice of utilizing, in large houses and palaces, four *iwans* (deep-vaulted reception halls)—one in the center of each side of the courtyard—was also adopted in many parts of the Islamic world in the 9th century. (See also ISLAMIC ARCHITECTURE.) Various interpretations of these traditions survive today in houses in Iran, Syria, North

Courtyard of a house in Cairo, Egypt.

Hindu house: looking from the men's reception space into the court.

Sunken courtyard in a Buddhist house, Sri Lanka.

Africa, and Spain. In Mamluk Cairo, the density of urban building led to the elevation of the *iwan* to the second-floor level, in the form of an arcaded gallery. The main reception room became a double *iwan* with a central space covered by a lantern, representing the residual form of a small courtyard. Large houses had two such reception rooms, one at ground level used by men and the other at right angles above it used by the women.

In Ottoman Turkey, cool weather for much of the year made courtyard life impractical; instead, the form of the Islamic courtyard house was retained but it was covered with a raised lantern or dome, surrounded by large *iwans* on two or four sides.

Late medieval houses in hot humid regions often utilized pierced masonry and wooden screens on the upper storeys to provide cross ventilation through the harem levels. Screens of this type in wood, enclosing projecting balconies, sometimes rose in tiers one above the other to give rise to the characteristic street scenery of many parts of the Islamic world.

The Hindu house

Strict religious rules determined almost every aspect of the traditional Hindu house. The permissible size of the house and the extent to which it was subdivided into zones—the most private areas being entered only by members of the family and people of equal or higher caste—depended on the caste of the inhabitants.

The kernel of a Brahmin house was the room in which the family possessions and the family shrine were kept; this sometimes doubled as the main bedroom—wherever possible, however, the shrine and the owner's sleeping room were separate spaces. In larger houses a third room was provided, for storing the large sacred vessels in which the rice and meat for ceremonies and weddings were kept. The roof of this section had to be as high, and preferably higher, than any of the surrounding roofs whether they were those of the same house or of neighboring houses. The doors of the shrine room and the owner's bedroom opened on the east side into the private living court which was regarded as a purified living area; entry by a foreigner or member of a lower caste rendered it polluted.

The men's reception room (*barsati* or *barasti*) faced north in the hottest climates to avoid the sun. It was often open on the outside but was separated from the private living quarters by a pair of doors. The court was square in plan and surrounded by rooms on all sides. In general the living space under the roofs was open to the court and separated from it only by a light wooden colonnade; the floors of these shaded spaces were raised 1 ft. 8 in.–2 ft. (50–60 cm.) above the pavement of

the court, and on this raised platform boys, girls, servants, and various other members of the family slept in various areas which were predetermined by custom.

Family prayers were said in the courtyard in front of the treasury shrine before sunrise and at sunset. The living court was used for all domestic activities, including washing and bathing, by the women and children of the family.

The Buddhist house

Buddhism in its pure form did not involve the worship of any god or spirit. Neither caste nor inequality between the sexes was recognized by it, and in many parts of the Buddhist world today this absence of hierarchy, dogma, and ritual is reflected in the design of the house. Buddhism survived in its purest form in Ceylon, where the late medieval houses are distinguished by their lack of division into zones or parts. A high wall enclosed the house, on the inside of which lay a colonnaded gallery under a tiled roof, usually extending around all four sides of a sunken courtyard.

Although the open plan emphasized the lack of segregation in the house, anthropologists believe that custom would still have led to a division into functional zones so that, for example, specific areas were probably used by different sections of the family for sleeping, for business, and for storage. It is not possible to establish the practices of 2,000 years ago, however, when in the early phase of Buddhism a great flexibility of use may have been deliberately introduced into the house.

The Chinese house

Archeological evidence suggests that Chinese houses had their origin in the prototype of a long, rectangular, hall-like building constructed on a raised platform or a timber framework with an infilling of brick and plaster. At an early period this was divided into two parallel rectangular structures separated by a square courtyard; the front structure, which could be opened using doors or screens on both of its long sides, became the reception hall for guests while the rear hall became the private living and sleeping space; it also contained the ancestors' shrine and served as the prayer room.

Eventually, by the period of the Tang dynasty, in the 7th to 5th centuries AD, houses had characteristically two zones: a formal courtyard in front, entered through an imposing gateway or doorway in a high wall, and a private living court in the rear.

As only thin walls of oiled silk, paper, or light sliding timber screens separated the rooms from the courtyard, houses in the north

Traditional courtyard house in Peking, China.

Interior of a guest room in a Japanese house. The beam and floor have tracks for movable screens.

Garden view of an old house in Kyoto, Japan. The veranda looks out onto the garden with its stepping stones leading to a fish pond.

Reconstruction of a Roman insula at Ostia, with tenements above shops (1st-3rd century AD).

of China were often extremely cold in winter. In some houses a raised seating area (*kwang*), made of masonry, sometimes had provision for the insertion of charcoal braziers underneath, but in general heating was limited to the use of small movable braziers. The inhabitants had to rely on padding clothing if they wanted to keep warm in winter. (See also CHINESE ARCHITECTURE.)

It was customary for Chinese houses to have a garden at the back and on at least one of the sides. Following the principle that man-made architecture was essentially composed of straight lines forming squares and rectangles, while nature was essentially curvilinear, the garden was made as irregular as possible and entrance doorways to it frequently took the form of pure circles. However small, the garden would normally contain water in the form of pools or a running stream—where this was not possible it might be represented symbolically by sand or gravel. Frequently the garden would be conceived as an ideal landscape scene, complete with literary allusions, reduced in scale so that even mountains and cliffs might be represented in a small space, as well as winding paths and bridges.

In larger houses the two-court nucleus in the center of the plan was extended sideways and sometimes longitudinally by further courts which allowed a subdivision of functions and the accommodation of an extended family of relatives and retainers. The outside edge of the house on the garden side might be made more irregular to allow the penetration of the garden between some of the rooms of the house, and separate pavilions for relaxing and entertainment were sometimes built in the garden itself. The pavilions often took on an exotic and fanciful architectural style and were linked by covered galleries with varied window openings and screens. In such a house the normal, ceramic-tiled roofs might be elaborated with green or turquoise glazed tiles, with dragons or phoenix finials on the overhanging eaves.

The Japanese house

Japanese houses were distinguished from the Chinese in that all the daily activities of the household occurred under one roof. The house was divided into two zones: at one end, usually the northern, there was a cooking zone with an earthern floor; at the other end, with the floor elevated on a timber framework and constructed of timber planking, was the living space. The external walls of the house were made up of areas of wooden boarding and light sliding screens of wood and/or paper. These were held in the timber post-and-beam framework which supported the heavy thatched or tiled roof.

Only the most important rooms of the house had a mat floor, made up of *tatami* mats approximately 6 ft. 6 in. x 3 ft. (2 x 1 m) in size, arranged in a traditional pattern. The remainder of the floors were made of wide planks. The rooms were divided from each other by lightweight sliding screens, usually with a thin timber framework covered with rice paper; these screens could be replaced by impressive painted screens during a time of ceremony or for an important visitor. In the heat of summer almost the whole of the wall surface could be removed to provide one continuous open space and thus provide the maximum cross ventilation. During sudden storms or to furnish insulation in winter, sliding screens of wood could be quickly put in position around the outer edge of the house. The cold winter was combated by the use of small charcoal braziers and by the wearing of thick padded clothing, as in China.

The main house was usually flanked by gardens surrounded by high walls. The gardens frequently followed Chinese precedent in representing landscapes in miniature. Very large houses were made up of a number of single house nuclei placed together; in particular, the living, reception, and visitor zones might spread across a large area producing an irregular profile on the garden side of the building.

Apartments

An apartment is a room or set of rooms for domestic use that is contained within a larger building. Such arrangements are known to have existed in republican and imperial Rome; the tenements at Ostia (1st to 3rd century AD) being the best preserved from this period. These were used to house the labor force that worked in the port of Ostia, and the repetitive cellular character of their planning and construction would seem well adapted to the provision of shelter for a large, undifferentiated group of social equals.

It was not only in cities that apartments were required. During the Middle Ages and in the early Modern period apartments were an inevitable part of the larger and more powerful houses throughout Europe. These households might contain as many as 500 apartments, and even a modest nobleman might have supported 50 or so. Members of the family, servants, visitors, and their servants all required accommodation, and this was provided either as a series of more or less similar chambers with independent access—the upper storeys being reached by internal stairs between each pair of rooms—or by rows of connected rooms. Each of these latter rooms could be occupied by one or more persons, usually servants or people of little account. More illustrious individuals would be given suites of rooms to themselves, or for themselves, their servants, and companions.

By the 17th century the French had developed a characteristic sequence of three rooms for this purpose: the antechamber, the chamber, and the closet. The antechamber was for those awaiting an audience and for public appearance; the chamber was for sleeping in and for more familiar entertainments; and the closet, more private than the bed, was for confidential transactions and for personal solitude. After the mid-17th century this arrangement was frequently supplemented by back stairs leading from the closet or chamber. There were many other arrangements of similar type.

The word apartment was first applied to these personal suites within grand houses, but it was only in the 18th century that the building type that would now be recognized as an apartment house reappears. Urban life in the preceding centuries had certainly involved the frequent subdivision and subletting of larger houses originally occupied by one family and the construction of tenements of repeating rooms (usually illegal and therefore difficult for the historian to trace). Suprisingly little is known of these.

Apartments for middle-class occupation were being built in Paris in the 1730s and blocks of buildings four or five storeys high were familiar by the end of the century throughout Europe. They soon developed a peculiar social structure of their own—a microcosm of the city as a whole, with the well-to-do occupying the second floor and perhaps part of the first floor, the respectable occupying the third floor, the hard pressed the fourth, and so on, with the social dregs in the basement and garrets.

With the growing fascination in the seamier side of city life in the 1830s and 1840s came more vivid descriptions of the living conditions of the margins of society, particularly in the Parisian *ouvriers* and the London rookeries. Multiple occupation, not just of houses but of rooms, was the rule. What offended social investigators and philanthropists alike, apart from the filth and the smells, was the apparent absence of any principle in the architectural organization of public and private space within the slums. Two classic types of apartment house developed as a way to supply the poor with housing that would encourage a clean, decent, and domestic way of life: the common-stair-access house and the gallery-access house. Both these types had existed before and had also been proposed for utopian schemes. Common-stair-access houses had been built by Robert Owen (1771–1858) at New Lanark, Scotland, in the early 19th century to house his mill workers, and gallery-access apartments had been put forward by Charles Fourier (1772–1837) for his Phalantery settlements, several versions of which were built later in the 19th century. But in the 1850s and

1860s the common-stair-access houses became the typical format for working-class tenements, while at the same time numerous gallery-access houses were also constructed. The English architect Henry Roberts had provided the specific models for both types in his Model Dwellings for Families in Bloomsbury, 1847, (gallery access) and his Model Houses for the Great Exhibition, 1851, (common stair access).

In the 1870s large numbers of specially built apartments began to appear for the first time in Britain. Most were urban but there were also suburban developments. The characteristic of the most advanced of these schemes was the incorporation of a variety of other services within the estate or building—communal gardens, leisure facilities, food delivery services, dining rooms, laundry services—as well as private accommodation; a practice which may have been suggested by the luxurious conversion in 1804 of the Albany in Piccadilly into serviced bachelor apartments. In any event, the apartment was able to offer something more than condensed housing; something that could not be obtained in private houses.

The opportunity of supplementing domestic rooms with these special facilities was developed much further in the U.S. and in the USSR. Already in the 1890s skyscraper towers of apartments around a core of elevators with a wide range of social facilities beneath were being built in New York. The sumptuous Ansonia, designed by Emile Duboy, for instance, had 19 storeys, 2,710 rooms, palm gardens, a laundry, a dairy, a restaurant to seat 1,000, billiard rooms, and a private catering service. A contemporary suburban counterpart of this is the self-contained condominium for senior citizens, young marrieds, the childless, singles, etc.; communities of equals within which there is an emphasis on public life.

After the revolution in the USSR, the constructivists and their allies worked out a number of propositions for communistic living. In some the family was no longer to be the basic social unit, its functions having been taken over by the state. An individual's social life was entirely public, and all social events could therefore be subtracted from the home and thrown open to the community. Nurseries, dining rooms, common rooms, and kitchens took the place of equivalent areas in the conventional house: only sleep and rest took place in private. Thus the apartment might be reduced to 100 sq. ft. (9 sq. m) of floor space for bed and personal storage, as in the Barshch and Vladimirov communal house project of 1929. The experiments with social condensers were short lived, however. Under Stalin more conventional policies soon prevailed.

Meanwhile, the apartment had been

Model houses, with common stair access, at Cowley Gardens, London, England, to a design first proposed at the Great Exhibition of 1851.

Tudor City, New York (1925-28): an elevator-access apartment complex of 12 brick-clad towers with tudor styling.

An apartment house (1913) opposite the Metropolitan Museum, New York. The 12 floors are incorporated into a Florentine Palazzo facade.

L'Unité d'Habitation, Marseilles, France (1951), by Le Corbusier. This apartment building has central corridor access.

Park Hill, Sheffield, England (1955-61), by J.L. Womersley: Gallery-access apartments.

Conjectural restoration of northern angle of the Great Court of Sargon's Palace at Khorsabad (742-705 BC).

Plan of Sargon's Palace, Khorsabad (742-705 BC).

adopted by the Modern Movement in Europe as the necessary and correct form for contemporary urban living. The arrangements in modern schemes were often similar to those devised in the 19th century—gallery-access slabs, common-stair-access blocks, elevator-access towers—but there were other types too. L'Unité d'Habitation realized by Le Corbusier (1887–1966) at Marseilles in 1955 had central corridor access to two-storey apartments on what is known as a scissor plan. Sandwiched in the middle of the block was a layer of shops, and on the roof a running track and various social facilities. The architectural organization was novel (though based on unbuilt Russian projects of the 1920s), as was the attempt to isolate the building and its occupants from the surroundings.

In contrast to this, the superficially similar schemes put forward in the 1950s and 1960s for slab blocks with "streets in the air," derived from ideas taken from Peter and Alison Smithson (b. 1923 and 1928), tried to recreate the social conditions of the traditional working-class streets they had replaced. The best known scheme of this type is Park Hill in Sheffield, England (1955–60), by J. L. Womersley (b. 1910).

Palaces

Palaces are the sumptuous homes of rulers—whether Egyptian pharaohs, French monarchs, or German bishops. As such, their functions extend beyond those of a simple house into accommodating the administration of territory and the exercise of authority. They also symbolize the concentration of political power and, since (unlike castles) they are unfortified, the existence of relative peace and stability. In their fulfillment of these purposes, all palaces can be interpreted as combinations of various functional elements—public rooms, the private apartments of the prince and his family, accommodation for functionaries and servants, storerooms, kitchens, and stables. Innovations in palace design are represented in the organization of these elements and by the attempts to achieve an image of pomp and circumstance. While distinguishable from houses and castles, palace architecture gives and receives inspiration from the domestic and military architecture of its epoch.

The Ancient Middle East and Mediterranean

The formality commonly associated with palace layout was present from earliest times. Evidence suggests that Egyptian palaces of the 3rd millennium BC were walled rectangular enclosures with a ceremonial gateway leading to a pavilion with a throne room and private apartments. Unlike their temples and tombs, the Egyptians always built their palaces of brick and timber and the remains of them are therefore scanty. Ruins at Amarna and Thebes, however, give us some idea of their splendor. The latter complex, built by Amenhotep III in the 14th century BC, clearly distinguishes between ceremonial spaces, of which the most important is a columned audience hall with a throne dais on the entrance axis and, lying at right angles to this, the suites of rooms for the prince and his harem women. Each suite is a complete dwelling repeating a unit of vestibule, throne room, bedroom, and bathroom. At Amarna, Amenhotep IV (1372–1350 BC) built a palace on two sides of a road. The official palace with courtyards and hypostyle halls is connected by a bridge with a "window of appearance" to the much smaller family palace.

The small temple palaces, which were used only during religious festivals, demonstrate Egyptian symmetry at its most rigid. A typical example is that built for Ramses II at Medinet Habu which has a main hypostyle hall in front of the throne room with the royal apartments clearly separated from the public spaces.

The remains of Assyrian palaces show the same clear division into functional areas, although the axial system of planning is not so clear. Each group of rooms is organized around an axis, but there is no overall control, the groups are simply added together. At Khorsabad, the Palace of Sargon II (742–705 BC) was built on a raised platform covering 25 acres (10 hectares). It consists of courtyards surrounded by rooms. The largest court is enclosed by temples and offices and, beyond this, a second ambassadors' court leads to a throne room with the royal apartments surrounding a third, smaller court. Thus, before they reached the throne, foreign emissaries had a long and imposing procession past carved stone walls and painted plaster murals reminding them of Assyrian power.

Nebuchadnezzar's brick palace (c. 600 BC) is made up of five courtyard complexes placed side by side. Each court has access to a throne room on its south side and residential quarters on the north. In one corner of the palace a line of massive vaults may have been the supporting structure for the hanging gardens of Babylon.

The theme of raising the ruler's palace on an artificial platform was taken up by the Persians, notably at Persepolis (begun 518 BC). Here, on a high terrace, Darius and Xerxes built square hypostyle audience halls and throne rooms flanked by ranges of smaller rooms. The whole complex is an assembly of several rectangular buildings but, although each one is axially planned and all the axes are similarly orientated, there is no attempt to impose a symmetrical relationship between them. Besides the scale and monumentality of the blocks, Persepolis is remarkable for the sculptural quality of the stairs leading to the podium, which was not surpassed for 2,000 years.

While the cultures of the Middle East were developing the palace on monumental and more or less symmetrical lines, the Minoan culture of Crete was exploring a different theme. The Palace at Knossos, dating from the 15th century BC, is an apparently random agglomeration of various building types growing outward from a single large courtyard. There are state rooms grouped on the first floor reached by a ceremonial staircase, but the organization of the whole complex, with numerous light wells and twisting passages between blocks, can only be described as labyrinthine. There is no precise delineation of the palace limits—the outer wall simply follows the projection and recession of the storerooms and servants' quarters on the edge of the complex. This deliberate informality is not an indicator of a lack of sophistication; the palace dwellers enjoyed bathrooms and toilets drained by a system of terra-cotta pipes.

The Romans

The next European palace builders, the Romans, did not adopt the Minoan precedent but looked further away for inspiration. Their palaces were planned in the formal, axial tradition of the Middle East. To this was added an engineering skill capable of spanning large spaces with concrete vaults, and an architectural vocabulary based on the Greek orders. Domitian's palace on the Palatine (the word palace is derived from the name of this Roman hill) is arranged around a series of arcaded courts. One half of the palace, the Domus Flavia, includes a vaulted audience chamber over 150 ft. (46 m) high which opens onto a peristyle from which the banqueting hall was reached. The private apartments were located in the Domus Augustana, a separate contiguous building at a lower level.

In the country, the Emperors constructed more extravagant projects. Hadrian's Villa at Tivoli and Maximian's Villa at Piazza Armerina in Sicily are both a series of freely disposed geometric conceits where the functional groups—the dining rooms, baths, and libraries—are careful symmetrical compositions, but there is no dominating axial relationship between all the parts and the surrounding landscape. In contrast, when the Emperor Diocletian built a palace at Split around 300 BC he adopted the Roman *castrum* as a model with very different results. The rectangular block is clearly quartered by two colonnaded streets—namely the *cardus* and *decumanus*—with the royal apartments facing south with a columned gallery overlooking the sea. The Emperor's octagonal mausoleum and a Temple of Jupiter are on a subsidiary east–west axis and the circular vestibule to the royal apartments is reached through a peristyle on the main north–south axis which has the earliest known columned arcade. The two

northern quarters of the block accommodated soldiers and officials.

In the troubled times following the fall of the Roman Empire, the feudal rulers were primarily concerned with building fortified residences. During the Middle Ages the castle, rather than the palace, was the symbol of secular authority. Reflecting contemporary military architecture, the few palaces that were built are notable for the introduction of the "Great Hall" around which other rooms were grouped. Halls are found at Charlemagne's palace at Ingelheim, in the 11th century at Goslar; and they culminated in the 14th-century timber hammer-beam roof of Westminster Hall. Toward the end of the feudal era the growing power of the cities found expression in the civic palaces of Italy and Flanders. Buildings such as the Cloth Hall in Ypres, the Palazzo Vecchio in Florence, and even the Palazzo Ducale in Venice are meeting places for council or senate, halls of justice and public archives rather than residences.

During this period the palace tradition was continued by the Arabs who, in buildings such as the 13th-century Alhambra at Granada, developed the Roman form of rooms arranged around columned peristyles containing pools and fountains. In spite of formal, axial planning they used a painted timber and plaster vocabulary of stalactite vaults and elaborate capitals to produce an environment of lightness and subtle elegance. 13th-century China also used the axial planning of the ancient Middle East to build strictly symmetrical complexes of open columned halls raised on low platforms.

The Renaissance

With the decline of feudalism it was the new merchant prince who sought to express his growing wealth and power through the erection of an imposing residence. The architects who built city palaces for the wealthy families of 15th-century Florence combined the form of the medieval town hall—three storeys surrounding a courtyard and itself derived from the Roman *insula*—with a grammar of architectural detail also based on Roman precedent. The earliest of these palaces, the Palazzo Medici by Michelozzo di Bartolommeo (1396–1472) built in 1444, presents a forbiddingly rusticated appearance to the street. The main rooms are on the second floor above the offices and stores, and look inward onto a shaded and quiet courtyard. The building extends over a city block and is crowned by a massive overhanging cornice. All the openings are regular, and a change in the rustication, together with carved string courses, defines each floor. The concern with articulation is carried further in the Palazzo Rucellai (1446–51) where, for the first time, Leone Battista Alberti (1404–72) uses three superimposed orders of pilasters, Doric, Ionic, and Corinthian, to subdivide the street facade. This established a building type that

The Palatine, Rome. View of Nero's Golden House, begun AD 64.

The Strozzi Palace, Florence, is a semi-fortified town house built in 1489 by Benedetto da Maiano. The outer walls are heavily rusticated, rising to an ornate classical cornice.

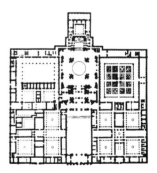

Plan of the Escorial near Madrid, Spain (1559–84), by Juan de Herrera.

Parts of facades of Roman Renaissance palaces. LEFT: Palazzo Farnese, Rome (begun 1534) by Antonio da Sangallo. RIGHT: Palazzo Massimo alle Colonne, Rome (1532–36), by Baldassare Peruzzi.

Overall view of the Escorial Palace. Built in yellow gray granite its vast plain and relatively unadorned facades measure 675 x 685 ft. (205 x 209 m).

Château of Chambord, Loire valley, France (1519–47).

was to accommodate many urban functions. For over 500 years, banks, offices, universities, and even department stores were based on the form of the Florentine palace.

In the 16th century, Roman architects developed the basic palazzo formula into complex and richly sculptured facades such as Raphael's Palazzo Vidoni-Caffarelli (1515), with its coupled Doric columns over a horizontally rusticated first floor. The grandest of Renaissance palaces, the Farnese, built between 1534 and 1540 by Giuliano da Sangallo (1445–1516) and Michelangelo Buonarroti (1475–1564) has a monumental gateway leading to the courtyard while the street facade is not articulated by pilasters but by the rhythm set up through highly modeled window aedicules with alternating circular and triangular pediments. Outside Rome, Andrea Palladio (1508–80) built palaces in the Vicenza which are notable for their elegance and clarity while, later, Guarino Guarini (1624–83) in the Palazzo Carignano in Turin (c. 1678) uses alternating convex and concave plans to give the street facade a Baroque interpretation.

Renaissance ideas traveled rapidly. The Palace of Charles V (started 1526), in the Alhambra, is unique for its huge circular court intended for bullfights, but otherwise it shows a full understanding of the Italian style. At the Escorial built for Charles' son, Philip II, Juan de Herrera (c. 1530–97) turned back to the Palace of Diocletian for inspiration. The palace sits on an artificial platform—another old idea—in the desolate hills north of Madrid. Constructed in an austere undecorated style, the building consists of an immense rectangle divided by 16 courtyards into a grid of college, monastery, and palace. It contains an early example of the "imperial" staircase which starts with one flight and divides at the first landing into two flights returning parallel to the first.

North of the Alps the Italian ideas were not so quickly established. The Chateau of Chambord (1519–47) built for Francis I is a medieval castle in plan, although the main block with a central spiral stair is noteworthy for its arrangement of rooms in suites instead of along a corridor. The detail is classical, although it is used in a quite unique way. The skyline, for example, is a fairy-tale composition of chimneys, cupolas, and pinnacles springing from a high-pitched roof.

Baroque palaces

It was in France, under Louis XIV, the greatest of the European absolute monarchs, that the palace reached its apogee. In Paris, where the old Louvre was rebuilt in courtyard form, the city palace achieved its most perfect form in Claude Perrault's (1613–88) east facade. Six hundred ft. (183 m) long, with slightly projecting center and corner pavilions, it consists of a two-storey colonnade raised on a plain, podium-like first floor. Unlike earlier work at the Louvre, there is no visible pitched roof, but a flat balustrade with a central pediment.

The palace that Louis XIV started to build in 1669 at Versailles was to become the prototype for most subsequent palaces in Europe—Blenheim in England, Schönbrunn in Austria, Caserta in Italy, and Aranjuez in Spain. The architects Louis Le Vau (1612–70) and Jules Hardouin Mansart (1646–1708) produced a structure over one-third of a mile (0.5 km) long that represents a fusion of architecture with painting, interior design, and landscape architecture. The palace, and therefore the monarch, stood at the center of a system of radiating paths which stretched out across the countryside, dominating nature in a way that was to be emulated in many city master plans such as Karlsruhe and L'Enfant's plan for Washington. On the urban side of the palace the wings enclose a three-sided *cour d'honneur* which contains stables, the chapel, and servants' quarters. The main rooms in the central block, the *corps de logis*, are on the second floor. There was no attempt to distinguish state rooms from the King's private apartments—the royal awakening was a most public affair.

The architects of lesser monarchs in emulating Versailles produced variations on the theme laid down by Louis' architects. At Stupinigi, in 1729, Filippo Juvarra (1678–1736) developed the *cour d'honneur* into an extended sequence of geometric spaces which culminates in a great domed salon. The architects of the minor German princes developed the staircase in a particularly original way. At Brüchsal the oval staircase dominates the palace and consists of flights of different curvature apparently floating on either side of a columned space. At Würzburg the same architect, Balthasar Neumann (1687–1753), and in the Belvedere at Vienna, Lucas von Hildebrandt (1668–1745), exploited the dramatic possibilities of the imperial staircase.

With the end of the 18th century and the overthrow of the absolute monarchies, palace building came to a halt. The palaces of the 19th century were exhibition halls—the Crystal Palace, the Galérie Des Machines. They were important for their use of new materials, glass and iron, but they did not fulfill any residential function.

This century, palaces seem to have returned to their origins in the Middle East. It is only in the oil-rich states of that part of the world that palace building is now taking place. The administrative functions of government are accommodated in office blocks, but there still exists a demand for the ceremonial residence. The palace for the Shah of Iran's sister, by Taliesin West Associates, is designed around a transparent, domed, air-conditioned garden which is surrounded by small domed apartments on a curving ramp. It is a 20th-century interpretation of a theme that is at least 3,000 years old.

Hotels

Ancient and medieval inns

The provision of accommodation for travelers was the main function of the ancient inns of the Roman world. The evidence is that these were planned like Roman villas, with two courtyards around which were arranged sleeping, eating, cooking, and stabling accommodation. The ancient Greeks, however, had no inns, instead the traveler lodged at private homes. Even today the same word is used in Greece for stranger and guest. The ancient Persians, however, built luxurious inns along their excellent highway systems.

In the ancient East, caravanserais developed along carrier routes for the shelter of caravans and travelers. These were watering places set about 8 mi. (12 km) apart, often fortified with thick encircling walls. Inside these walls arcades surrounded a central court, with stables and stores at ground level and sleeping accommodation above. The upper arcades were often separately domed, the corners sometimes raised as watchtowers which, together with the heavily emphasized gateway, lent the whole structure a fort-like appearance. Khans (similar structures but smaller in scale) developed in villages and market towns, and frequently came to be used by a particular trade. The silk khan at Bursa in Turkey, is a finely preserved example.

In Europe the enclosed arcaded form recurred in monasteries, which were effectively the only form of accommodation available for the traveler in the early Middle Ages. Free board and lodging for wayfarers was considered to be a Christian duty, although a voluntary contribution according to the traveler's means was generally expected. The monastery provided rooms with varying degrees of comfort, depending on the social status of the guest.

Commercial inns were introduced by the Romans into Britain. The *taberna,* where the legionaires and officials used to drink, was separate from the *caupona,* where sleeping accommodation was offered. Some of the Roman inns, as at Bath and Silchester in England, had a system of baths, consisting of a series of steam-heated rooms of increasing temperature, a massage room, and cold plunge. The Roman baths, or *thermae*, provided libraries and eating, sleeping, and rest facilities, and were therefore the precursors of the European spa movement of the 19th century. Similar forms occurred throughout the world; in the Islamic East baths called *hammans* were attached to mosques, and these later developed into spas in the 16th

century, while similar baths existed in Japan.

Following the Norman Conquest in Britain, towns developed and trade fairs became popular, attracting merchants and visitors from other regions. Pilgrimages also became fashionable. The monasteries could not cope with the increased demand for accommodation and they began to build separate lodging houses called "inns." The Crusades of the 11th century substantially increased the number of travelers in Europe and abroad, and the Knights Hospitalers created shelters and hospices for Crusaders and pilgrims to the Holy Land. In some cases, tokens were given to guildsmen and knights which, when matched, ensured reciprocal accommodation throughout Europe. In Coventry in England, the Guild of Merchants built lodgings for pilgrims in 1425.

Noblemen at this period were traditionally accommodated with their entire retinue in the host's palace. As this custom increased, special lodging houses were built at the gate of the estate to house the servants. These came to offer accommodation for travelers when unoccupied, as did the manor house when vacant, both in the country and in town. In London, the great houses of the nobility also served as hotels for visiting nobles; one such example was the Savoy Palace of the Earl of Richmond, on the Strand. However, as the wealth that upheld this hospitality declined, some of the private houses became commercial inns.

The dissolution of the monasteries by Henry VIII in the period 1536–40 gave a further impetus to the development of commercial inns. Another stimulus was provided by the increase in coach travel from the 16th century onward. By 1576, England had about 6,000 inns. Medieval inns in England, Spain, and Germany were usually planned on three or four sides of a courtyard with stabling and

Front view of the Palace of Versailles, Poissy, France (1669–85), by Louis Le Vau and Jules Hardouin Mansart.

The George Inn at Southwark, London, England; a surviving example of a medieval coaching inn. The present structure dates from about 1676.

Silk Khan at Bursa in Turkey, dating from the 15th century.

stores at ground floor and sleeping quarters accessed by timber balustraded galleries above. The George at Southwark, London, still in existence, is a fine example of a galleried coaching inn. In 1473 it had 13 guest rooms, each with three beds, a table, and benches.

Travelers at this time usually ate whatever provisions they carried with them or dined with the innkeeper in his kitchen. Only later were the rich served in their rooms, and this tradition lasted right up to the 19th century. When liquor was sold this was usually drunk in the kitchen, where it was stored. Slowly the "tap room" came to be separated off with a low partition or rail; the origins of the bar counter.

In the years from the fall of Rome to the time of the Reformation, European inns gradually developed from small, uncomfortable buildings to larger, more hospitable structures.

The European "grand luxe" hotel

In the 17th century it became fashionable for young British gentlemen to make the "Grand Tour" of Europe, and as a result luxurious, as opposed to purely expedient accommodation developed. By the 1760s Dessin's Hotel in Calais, France, was reputed to have a theater, workshops, ballrooms, as well as stabling accommodation. Similar was the Rotes Haus in Frankfurt, Germany (1767), and the Royal Hotel in Plymouth, England (1811).

The spa movement, which reached its zenith of popularity in the mid-19th century brought about a marked change in hotel development throughout Europe. Although European spas had their origins in the 16th century, when the Belgian town of Spa first became fashionable, it was only later in the 19th century that fashionable society went to "take the waters" at such resorts as Baden Baden in Germany, Bad Gastein in Austria, and Vichy in France. Palatial resort hotels developed on a "grand luxe" level of comfort with private suites of rooms, as at the Hotel Byron and the Hotel Interlaken in Switzerland. With the new type of clientele, entertainments were provided, and balls and concerts became an intrinsic part of hotel life.

Parallel to the development of these palatial resort hotels, where the rich could live in their accustomed style, where great banquets could be served, and international meetings held, there also emerged at this time the less lavish commercial hotel. Commercial hotels were often built near railroad terminals, and catered specifically for the needs of traveling salesmen. Special facilities such as writing desks, stockrooms, mail delivery and collection points were provided. In London between 1840–65 hotels were built by the railroad companies at the terminals of Euston, Pad-dington, Charing Cross, and St Pancras to accommodate the ever-increasing number of railroad travelers. These hotels were generally less luxurious and less expensive than the "grand luxe" type of hotel. Similar hotels soon appeared in every city in Europe, as well as in the U.S. and Canada.

U.S. developments

Hotel development in the U.S. followed very similar lines to that in Europe. Inns started in seaports in the 17th century, then, as people moved inland, inns were built along the rivers and the post roads. One of the earliest inns was the Blue Anchor in Philadelphia, where William Penn (1644–1718) was welcomed on his arrival from England in 1682.

As roads improved, so the inns multiplied and improved. In form they were similar to the European model with communal sleeping accommodation above a large ground-floor communal room which served as bar, meeting room, and eating place. In New York in the late 17th century, the Dutch Mayor built special accommodation to house the continual stream of settlers that he was expected to welcome; a hospitable tradition previously adopted by the European gentry.

Spa resort hotels in the U.S. developed much earlier than their European counterparts. One was built at Yellow Springs, Pennsylvania, after the mineral springs were discovered in 1722, and a hotel at York Sulphur Springs opened in 1790. By 1830, Saratoga Springs, New York, had become the most fashionable watering place in the U.S. and other spa resorts followed.

The Tremont Hotel in Boston (1829), designed by Isaiah Rogers (1800–69) was at the time the most modern and luxurious hotel in the world. It offered private, lockable bedrooms, introduced the idea of bellboys and desk clerks, and was the first hotel with indoor toilets. The hotel occupied a full city block and had an impressive Greek portico.

The Tremont was emulated immediately throughout the U.S. where cities vied with each other to construct the most prestigious hotel. The Astor Hotel opened in New York in 1836 with 309 rooms, and was two floors higher than the Tremont. The San Francisco Palace Hotel was built in 1875 with 775 rooms, at a cost of 5 million dollars, while the United States Hotel in Saratoga Springs boasted 1,000 rooms.

The city hotel became the focal point of social activities for a growing middle class. Winter gardens, conservatories, piazzas, and bridal suites were some of the more elaborate facilities provided. The development of the steel frame and the elevator allowed hotels to increase in size. It was essential that strict fire precautions were taken, and the Palace Hotel (1875) was one of the first to have its own fire

Marlborough/Blenheim Hotel, Atlantic City, New Jersey (1905), by Price and McLanahan.

Port Grimaud, France (1966), by Francois Spoerry.

The Hotel Fontainebleau, Miami Beach, Florida (1955), by Morris Lapidus. An extravagant resort hotel.

Regency Hyatt Hotel, San Francisco (1973), by John Portman. View of the internal lobby space.

hose system, with reels on each floor, a tank on the roof with 130,000 gallons (590,980 liters) of water. It also pioneered the idea of building its stairs in vertical brick compartments as smoke checks. Despite all these precautions the hotel burned down in 1906.

The development of services radically affected the planning of hotels. The Regent Hotel at St Leonards was one of the first to have toilets on each floor, while the Lindell Hotel in St Louis (1863) had running water in all rooms and bathrooms in 14 suites. Central heating was first installed in the Eastern Exchange in Boston in 1846. Passenger elevators were installed in the 5th Ave. Hotel, New York in 1859, and electricity came into use in the 1880s. The introduction of these services meant the planning of hotels with vertical stacks, and radiating corridors. The Statler Hotel in Buffalo (1908) was the first hotel to offer a bathroom in each room.

The Statler hotels were a type of hotel designed specifically to appeal to the commercial traveler. From the 1860s to the early 1920s almost all hotels were built near railroad stations. The Statler Hotel in Buffalo the first modern commercial hotel, the brainchild of Ellsworth M. Statler, who developed the most successful chain operation in the history of hotel development. The Statler hotels were sold to Hilton in 1954.

Europe was behind the U.S. in services development. In 1889 the Savoy led the field in Britain by installing 70 bathrooms fed from its own well, and its own generator to ensure a continuous supply of electricity. By 1930 the basic mechanical engineering for the modern hotel had been established: the plumbing and heating services, elevators, lighting, telephones, and mechanical ventilation.

The 20th century

Just as the advent of the coach, and then the railroads, played a major part in the development of hotels, so the introduction of the automobile and of mass air travel has brought about a dramatic increase in the number and variety of hotels. Airport hotels have been built at almost every major airport terminal in the world. At least 50 hotels and motels operate near Chicago's O'Hare airport. These hotels are particularly convenient for transit and other passengers, as they often provide courtesy cars to and from the airport.

The second effect of the air age on hotel development has been the increase in tourist resort hotels, particularly in response to demands of package tour operators. In some parts of Europe, and particularly in those countries bordering the Mediterranean, tourism may rank as the second- or third-largest source of national income. In many Mediterranean regions, architects are now becoming aware of the need to try and preserve local

architectural features; the unique character of each locality. Resort towns, such as Port Grimaud in France, as well as hotels in Morocco, Algeria, and Tunisia, have established low-rise developments using traditional forms. The El Jerba Hotel in Tunisia is an expansive covered walk, lined with small shops and enclosed spaces formed by the public rooms—all combining to create the scale and ambience of an Arab village. The Club Mediterrané, with its fashionable villages of reed huts and beach cabins, has similarly provided a solution to the problems of intrusive structures and overdevelopment.

In the U.S., resort hotels have been developed along coastal regions, and particularly along the Florida coast. Here the tourist resort hotel developed a flamboyant architectural vernacular of its own. Architects such as Morris Lapidus (b. 1902), designer of the Eden Roc and Fontainebleau in Miami Beach, elevated the resort hotel to a level of fantasy to include palatial lobbies, banquet halls, gymnasiums, casinos, nightclubs, swimming pools, and sumptuous restaurants. The Fontainebleau set the trend for the overdevelopment of Miami Beach's Collins Avenue. In 1976 a bankrupt Fontainebleau was sold at auction.

The growth of automobile use created the motel, which first appeared in the U.S. at St Louis Obispo in 1924. By the 1930s the motel had become an established form of accommodation in the U.S. although it was only introduced in Europe in the 1950s. In its initial stage the motel offered very little in the way of luxury, but for the traveler who wanted cheaper, informal, standardized accommodation the motel provided the answer. Today most of the motel chains in the U.S. offer a standard of comfort comparable to the more luxurious resort hotels. The modern form for many motels and hotels consists of a multistorey bedroom block and an expansive low-rise structure, which accommodates public rooms, administration, and services.

In any discussion of urban hotels, mention must be made of the American architect John Portman (b. 1924) who set a whole new pattern for hotel design with his Regency Hyatt Hotel in Atlanta, Georgia (1972). The building features an internal lobby which pierces the entire height of the building, crowned with a skylight and revolving roof restaurant. All bedrooms overlook this grand lobby which contains a cocktail bar, news stand, and tree-surrounded seating area. Glass elevators move up the lobby shaft, depositing visitors on the appropriate balcony floor for their bedrooms, or taking them up to the restaurant. Portman's idea was to bring the city into the hotel lobby and to create a social focus there. This pattern has now been successfully repeated in many hotels throughout the U.S. by both Portman and others.

Industrial

Factories

A factory implies a certain level of organization. The first real evidence of this after the collapse of the Roman Empire can be identified by the growth of the monastic institutions in Europe from the 12th century. Monastic organization was rigidly structured, teams of monks being allocated specific tasks. It was no accident that when the reaction came to what was little short of monastic rule in some parts of Europe, the great workshops were seized and put to a similar secular use as manufacturing centers. A later example was the Montgolfier family's use of the Abbey of Fontenay in Burgundy to develop their papermaking industry in the 17th century.

The monastic orders, not being so preoccupied with defense, had developed the basilican building type using locally available materials, stone, and timber, to provide generously lit and well-ventilated work areas, not unlike the abbey buildings themselves.

The influence of the watermill

The first mills, in Asia Minor, employed human and animal power as far back as 5000 BC, to split grain from the wild grasses. The vertical millwheel was probably a Persian invention for irrigation, a form that can still be seen today. The major technical breakthrough, however, was in the hydraulic engineering theories of Archimdes (287–212 BC). He not only developed the Archimedes screw for raising water, but also horizontal gearing for power transmission.

The overshot waterwheel was developed in Roman times, producing more power in proportion to the volume of water required than undershot types, but this innovation required sophisticated hydraulic engineering. The dams and weirs to raise the level of the water made river navigation virtually impossible in many areas until the lock was developed in France in the 17th century. The first recorded overshot wheel was in the *agora* at Athens dating from the 4th century AD, powering two sets of stones through reduction gearing. Later, a multiple mill was built by the Romans at Barbegal in France. It was built on a stream running down a steep hill; a double mill race drove eight pairs of overshot wheels, each driven from the tail race of the other.

Mills were not confined to grinding grain. A horizontal mill dating from AD 31 has been discovered in China, used to power bellows for a forge. The Romans used edge mills (where a pair of vertically disposed stones rotate over a circular trough) for grinding olive oil. Cloth mills were recorded in the 13th century and continued with only minor changes until steam-powered machinery replaced them. The iron industry tended to be located away from communities prior to the Industrial Revolution because of its reliance on waterpower and charcoal. Waterpowered tilt hammers, used for forging, were first recorded in Germany in 1010. Waterwheels were also in use until the end of the 18th century for powering furnace fans, wiredrawing machines, and machinery for boring out cannon barrels.

Almost 6,000 mills were recorded in England in William the Conqueror's Domesday Book of 1066, although some were animal powered. The mill of the Middle Ages was a simple, rectangular building, often built in the style of domestic accommodation; timber frames infilled with wattle and daub, with a horizontal shaft from the waterwheel piercing one wall to drive the mill stones through gearing. It was the millers who were the first engineers, pioneering developments in shaft drive, gearing, belt drive, and mechanical handling equipment; they had to manipulate heavy loads in confined spaces.

Mill structures developed as strong timber frames clad with weather boarding, but their proximity to running water meant that they were often built on marshes, on timber pile foundations. Mills seldom survived wet rot for more than two centuries. Brick mills did not appear in Britain until the 17th century, and then were not firmly adopted until well into the next.

Before the 18th century large buildings, where a substantial number of workers concentrated around some manufacturing activity, began to appear in many countries. These buildings, when the production process involved the use of machinery, took advantage of some concentrated power source, which in the early stages was invariably waterpower. The French royal workshops were established by Jean Baptiste Colbert (1619–83) in the 1660s; the Gobelins tapestry factories near Reims are a good example. The tobacco factory at Seville built between 1728–70 grew to become a vast production area measuring 615 x 480 ft. (187 x 146 m).

British mills

The predecessor of nearly all the British, and later of the continental and American mills, was a silk factory built near Derby by John Lumbe between 1718–22. It consisted of a five-storey building measuring 110 x 39 ft. (34 x 12 m), with 468 windows. The 18 ft. (5 m) waterwheel activated 26,000 machine wheels in this building when 300 people were employed.

Technical developments in spinning technology, such as James Hargreaves' (d. 1778) invention of the "spinning jenny," Richard Arkwright's (1732–92) water frame (1769–75), and Edmund Cartwright's (1743–1823) loom of 1785, contributed to the rapid growth of the textile industry. Many of the British 18th-century factories were built for this type of manufacture. Arkwright, Strutt, and Need's mill at Cramford was the first cotton mill and was built in 1771.

Until the introduction of iron, these buildings had internal structures of heavy wooden beams, joists, columns, and trusses. The wider span afforded by the timber trussed-roof construction provided an attic for the spinning machinery. The lower floors were used for other manufacturing activities requiring mechanical power from the horizontal shafting, and the remainder of the space was used for storage. Since they were lit by naked flames, and inflammable lubricants were used on the machinery, these buildings were a great fire hazard. The spectacular burning in 1791 of the Albion Flour Mill in London, designed by Samuel Wyatt, is just one of the many disasters that occurred.

Two interrelated influences were instrumental in design innovation: the need to reduce fire risk, and the adoption of steam power.

The first use of cast-iron columns was at Calver Mill in Derbyshire, England, in 1785; this building still used timber beams and thick masonry load-bearing walls. William Strutt (1756–1830), in partnership with Richard Arkwright, erected a six-storey mill at Derby in 1792–93, described as "fireproof," with iron columns and brick-arched floors sprung from plaster-encased timber beams. The first completely iron-frame mill was built by Charles Bage at Shrewsbury in 1796. It employed cast-iron columns and beams which supported flat brick arches which formed the floors.

Thomas Newcomen (1663–1729) invented a rudimentary, steam-assisted, atmospheric pumping engine in 1712, but it was not until Boulton and Watt perfected steam operation with rotary motion in 1783 that steam power became an industrial reality. The waterwheel was limited in its power production to about 20 horsepower. William Strutt's iron-framed mill at Belper of 1804, with its huge wheel and full exploitation of horizontal and vertical shafting, was the ultimate in this development. Steam power was revolutionary because the factory was no longer tied to a site determined by a natural power source, but the early steam-powered mills still tended to be built by rivers for reasons of coal supply. Strutt's Derby mill of 1792 was steam powered, with a centrally positioned engine house to reduce shaft length. Bage's iron-framed mill was designed for steam power with the engine house at one end. The restricted length of shafting through power loss was well illustrated when this mill needed to be expanded two years after the initial completion—another engine house had to be installed at the opposite end of the extension, with its own independent shafting.

It was not until Boulton and Watt in 1801

constructed the mill in Salford for Philips and Lee that iron and steam technology were truly integrated. The seven-storey mill, 140 x 42 ft. (43 x 13 m), employed 9 in. (225 mm) diameter, hollow, cast-iron columns for all floors, which were also used for heating. The I-section cast-iron beams carried the brick arches of the floors. Heavy floor loadings were possible throughout the building.

A large number of mills were built on this type of model, and they were much admired by foreign visitors, among them the Frenchman Dupin who wrote on the superiority of British industry after the Napoleonic War which ended in 1815, and K. F. Schinkel who visited Britain in 1826 for the Prussian government.

Working conditions in these early factories were often extremely bad. The work force—a large proportion of which consisted of children—worked long hours in badly ventilated and badly lit buildings, close to dangerous machinery. In the early years of the 19th century, visionary reforms were proposed by writers such as Robert Owen in Britain, author of *A New View of Society* (1813–16), and Charles Fournier in France, author of *Le Nouveau Monde Industriel*. Both these reformers proposed that the responsibilities of mill owners extended beyond just paying wages, and that they should provide housing, schools, and many other facilities for their workers as well as some involvement in running the enterprise. Owen experimented with his ideas after buying the New Lanark Mill at the age of 28. By 1800 he employed about 180 people who benefited from cooperative housing, education for children, encouragement of trade unions, and so on. Many reformers and philanthropists, such as André Godin and Titus Salt, were to follow with further practical experiments in opposition to upholders of the status quo such as Dr

Late 18th-century silk mill at Malmesbury, Wiltshire, England.

Stanley Mill, Gloucestershire, England (1813). Interior view of cast-iron columns and framing incorporating supports for shafting and machinery.

Titus Salt's mill at Saltaire, Yorkshire, England (1853), by William Fairbairn and architects. Lockwood and Mason.

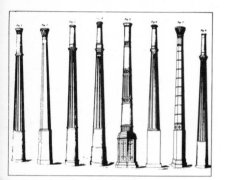

Ornamental chimney designs from a mid 19th-century bricklayer's treatise.

The Duke of Devonshire's iron and steel works at Barrow-in-Furness, built in the 1870s. Apart from a large number of blast furnaces the complex included docks for barges and ships and extensive railroad yards.

Andrew Ure who defended the factory system as it was in his *Philosophy of Manufacturers,* published in the 1830s.

Many innovations in structure and power transmission occurred before 1850. Cast-iron beams and columns were developed with more structurally efficient sections, investigation into fireproofing iron structures was initiated and the use of wrought-iron joists and beams, which began in the 1830s, became commonplace, often combined with lighter arches between them, as in William Fairbairn's floors for his sugar refinery (1844–45), which had arches of boiler plate topped with concrete.

During this period Fairbairn established that the main efficiency problem with the stationary steam engine was the Watt vertical boiler. Fairbairn introduced at this refinery the horizontal multitubed boiler, later adopted worldwide as the Lancashire boiler.

Most textile mills of this area were based on a rectangular structural grid of between 9 and 12 ft. (3 and 4 m), and were four to seven storeys high. Apart from the influence of power transmission, the need for natural lighting and ventilation kept the proportions long and narrow, seldom exceeding two lines of machinery across the width. It was those great innovators Matthew Boulton (1728–1809) and James Watt (1736–1819) who pioneered gas lighting in factories. Their chief engineer, William Murdock (1754–1839), installed gas lighting, perfected from Philippe Lebon's experiments in Belgium, at their ironworks in Birmingham. Their first major installation was in the mill for Philips and Lee in 1801, and by 1804 they offered a commercial installation service to mill owners.

Early textile factories were envisaged as massive machines. The structural frame of the building which carried all the floors also served as a support for all the shafting that transmitted power to the various looms and other machines from the central waterwheel or steam engine. In many cases, the cast-iron columns had brackets and bearing supports cast into them to hold this machinery in place.

One factory complex stands out in the early history of the factory as a coordinated power and service package—Saltaire. Built by Titus Salt, it was a model factory in the country, with a model village for the work force—much as Robert Owen had done at New Lanark 50 years before. The mill complex was designed by Fairbairn in 1853, with the architects Lockwood and Mason. There were two mill blocks disposed about a central entrance, each five storeys high and flanked with its own engine house. The attic was high enough to clear the entrance arch and engine houses, and ran 550 ft. (168 m), the whole length of the two buildings; the longest room in Europe at the time. The cavity walls were revolutionary in that they drew cool air in at floor level;

the hollow brick arches supported on wrought-iron joists drew hot, moist air out through ceiling vents to the outside. The spans were 27 ft. 6 in. (8.4 m) and 22 ft. 10 in. (6.9 m), unequal due to the demands of the machinery, an early instance where machine layout generated building design. Another innovation was Fairbairn's use of rolled-iron angle to build trussed girders spanning the whole 50 ft. 4 in. (15.3 m) width of the attic without intermediate support, and carrying iron purlins, in turn supporting extensive roof lights.

Multistorey construction was not used for the vast weaving shed. The demand for even daylighting and Salt's preoccupation with better working conditions resulted in the prototype for many factories to the present day. North-light trusses were used over a 36 ft. (11 m) span in one axis and 18 ft. (5.5 m) in the other. Integral cast-iron gutters formed the junction between cast-iron columns and the trusses. All the drive shafting was restricted to an undercroft, the machinery being powered by belt drive through slots in the floor. A large ornamented chimney belonging to the steam engine plant was made an integral part of the composition of this complex. Chimneys were used by the designers of many mills as an opportunity for architectural embellishment. The Tower Works, built at Leeds in about 1900, had a chimney in the style of Giotto's campanile in Florence.

Single-storey textile mills and factory sheds had existed previously. Marshall's Temple Mill at Leeds (1838–40), designed in the Egyptian style by Ignatius Bonomi, is an interesting example. It consisted of a hall of slender columns supporting cross vaults, each with a central skylight. The domes which were covered in waterproofing drained through the columns. They were covered with earth and grassed over, and sheep were reared to keep the grass under control. This elaborate roof served the purpose of helping to maintain the correct level of humidity for flax weaving.

Brick or masonry walls supporting trusses in cast or wrought iron, or timber, were often used to make enclosures for production spaces for heavy manufacturing processes such as those found in foundries and iron-rolling mills.

The roof of Maudsley and Field's machine workshop built in 1825 was made of cast iron. It collapsed soon after it was erected, probably because it lacked horizontal ties, but this was not widely reported. It continued to be illustrated in European construction manuals for many years as an example of an elegant structure in that material.

Light enclosures using an iron or wooden framework with a cladding of boarding or corrugated iron had, by the mid-century, become a common form for factory sheds.

A typical, well-designed multistorey mill of the 1880s in Britain had many features in common with those built early in the century, but there were some refinements. Floor structures of wrought-iron joists carried arches of concrete, but the joints were normally cased with fire-resistant material. Flat roofs built in a similar way to the floors, but covered in waterproof material, were not unusual. The external masonry walls would often have piers incorporated in them to carry the loads of the beams, and to make it possible to employ wider openings for the windows, which gave more light and made deeper plans possible. Stair towers would be designed to project from the main structure so as not to create awkward corners in the plan. Doors separated the stairs from each level. Toilets and other sanitary facilities were accommodated on half landings with good ventilation, and in many cases fire hydrants would be provided. Fire escapes became a common adjunct, and in many cases buildings provided for heating and ventilation. These buildings had regular grids of columns whose modules were arranged to conform as far as possible to the dimensions of the machinery, and to space occupied by processes, within the limits of economic spans. In some areas, standard modules were used for buildings serving many different purposes, leading to the standardization of components and ease of planning.

European factories

In Europe, British industrial processes and techniques had a strong impact on the design of factories. William Wilkinson (1738–1808), the younger brother of John Wilkinson (1728–1808), helped establish the great royal foundry complex in France at Le Creusot between 1779–85. These buildings consisted of double and single-storey volumes arranged formally around courtyards. European mechanized cotton manufacture began in Germany in 1784 using Arkwright's methods, introduced at Ratingen near Düsseldorf by J. G. Brugelmann.

Claude-Nicolas Ledoux (1736–1806), who became inspector of the Royal Saltworks in 1771, began to build a new Saltworks plant at Chaux in 1776. This complex, based on a semicircle of buildings, was never finished but its famous plan survives, along with an expansion of the project to the level of a large visionary assembly of interrelated buildings on or near the site, serving many industrial housing and social functions. Ledoux also produced schemes for other types of industrial buildings including large foundries. In many cases early European factory complexes, often funded by the State, were more impressively and formally planned than those in Britain.

Le Grand Hornu, near Mons in Belgium, was designed by Bruno Renard, a pupil of

Percier and Fontaine, and was begun in 1822. This factory complex consisted of buildings arranged in a formal rectangle, forming a courtyard, with two semicircular ends measuring 460 x 260 ft. (140 x 79 m). Workers' cottages and other facilities were planned nearby.

European engineers developed incombustible and fireproof framing techniques, similar in many respects to those pioneered in Britain. Roof framing for sheds and workshops followed similar evolution to those in Britain and the U.S. Camille Polonceau (1813–59) invented a simple trussed-rafter truss in 1837 which was very similar in form to those developed by A. Fink in the 1840s. These were first used in railroad-shed structures, but quickly found a wide application in factory buildings of moderate span. They could be made in timber with iron tie rods, but they were often made entirely of iron. Large areas of unencumbered floor space were often required in buildings associated with heavy industry, and many engineers rose to the occasion by providing innovative structures. The Galérie des Machines designed by Ferdinand Dutert and Victor Contamin between 1887–89 in Paris for the International Exhibition of 1889, although not strictly an industrial building, shows how far these techniques had developed by the end of the century. In this vast building, designed to exhibit large machines, three-pin portal frames spanned 377 ft. (115 m).

The Menier Chocolate Mill at Noisiel-sur-Marne, designed by Jules Saulnier (1817–81) between 1871–72, broke with the tradition of load-bearing external walls. Instead Saulnier designed this four-storey building—which straddles the river like a bridge—on four massive piers, as a complete freestanding cage. The structure was made up of riveted wrought-iron sections with cast-iron columns and an extensive system of diagonal bracing which is visible on the facades. The non-load-bearing external walls were enclosed in carefully executed decorative brickwork with a repetitive motif of cocoa flowers and the Menier monogram. The machines in the factory were driven by three turbines which had become a viable and more efficient alternative to the waterwheel.

Many of the developments in turbines had taken place in France from 1826 when Poncelet invented his inward-flow turbine. Other developments included the outward-flow turbine designed by Benoir Fourneyron (1802–67), and Jonval's axial-flow turbine, introduced in 1843.

Multistoreyed factories of skeletal construction became fairly common in France and other European countries. The metal framing, often consisting of wrought-iron lattice beams and diagonal elements, was in many cases exposed outside the enclosing

Drying shed at Güell Textile Mill, Santa Colonia, near Barcelona, Spain (late 19th-century).

Menier Chocolate Mill, Noisiel-sur-Marne, France (1871–72), by Jules Saulnier. A pioneer framed structure with non-load-bearing walls.

Partial view of the Royal Saltworks, Chaux, France (1776–79), by Claude-Nicolas Ledoux.

Templeton's Carpet Factory, Glasgow, Scotland (1889–92), by William Leiper. Glazed brickwork in Venetian Gothic.

Fagus Shoe-Last Factory at Alfeld-an-der-Leine, Germany (1911), by Walter Gropius and Adolf Meyer. The external walls consist of glass screens between the columns of the frame.

walls. A good example of a building of this type is the Usine de la Société Urbaine d'Air Comprimé in Paris, built by the engineer Joseph Leclaire in 1891.

The great majority of European factory buildings, as in other parts of the world, were utilitarian in design, but occasionally these buildings were elaborately ornamented as was the Benedictine liqueur factory at Fécamp (1893–1900), designed by C. Albert in a French Gothic and Renaissance style.

Toward the end of the century reinforced concrete began to be used for factory buildings in Europe. A pioneer in this field was François Hennébique (1842–1921) who built the large Charles VI spinning mill at Tourcoing in 1895. The tradition of reinforced-concrete factory building in Europe soon became firmly established. At first the forms used were similar to the repetitive post-and-beam systems typical of wood and metal construction, but forms based on mushroom columns and slab construction introduced before 1910 by C. A. P. Turner in the U.S. and Robert Maillart (1872–1940) in Switzerland were soon to be included in the vocabulary of reinforced-concrete construction along with other forms appropriate to this new material. Examples of these include the Esders sewing workshop in Paris (1919), designed by Auguste Perret (1874–1954); the Van Nelle Tobacco Factory (1828–29), designed by Johannes Andreas Brinkman (1902–49) and L. C. van der Vlugt (1894–1936); the Fiat Works in Turin (1927), designed by Matté Trucco; and the Boots factory at Beeston, near Nottingham, England (1932), designed by Owen Williams (b. 1890).

The AEG Turbine Factory at Berlin (1908–09), designed by Peter Behrens (1868–1940), and the Fagus Factory at Alfeld (1911), designed by Walter Gropius (1883–1969) and Adolf Meyer (1881–1929), were both forward-looking buildings combining symmetrical and severe repetitive compositions with large areas of glass. These buildings were to become very influential to European Modern Movement architects. Before World War I, these avant-garde architects began to admire the forms of American industrial buildings, often built by little known engineers, and these were published as examples of the direction architects should take.

Among factories designed by European architects, mention should be made of Hans Poelzig's (1869–1936) monumental brick chemical works at Luban in Germany (1911–12), Erich Mendelsohn's (1887–1953) hat factory at Luckenwalde (1921–23), with its angular shapes, and Alvar Aalto's (b. 1898) cellulose factory at Sunila, built in 1936–39.

Factories in the U.S.

In the U.S. the adoption of iron for factory buildings was slow. Spinning machinery was first introduced in 1787 at Beverley, Mass., and the first power loom was used by Francis Cabot Lowell in 1814 at Waltham, Mass. Initially, industrialization was limited to the northeastern states. In 1810 there were 54 mills in Massachussetts, 26 in Rhode Island, 14 in Connecticut, and 12 in New York State.

Until 1810 nearly all American mills were timber-framed and clapboarded structures, normally not exceeding three storeys. The first cotton mill in the U.S. was the Slater Mill at Pawtucket, Rhode Island, built in 1793. External walls in stone began to be used in about 1810—an early example of this form was the first Georgia mill at Smithfield, Rhode Island, built in 1812. This mill was totally rebuilt in 1853 but continued the tradition of building an internal timber frame with stone external walls. Many American mill owners built houses for their workers, especially when their mills were built at a distance from settlements in order to take advantage of waterpower.

Ingenious labor-saving techniques were a common concern of American mill designers. As early as 1783, Oliver Evans (1755–1819) built a totally mechanized grain mill at Redley Creek, Delaware, in which all the products were handled by mechanical power or gravity and passed from one process to another by systems of chain bucket conveyors, chutes, Archimedes screws, and other methods. This early precursor of automated factories was published by Evans in his *Young Millwright and Millers Guide* and was much admired by British and European mill builders including Thomas Telford (1757–1834), who advocated its use in his unpublished work on grain milling.

Rapid industrialization in the U.S. led to the emergence of vast industrial complexes, often exceeding European examples in size. The Bay State Mills at Lawrence, Mass., of 1846, consists of a large range of buildings including three nine-storey blocks in a row, each 200 ft. (61 m) long. In many cases these large enterprises did not restrict themselves to one type of production. The Amoskeag Mill at Manchester, New Hampshire, made textiles as well as heavy industrial machinery. The first iron columns to be used in an American textile mill date from 1846 at the Metacomet No 6 Mill at Fall River, Mass.

The Pemberton Mill at Lawrence, Mass., built in 1853, had hollow cast-iron columns supporting timber beams and joists. This building collapsed in 1860, killing 200 people through a failure in the columns which were defectively cast. Disasters of this type helped to establish stricter controls on building, and insurance companies played a major role in effecting these changes. After this period, reliable methods of iron framing became common in multistorey factories. In the U.S.

there were significant innovations in structures associated with industrial processes. James Bogardus' Shot Tower, built in New York in 1855, was an eight-storey tower. octagonal in plan, using a load-bearing, external, cast-iron frame infilled with thin, non-load-bearing, brick panels. Because of bad soil conditions, a light structure was necessary and a freestanding skeleton frame was proposed and built.

The cast-iron fronts and building frames marketed by Bogardus, Badger, and others from the 1850s onward had an impact on the design and construction of industrial buildings by introducing nonmasonry incombustible walls in multistorey structures together with other innovations. Large grain elevators posed problems to designers of industrial buildings and after the mid-century many patents were obtained for improved and more economical methods of construction. G. H. Johnson, for example, patented systems of reinforced brickwork in 1862 and 1869 for these structures. Toward the end of the century they were being built in reinforced concrete.

American factory designers helped to make important advances in the design of roof trusses for factory sheds from the 1840s onward. These structural elements, first made out of wood and later iron, were in most cases derived from variations on truss systems developed for railroad structures. By the end of the century there were over 60 different types of roof frame in use.

The use of reinforced concrete as a structural material in place of wood and iron was pioneered in the early 1880s by E. L. Ransome (1844–1917) in California, where iron was expensive. In 1885 he built a mill for Starr and Company at Wheatport, in California, with a complete reinforced-concrete structure. This was the first of many buildings built by Ransome and his company throughout the U.S. that exploited the opportunities offered by this new material. At the Borax Works at Alameda, California, he cast beams, slabs, and joists as homogeneous elements in 1889. In the New Jersey factory on the Pacific Coast, Borax Company at Bayonne (built 1897–98), Ransome brought concrete frame onto the facade, breaking away from the tradition of small windows which are required in other types of masonry construction. Albert Kahn's Packard No 10 building in Detroit, Michigan, built in 1905, is another example of this type. The use of concrete construction spread to many types of industrial buildings throughout the U.S. Before World War I, in the areas around the Great Lakes, large industrial complexes contained factories with unadorned, repetitive, concrete frames and there began to appear massive batteries of grain and cement silos on the skyline.

The electric motor

From the later years of the 19th century, the influences on factory design were: increased speed of design and construction; reduced costs; increased productivity through attention to materials flow and machine layout; and the ability to alter production technology.

While the Industrial Revolution in Britain stimulated rapid technological development, innovation was proceeding rapidly in Europe and the U.S. The elimination of the dependence on shaft and belt drive from a central power source virtually led to the modern factory as we know it. Michael Faraday (1791–1867) had discovered the principle of the electric motor in 1821, and the dynamo by 1831, but substantial power generation was not practical until the advent of Gramme's direct current generator in 1870. This was developed by R. E. B. Crompton (1845–1940) in Britain, and Werner von Siemens (1816–92) in Germany. Crompton, famous for his steam trains, had pioneered electric lighting in his cousin's cast-iron foundry at Stanton in 1877, to enable the operation of a night shift. This was a great improvement over the gas lamp, opening the way for deep-plan buildings. From the 1870s, Crompton developed electric power rapidly for industrial use, joining with Edison in 1883 to form the United Electric Light Company in the U.S. By the mid-1880s, electric motors started to replace steam engines, at first on shaft drive systems. The rapid development of a range of small motors opened up a new area of planning flexibility; plant layout and materials handling were established as major contributors to production cost reductions.

The interwar years in the U.S.

While Gropius and Meyer and the other Bauhaus pioneers searched in the late 1920s for their "complete and inseparable work of art," architects and engineers in the U.S. were concentrating on designing factories for the needs of rapidly expanding mass-production industries. Management sciences, such as operations and methods analysis were well known by 1920, typically in the Ford organization, and during the 1920s work study and factory layout became highly developed. The relationship between the factory structure and bay size to materials flow and production organization was clearly understood, with architects like Albert Kahn (1869–1942) pioneering integrated industrial design.

Where the typical factory of the turn of the century had been dominated by the unidirectional nature of the structural frame, due to overhead gantry cranes, the use of north-light trusses, or pitched Warren or Fink trussed construction, a new awareness of the benefits from improved mechanical handling techni-

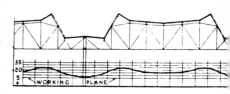

Section through the press shop at Chrysler Corp., Detroit, Michigan (1936) by Albert Kahn Associates, showing interior lighting levels produced by monitor roofs.

York Shipley Factory at Basildon, England (1962), showing wide-span monitor roof construction.

Production lines inside a modern artificially lit, deep-plan factory.

Typical deep-plan factory of the current American pattern with unit environmental control mounted on roof, central boiler plant, and room for expansion.

ques demanded new developments in steel frame design. The mass-production lines employed overhead conveyors, and numerous piped services; and the development of forklift trucks in the early 1930s required unrestricted floor surfaces. Freedom of production layout required deep buildings and wide spans. Lattice steel construction technology had progressed rapidly with bridge design, and it was to bridge-type trusses that Albert Kahn turned for his wide-span monitor roofs of the 1930s.

Kahn had been concerned about the uneven light from the north-light roof form, which produced pronounced shadows behind machinery. The monitor roof introduced a controlled amount of south light opposite the north light, resulting in a greatly improved overall lighting level without producing too much insolation or heat loss. This form was used for the Harrison Radiator Division of General Motors, and for the press shop for De Soto, part of Chrysler at Detroit, in 1936. Perhaps the greatest demonstration of the steel truss and monitor construction was Kahn's hangar for the Martin Aircraft Company in Baltimore in 1937. A column-free area of 500 x 330 ft. (150 x 100 m) was achieved by 30 ft. (9 m) deep trusses spanning the longest dimension, bridged alternately top and bottom to form a monitor roof.

The lessons learned in flat-truss construction were used by architects and construction companies for the U.S. factory building boom in the 1940s to support the war effort. The Austin Company of Cleveland built 300 ft. (91 m) span trusses for the Boeing bomber factory at Wichita, Kansas, supporting 10 ton (9,090 kg) capacity underslung cranes. The same company built a clear-span factory for the Singer Sewing Machine Company at Finderne, New Jersey, designed for rapid services rearrangement to suit alterations of the workbench layout; the services ran within the roof trusses. But the major departure which heralded much of the factory development in the U.S. in the 1950s and 1960s was the deep-plan space which incorporated no roof lighting and was fully air conditioned. The windowless, air-conditioned box had arrived.

Current developments

After World War II, two main requirements dominated factory design in Europe and the U.S.: firstly the provision of flexible space for optimizing production layout and materials flow, and secondly low-cost speed of erection. In some cases these goals proved mutually exclusive, with the low-cost steel or concrete portal frame being calculated by the plastic theory of design to contain so little material that although capital cost was reduced, so was the structure's flexibility to accept service loading. This form of structure also returned to the unidirectional emphasis of half a century before. Concrete ribbed slab and column construction was carried to a high level of structural efficiency by Pier Luigi Nervi (b. 1891) in the State tobacco factory at Bologna in Italy in 1949–50.

Considerable advances were made in the 1960s in construction systems employing both steel and concrete. The monitor roof continued to be favored as an alternative to the deep-plan, air-conditioned factories. Developments in three-dimensional steel structures, or space frames, advocated enthusiastically by Konrad Wachsmann (b. 1901) and others, allowed very large areas to be covered with only perimeter support. This, however, could only be achieved at a high cost, and although theoretically it offered a limitless choice for routing services, in practice the numerous structural members proved to be a constraint. Equally, the experiments with thin concrete shell forms, begun in 1920 by the engineering firm Dykerhoff and Widmann—who built many shell concrete factory roofs and other structures after World War I—were not sufficiently adaptable in many cases for the servicing needs of modern industry operating in conditions of increasing change.

The oil crisis of 1973 imposed another influence on factory designs, perhaps currently the most important one—that of energy conservation. There are currently numerous schemes attempting the recovery of process heat for reuse in the factory, but it is becoming clear that the scale of the manufacturing operation has to be large to justify the capital cost against current savings. The energy crisis has reopened the arguments for and against the deep-plan factory with an artificial environment, or the naturally lit plan types. There is a desire to return to smaller factories with more natural environments, due to experience in the 1960s of centralized plants being closed for long periods through strikes, and increasing evidence that both improved productivity and labor relations result from care being taken in the design of the work place. It is also being accepted that full automation is only economic in certain well-defined areas of production, such as welding car bodies, and that developments in mechanical handling techniques, such as wire-guided automatic work carriers, can be successfully integrated with personnel to increase productivity and improve job satisfaction.

Warehouses

"Warehouse" is a generic term for three distinct storage functions. A warehouse can be used like a tank, to even out peaks of production and continuous consumption, such as a harvest, or continuous production and irregular demand. Sufficient quantities can be metered out to control the selling price, and considerable economies result from continuity of production in industry. Warehouses can equally be transshipment and redistribution points; collecting diverse material from many production locations, sorting and sometimes packaging it before redistribution to customers for reprocessing or sale. This operation was typical of the great mercantilist companies in the Middle Ages. Warehouses can also be used as repositories, where storage area or volume is charged to companies short of their own space. A warehouse therefore has to provide volume with security, together with easy access and handling for the stored material. Demand for space dictates the size, but the current construction and handling technologies generate the form the building takes.

Storage in early civilizations

Commercial storage played an important part in the growth of intercommunity trade in southern Europe and Asia Minor, and there is evidence of substantial storage complexes from the Minoan and Mycenean cultures. The early Mycenean *megaron,* from about 2000 BC, has been found in round and rectangular form, in groups surrounded by palisades. In a late Minoan building at Ninos (c. 1500 BC), evidence has been found of groups of individual storehouses centralized under one roof. The building was 100 ft. (30 m) square, with a cellular construction of closely spaced storage chambers.

The expansion of the Greek Empire in the 4th century BC required the development of a storage and distribution network to supply the military machine. In 330 BC a very large warehouse was constructed at Piraeus, the main base for the Greek navy. The building measured 405 x 55 ft. (124 x 17 m), with a gabled roof supported by two rows of columns 35 ft. (10 m) high. We can only surmise the construction methods that were employed, but it is likely that heavy timber frames would have spanned from the masonry external walls across the columns—similar to the larger temples.

As the Roman Empire supplanted the Greek, the extent of their conquests and settlements posed a considerable logistic problem. A highly organized, centralized distribution system evolved based on the ports around Rome, with redistribution centers at the principal ports and cities of the Empire. The Romans had access to a wide range of construction techniques, including concrete and wide-spanning timber trusses. A very large granary, measuring 285 x 160 ft. (87 x 49 m), was built in 193 BC at Porticus Aemilia in Rome. This was a vaulted structure, disposed on three levels down a slope, and constructed entirely of concrete. Still larger warehouses were erected at Ostia, the principal supply port. The most important commercial buildings were three public warehouses (*horrea*) that were used to store goods after discharge from seagoing vessels, prior to reshipment up the River Tiber. The *horrea* were large, enclosed, rectangular buildings, with storage chambers opening off a colonnaded courtyard. Construction was typically concrete vaulting. The barrel vaults of a warehouse at Tivoli were over 30 ft. (9 m) high; nothing like them occurred again until the 19th century.

Medieval and early mercantile warehouses

In many parts of Europe, the feudal organization operated on the tithe principle, each artisan or serf providing one-tenth of his output, and large buildings were required to store these dues. The tithe barn is the forerunner of the pre-Industrial Revolution warehouse—with their massive timber construction these buildings survived many centuries. The tithe barn at Great Coxwell in Oxfordshire, England, still survives. Built on a stone foundation the store is 152 ft. (46 m) long, 44 ft. (13 m) wide and 48 ft. (15 m) high to the apex of the pitched roof. Heavily framed roof construction provided a clear span across the width and bore onto thick stone walls, the lateral thrust being absorbed by substantial buttresses.

As trade developed during the medieval period, alliances were established between several cities in northwest Europe, in an attempt to resist economic and military harassment. The greatest of these organizations was the Hanseatic League, which at the height of its power controlled 80 cities. The records indicate that substantial warehouses were built in the cities of the League; these were timber framed in the north German ports but sometimes of brick construction in Flanders and Brabant. Because of restricted quay area, these warehouses developed as multistorey buildings. At first manual hoists were used, but by the end of the 15th century cranes powered by horsedriven treadmills were in use. They comprised beams mounted as jibs on stout timber turntables to slew loads from vessels' holds into the upper storage levels of the adjoining warehouses.

While the mainland of Europe suffered from the devastation of the Thirty Years War and the trade leagues' disintegration, the British economy continued to develop due to the stimulus of trade with the new colonies across the Atlantic. The watermill provided prototypes for the warehouses of the late 18th century, with their massive timber construction and clear spanning storage lofts, supplied by hoists cantilevered out from the gable (the *locum*) and powered via a crown

The Dutch East India Company warehouses and timber wharf. The monumental symmetrical five-storey brick building in the background has regularly spaced loading doors flanked by windows. Goods delivered to the wharf would be hoisted for storage at the appropriate level. This type of arrangement remained common until the end of the 19th century.

Warehouses around the basin of the Albert Dock, Liverpool, England (1845), by Jesse Hartley.

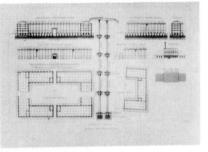

Plan, elevations, and sections of the Quadrangular Store House in Sheerness, England (1827), by Edward Holl.

Mid 19th-century warehouse in the City of London, England, showing loading doors at various levels and crane.

wheel and pinion from the main drive shaft of the waterwheel. Because these mills had to be built adjacent to running water they were often founded in marshy soil: the consequent problem of rotted structures led to the development first of massive masonry footings, and eventually to stone or brick walls.

Developments in the Industrial Revolution

There was little difference between the early factories and warehouses—both were multistoreyed, with timber-framed floors and roof construction bearing onto massive walls of local stone or brick. Their main disadvantage was that they were very inflammable. The answer lay in cast iron.

Charles Bage's mill at Shrewsbury, England (1796), was the first to employ cast iron for both columns and floor beams; Boulton and Watt perfected the technique in their Salford mill of 1801. Their cast-iron framework employed hollow cast-iron columns of 9 in. (225 mm) diameter and cast-iron I-section beams from which brick arches were sprung. Equally heavy storage was possible on all floors.

In Britain, the expansion of commerce demanded the creation of well-organized warehousing, especially in seaports. In London the construction of enclosed dock basins, with locks to remove the influence of tides, was begun in 1800. These basins usually had quays on all sides with warehouses opposite them for storing goods. In other places along the river, warehouses continued to be built directly on the waterfront. Examples include the West India Dock by Jesse Hartley, begun in 1800, and St Katherine's Dock by Philip Hardwick (1792–1870), begun in 1827. Many of these buildings were built with cast-iron columns and beams with brick arches, on the model of the textile mills. Where, as in St Katherine's Dock, they were built directly over the quay, the lower storey is treated as a colonnaded loggia to facilitate the unloading of goods. In other cases, such as at the London Dock, the space under the quay was vaulted with brick domes or cross vaults on cast-iron columns to provide further storage. The Albert Dock at Liverpool, built by Jesse Hartley in 1845, is a further illustration of an extensive dock basin surrounded by monumental and severe five-storey dock warehouses, whose only decorative features are massive Doric cast-iron columns on the ground-floor loggia over the quay.

Large warehouses were built in other countries, and these were often similar in most respects to those which evolved in Britain, although the use of iron framing was less extensive—as in the Packhof in Berlin, by Karl Friedrich Schinkel (1781–1841).

In Britain, John Rennie (1761–1821) advocated the use of iron framing for naval stores and warehouses in 1807 and when a large complex—the Quadrangular Storehouse at Sheerness—was built by Edward Holl in 1827, a special structural system was evolved to cope with the very bad bearing capacity of the local soil. Instead of heavy brick arches between cast-iron beams, Holl made his floors out of much lighter Yorkshire flagstones supported on joists between the beams.

In 1845, William Fairbairn (1789–1874) advanced the design of the iron frame by replacing the brick arches with thin, vaulted iron plates filled to floor level with concrete for his eight-storey sugar refinery and warehouse, while at a later date he introduced wrought-iron beams. He still, however, relied on massive masonry walls for lateral support. In 1837 the Lorillard warehouse was erected in Gold Street, New York, with cast-iron piers and lintels in the external walls.

In 1856, Colonel Greene designed a boat store, again at the Sheerness shipyard, in which the iron frame, consisting of cast-iron columns and wrought-iron beams, was also the supporting element of the facade which no longer relied on massive masonry walls to stabilize the building frame. The facades are made up of wide horizontal strips of glazing over corrugated-iron spandril panels between columns. This innovation in design made it possible to introduce more light into interiors, making deeper building practical.

When warehouses were built close to commercial and office districts they were often embellished with architectural features based mainly on Gothic and Classical forms, and it was not uncommon for the same building to combine administrative and storage functions.

The reputation of iron-framed warehouses was severely tarnished by a great fire in Cubitts Building Yard in London in the 1850s, when a large iron-framed warehouse filled with timber and joinery burned down and collapsed very rapidly, while a timber-framed building which also caught fire remained with its roof. From this time onward a better understanding of fire protection techniques began, and warehouses were compartmented internally to contain fires to manageable areas.

In the U.S., the work of James Bogardus (1800–87) and other iron-founders such as Daniel Badger advanced the use of iron structures. Many prefabricated buildings and facades were intended for warehousing. For his warehouse for Harper Brothers in New York (1854), Bogardus designed a cast-iron frame that was completely self-supporting. The design was an early example of true system building, with standard components produced for easy transportation and rapid assembly in the developing ports and cities in the U.S. Bogardus' pamphlet *Cast Iron Buildings* (1856) had a major influence on industrial building design, and by the beginning of the next decade, systems of the type he initiated were firmly established. Fine specimens of these warehouses were built on the waterfront at St Louis and in other cities. In 1865 grain elevators using cast-iron framing were erected in New York by Daniel Badger, a rival of Bogardus. In the same year, Hippolyte Fontaine built a large six-storey warehouse at the St Ouen Docks in Paris. This large building, measuring 630 x 82 ft. (192 x 25 m), had its external walls supported by cast-iron columns and wrought-iron beams. It was therefore an early example of a substantial iron-framed building with non-load-bearing external walls. Stability was ensured by rigid column-and-beam connections.

The form of the warehouse did not change significantly until the 20th century. There was a tendency toward deeper buildings after the introduction of electric light on a commercial scale in the 1880s. Cast-iron cranes, fitted to the slewing brackets on the face of the warehouses, still lifted material into loading doors disposed in a vertical line on every floor. The safety elevator, developed in the 1850s by Elisha Otis (1811–61), did not make an industrial impact until the 1880s, when electric motors also became available for powering cranes.

A major development in the construction of warehouses, but still adhering to their traditional multistorey form, was the introduction of reinforced concrete.

In the U.S., many large warehouses and storage building complexes were built in the major manufacturing centers to store goods and prepare them for distribution. The use of reinforced concrete on a large scale was introduced by E. L. Ransome (1844–1917) in his storage buildings for the Arctic Oil Works at San Francisco in 1884. Soon after the turn of the century, concrete was being widely used in a variety of structures associated with storage functions. Large grain elevators were built at major transshipment points, especially along the shores of the Great Lakes. The eight-storey warehouses designed by R. E. Schmidt, Garden, and Martin for the Montgomery and Ward Company in Chicago (1908) are a good illustration of an extensive warehouse complex. Here windows are larger than in most masonry precedents, allowing for deeper, better lit plans. These vast and sometimes stark storage buildings were often built around special docks for railroad cars, to facilitate loading and unloading. In some cases rail tracks occupied areas within the first-floor areas of these buildings.

The use of beamless floor slabs was initiated by C. A. P. Turner in the U.S. in 1908, and by Robert Maillart (1872–1940) in Switzerland in 1910. Beamless floors were more efficient structurally as load-bearing surfaces than their trabeated predecessors since they created spaces without using downstanding beams which restricted storage and consequently reduced the useful light of each storey. Maillart's warehouse in Zurich (1910) is an early example of this type of structure, which gained popularity in Europe and the U.S. between the wars.

Single-storey storage sheds

While developments were taking place in the evolution of framing techniques for multistorey warehouse buildings, single-storey sheds were also receiving attention from designers, who often adapted structural forms developed for bridges, factories, and other industrial buildings. Improvements in truss design and roof framing, first in timber and later in iron, resulted in considerably cheaper and more secure structures.

Corrugated iron, a major innovation in the development of building materials, was originally designed specifically for the construction of storage sheds. Henry Robinson Palmer (1795–1844), while working as engineer to the London Dock Company, was faced with the task of providing extra storage space on the quayside. In 1829 he developed and patented the use of corrugated sheets of wrought iron which were manufactured by Richard Walker. Palmer's sheds consisted of corrugated-iron vaults supported by cast-iron gutters carried on columns. This type of structure was considerably more economical than structures framed and enclosed in other ways. With the development of galvanizing in the late 1830s, this material quickly became popular for roofing and enclosing walls of single-storey warehouse buildings. This popularity has continued to the present day, and a large variety of different corrugated sheet materials are now available. From the 1840s onward numerous manufacturers in Britain, such as E. T. Bellhouse of Manchester, Morewood and Company, J. H. Porter, and Charles D. Young of Edinburgh and London, marketed prefabricated corrugated-iron sheds throughout the world. The California gold rush of 1849 provided one of the many markets for their buildings. An early American firm in this field was Marshall Lefferts and Brother who were established in New York in the early 1850s.

Developments in mechanical refrigeration which began in the mid-19th century led to a new kind of storage building—the refrigerated warehouse. They are normally sealed, compact, highly insulated volumes with a minimum of openings to reduce heat loss. By the 1880s the rapidly growing international trade in frozen meat, and later in frozen vegetables, helped expand the demand for buildings of this type.

Robert Maillart's concrete warehouse in Zurich, Switzerland (1910).

Early 20th-century Dutch dock warehouse. Machine rooms for hoists interrupt the skyline. These deliver goods to the various levels within the building, which is artificially lit.

Corrugated-iron warehouse shed advertised in Marshall, Lefferts, and Brother's catalog of 1854. This New York firm produced sheds with or without sides, in any length, and with spans up to 30 ft. (9 m).

Single-storey dockside warehousing.

A reach truck working in a warehouse with a pallet racking arrangement.

A maze of steel structural hollow sections forming the Eastern Electricity Board's new computerized "superstore" warehouse at Waltham Cross, Hertfordshire, England.

The development of unit load handling

Battery electric platform trucks were used in multistorey warehouses from about 1910, but although they moved between floors by elevator, they still had to be loaded and discharged manually. Some crude lifting machines were in use in the 1920s, but the real breakthrough did not occur until Sears introduced their Model L in the U.S. in 1930. This was a forklift truck as we know it today, with hydraulic lifting and mast tilting, battery electric drive, and rear steering. From that date, the days of the multistorey warehouse were numbered.

During World War II palletized handling and stock control techniques were developed in order to deal with the transportation and storage of vast tonnages of material, and after the war these techniques were firmly established in the U.S., Europe, and Britain. Old single-storey factories were pressed into service as warehouses, but it was soon recognized that much space and handling effort was being wasted by the intrusion of columns and the pitched roof structure. Developments in factory design, particularly in the U.S. in the 1940s with wide-span welded trusses at economic prices, led in the 1950s to specially built warehouses, specifically designed to accommodate the newly introduced pallet racking systems which greatly improved selectivity and the use of built volume. The structural grid was a balance between the capital cost of wide spans and the necessity to space the columns in multiples of the storage racking and forklift-handling aisles.

In the 1960s, the widespread replacement of small retail outlets with supermarkets operated by a few large companies increased the pressure to centralize warehousing and distribution facilities. But the increase in area demand coincided with rapidly increasing land prices. New handling methods were required to store vertically rather than horizontally. Computerized stock control had already been developed to reduce order/delivery cycle times: the concept of automating the lifting devices was the logical development, eliminating the problem of operators confined to inhospitable working conditions. Forklift technology had been exceeded with the demand for storage to heights of more than 25 ft. (8 m); it was crane manufacturers who were the innovators. The stacker crane, which revolutionized high-density storage, involves a rigid vertical mast structure with a pallet-carrying fork assembly. The mast runs on a single bottom rail, stabilized by another at the top, powered by its own traction motor. The Otis Company in the U.S. were pioneers in the field of computer-controlled stacker cranes, although development was so rapid in this field that there was clearly parallel design work by other companies.

The first automated warehouse to attract widespread attention was for the Brunswig Drug Company in Los Angeles. Built in 1960, it was rapidly made redundant by an overall change in company policy. The next major development—which was really the precursor of subsequent automated warehouses in the sophistication of its control systems, handling plant, and the fact that the whole warehouse was chilled—was for the kitchens of Sara Lee at Chicago, a cake manufacturer. More than 100 product lines were handled automatically.

In Europe, automated warehouse development was initially confined to countries with very high land and labor costs: Sweden and Switzerland. A 63 ft. (20 m) high warehouse for a paper company in Lausanne, Switzerland, was the first to integrate the structure of the pallet racking with support for the roof, the walls, and the stacker crane's guide rails. This innovation was quickly copied in the U.S., Sweden, Germany, and Britain. A warehouse for the Dr Maag Pharmaceutical Company in Germany successfully overcame the constructional and operating problems experienced in some of the early integral building/rack structures from the very tight tolerances required by the electronic control of the machinery: the 56 ft. (17 m) high rack structure was made of vertical, precast concrete elements. The 63 ft. (20 m) high automated warehouse for Pressed Steel Fisher at West Bromwich in Birmingham, England, was innovatory in that the whole volume was fully air conditioned, with humidity control, for storing unpainted car bodies. From the mid-1960s the continuing trend for higher land prices in Europe stimulated the development of automated warehouses to over 100 ft. (30 m) storage height, while in the U.S. control systems became more reliable with the warehouse height leveling off at about 63 ft. (20 m).

Developments in the U.S. and Europe in the late 1960s concentrated on increasing the lifting height, using forklift technology. As a result three distinct building types have evolved, each with different machinery: warehouses of about 25 ft. (8 m) clear height for forklifts and reach trucks; warehouses of about 40 ft. (12 m) for turret trucks (forklift-type machines that do not turn in the aisle between the racking to place the load, but handle it from side to side in an aisle just wider than the pallet); and automated warehouses for higher lifting. The first two types do not employ integral pallet racking, permitting flexibility of layout. The column centers are normally spaced to allow racking to be economically installed for either system. Wide spans can be cheaply provided through advances in structural steel and reinforced-concrete design, principally by the two- and three-pin portal frame. First demonstrated in the structure of the Halle des Machines at Paris (1889), this form has been perfected by the plastic theory of design to maximize strength for the minimum of material. Pioneer forms of this type of structure, used on a commercial scale as "standard" buildings, were "Butler" in the U.S. and "Conder" in Britain.

Current warehouse innovations aim to reduce energy costs; for example, the use of loading dock shelters, where the box-bodied truck forms a hermetic seal with the face of the warehouse. Other developments continue to refine mechanical handling techniques for ease of control and improved reliability and flexibility of operation. Perhaps the greatest innovation of the early 1970s was the minicomputer, offering automated control for small installations at a comparatively low cost and retailing systems completely integrated with centralized storage. Existing warehouses are sometimes modified by converting them into sealed secure volumes by closing off their windows and installing air conditioning.

Future trends must include the eventual elimination of the stockholding warehouse, except at a small local scale. In its place real time computer control for ordering, stock control, and manufacturing programing will enable the rapid distribution of required goods directly from smaller factories to nearby outlets. Most of the storage capacity would then be contained within the distribution network.

A turret truck operating in a warehouse with narrow aisle racking.

Commercial and administrative

Oriel Chambers, Liverpool, England (1864), by Peter Ellis, with cast-iron facades and large plate glass bay windows foreshadowing the form of Chicago buildings of the 1880s.

Leiter Building II in Chicago (1891) by William Le Baron Jenney. This building represented an advance in structural techniques with the iron framing extended to the outer walls.

Offices

The office as a distinctive building type is a late 19th-century phenomenon. Before that time the office function was an adjunct of government, trading, or manufacturing. It was accommodated in the merchant's house, the palace, or the place of production. It was unusual for a building to be designed especially for office use; Somerset House in London, England, built in 1789 for administrative offices of government, stands out as the exception. However, as manufacturing changed from a craft to an industrial base, there was a greater demand for administrative control and the keeping of records, and the office function grew rapidly. As organizations became more complex and the amount of transactions multiplied toward the end of the 19th century, the demand for specially designated office space increased. Since the 1950s offices have again developed dramatically and now provide a sophisticated environment for a wide range of functions.

Office work is generally concerned with the receiving, recording, arranging, and giving of information, and the safeguarding of assets by ensuring that the cash and stock representing the value of the business are fully accounted for. Offices exist wherever record keeping and the exchange of information is coordinated or performed.

Innovations in office design have occurred both in the form of buildings and in their interior environment and planning. Developments in communications and information production and handling equipment have had a dramatic effect on the scale and organization of the office function. This in turn has created demands on the type of building and environment required. In the late 19th century a large office may have employed no more than 50 staff in "counting rooms," as an adjunct to the manufacturing plant, whereas today up to 5,000 staff may be employed in a single corporate headquarters.

Innovations in the function and organization of offices

During the early part of the Industrial Revolution the small numbers of artisans employed by manufacturing establishments began to be replaced by much larger work forces. As production increased, so also the supporting functions of financing, insuring, and exchanging goods, and the administrative function of directing, controlling, and accounting for the factory output developed. Although the accounting and managerial functions remained attached to the plant, the trading and financing functions were separated from the factory and occupied the most central locations of cities. The first major clustering of specialized office accommodation in London, England, occurred with the location of insurance firms around the Strand and in Fleet Street in the 1830s. By 1850 distinctive office building types began to appear, such as Oriel Chambers built in 1864 in Liverpool, and exchanges such as the London Coal Exchange of 1849.

The speed with which business could be transacted was increased by a series of inventions. The first patent for a typewriter was granted in England in 1714, but it was not until 1868 that a practical commercial model was developed by Remington. In 1837 Cooke and Wheatstone invented the needle telegraph, and by 1868 almost all major American cities had telegraphic links. The duplicator, developed by Gestetner in 1861, further increased the flow of paper. The most important invention for the speed and ease of communication was the telephone. Alexander Graham Bell (1847–1922) patented the telephone in 1876 and by 1900 it was an indispensable tool.

By the end of the 19th century more or less instantaneous contact was possible between the manufacturing plants or warehouses and the financial or commodity exchanges. The office function was no longer restricted to being close to the function it was servicing. A further concentration of office building took place: in London between 1867 and 1886 new "office streets" such as Victoria Street in 1871 and Shaftesbury Avenue were constructed. In New York, offices began to be located in downtown Manhattan during the 1880s. Chicago's Central Business District, within the loop of elevated railroads, also emerged in the 1880s. The development of the office as a distinctive building form continued: large firms began to cluster all their office functions together in the cities, in prototypes of the corporate headquarters.

The 19th-century office was a small room where no more than three or four clerks might work, or where the individual went for solitude from the noise of the production area or the exchange. The Leeds Corn Exchange in England (1863), for example, had a central oval dealing floor overlooked by small rooms accessed by galleries. As industry and business developed, these types of offices were superseded by the "general office" or "pool." These housed up to 100 clerks in large, often toplit rooms surrounded by small individual rooms for managers. Two examples of this arrangement are Frank Lloyd Wright's Larkin Building in Buffalo, New York, built in 1904, and the Lever Brothers Port Sunlight Building in Cheshire, England, built in 1914.

Between 1920 and 1950 the use of machines for collating, storing, and reproducing information increased dramatically. New office occupations arose as the traditional role of the clerk was taken over by machines which could be managed by workers with a lower level of skills. At the same time the number of women in office work began to increase.

Offices became more organized and took on similarities to the factories they served. Departments were formed with defined functions which were part of a sequence of information flow.

IBM's first automatic digital computer was completed in 1944 and by the early 1960s computers began to be applied to office activities. The number of computers in use in offices is at present rapidly increasing with the development of minicomputers and the latest microprocessors. The impact of the computer has been to encourage the further systematization of office work flow, to restore the need for skilled office workers, and to enlarge the management and planning role of office organizations. Automatic data processing has freed time for thinking and planning. The office function today is less concerned with processing paper than with face-to-face communication and the exchange of ideas.

Technical innovations and building form

As office work expanded, greater pressures were placed on the land available in central business districts, and development began to go upward. Some of the most successful early tall buildings are to be found in Chicago. The first central city buildings were limited to 10 storeys, due to the need to strengthen the normal 12 in. (300 mm) thick masonry wall for every additional floor above two storeys.

It was not until the 1880s, with the development of the cast-iron frame and the passenger elevator, that the skyscraper became an established building form. In 1884 the Home Insurance Building of Chicago, designed by William Le Baron Jenney (1832–1907), was completed, using a steel and iron skeleton frame. This was followed by a number of glass and skeleton-framed buildings which created the original Chicago style. The steel frame, clad with masonry fire protection, allowed for a uniform thickness of external walls. This in turn provided more usable office space and increased the size of the offices. The technology was available for the city to build upward, providing goods and people could be transported rapidly and efficiently to the upper floors. The first commercial elevator was installed by Otis in New York in 1857. In the initial stages elevators were used mainly for hotels, and it was not until 1873 that the first elevator was installed in an office for the Western Union in New York. By 1887 electrically operated elevators were introduced and limitations on the height of buildings were finally lifted. The "skyscraper" was born. The progression upward was rapid: in New York the Metropolitan Building of 11 storeys in 1892 was followed by the Singer Building of 47 storeys in 1908, the Woolworth Building of 60 storeys in 1913, the

102 storeys of the Empire State in 1931, and in 1971 by the towering World Trade Center of 110 storeys.

The space provided was anonymous serviced space built speculatively to be rented off to the highest bidder. The Schiller Building in Chicago of 1891, designed by Dankmar Adler (1844–1900) and Louis Sullivan (1856–1924), was intended to provide either hotel or office accommodation, either use requiring a lobby or reception area, corridors lined with rooms, stairs, elevators, and bathroom facilities. Today an office building can be equivalent to a small city in terms of space provision. The World Trade Center has over 20 acres (8 hectares) of office space, and can accommodate 50,000 workers. In New York during the early 1960s a new profession was formed to plan, manage, and adapt the stock of speculative space. These firms of space planners, from their experience of designing the interface of office organization and their buildings, have begun to influence the design and quality of office shells.

In Europe tall office buildings did not appear until after 1945 due to height restrictions. Some early examples are the Pirelli Building in Milan (1961) and the Shell Center in London (1962).

Although buildings were able to develop upward, they were constrained in the depth of space by the need for natural light and ventilation. In 1870 Swan and Edison invented the incandescent lamp and by 1907 tungsten was introduced for commercial purposes. It was not, however, until 1938 that fluorescent tubes were introduced commercially by Westinghouse and GEC. Fluorescent lighting reduced energy consumption, glare, and above all, heat build up. This allowed the continuous use of artificial lighting, an increase in the depth of buildings,

Distant view of BMW Headquarters, Munich, Germany (1973), by Professor Karl Schwanzer.

PSFS Building, Philadelphia (1932) by Howe and Lescaze. One of the first modern air-conditioned buildings independent of period styling.

Union Carbide Headquarters, New York (1959) by Skidmore, Owings, and Merrill.

and reduced the problems of air conditioning.

The principles of conditioning the environment were well known in the 19th century and used extensively in industrial buildings and ships. Apart from isolated examples, such as Frank Lloyd Wright's Larkin Building of 1906, which was specifically designed for a noisy and dirty location, the air conditioning of office buildings was not considered to be an economic proposition until the 1920s. The 1928 Milam Building in San Antonio, Texas, is the earliest air-conditioned building given over entirely to offices. The Howe and Lescaze PSFS Building in Philadelphia was completed in 1932 and was probably the first truly modern, air-conditioned office slab, and is a pointer to the New York developments of the 1950s. The typical office floor plan is interesting in that although offices are organized around a central corridor, there is an inner bank of offices relying on artificial lighting and an outer bank of smaller perimeter offices. The day-to-day use of air conditioning for commercial office buildings began to take effect after World War II. By then, the level of technology had made it practical to localize input and extract so that the building interior could be subdivided into cellular spaces as required. The buildings of Skidmore, Owings, and Merrill provided space that was uninterrupted by columns, service cores, and fixed partitions, and no longer dependent on the perimeter for its services. The Pepsi Cola Building, New York (1959), the Inland Steel Building, Chicago (1957), and the CIS Building, Manchester, England (1962), are examples of the flexible uninterrupted space that architects were striving for.

Innovations in the design of office interiors

From the beginning of the 19th century until the end of World War II office buildings and organizations increased in size and in the amount of paper needing to be processed, but there was no fundamental change in the nature of the interior environment. The typical prewar office interior consisted of heavy desks, butted together in clerical areas, and glass-partitioned cubicles. The introduction of computers in the 1950s reduced much of the repetitive clerical work; work flow became less critical and working groups became smaller and more project orientated, adapting as the workload and the function of the organization changed. The demand for organizational flexibility resulted in buildings being designed so that ceilings, service points, partitions, and sometimes even furniture, were dimensionally coordinated, of the same style, and easily adaptable. Skidmore, Owings, and Merrill's Union Carbide Headquarters (1959) reflected a systems approach with total modularity and interchangeability throughout.

In 1963 Robert Probst, in association with George Nelson, brought out the prototype of Action Office Furniture. Probst, perceiving both the limitations of continuously moving partitions and the changing nature of office work, conceived of a system of screen-hung work surfaces which could be adapted to a variety of styles of work and management. Action Office provided the privacy of the individual room, with ease of adaptation, and the possibility of interaction. In 1968, Action Office 2 was launched on the commercial market, interior layouts became more relaxed and the number of enclosed offices was reduced. Furniture began to be marketed as a coordinated system rather than as individual desks, filing cabinets, and chairs. The production of furniture was big business and large industrial organizations entered the market, such as Westinghouse and Steelcase in the U.S., and Strafor and Olivetti in Europe. Furniture has taken on many of the traditional roles of the building shell by carrying cables (Voko), supporting lighting (Westinghouse), providing privacy (Marcatre), and allowing the character of finishes to be varied.

Up to the early 1950s the impact of air conditioning and the ability to build deep spaces which were not dependent on the perimeter for light and ventilation had only been used in a few buildings. The speculative office building of the U.S. in the early 1950s began to break away from the constraints of the 25 ft. (8 m) depth from window to core or corridor, and started to provide deeper spaces and a central core. In Britain, speculative office forms continued to be dictated by the demands of natural light and individual offices, and the shallow-depth office block persisted. In Germany, however, where the tradition for custom-designed office blocks was stronger, a new generation of building shells appeared. The Quickborner Team in the late 1950s developed a new approach to office planning through analyzing individual needs and work relationships. Office landscaping, or Bürolandschaft as it was termed, proposed a free-form layout in deep open spaces which, it was claimed, could improve organizational efficiency by facilitating work flow and ease of interaction. The resultant buildings were two or three storeys high with floor areas at least 70 ft. (22 m) in depth, and normally with an off-center core to allow for a zone of cellular spaces. The Nino Building in Nordhorm (1963) is an early German example, and the experimental offices of 1969, for the Department of the Environment at Kew, pioneered the concept in England.

In the U.S. the European tradition for cellular offices has never been as pronounced. The regular rows of desks of the "general office" naturally developed into open planning, as reflected in the Union Carbide Building of 1959 by Skidmore, Owings, and Merrill.

The emphasis of postwar office development was on the image-conscious corporate headquarters. Telecommunications and transportation had become so sophisticated that large corporations felt able to leave the security of physical proximity to their suppliers, clients, and competitors, and often moved to "green field sites." With the removal of constraints on site and land values, buildings could spread outward instead of upward. Buildings began to reflect their natural settings, and office work took on a new quality, as for example in the Connecticut General Building, completed in 1957.

Present developments and future trends

The euphoria of the 1960s for fully air-conditioned, deep, open-planned space, uninterrupted by walls and individual offices has been tempered by experience. The realities of planning and managing single floors with up to 500 staff in one space, and the realization that some activities or individuals might require offices for reasons of privacy or status, have stimulated designers to look at alternative plan shapes. Buildings are reverting to medium-depth space of 48–52 ft. (15–16 m), where either cellular or open-planned layouts can be accommodated. This is described in Germany as "reversible space." As with the domestic scale of 19th-century offices, the building structure again begins to manage and regulate activities.

In 1973 an outstanding innovation in office design was achieved in Herman Hertzberger's Centraal Beheer Cooperative Insurance Headquarters in Apeldoorn, Holland. Hertzberger has taken the undifferentiated character of Bürolandschaft as his starting point and then proceeded to bring the building shell back into use. Hertzberger and his client argued that the building should provide easily definable space for small working groups and reflect the better level of education and environmental awareness of office workers. The building is a honeycomb of constant-sized spaces, defined by structure and punctuated by voids, within which the individual can manipulate furniture, decoration, and lighting. Centraal Beheer may well reflect the future of office building design.

Office organizations are also in flux. The repetitive data-processing functions can be handled by computer, leaving the office worker to plan and manage. Greater emphasis is placed on meetings, working in small project groups, and interaction with outsiders. Organizations are becoming less hierarchical, more open and participative. The design implications of such changes are to provide buildings that are open yet also afford small group spaces. Trends in office design are moving away from undifferentiated space to multifunctional buildings which provide a variety of different-sized and interconnected spaces. The character of spaces is differentiated in plan by structure and walls, and in section by changes of level and voids. Lighting is becoming more variable, relying on task lighting associated with the work station to provide for specific requirements.

(See also SERVICES.)

Skyscrapers

No other building type delights in such an evocative name, nor one so free of functional connotation than the skyscraper. Whatever else a skyscraper might stand for it stands, above all, for technical confidence, for mastery of technical means, and it has done so since the patrician families of medieval Italy raised their personal fortified towers over San Gimignano and Bologna in the 13th century.

The origins of exceptionally tall buildings are, of course, older still. The tower of Babel is often cited as a key antecedent, and it must be admitted that the actual monuments that inform the myth of Babel are remarkable for their technology (mainly of brick construction) and for their effect in punctuating relentlessly hot and barren landscape. In a similar way, the Gothic cathedrals of 13th-century Europe loomed over huddled towns. The aim in both instances was to reach heaven, whether literally or metaphorically.

Subsequently, the skyscraper has become a conventional vehicle for more frankly secular aspirations which have to do with erecting permanent marks of human ingenuity and daring. Rational justifications of the basic skyscraper concept, whether found in the logic of real estate, or of management theory, have often obscured but never extinguished the fundamental urge to create man-made landmarks.

Origins of the modern skyscraper

Three vital technical developments have paved the way for modern skyscrapers. The first was the rapid development of structural engineering from the 1830s onward. Laws of statics, discovered empirically, were resolved into formulas which described the behavior of a whole class of structures (i.e. columns, trusses). (See STRUCTURAL THEORY.) In this way, for the first time, generalized laws and principles could be applied to different classes of forms and materials, and could explain behavior independent of experience with a specific form. The ability to reliably predict the behavior of forms greatly reduced the dependence on past experience that had governed building over the previous 5,000 years. Any form to which the laws of statics may be applied can be realized. Thus, detailed calculation was sufficient to persuade the French

Interior view of open-plan Burolandschaft office in the Osram Headquarters, Munich, Germany (c. 1960).

John Deere Offices in Moline, Illinois (1964), by Eero Saarinen. Built in a landscaped park in the open countryside.

Centraal Beheer Cooperative Headquarters in Apeldoorn, Holland (1973), by Herman Hertzberger.

San Gimignano, Siena, Italy. Thirteen of the medieval towers still dominate the town.

Reliance Building, Chicago
(1894-95), by Daniel Hudson
Burnham and John Wellborn Root:
an early Chicago skyscraper with a
skeleton frame.

Daily News Building, New York
(1930), by Raymond Hood. An early
multi-functional skyscraper.

R.C.A. Building in the Rockefeller
Center, New York (1931–32) by
Corbett, Harrison and McMurray.
Part of a large complex of 14
buildings.

government to allow Gustave Eiffel to build his 1,010 ft. (300 m) high tower, which, in 1889, was the tallest structure in the world.

The second development was the making of a clear differentiation between the supporting skeleton of a building and its enclosing skin. Although this distinction was made to a limited extent in Gothic architecture, and in the domestic architecture of several cultures, it was not part of the Renaissance load-bearing masonry tradition which informed most of what was considered architecture in the mid-19th century. Hence the popularity of the Gothic style for many early skyscrapers, notably the 1913 Woolworth Building with a height of 792 ft. (241 m) and the 1923 Chicago Tribune Building with a height of 450 ft. (137 m).

The last tall building in load-bearing masonry construction was the Monadnock Building (1891) in Chicago. Sixteen storeys high with brick walls 7 ft. (2 m) thick at the base, it stands close by the first skeleton-frame skyscraper, the 10-storey Home Insurance Company (1885) designed by William Le Baron Jenney (1832–1907). The crucial virtues of the framed building were an improved strength to weight ratio, with a consequent relative economy of foundations, "fireproof" construction to the extent that floors were no longer supported on timber, and speed of erection.

The third development was the invention of a safety device by Elisha Graves Otis (1811–61) to prevent elevators from falling. He demonstrated his safety elevator at the 1853 Crystal Palace Exhibition in New York, and by 1872 more than 2,000 of them were in service. The architecture of hotels, apartments, office buildings, and department stores evolved rapidly in response to the potential of the steam-powered elevator. Electric drive, push-button control, and speeds of 700 ft./min. (212 m/min.) were commonplace by the turn of the century.

Chicago

The fire which destroyed a large part of Chicago in 1871 effectively produced the first modern city in the world. After the fire, new building within the area of the Loop had to be of fireproof construction and, as fireproof construction was too expensive for most residential building, this had the effect of producing a downtown area which was used almost entirely for commercial and administrative purposes. This specialization in turn pushed up land values and created considerable pressure to build the largest possible building on any particular site. The result was the cluster of skyscrapers built within the area of the Loop in the decade 1885–95.

The Chicago architects, William Le Baron Jenney (1832–1907)—who was an engineer as well as an architect—Burnham and Root, Holabird and Roche, and others, wanted to break down distinctions between architects and engineers, and to evolve an architecture depending for its effect on mass and proportion rather than on ornament. The Reliance Building, a slender 15-storey tower of 1895, designed by D. H. Burnham and J. W. Root, epitomizes the Chicago skyscraper. Finished in white glazed terra-cotta panels and glass, the tower sports no cornice and has a freely subdivisible floor plan served by a bank of elevators on one side.

An important aspect of the Chicago School was the way in which novel forms and plan types were evolved for departmental stores, apartments, and hotels, all based on the technical potential of frame construction, innovative plumbing and drainage, electric lighting, and demountable partitions. Chicago has also produced at least two unbuilt but important skyscraper projects. In 1891 Adler and Sullivan proposed a 36-storey skyscraper, the first with a system of setbacks which anticipated the requirements of the New York City Zoning Law of 1916, on a downtown site. In 1956, Sullivan's pupil, Frank Lloyd Wright (1869–1959), designed the Mile-High—a skyscraper of 528 floors.

The Chicago School also had a strong influence on the development of architecture elsewhere, and particularly in Europe. Germans who visited Chicago in the 1920s to see Frank Lloyd Wright's houses, stayed to marvel at the earlier commercial buildings. Sigfried Giedion (1893–1968), the architectural historian, saw the Reliance Building as a possible source for Mies van der Rohe's studies (executed in 1919 and 1921) for 30-storey, glass-clad towers.

Up until 1960, it was generally accepted that the steel frame was the only suitable form of construction for skyscrapers. European developments in reinforced concrete were, however, helped by the relative economy of concrete outside the U.S., Canada, and Japan, and some tall buildings have been constructed in that material. It is therefore somewhat ironic that Water Tower Place in Chicago will be the world's tallest concrete skyscraper at 860 ft. (262 m), with 7 floors of shopping, 19 floors of hotel, and 40 floors of apartments.

New York

One of the greatest skyscrapers is the Woolworth Building (1913) on New York's lower Broadway designed by Cass Gilbert (1859–1934). This technically advanced building of 55 storeys, clad in terra-cotta, is vaguely Gothic in style. What became known as "Woolworth Gothic" was the style that launched Raymond Hood (1881–1934) as the master of skyscraper design in the 1920s and

1930s. He won the Chicago Tribune competition of 1922 with a 34-storey structure reaching a height of 450 ft. (137 m) which defeated entries by many notable European modernists.

The logic of building high in New York was only partly due to real estate values. There was also a positive desire for the density of activity which resulted from building skyscrapers close to each other. Some of the architectural consequences of such desires were most clearly worked out by the draftsman Hugh Ferris, particularly in the *Metropolis of Tomorrow* which he published in 1929.

The concept of multifunctional buildings, sometimes with nonutilitarian interiors, began to materialize in the late 1920s. Raymond Hood's Daily News Building of 1930 had a 50 ft. (15.2 m) high circular lobby in black glass containing only a spotlit 10 ft. (3 m) globe. Only the radio transmission possibilities of the 102-storey Empire State Building made it economic. Designed in 1929 by Shreve, Lamb, and Harmon, the 1,044 ft. (318 m) building remained unsurpassed until 1973 when the twin towers of the World Trade Center (1,350 ft./411 m) were completed. The 1931 Downtown Athletic Club by Starrett and Van Vleck has the most startling amalgam of uses, including a miniature golf course on the 7th floor and a swimming pool on the 12th floor. Less whimsical and far larger in scale is Rockefeller Center in midtown Manhattan, by Hood, Harrison, and others. This complex of 14 buildings, built between 1931–39, anticipated today's multifunctional skyscrapers in many ways. It contains the famous 6,200-seat Radio City Music Hall and the 70-storey RCA building, as well as an open-air ice rink. Below street level, a continuous shop-lined concourse links together all the separate buildings, while the upper levels accommodate several bars and restaurants, often with their own roof gardens and observation decks.

Postwar skyscrapers

The United Nations Building of 1950 ushered in a new era in skyscraper construction. Although not exceptionally tall, it was the first tall building clad mainly in glass curtain walls. The technical problems, which were highlighted by the use of the curtain wall, spurred industry to develop air-conditioning systems, scalants, and the technology of thin metal claddings in general. Despite the basic technical shortcomings of the curtain wall, hardheaded clients were no match for the fanciful logic of industry and art in unison. Ingenious economic arguments were used, and these bought time until the curtain wall was the cheapest way to clad tall buildings. Lever House (1952) by Skidmore, Owings,

and Merrill and the Seagram Building (1958) by Mies van der Rohe and Philip Johnson, both situated in New York, were representative buildings of this period.

Recent developments

Three-dimensional structure is the key structural concept of today's skyscrapers. The Empire State Building was designed as a series of plane frames, while today's buildings are designed as three-dimentional "tubular" cantilever structures. The core of the building constitutes an inner tube, and the outer columns and spandrels constitute another tube. The tube concept is attributed to Fazlur Kahn, the structural engineer for Chicago's John Hancock Tower, a 100-storey tapered tube with external diagonal bracing 1,127 ft. (344 m) high, completed in 1968. The Hancock Tower was swiftly followed by the N.Y. World Trade Center of 1973, engineered by Leslie Robertson, which was 1,350 ft. (411 m) high. These twin tubular structures were the first to use sky lobbies, which allow elevators to travel above each other in the same shafts and save vital core space.

The tapering form of the Hancock Tower, besides being more resistant to wind pressure, also corresponded in section with the different floor areas which were required at different heights in the building, and this innovation has been taken further in the design of a number of skyscrapers whose sections taper in curves in one direction but not the other. The First National Bank Building, built in Chicago in 1969, is an early example of this type of skyscraper. Some skyscrapers, such as the Hancock Tower in Boston, are now being subjected to additional loading on their upper floors in order to increase their inertia and so reduce their movement in strong winds.

The highest skyscraper at the present time is Fazlur Kahn's giant Sears Tower in Chicago, completed in 1974, which reached a height of 1,450 ft. (442 m). The concept of the building is a "bundle" of nine perforated square tubes, two of which terminate at the 50th floor, two at the 66th floor, three at the 90th floor, and two of which go the full height of 110 storeys. As Fazlur Kahn himself says "today without any real difficulty, we could build a 190-storey building. Whether we will and how the city will handle it is not an engineering question, it is a social question."

Shops, stores, and shopping centers

The open market, composed of temporary stalls, is the earliest known arrangement for buying and selling goods. The Greek *agora*, the center of the city, was both a public meeting place and market. It was surrounded by public buildings with arcades under which permanent shops were built as early as the 5th

The United Nations Building, New York (1950) pioneered the use of curtain walling on a tall building.

John Hancock Tower, Chicago (1968) by Skidmore, Owings, and Merrill (engineer Fazlur Kahn). A 100-storey braced tube structure.

century BC, and even at this time both market and shops were zoned for the sale of different kinds of merchandise. The shop itself was the space between the warehouse or workshop and the street where the purchaser stood to do his business.

The Roman forum was originally occupied by a market and shops, but during the Empire civic and religious buildings were sited here and shops were regrouped elsewhere in specially built market places. Covered markets date from this time; the one at Pompeii, built in the 1st century AD, was typical with walls decorated with mythological subjects and pictures of the goods on sale. Trajan's markets in Rome, built in the early 2nd century AD, were the most ambitious and architecturally complex examples of this general type. Each *taberna*, or individual shop, had a sales counter toward the front and storage behind, and sometimes above, and this basic arrangement continued to be used throughout Europe until the end of the 17th century.

The Moslem and Asian pattern for selling was established in ancient times and has continued with little alteration to the present day. The bazaar grouped together shops and workshops by trades along passages with connections between them. Roofs varied from straw mats to stone vaults. Instead of counters, shops had low platforms where artisans sat working or shopkeepers, surrounded by merchandise, bartered with their customers.

In the West, shops remained open-fronted until the beginning of the 18th century, although the customer was often protected by a canopy or the overhanging upper storey of the building. The first enclosed and glazed shopfronts appeared in Holland in the late 17th century; in France they appeared by about 1700, and in England by about 1736. In all these early examples, the windows were glazed in small panes, as larger ones were not yet available. Plate glass was introduced at the beginning of the 19th century; the biggest sheets available were 8 ft. (2.5 m) by 4 ft. (1.2 m) and, in the larger shops, windows were often arranged in series separated by classically detailed cast-iron columns.

Shopping arcades

Arcaded streets sheltering shoppers were common throughout Europe in medieval times, and stalls and shops under a common roof were introduced in the Royal Exchange in London in 1566. However, the shopping street entirely protected by a continuous roof did not appear until the end of the 18th century. Twenty such streets, usually with pitched glazed roofs, were built in Paris between 1790 and 1860, and of these the Galérie d'Orléans, finished in 1830, was the most grand and had a glazed barrel roof.

In England, the best examples of this type

Market hall at Lyons, France (1909-13) by Tony Garnier.

Rue de Rivoli, Paris, France (1811): a fashionable arcaded shopping street.

were the Royal Opera Arcade, finished in 1818 by John Nash (1752–1835) and Humphrey Repton (1752–1818), and the Burlington Arcade, finished in 1819 by Samuel Ware, both in London. Also planned for London but never built was Joseph Paxton's Great Victorian Way—it was to have been 10 miles (16 km) long, the glazed roof would have been 108 ft. (33 m) high and it would have been served by a regular carriageway with sidewalks, as well as an underground railroad. The finest of all arcades, the Galleria Vittorio Emanuele II in Milan, finished in 1867 by Giuseppe Mengoni (1829–77), had a glazed barrel roof 137 ft. (42 m) high, and the arms of the cruciform plan met in a domed octagon 128 ft. (39 m) in diameter.

The finest American example was in Cleveland, Ohio, finished in 1890 by Eisenmann and Smith, and had a pitched glazed roof and four upper levels of individual shops, served by iron balconies and grand staircases. In Moscow the GUM building, finished in 1893 by Pomeranzev, combined features of the Eastern bazaar with those of the arcade; it was planned on an equal grid covering 16 blocks and had shops on four floors.

Department stores

The first large shops or specialized stores were built in Paris after the Revolution; they were known as *magasins de nouveautés* and sold women's clothing, fabrics, and accessories. Similar shops appeared in England in the 1830s, the most notable being Kendall Milne in Manchester, started in 1831, where all the goods were visibly priced. The first true department store, that is one with a range of different departments, was the Bon Marché in Paris which was organized along such lines in 1852 by Aristide Boucicaut. In 1869 he commissioned a new building from M.A.La-

Burlington Arcade, London, England (1819) by Samuel Ware, housing a complex of 72 shops.

planche which was planned around a sensational curving staircase to entice shoppers upward. Later extensions in glass and iron were by Louis-Auguste Boileau (1812–96) and Gustave Eiffel (1832–1923). Despite the fire risks, such features as open staircases, domed skylights, and exposed iron frames were common in many of the department stores which followed, especially in France.

In America, the design of the building for E. V. Haughwout in New York of 1857 by J. P. Gaynor was very advanced, although the shop itself specialized in high-class furnishings and was not therefore a department store. It was planned around a central well with a large rooflight and was the first building with a passenger elevator, installed by Elisha Otis. On the upper floors were workrooms and storage space.

John Wanamaker (1838–1922) started the first American department store in Philadelphia in the late 1860s but neither his new building of 1875, nor others which followed in New York, Chicago, and elsewhere, were of much architectural interest. The one great exception to this was the Schlesinger and Mayer store, now Carson Pirie Scott, in Chicago, built in 1899 and extended in 1904, which was the last major work of Louis Sullivan (1856–1924). The handsome proportions of the steel frame are directly expressed on the street elevations, and internally they create a light and airy atmosphere on all floors.

In Europe, generally, considerable attention was given to the facades of department stores. In Berlin, the sculptural use of granite and glass on the front of the Wertheim store built in 1896–1904 by Alfred Messel (1853–1909) was in sharp contrast to the large areas of continuous glazing, the first fully developed curtain walling, on the Hermann Tietz store (1898) by Sehring and Lachmann. In Brussels, L'Innovation had a splendid Art Nouveau front in iron and glass by Victor Horta (1861–1947), finished in 1901 and sadly burned down in 1966. In London, the most interesting facades were those of Harrods, a buff terracotta version of the French Renaissance, finished in 1905 by Stevens and Hunt, and Selfridges, which was started in 1908 by Frank Atkinson with Daniel Burnham (1846–1912) as consultant architect, and is imperial Roman in style. The Modern Movement's answer to facades of such grandeur were those which Erich Mendelsohn (1887–1953) designed for the Schocken stores in Chemnitz and Stuttgart, of 1926 and 1928 respectively. In both these buildings the structural and expressive qualities of reinforced concrete were exploited to the full.

Shopping centers

All these examples were in downtown loca-

GUM Department Store interior, Moscow, USSR (1888-93) by A.N. Pomeranzev. Originally a series of separate shops in 16 blocks, linked by glass arcades.

tions and basically relied on shoppers arriving on foot, but by 1930 in the U.S. the increasing use of private automobiles had begun to suggest a whole range of different possibilities. In California a developer, A. W. Ross, had in 1921 purchased 18 acres (7 hectares) of open land along Wilshire Boulevard to the west of Los Angeles, as it then was, and easily accessible from the fast-growing communities of Beverly Hills and Hollywood. This land ran for a mile (1.6 km) along the Boulevard, and by 1928 there had been so much commercial development along this section that it became known locally as the "Miracle Mile." Many of the individual developments incorporated a variety of functions—shops, offices, and very often, a movie theater—but what they all had in common was a parking lot to the rear, often with direct access into the development, as well as the usual type of entrance directly off the boulevard.

An early example of a department store with this double access arrangement, on Wilshire Boulevard but not part of Miracle Mile, was Bullocks-Wilshire of 1928 by John and Donald Parkinson. Later examples of this type made for greater use of modern materials and environmental control systems; for instance, the Sears Roebuck Building on Pico Boulevard, Los Angeles, built by John S. Redden and John G. Raben in 1939, had a considerable amount of parking on the roof with stairs and an escalator leading down into the windowless sales floors below.

It was not long before the importance of automobile access was generally accepted, with serious consequences even for small individual shops. Downtown, where it was difficult to arrange parking, no longer seemed such an attractive place to be; a through route with lots of passing traffic and space for

Aerial view of Northland Center, Detroit (1954), by Victor Gruen and Associates. A group of stores and shops surrounded by acres of parking lots. A typical American suburban shopping center.

Au Bon Marché store interior, Paris, France (1852–76), by Louis-Auguste Boileau and others. The various shopping levels are linked together by glass-covered lightwells and monumental staircases.

The Galleria, Houston, Texas (1970), by Neuhaus and Taylor. A three-storey enclosed mall is the focus of this shopping center.

The interior of the Eaton Center in Toronto, Canada (1976), by Craig, Zeidler, and Strong. Part of a large downtown block redeveloped into a shopping center on various levels with automobile parking facilities.

parking seemed far better. Some facilities were even designed in such a way that goods could be purchased without getting out of the automobile. Examples of the "strip" environment which these developments resulted in can still be found on highways around most American towns and cities.

Strip development on the American pattern is comparatively rare in Europe, largely due to stricter planning controls. Instead, the demand for a high degree of vehicular access to shopping facilities has had to be dealt with largely within the context of existing downtown areas. In some cases this was facilitated by the large amount of reconstruction which was necessary after World War II. In England, for instance, the centers of Plymouth and Coventry were rebuilt during the 1950s with the primary intention of separating vehicular and pedestrian circulation. In both cases, many of the shops faced into traffic-free precincts served by multistorey parking structures. Elsewhere, the same intentions were pursued in a more piecemeal manner, and in many cases existing shopping streets were transformed into pedestrian precincts by the exclusion of vehicular traffic.

In the U.S., more recently, there has been a definite tendency away from strip development as such and back toward the advantages once enjoyed by downtown shopping: choice and variety in a very small area. The result has been shopping centers, sometimes of enormous size, some located on surburban sites with parking at ground level, and others located on urban sites with parking structures. Houston, Texas, has examples of both. Town and Country Village, started in 1965, is in a surburban location and consists of 120 shops grouped to create a village atmosphere. In addition, there are offices, a hotel, several restaurants, four theaters—movie and live— and an ice-skating rink. The whole development covers 120 acres (49 hectares) and there are 17 entrances and exits for easy access. The Galleria, built in 1970 by Neuhaus and Taylor, is also in a surburban location but is more urban in form. Specialty shops are arranged on three levels along either side of a wide mall with a glazed roof and skating rink below. In addition there are offices, a department store, a hotel, three restaurants, an athletic club, ten indoor tennis courts, and a medical clinic, all within the same building complex. There is parking for 7,000 automobiles. The Houston Center, begun in 1972, when completed in about 30 years' time will be the largest continuous mixed commercial development in the world, with a capacity of 40,000 automobiles. In Montreal, Place Ville Marie, completed in 1966 (I. M. Pei and Associates) and Place Bonaventure completed in 1967 (Affleck, Desbarats, Dimakopoulos, Lebensold, Sise) are examples of large com-

mercial developments which, besides their parking facilities, also connect into a continuous system of underground walkways.

Banks

Bank buildings only developed as a specific type toward the end of the 18th century. European banking grew steadily from the 14th century, when consequent to the increase in interregional trade and the rise of wealthy mercantile family dynasties, letters of credit and money began to be exchangeable for goods. At the same time the attitudes toward usury changed. At first the wealthy banking families (Medici, Chigi, Fuggers, De La Poles) carried out their affairs from their multifunctional palaces, and the only reference to any special accommodation for banking was the *Camera dell Tasca*—the money bag room. Subsequently exchanges developed with large covered floor areas to accommodate the regular goods fairs which took place in commercial cities. These developed as columned halls or open loggias underneath town halls. The Loggia dei Mercanti at Bologna (1382), the Taula dei Canvi at Barcelona (1383), and the Royal Exchange in London (1566), are examples. At Bruges the activity took place in a cloistered square surrounded by buildings erected by mercantile families. In both the exchanges and the private palaces the ground floor became increasingly penetrable by the public.

Banking practice had become established by the middle of the 17th century. Banks were privately owned and like many commercial activities banking was carried out from private residences. The Bank of Jones, Loyd and Company of London was the residence of Mr Loyd in 1796 and for a time in 1724 the Bank of England, founded in 1694, was housed in the home of the Bank's head, Sir John Hublon.

After the English Joint Stock Legislation in the 1820s and 1830s, banks grew rapidly as a public service. Banks came to reflect the collective aspirations of a capitalist society. The power and stability associated with classical models produced a commercial classicism based on the more monumental Greek, Roman, and Renaissance forms. The imagery went hand in hand with the models which accommodated early banking—the Palace and the Exchange. The development of a specific functional program began to modify the models. Already the ground floors of bankers' houses had grown to accommodate the continuing increase in public and staff, and the counter became a necessity to separate the two. Lighting these increasingly large floor areas became an acute problem. Windows grew and developed into glazed, arcaded, ground floors along with other commercial enterprises. But in banks this tendency was restricted for reasons of security—both in reality and for appearance sake.

The internal courtyard of the palazzo, utilizing the developing technology of glass infilling an iron skeleton, provided the solution. Already the exchanges had adopted this form of construction—the Paris Bourse (1763 and 1808), the London Stock Exchange (1801), the St Petersburg Exchange (1804), and finally the Berlages Amsterdam Exchange (1898–1903). Furthermore, the Renaissance palazzo was ideal for the new long facades of city blocks, being capable of horizontal extension through repetition. The increasing need for private, cellular offices could be satisfied by arranging these along corridors which, to be fully utilized, had to be double banked with offices looking into internal light wells. These light wells when glazed over at first-floor level served as public top-lit halls, continuing the traditional organization of public first floors with more private accommodation above.

The Bank of England designed by Sir John Soane (1753–1837), occupying a large city block, was both the earliest and largest example of a new type of building for banking and came to serve as the standard model. The first bank building of 1732 was a large Palladian basilican plan. This was enlarged by Tyler in 1765, who designed a top-lit rotunda with radiating halls (for which there were Roman models). It was Soane who mastered the solution with his series of interrelated halls (1788–1823), top lit with skylights in lanterns and drums, arranged behind a giant blind external wall. Soane also eliminated all timber in the bank's construction after 1792 to minimize fire risk. In the later design for a branch bank in Liverpool by Charles Robert Cockerell (1788–1863), the vaults were constructed entirely of masonry.

Bank buildings began to increase in number after the turn of the century. The Royal Bank of Scotland in Glasgow (1827) and the Commercial Bank of Scotland in Edinburgh are large, fully pedimented structures, deep in plan. The London and Westminster Bank (1823) was typical of many that followed Soane's model. Internally, Sir William Tite (1798–1873) designed a large, rectangular, top-lit hall with a giant Corinthian order running through two floors supporting the lantern. A hydraulic lift was installed to raise and lower the heavy cash and plate boxes from the vaults. The Westminster Bank at Bishopsgate, designed by John Gibson (1817–92) in 1862, is a fine preserved example of an early banking hall, with three oval glazed domes and colonnades of Corinthian columns.

The basic form remained unchanged throughout the 19th and early 20th centuries. The Postal Savings Bank in Vienna (1904–06), designed by Otto Wagner (1841–1918), is one of the few not adhering to the classical tradition; nevertheless in plan it conforms to the type. The roof is developed into a complete double skin of steel and glass. Wagner was innovative in his placing of the banking hall on the second floor, leaving the first floor free for the circulation of post. The Crédit Lyonnaise, Paris (1908), where the glass and iron is highly wrought with Louis Quatorze motifs, has the same plan. A more recent example is Helsinki Pension Bank (1952) designed by Alvar Aalto (1898–1976).

In the early 19th century in England and the U.S., branch banks grew rapidly and began holding small savings. They were small and often occupied narrow, terraced sites in main shopping streets. The pedimented temple front was the classical model adopted. At the Bank of England branches in Liverpool, Manchester, and Bristol (1844–46), Cockerell developed a solution which effectively employed a false facade containing three floors of offices in the temple facade arcade, behind which was hidden the top-lit banking hall.

American banks developed along very similar lines. Benjamin Latrobe (1764–1820), a pupil of Cockerell's, designed the Bank of Philadelphia in 1798 as a temple with a freestanding Ionic portico, complete with entablature, pediment, and steps up to the main floor. The cella of the temple was domed with a skylight to form the banking hall. Similar, but more monumental, was the Grecian-style Second Bank of the United States (1818) designed by William Strickland (1788–1854). The growing use of cast iron in American buildings in the first half of the 18th century was reflected in early bank buildings. The Miners Bank, Pottsville, Pennsylvania (1829), contained cast iron in its facade, and the Continental Bank in New York (1856) had an iron frame.

Until the mid-20th century, the majority of banks remained within the classical tradition. The Westminster Bank in London's Threadneedle Street (1930) was almost an exact copy of the Palazzo Massimo in Rome designed by Baldassare Peruzzi (1481–1536). Similarly, in Europe, banks were treated as exercises in the hybrid styles of the 19th century—as in the eclectic design of the Italianate Kreditanstalt, Zurich (1873), and the Baroque Wechselbank in Munich (1895).

The only significant development during the 19th century was in the increase in security. Locks, with the key acting directly on a bolt, had long existed and were vulnerable. Joseph Bramah's lock with circular tumblers (1784) was the first to give real security. This was widely used in banks until Linus Yale developed the pin tumbler cylinder lock in 1851, followed by his combination lock which became standard for bank vaults and safes, as has the time lock which was developed in the

Midland and International Bank, Throgmorton Street, London, England (1871) by William Burnet. Formerly the Ottoman Bank, this is a typically Victorian building in its Italianate style.

Bank of England, London, England (1799), by Sir John Soane. Interior of one of several domed banking halls in the building.

Main banking hall of the Postal Savings Bank in Vienna, Austria (1904–06), by Otto Wagner.

U.S. During the same period, fireproofing of safeboxes developed, initially in France in the 1820s. These had two skins of iron filled with heat-resisting material. Basic construction techniques remained unchanged and improvements in material technology made these increasingly effective; consequently by mid-century they were in general use.

Architecturally, banks developed no further. As banking became more widespread, and with the introduction of automated accounting in the 1920s, the amount of clerical work increased disproportionately to the work in the banking hall. Central bank headquarters became no different from standard office premises. Banks patronized the contemporary international modern idiom of postwar architecture and commissioned some of the better examples. Skidmore, Owings, and Merrill's Chase Manhattan Bank and Manufacturers Hanover Trust (1953) in New York are examples. In England, the National Westminster's new headquarters, undistinguished architecturally, will nevertheless be the highest building in London. With the acceptance of deep, permanently lit spaces, banking halls, large or small, differed only in their furnishings from other commercial enterprises. There is one interesting difference in the fitting out of banking halls. In Europe money is handled only by a cashier—the main floor being furnished with desks, the counter acting only as a space divider. In England and the U.S., however, all counter staff handle money, and the need for secure partitioning between public and staff areas has developed. These take the form of steel grills, sometimes fitted with bullet-proof glass.

During the 1960s, banks introduced mechanized cash dispensers and started to experiment with fully automated cashiers. Owing to their frequent failure, their introduction has been slower than expected. Nevertheless, the development of bank services as a standard clerical operation, the acceptance of permanent artificial lighting of deep spaces, and the likely growth of mechanized cashiers have already rendered obsolete that special architecture that grew out of the person-to-person nature of early banking.

Exhibition buildings

The history of exhibitions goes back many centuries. One of the earliest records of an organized exhibition can be found in the Bible in the Book of Esther which records that the Persian King Xerxes in 486 BC "shewed the riches of his glorius kingdom" in his palace at Shushan. Later in the Middle Ages, huge fairs were held in the principal European capitals and formed an important part of the economic system of the time. In the U.S. and Europe the word "fair" is still used to describe a commercial exhibition. The medieval fairs

were not held in specially designed buildings, but consisted of temporary stalls and displays of merchandise arranged according to different classifications, very similar to a modern trade fair. Today, such a display would be called a "commodity fair" as the goods were sold on the spot.

Trade fairs

During the 18th century, as the medieval fair declined, the "trade fair" developed to stimulate industrial growth. The new form of exhibition, where goods and products were displayed rather than sold directly to individual customers, was first exploited by the Society of Arts in England as early as 1756; and in 1761 the first organized industrial exhibition was held in a London warehouse adjoining the Society's headquarters.

The idea of specially organized displays of machinery and products spread to Europe and in 1798 the first large-scale national industrial exhibition took place in Paris. In 1798, Baron de Neufchâteau organized a national exhibition in temporary buildings erected on the banks of the River Seine as a means of encouraging French industry to compete for world markets. The early years of the 19th century saw several such national exhibitions in France and other European countries, as well as in the U.S.—all in existing buildings such as the Louvre in Paris or temporary structures of little interest, which have long since disappeared.

In 1845 the Society of Arts in London decided to hold a national industrial exhibition, which was eventually to change the whole history of exhibitions. Prince Albert's interest was gained and after a disappointing start the advice of Henry Cole, an enthusiastic and gifted civil servant, was sought. In the meantime, the Society held small annual exhibitions in London, and in Birmingham a permanent exhibition building, known as Bingley Hall, was built in 1850. This was the first specially built exhibition building in England and, as it is still in regular use, it can fairly claim to be the oldest exhibition building in the world.

The international buildings of the 19th century

Following a visit to Birmingham by the Prince Consort and a visit by Henry Cole to the Paris Exposition of 1849, it was decided that the "Great Exhibition of Industry of all Nations" should be held in London in 1851. Thus the first international exhibition was born. As no existing building was large enough to house the proposed exhibition, a part of Hyde Park in London was allocated as the site for a temporary building. Joseph Paxton (1801–65) was the designer for this

Helsinki Pension Bank, Finland (1952), by Alvar Aalto. The building houses a banking hall as well as administrative offices.

unique structure, which was named the Crystal Palace. It was remarkable not only for its scale, but because it was completely prefabricated using iron, glass, and wood components.

The 1851 Exhibition started a wave of similar enterprises in other parts of the world; New York, Paris, Vienna, Philadelphia, and Chicago all staged international exhibitions in specially designed settings between 1853 and 1900. The Chicago Exhibition of 1893 extended over nearly 700 acres (283 hectares) including 200 acres (81 hectares) of buildings. The Chicago Exhibition attracted 21 million visitors, while more than 48 million visited the 1900 Paris Exhibition. Such enormous undertakings often resulted in spectacular and novel buildings of which no details now exist; and designers soon realized that new ideas could be explored in buildings which had only a limited life. Some of these unique structures have survived, but others have been demolished.

The Crystal Palace, designed to last for only six months, was reerected at Sydenham in London, and survived until it burned down in 1936. The tower designed by the French engineer Gustave Eiffel (1832–1923) as the temporary centerpiece of the Paris 1889 Exhibition, still stands over 1,000 ft. (305 m) high and is one of the most remarkable engineering structures of the world. The Galérie des Machines (1889), designed by Ferdinand Dutert (1845–1906) and Victor Contamin (1840–98) with steel portal frames and an uninterrupted span of 375 ft. (114 m), was a remarkable example of innovation in building. It has since been demolished. Three buildings remain from the Paris Exhibition of 1900: the Grand Palais, the Petit Palais, which became the Paris Museum of Fine Arts, and the Alexander II Bridge linking the Champs Elysées with the Esplanade des Invalides.

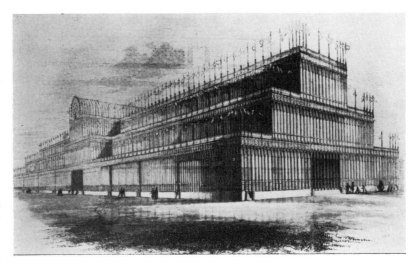

The Crystal Palace, London, England (1851), designed by Joseph Paxton. Exterior view.

Exhibitions in the early 20th century

The early years of the 20th century were marked by a number of major exhibitions in the U.S., including St Louis in 1904 and Buffalo, New York, where President McKinley was assassinated while attending a public reception. The St Louis Exhibition covered an incredible 1,272 acres (515 hectares). Everything was planned on a vast scale; the agricultural building alone covered 20 acres (8 hectares), and a tour of all the exhibits involved a 9 mile (14 km) walk. The exhibition was used to develop the new invention of radio transmission, and automatic telephone exchanges and teleprinters were on display.

Up to the outbreak of World War I, many major exhibitions were held in Europe and the U.S., and architects and engineers used these as opportunities to develop new building materials and techniques. New scientific developments and inventions such as electric lighting, motion pictures, automobiles, aeronautics, and radio transmission were all first seen by the general public at these international exhibitions, which became recognized showplaces for scientific and technical innovation.

The first major exhibition after World War I was the British Empire Exhibition held at Wembley in 1924 and 1925. Most of the buildings were temporary, Neo-Classical in style, and mediocre in design. The only exception was the great concrete stadium seating 125,000 people, designed by engineer Sir Owen Williams (b. 1890) and architect Maxwell Ayrton, which is still the scene of many major sports events and rock music concerts. In 1928, the International Exhibitions Bureau was set up in Paris and a convention signed, which was intended to control and to lay down the basic guidelines for the organization and operation of international exhibitions.

In the years before World War II, regular international exhibitions or "world fairs" were promoted, the most important being those at Brussels in 1935, Paris in 1937 (which included Picasso's famous mural painting *Guernica* in the Spanish Pavilion), San Francisco-Oakland Bay Bridge, the Golden Gate Bridge, and Treasure Island. This was an artificial island 200 acres (81 hectares) in area where the two bridges met. The exhibition also introduced fluorescent lighting.

The New York World Fair, the largest exhibition ever held, was built on the site of a 3 mile (5 km) long garbage dump near New York, now called Flushing Meadow Park—a remarkable example of land reclamation. The New York City building subsequently became the first home of the United Nations which was set up in 1945.

Interior of the Crystal Palace, London, which was 1,848 ft. (563 m) long and 108 ft. (33 m) high and contained industrial exhibits and work from the whole of the British Empire.

Aerial view of the site of the Chicago Exhibition of 1893.

The Paris Exhibition of 1937: general view from the Trocadero.

The 1930 Exhibition in Stockholm, Sweden, designed by Gunnar Asplund. Detail of glass-walled pavilion.

The Atomium which was the centerpiece of the 1958 Brussels International Exhibition.

Post-World War II developments

The first major exhibition after World War II was a national one: the Festival of Britain which, in 1951, celebrated the centenary of the 1851 Exhibition. The centerpiece was the South Bank Exhibition with Sir Hugh Casson (b. 1913) as coordinator, featuring such important buildings as the Royal Festival Hall built in 1950 (architects Sir Robert H. Matthew (b. 1906) and Sir Leslie Martin (b. 1908)), undoubtedly one of the finest concert halls in the world and the model for many buildings in Europe and the U.S., and the Dome of Discovery (architect Ralph Tubbs), the largest dome in the world at that time, constructed of aluminum and since demolished. This exhibition was notable for its overall design standards and landscaping.

The first postwar international exhibition was held in Brussels in 1958 and the central feature was the Atomium, based on the enlarged atomic structure of a metal crystal. This structure still stands although, apart from the permanent buildings erected by the Belgian Government, the other national and international buildings have been demolished. Some of the buildings were reerected elsewhere; the German Pavilion, designed by Egon Eiermann (b. 1904) was rebuilt as a school in Germany and the British Industry Pavilion, designed by Edward D. Mills (b. 1915) was rebuilt in Hilversun, Holland, as a permanent exhibition center. Considerable use was made of many new building techniques; curtain walls, prestressed concrete, and welded steel all played an important role in the building's construction.

Other postwar international exhibitions included New York in 1964–65, Montreal in 1967, and Osaka, Japan, in 1970. The two most outstanding innovations were to be found at the Montreal Expo: the United States Pavilion was in the form of a 250 ft. (76 m) diameter geodesic dome designed by Richard Buckminster Fuller (b. 1895), and the German Pavilion by Rolf Gutbrod and Frei Otto (b. 1925) was a vast tentlike structure suspended by prestressed cables from eight tubular masts, a forerunner of the highly innovatory suspended structures since designed by Otto.

Osaka 1970 may be the last of the great international exhibitions, the next is scheduled for 1982 in Barcelona, but many people belive that rapidly changing world conditions may mean that Expo 1982 may never happen. However, exhibition buildings will continue to be built for permanent exhibitions and these will owe much to the experimental work that has been done throughout the world on buildings that were, with a few notable exceptions, intended to be only temporary. In fact, most of the world's exhibition centers grew from buildings first built as temporary international exhibitions.

Fuji Group Pavilion by Mutaka Murata at Expo '70 in Osaka, Japan. A double-skin inflated structure with 16 closed tubes bent to a horseshoe shape to enclose the exhibition space.

Municipal Exhibition Hall in Kitakyushu, Japan (1977): a cable-stayed enclosure by Arata Isozaki.

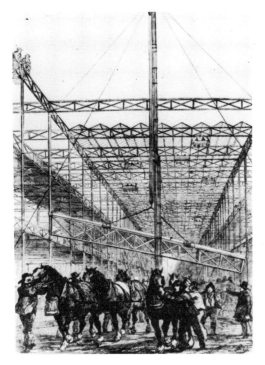

LEFT: Raising the girders of the Crystal Palace (1851) RIGHT: Erection of the transept arches of the Crystal Palace.

Osaka, Expo 70: Canadian Pavilion by Erickson and Massey.

Osaka, Expo 70: Mitsui Group Pavilion by Takamitsu Azuma.

Government

Doge's Palace, Venice, Italy
(1309–1442). The largest and most
prominent Venetian civic building.

Town Hall at Brussels, Belgium
(1401–55). A large late medieval
civic building with an imposing
spire, rivaling contemporary
ecclesiastical buildings.

Palace of Versailles from garden
(1669–85), by Louis le Vau and
Jules Hardouin Mansart.

Civic buildings

The civic building, as an expression of the collective, began when administration ceased to be the direct prerogative of a tyrant. This is, of course, a partial definition as the administrative process and the structures it requires is much the same under any ideology. But the involvement of the ruled in the process of decision making produces structures, social and physical, which are participatory and are also the expression of civic pride and consciousness.

Personal rulers administered (and in many countries still administer) summary justice and controlled their societies by simple physical force. They found, and still find, a palace or a fortress an appropriate architectural setting. This physical image of power still remains with us, transferred perhaps, in the case of the palace, to the town hall or parliament building.

With the birth of the Greek democracies, the process of civic consultation, persuasion, and confrontation began. The Athenian citizen assembly took place on the Pynx, a natural amphitheater below the Acropolis. Here the great speeches were made, on a bare hillside, under the temples. The city council of delegates, first representing the original tribes and later the city sectors, met in the *bouleterion*, next to the *agora* or market. It was a small building with a semicircle of tiered seating facing the chairman's dais, and this arrangement still constitutes our civil assemblies. The *bouleterion* was never a prominent formal element in Greek cities. It was placed near the center, the *agora,* but neither in scale nor formal complexity was it ever dominating. In this it reflected the relatively small role of the state in the citizens' lives. These two forms of assembly, the popular and the delegate, still persist into our own day as assembly hall and council chamber.

The Romans established a rule of law, backed by force. Administration was handled by appointees from the center, so that no new participatory administrative building type evolved. Their halls of justice, the *basilicas,* were, however, the prototypes of most subsequent covered public spaces. The requirements of a courtroom, a large space with a broad apse for the judge's seat, work very well for public assemblies, and the plan with its nave and aisles is still often used. *Basilicas* were directly adopted by the early Christians as a suitable form for churches. The early church was primarily an assembly, where it was essential to hear the message, and to see the speaker. In the *basilicas* the Romans developed a building type of considerable grandeur, which could take its place in the city along with the baths and the temples, as an urban element.

The medieval period saw the reestablishment of the city as an independent entity, and the town hall became an integral and important focal point of the city fabric. The free cities were run by oligarchies of trade guilds and the city was divided into areas dominated by a particular activity. Each area had its guild hall, where the guild held its meetings, and these buildings, together with the parish churches, provided a focal point of local life. The system still exists in Venice.

The 16th–18th centuries were times of increasing nationalism and centralization. Civic buildings were the palaces of the powerful and Louis XIV was the exemplar of his period. He built extravagantly and established at Versailles in 1669 an extraordinary example which was to be emulated throughout Europe and the Americas. Versailles was not only the home and office of the king, but also the home of the attendant court and an army of bureaucrats who ran the various ministries. As most decision making was personal, and ministers depended entirely on the king's presence and favor, their staffs were similarly centralized and concentrated. (see PALACES.)

The free cities of the Middle Ages were soon destroyed by the authoritarianism of central governments. One of the first such cities, Florence, was also one of the first to succumb to tyranny. In 1560 the Medicis built the first modern block, the Uffizi, to contain their bureaucracy, connected by a half-mile (804 m) secret passage to their Pitti Palace.

The 19th century

The French Revolution at the close of the 18th century set loose an ideology which is still being worked out in both social terms and in terms of civic architecture. The revolutionary assembly followed the classical pattern; a semicircle of tiered seats which could accommodate a wide range of opinion. The Assembly, symbol of the representation of the people by the people, was the essence of both the French and American Revolutions. In the U.S. the puritan tradition of town assembly and devolved civic administration was both a contrast and an example to contemporary Europe. In France, when the Second Republic was set up in 1848 (only to become the Second Empire a few years later) it established a system of local government. It was later clothed and formalized by Georges-Eugène Haussmann (1809–91) and remains a model today.

Haussmann's 17 years as Prefect of the Seine (1853–70) enabled him to establish the form and responsibilities of local government, and to provide its prototypical architectural forms. Each urban sector was focused by the provision of a town hall, and a network of related police stations was established. Some functions were centralized: the provision of clean water and sewerage, the establishment and maintenance of parks, the planning and building of roads. Through his system of planning control and consultation, he produced the most beautiful city in the world, with very little direct intervention, and at little cost to the community. It is an example of the force of a book of rules which were, however, simple, elementary, and easily understood. He was fortunate that in his time there existed a generally agreed architectural aesthetic. It was carefully controlled, monitored, and suitably rewarded through a small, powerfully entrenched academy at the Ecole des Beaux Arts. This school provided the prestige, intellectual force, and leadership in architectural matters until 1914.

Haussmann restructured Paris, provided markets, fire stations, slaughter houses, roads, canals, and a major world exhibition in 1867. Each initiative was the occasion for a civic building, a monument; for a reinforcement of the beauty and coherence of the city. Since the Revolution of 1789 the reality of communal responsibility for every aspect of urban life has inexorably come about. The royal collections became the core of national museums; the local landowners' palaces became, in due course, civic structures. This process continues today with the conversion of magnates' houses into museums, as in the recent transformation of the Carnegie house in New York to the Cooper-Hewitt Museum, part of the Smithsonian Institute.

The 19th century had command of several architectural languages, all rich in meaning and rhetoric, which it applied with enthusiasm to a wide range of quite new problems, in scale and in building types. The first problem for architects was the drastic change in the scale of the building process during this period. Though large buildings had been built previously they were always exceptional—palaces or cathedrals. After 1820, large buildings became commonplace in cities as opera houses, railroad stations, covered markets, and the like were erected in the heart of old cities. The languages of eclectic architecture were extended, or compounded, to deal with these buildings. In France, the classic tradition, always capable of a grand statement, produced a series of triumphant solutions: the Paris Opéra, Les Halles, the Bibliothèque Nationale. They were grand in scale and conception, and were carefully placed in the city to maximize their civic and visual importance.

In England another method was advanced: aggregation. Here there was a longstanding mistrust of megalomanic architecture, which was closely associated with French centralism and Empire. Large buildings like the National Gallery, the Law Courts, or St Pancras Station all pretend to be an agglomeration of smaller buildings, with a variety of elements in juxtaposition to create a romantic, picturesque effect. The inspiration was, of course, in the varied skylines of northern Gothic building which coalesced into homogeneous cities. Within such complex organisms each house, hall, or church retained, indeed proclaimed, its individual identity. Out of this spirit of mutual competition, of small-scale challenge and response, confrontation, and emulation, there developed an unparalleled density of visual and tactile experience. The understanding of the need for detail and complexity is a constant throughout the 19th century, and it makes the buildings both readable (in the sense of structuring our understanding of the purpose and use of the building) and somewhat endearing.

In the U.S. during the same period there was also great concern at the rapid and uncontrolled growth of many cities. Frederick Law Olmsted (1822–1903) saw parks, parkways, and park systems as a way of relieving high densities and overcrowding while, at the same time, giving some sense of order, in spatial terms, to different areas within cities. Probably his most famous undertaking in this respect was Central Park in New York, designed in 1958 in partnership with Calvert Vaux (1824–95). Olmsted and his followers also saw these open spaces as being suitable, in many cases, for locating civic buildings. This combination of open spaces and civic buildings was to provide the basis for the City Beautiful Movement, which was first formulated in the 1890s and has continued to be influential throughout much of this century. Boulevards, parks, and squares were used as a means of creating a sense of order on a large scale, then within the more important of these spaces were located civic buildings, visible and accessible to all. Daniel Burnham (1846–1912) was the leader of this movement and executed proposals for many cities in the U.S. The most complete and ambitious of these proposals were contained in his *Plan of Chicago*, published in 1909. Many later developments such as the Civic Center at Los Angeles, first projected in 1939, are based, at least in part, on the same approach.

The 20th century

The Modern Movement, on which our current architecture is based, sought its poetry through control of, and involvement in, the industrial process. This essentially hopeful ideology is still with us, although now often questioned. It was a theory all too convenient for both bureaucrats and businessmen, because it discounted the spiritual elements, and set aside the humanizing (and expensive) qualities of craft tradition. Since World War II, the main function of civic buildings have therefore been largely lost. The representation of, and contact with, the community which

Uffizi Palace, Florence, Italy (1560–71), by Giorgio Vasari. Built by the Medicis to house city offices.

Section and elevation of the Capitol in Washington D.C. (1856–64). The large iron-framed dome was designed by Thomas U. Walter.

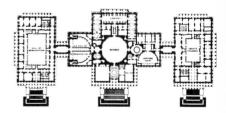

Plan of the U.S. Capitol. The building, which is in the Classical style, has an imposing grandeur with its large dome dwarfing the three-storey wings around it.

Victoria Tower, Houses of
Parliament, London, England
(1835–36), by Sir Charles Barry.

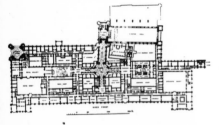

Plan of the Houses of Parliament in
London, England. The effect of the
regularity of the plan is reduced
externally by the asymmetrical
arrangements of the towers.

View of one of the wings of the
UNESCO Secretariat, Paris, France
(1957–58), by Marcel Breuer and
others.

the building is intended to have, is set aside in the name of efficiency, function, and economy.

Furthermore, new government buildings today tend to be built out of the city itself, on open campus sites. This fashion, based on Garden City propaganda of the 1890s, is exemplified in the creation of the major monumental ensembles of the British Commonwealth: New Delhi; the Union Buildings in Pretoria; Canberra; and Ottawa. Here the monuments take on a life of their own, responsible only to themselves, with wide, empty spaces around, far removed from any urban activity.

The major postwar group of comparable quality is Le Corbusier's Government Center at Chandigarh in India. Here one of this century's great architects produced a solution very similar to the New Delhi of Sir Herbert Baker (1862–1946) and Sir Edwin Lutyens (1869–1944). Vast, searing spaces between the primary buildings ensure an inevitable sterility in use. Le Corbusier was, however, supremely conscious of architectural language and his buildings at Chandigarh provide many clues for the future. He understood the primary need for focus and identification, and so developed a series of forms to signal entrances, hierarchies, and uses. Such forms are, of course, the basic language of Classical architecture. The orders carry appropriate meanings about function and use as well as such combinations as porches, colonnades, and balconies—being well understood symbols for entrances, indications of route, and hierarchy of importance.

Le Corbusier developed the grand entrance to his buildings, inventing forms equivalent in monumental quality to classical prototypes, to draw the eye and establish a hierarchy of route. In his Secretariat Building, an ordinary office building in most respects, he structures the facade with double-scale elements which indicate the ministerial suites; a hierarchy of importance. The Parliament Building, too, expresses its constituent elements simply and clearly—an easily read and impressively scaled entrance portico, the council chamber signaling its presence above the surrounding service spaces. The forms are new, but the method is classical.

Also classical is the device of the grand scuptural monument—in Chandigarh the "Open Hand"—to lock the spaces visually together, and to provide the iconography essential to any urban ensemble. Civic monuments act in two quite separate dimensions: as urban funiture and also as reminders of the heroic past or the hopefully heroic future.

The most spectacular new government ensemble is Brasilia, a new capital set 800 mi. (1,280 km) in the heartland of Brazil. The intention of the new capital is to focus the country's attention inward to its own resources and development. The plan, by Lucio Costa (b. 1902) is a Beaux Arts masterpiece of cross axes and major monumentality. Like Chadigarh, the government sector is too grandly spaced. It is a city for automobiles, for speed, and all the effects are best observed from a vehicle rather than as a pedestrian. The city is nonetheless the product of a heroic vision, and the forms invented by Oscar Niemeyer (b. 1907) have an elementary geometric power which perfectly matches the vision.

The League of Nations Building in Geneva, Switzerland, was the result of a famous and controversial architectural competition in 1927, which established modern architecture as a viable alternative style. The completed building, in the white stripped classic style of bureaucratic buildings everywhere, conforms to the image of an isolated building camping in a park, which became the hallmark of postwar planning. It was only the generosity and intelligence of the Rockefellers (who also owned much surrounding land) that brought the United Nations Building into the urban context of eastside Manhattan. The support structure of a great metropolis sustains the organization without effort. The building, though much watered down from Le Corbusier's conception by Wallace K. Harrison, one of the architects of the Rockefeller Center, fits into the city with ease, and has helped regenerate a decayed neighborhood. The United Nations Building was a development triumph, and it illustrates a prime function of the civic building to sustain its neighborhood, to provide the necessary dream, the psychic energy of the city.

When the United Nations came to build in Paris it chose Marcel Breuer (b. 1902) as the architect, and his UNESCO Building is a paradigm of postwar building. It has a very high quality of finish and the organization of the plan, based on an early Le Corbusier design, is elegant and clear. First-rate collaborators like Pier Luigi Nervi (b. 1891), Pablo Picasso (1881–1973), Joan Miró (b. 1893), Henry Moore (b. 1898), and Isamu Nogachi (b. 1904) enabled Breuer to create one of the very few convincing civic buildings of the century. Its tragedy is that, coming at the beginning of the world's greatest building boom, its forms were instantly copied and debased around the world.

The few buildings of real quality built anywhere in the world by the public sector since World War II are almost invariably designed by independent architects working in conditions of competition very similar to those of the 19th century. Such buildings are exceptional and are the result of architectural competitions or chance appointments, for bureaucracies have generally been unwilling to take the responsibility for patronage.

The growth of large design offices within local and central government is also a recipe for a loss of architectural quality. Such organizations are concerned with committee decision making and are fundamentally irresponsible, and unresponsive. No one can be blamed, or even singled out for praise. Nevertheless there are some hopeful trends. Firstly, there is an awareness of the value of our existing buildings and monuments. Secondly, there is a slowly developing language of modern architecture which is in great need of invention and experience. Here the problem is to overcome the anonymity of industrial systems, without losing the economic benefits. It is here that civic buildings have a major role to play in the development of a meaningful and responsive architecture. An example like Boston City Hall, designed by Kallmann, McKinnell, and Knowles, introverted and self-regarding as it is, nevertheless makes an attempt at complexity of form and association which is very welcome. Thirdly the testing of postwar building technology is now almost complete. We have tried almost everything, every method, every material, and can learn some lessons from that. Fourthly, the radical changes in the energy equation will bring a need for higher quality and more intelligent buildings.

Civic buildings should by definition be exemplary. We still have a tradition of architectural competitions for particularly important buildings. It remains our best hope for an elevating, ebullient civic environment.

The Assembly Building Chandigarh, capital of the Punjab, India. Completed 1961 by Le Corbusier.

City Hall, Boston, Massachusetts (1962-69), by Kallmann, McKinnell, and Knowles.

National Congress Group in Brasilia (1960) by Oscar Niemeyer, housing administrative offices and the Assembly Building.

Educational and research

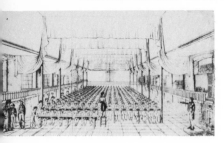

Central School of the National Society, London, England: early 19th century.

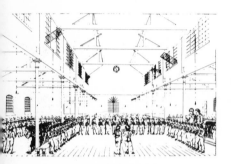

Central School of British and Foreign School Society, Southwark, London, early 19th-century.

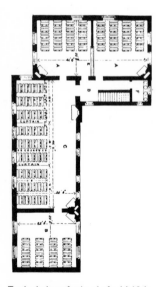

Typical plan of school of mid-19th century showing galleries and separate schoolroom.

Schools

Every civilization in history is known to have had a system of education, but in very little evidence remains of any specialized school buildings. Even in our own civilization the physical remains of medieval schools are extremely scanty. Teaching methods were probably oral rather than by written exercises, and very little provision was therefore required by way of equipment or buildings. Many of the buildings used were not designed for schooling.

The first development of importance occurred in the 14th and 15th centuries with the emergence of religious colleges whose activities included education. Winchester (1382) and Eton (1440) in England are two surviving examples. The school at Winchester consisted of a single room some 45 x 29 x 15 ft. (14 x 9 x 4.5 m) built of stone walls with no fireplace. Stepped seats within the window reveals were provided for monitors to oversee the pupils. The 16th century saw the establishment of many new schools, the prototype for which was St Pauls School, London, founded by John Colet, the then Dean of St Pauls. Erasmus (1466–1536), who was a pupil, describes the school as consisting of four chambers; two were schoolrooms divided by a curtain, one a chapel, the other an entrance.

By the 17th century there was something approaching a national system of schools in most European countries. The majority of school buildings, however, still consisted of a single room. They were unpretentious in appearance and differed little from the houses of the period. Yet in 1660 Charles Hoole, an early writer on education, proposed a three-storey school building for 500 scholars. This was to contain, on the middle floor, a schoolroom divisible by folding partitions into six classrooms with a desk for every pupil; the upper and lower floors to contain further classrooms, a library, and gallery. In the 18th century, attention was turned to the education of the poor with the founding of charitable schools. Blue Coat School, Westminster (1709), is typical, consisting only of a single room 42 x 30 ft. (13 x 9 m) for 86 boys and girls.

The 19th century saw the development of mass education in Europe. In England, two educationalists, Joseph Lancaster (1778–1838) and Andrew Bell (1753–1832), proposed model schools which were to become the prototypes for the schools that attempted to provide an education for the enormous numbers of poor children inhabiting the manufacturing towns. The Lancastrian Schools, following Lancaster's model of 1811, consisted of a single large schoolroom 70 x 32 ft. (21 x 10 m) to accommodate 320 children seated in 20 rows of benches facing the master at one end. The floor of the schoolroom

sloped slightly toward the front so that each child could be seen. An aisle 5 ft. (1.5 m) wide was left around the perimeter in which children stood in semicircles to go through their lessons with a monitor. Windowsills were high, leaving walls free for lessons, and baize curtains were hung from the exposed roof structure to keep down the level of sound. The Madras Schools, as Andrew Bell's model became known following his experiment with the form in India, was somewhat freer in plan and specification. Its model Central School in London consisted of two rooms, one for 600 boys, the other for 400 girls, each with cast-iron columns supporting a trussed roof structure. A single row of benches was placed around the perimeter leaving the central space empty for groups of children to stand in squares around their monitors.

This principle of teaching by monitors was challenged in 1820 when Samuel Wilderspin opened a model school for infants which included a smaller room off the main schoolroom for pupils to be taught directly by the master. Wilderspin later extended this principle to the schoolroom itself, with the innovation of raked seating from which all the children could see the master. The system, which had originated in Holland, relied upon the use of pupil-teachers to teach classes, and the typical plan of a school became that of a long narrow schoolroom 65 x 18 ft. (20 x 5.5 m) accommodating 120 children seated at desks on raked galleries facing across the width of the schoolroom. The children could be divided into groups by drawing curtains across the room. A small classroom 13 x 20 ft. (4 x 6 m) opened off the schoolroom for "object" lessons. The style of these buildings, of which several remain, varied from plain Tudor to plain Gothic.

In the 19th century it was Germany that led the world in education and consequently in the design of schools. King Frederick II (1712–86) had established the principle of compulsory school attendance as early as 1763. Teaching was, without exception, in classes of less than 60 pupils. The design principles for schools had been rigorously researched, the fundamental principle being that of daylighting. By 1850 a typical elementary school consisted of a set of identical classrooms arranged on two or three storeys that were planned compactly around two staircases, one for each sex. It was established that 27 ft. (8 m) was the maximum distance at which a child could read a blackboard, and that 21 ft. (6.4 m) was the maximum economic span to avoid columns. The classroom size was therefore 30 x 21 x 13 ft. (9 x 6.4 x 4 m). Windows were invariably on the pupils' left-hand side when facing the blackboard to avoid shadows being cast on the pupils' work. Secondary schools, of which by this time Germany had a completely developed system, were designed on corridor plans. Classrooms for as many as 900 pupils were supplemented by specialized spaces: laboratories, drawing rooms, and rooms with special equipment for mathematics and natural philosophy. They also included a hall in which the majority of the school population could congregate, and a gymnasium building. These schools were usually grand and built in a Neo-Classical style, much influenced by Karl Schinkel (1781–1841).

In the U.S., Henry Barnard (1811–1900), an educator from Boston, had published a range of school plans which had become widely accepted. Educators largely followed the German model but had reservations about a total commitment to teaching solely in classes. Many schools were built with sliding partitions between classrooms such that they could be opened up to form a large space for the simultaneous teaching of larger numbers. The American plans were ingenious, but paid less attention than those of the Germans to matters of daylight, and more to matters of economy. The buildings tended to be solid, compact in plan, and somewhat plain externally.

In England E. R. Robson (1835–1917), the architect to the London School Board, published in 1874 a comprehensive survey of contemporary school design in his book *School Architecture,* and consequently established the prototypical plans for English schools. They, like the American schools, followed the German model of classrooms, but also included on each floor a large schoolroom or hall with raked seating at one end for teaching in larger groups. This plan would be repeated on as many as four floors in schools containing as many as 1,500 children. Robson's schools paid considerable attention to environmental design and equipment, incorporating ducted ventilation systems and the innovation of the locker desk. In his search for an architectural style which suited the needs of daylighting and economy, Robson developed and established the Queen Anne Style as the accepted style for school buildings in his time.

By the end of the 19th century there was increasing concern for health standards in the seriously polluted cities of this period. Daylight, sunlight, and above all ventilation, were seen to be the solutions at a time when surburban development was beginning. One innovation of this movement was the "open-air" school, the first of which was built at Charlottenburg, near Berlin, in 1904. Such schools consisted of single-storey pavilion classrooms in which the complete side of a classroom could be opened—the concept possibly inspired by the design of isolation hospitals of the time. The more lasting outcome, however, was the introduction of cross ventilation as a principle in school design. The

earliest designs were for schools in which each classroom was a pavilion attached to a connecting corridor. The principle of daylighting from the left was replaced with that of daylighting from both sides of the room. Later designs had quadrangular plans with classrooms and a hall on one side of an open corridor which enclosed a courtyard. Hygiene had obviously distracted architects from the problems of the appropriate style for school buildings which in this period became plain and parsimonious.

The ideas of hygiene and economy continued to dominate school design for the next 40 years. Sunlight became an important ingredient, and the "finger plan school" became the norm: parallel rows of south-facing, highly glazed classrooms linked by long corridors. In the 1930s, under the influence of the Modern Movement, the style of school buildings changed to one of flat roofs, steel frames, and metal windows, but the pattern of regimented and identical classrooms persisted. In an era of depression and economy, the Cambridgeshire Village Colleges in England were an isolated architectural and educational innovation. These were community schools, offering education and culture to children and adults alike. The most famous, at Impington, was designed by Walter Gropius (1883–1969) together with Maxwell Fry (b. 1899) in 1939. The college consists of a hall with two wings, one containing classrooms, the other, curving around a promenade, contains a library, recreation spaces, and common rooms. Constructed of brick and glass, its appearance was "modern" yet not industrial.

For some time after World War II school design continued essentially unaltered—that is in the "finger plan" form. The recent advances in environmental science led architects to concentrate even more on refining aspects of lighting and ventilation. In the U.S. architects developed linear buildings with uni-, bi-, or even tri-lateral lighting.

More serious innovations in school design appeared in the late 1940s, and they sprang from a development, or rather a belated acceptance, of "child-centered" educational ideas. A far deeper understanding between educationalist and architect existed at this time than had ever prevailed before, and particularly so in Hertfordshire, England, where these innovations originated. The schools designed were relatively compact, with classrooms designed as multipurpose workrooms, clustered in groups around shared facilities. They were constructed of factory-prefabricated, standardized components: lightweight steel frames with timber infill panels. The light, domestic, and informal character of these schools ideally reflected the new freer and varied educational ideas within the schools.

These ideas on design and prefabrication

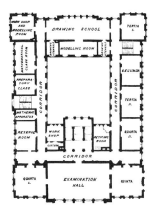

Plan and elevation of the Gymnasium at Liegnitz, Germany.

Gymnasium at Liegnitz, Germany, (1867). A typical German school of the mid-19th century.

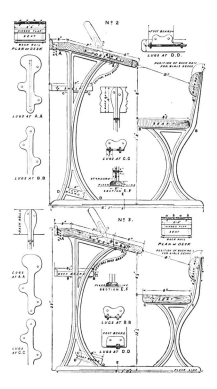

Desks for graded schools (1874).

Primary School at Cheshunt,
Hertfordshire, England (1947).
One of the system-built, prefabricated
schools developed after World War I.

Fodrea Community School,
Columbus, Indiana (1973), by
Caudill, Rowlett, and Scott.

were adopted and developed by the British Ministry of Education and spread rapidly in primary school design. Consortiums of local authorities were formed to sponsor prefabricated school construction systems. the Ministry architects researched and developed design ideas in model schools. Their Amersham Junior School of 1958 soon became typical. Instead of classrooms there were a series of interrelated and varied activity spaces; defined areas had vanished with the more informal use and the demand for more usable space. These ideas reached their full development in the Eveline Lowe Primary School of 1966. Here the teaching spaces became a continuous series of semi-open bays for particular activities—painting and craftwork, reading and discovery—with carpeted stepped spaces for story telling. School furniture design had also developed: it was now designed to stack and fit together to match the freedom and flexibility of use of space. Internally, these schools, with their varied spaces and a child-size scale that offered freedom of movement yet maintained a sense of enclosure, were some of the most sophisticated and sympathetic of modern architecture. Externally, they were somewhat incoherent.

Innovation in secondary schools in England came more slowly. The range of spaces specifically designed for particular activities increased as curricula became more diversified, but formal and rigid planning around a structuring circulation system continued. The first significant developments were associated with attempts to ameliorate the social problems of scale in large comprehensive schools. Spaces were grouped into faculty blocks or "house" blocks and the schools designed on campus principles. In the 1960s, following the publication of the Ministry of Education's development project for the Arnold School, Nottingham, ideas of open or semi-open planning spread to secondary schools. Their plans became amorphous in a similar manner to those of primary schools; the rigid, structured, and institutional atmosphere deliberately exchanged for one of freedom, informality, and variety.

Innovation in school design in the U.S. took a significantly different direction, one based more on the possibilities of technology, and particularly the attractions of physical flexibility, and less on new educational ideas. One important innovation came in 1952 with the Hillsdale High School, San Mateo, California, designed by John Lyon Reid. This was the first deep-plan school. It abandoned the constraints and the costs of linear buildings to contain its teaching space in a top-lit and air-conditioned single-storey steel structure measuring 196 x 430 ft. (60 x 131 m). The classrooms, otherwise quite conventional, were formed internally with relocatable partitions. Another innovation introduced prefabrication to American school building. The concept, developed by Ezra Ehrenkrantz, was for long-span, steel roof structures incorporating lighting and air conditioning integrated with a demountable partitioning system. The implications of this system were toward deep-plan and open-plan schools. And this was the direction that school design took. Schools were designed on pure open-planning principles with large carpeted areas subdivided by screens or demountable partitions and with a controlled air-conditioned environment. In the affluence of the 1960s the range and variety of facilities included in schools, now often having as many as 4,000 pupils in a high school, would be expected to include extensive workshops and laboratories, sports halls, swimming pools, and auditoriums.

The most recent innovations in schools have followed the revival of the concept of the community school. As a consequence, secondary schools have grown even larger in size, with the addition of libraries, sports centers, and theaters to their facilities, and these valuable resources are shared with the local community. The definition of the school as a building as well as an institution may be disappearing into history.

Universities

The first universities were founded in Italy in the 11th and early 12th centuries. Up to that time higher education was the prerogative of monasteries and cathedrals. The universities arose to teach secular subjects, and were the outcome of a civilization which, by the 13th century, required professional and scholastic expertise. The earliest seem to be Salerno in southern Italy, which was famous for medicine, and Bologna in northern Italy, renowned for law. The University of Paris was founded around the same time. A university was originally called a "studium" or "studium generale," the word "universitas" referring to the body of teachers who were licensed to teach. The university needed no more than suitable classrooms, and these, for the small numbers of students at this period, were located in private houses.

The first specific university buildings were colleges designed to provide living accommodation rather than teaching space. The first records of university colleges are at Paris, the earliest being that of the Hôtel Dieu in 1180. The Sorbonne was originally a college founded by Robert de Sorbon in 1256. The University of Oxford, England, had its first college, Merton, founded by Walter de Merton in 1264 and incorporated in the university in 1274. The University of Cambridge, England, was established following a migration of students from Oxford in 1209, and a subsequent migration from Paris in 1229, and had its first college, Peterhouse, founded in 1280 by a Bishop of Ely. The English colleges, however, were not intended for students but for teachers, and were in fact very similar to religious colleges of priests serving in a church. The students, many of them foreign, formed *hospitia,* or hostels, to accommodate and protect themselves.

The Old Court, Corpus Christi College, Cambridge (1352), is the best surviving example of a 13th-century college. In its original form it consisted of a hall, kitchens, offices, a masters' lodge, and chambers which were located in a two-storey building around an irregular quadrangle or court. The chambers each consisted of a large room together with four small cubicles. The cubicles were for study, the large room for sleeping. The most important prototypical college was, however, New College in Oxford, which was founded by William of Wykeham in 1379. It is both the first English college where staff and students lived together inside the college, and the first university building aimed at unified monumental architecture. The college, built around a central quadrangle, consisted of members' chambers, a chapel and hall, and a gatetower to the west. The structure was originally of two storeys. New College became, in academic constitution and architectural form, the model for successive colleges.

At about this time the first buildings specifically designed for lecturing appeared. The activities of the universities had outgrown the capacity of private rooms, but their demands for space were not large. Cambridge built its Divinity School some time between 1350 and 1400, and during the 15th century a library and buildings for the other subjects were erected—for canon law, civil law, and philosophy—the whole forming the small quadrangle known as "Old Schools." Oxford similarly built a quadrangle but on a far grander scale (1426–80). The Grandes Ecoles at Caen (c. 1436), in a building 250 ft. (75.5 m) long, had a Salle des Arts and Ecole du Droit on the first floor, a Salle de Théologie and Classe de Médecine on the second, and a library on the third. Krakow University, Poland (c. 1400), was built around a cloistered courtyard, with lecture rooms on the first floor and larger halls and living quarters for masters on the second.

The later 14th and 15th centuries were great university-founding periods. During the 13th century there were only eight universities in existence—Bologna in Italy; Paris, Toulouse, and Montpelier in France; Palencia and Salamanca in Spain; and Oxford and Cambridge in England. Thereafter, universities were founded at Prague, Krakow, Vienna, and Heidelberg, and seven more were established in Germany, twelve more in France, and three in Scotland. (The number and size of the colleges at the existing uni-

Perspective of Trinity College, Cambridge, England (1535) showing buildings centered around the Great Court.

The Rotunda, University of Virginia, Charlottesville (1822–26), by Thomas Jefferson.

versities also increased.) The forms of these universities varied. Many of them were built on the peripheries of the towns.

In Italy, in the 16th century, the scale and monumentality of university buildings increased enormously, and the common form adopted was that of the Renaissance palace. (The English college had adopted something of the form of the religious college and country house.) They consisted of a large square block with a square central courtyard with cloistered galleries on one or two floors. Jacopo Sansovino (1486–1570) built in Padua the University of the Venetian Domains, the courtyard of which was completed in 1547. The Archiginnasio (1565) was built to house Bologna University. Such monumental and unified buildings for universities were unknown at this time in Britain.

The small-scale collegiate building around a quadrangle remained the pattern of building at Oxford and Cambridge, and the serene academic atmosphere produced the ideal model for collegiate developments in American colonies. But none of the nine colonial colleges which were chartered between 1636 and 1780, and which represented the seeds of American universities, had the money to produce such an environment. Like Old College, Harvard (1638), they were plain, individual structures sited on open land. Harvard's Stoughton Hall (1698) became a model for future dormitories, with bedrooms on four floors, each room having cross ventilation, a fireplace, and a private study.

In 1807 William Wilkins (1778–1839) demonstrated a new form of college layout in his design of Downing College, Cambridge. His design incorporated the discoveries of 18th-century English landscape and became the prototype for the U.S. campus. Instead of the quadrangle format it consisted of separate buildings and pavilions, all in pure Neo-Classical Greek style, around three sides of a lawn. The first realized campus plan in the U.S. was that prepared in 1813 by John Jacques Ramee for Union College, Schenec-

Portico of University College, London, England (1827–28), by William Wilkins.

McGill University, Montreal, Canada.

School of Architecture (Crown Hall) at the Illinois Institute of Technology (1955) by Mies van der Rohe who designed the whole campus as a series of steel and glass rectangular pavilions.

tady, New York. The most important plan was that by Thomas Jefferson for his University of Virginia at Charlottesville, begun in 1817. With a Neo-Classical architecture, Jefferson produced a design which accommodated the activities of the university in a rational and functional manner. The plan was arranged around a rectangular open space, with ten pavilions each containing the living spaces and a teaching hall for the professor of each school of study. These pavilions were linked by colonnades, off which opened single-storey students' rooms. Behind the pavilions were gardens, and between them paths for the servicing of buildings, all enclosed by a further line of student quarters. At one end of the open space, in a rotunda, stood the library. Each of the buildings was designed differently to furnish examples for architectural teaching. In Jefferson's words "such a plan would afford that quiet retirement so friendly to study."

The 19th century saw a great upsurge in university building, due no doubt to the appetites of industry and commerce for expertise. Science had begun to appear in universities in the 17th century; in Italy it was absorbed by the existing institutions, while in other countries new specialized universities were formed—the Ecole Polytechnique in Paris in 1794, and the Vienna Technische Hochschule in 1815. These new foundations produced a single building complex, housing administrative and academic accommodation with inbedded lecture theaters for the various disciplines. Such new universities or colleges were established in most major cities of Europe, their architectural arrangement reaching the grandest expression in Nenot's building for the Sorbonne, Paris (1895). One prototypical form for these universities was that established by William Wilkins in his design for London University (now University College) of 1827–8. It consisted of a grand central Neo-Classical block with portico and dome raised on a podium and two monumental wings enclosing an open space on three sides. The organization of the building bore no relation to the academic activities it contained.

By the end of the 19th century the university curriculum had broadened, and university buildings included a wide range of specialized spaces. The lecture theater had appeared as a new and distinctive element in all university teaching buildings as this form of instruction became the norm. The lecture theater has its origins in medicine as theaters for the demonstration of anatomical dissections. Sir Christopher Wren (1632–1723) built one in 1689 for the Royal College of Physicians which consisted of a regular 16-sided space of 40 ft. (12 m) internal diameter with steeply raked seats and a demonstration table in the center. The theater built at the

Royal Institution in London (c. 1800) for the demonstration of scientific experiments was semicircular in plan, measuring some 60 x 40 x 30 ft. (18 x 12 x 10 m) and with raked seating. Here, the principle of direct sight and sound paths between audience and speaker was established, and in 1909 Guadet published a range of lecture theater plan forms for various academic subjects. The idea of research as one of the basic functions of a university had also become established by the end of the 19th century, and the scientific laboratory and workshop became part of the normal range of university buildings. The form of student housing also changed. Many of the newly founded universities were local and hence not residential. Of the others, few contained the collegiate pattern of combined residential and academic life. Instead, the student dormitory and the hall of residence established themselves as university building types, with the principle of students living outside the college in lodgings.

The form of the 19th-century university was essentially monumental and historicist. Contiguous Gothic structures around quadrangles, still inspired by Oxford and Cambridge, was one form. The University of Chicago designed by Henry Ives Cobb (1859–1931) in 1893, is the finest example in the U.S: its enclosed spaces formed by sculptured Gothic walls connected by towers and gateways emulate the character of its 15th-century precedents. The Neo-Classical campus plan designed around vistas remained another form, as in the University of California at Berkeley and the University of Birmingham in England. Both were conceived in monumental master plans which were never realized. Beginning in the late 1920s university buildings began to free themselves from historicist if not monumental influences. Perhaps the most significant plan was that of Mies van der Rohe (1886–1969) for the Illinois Institute of Technology.

Following World War II, the demand for university education increased dramatically, the existing universities expanded, and entirely new universities were established. The design of these was based on a more functional attitude toward the activities of a university, and reflected an academic desire to break down the barriers between schools of study. The need for growth and change was taken very seriously. The planning concepts very much followed those prevailing in town planning, tending away from the campus plan to more compact layouts, to the segregation of traffic and pedestrians, and to the use of structuring devices for the overall organization of the university. In England, between 1958 and 1960 the construction of seven new universities began. The variety of adopted forms exemplified different planning concepts. Sussex was designed as a series of precincts of individual buildings set in parkland; York as a series of buildings linked by a covered way arranged around a lake: Essex, designed on a linear plan, extendible at both ends—like Simon Fraser University, Vancouver; and Loughborough designed not as a group of buildings but as a single building type with a universal space for teaching and research buildings, the whole structured by a disciplining grid. The Free University of Berlin, designed c. 1960 by Candilis, Josic, and Woods, gained considerable attention as a prototype. It is organized by a grid of pedestrian routes connecting low-rise buildings in a very compact and highly interconnected form. The buildings are serviced by underground passages, and the whole complex is deliberately unfocused and unmonumental.

During this period each of the constituent building types of the university became the subject of research and their design evolved into highly specialized forms. Student residential buildings became more independent, the study bedroom closely designed around the imagined activities of a student. In 1921 Le Corbusier (1887–1966) produced a scheme which provided each student with the experience of an attic studio but his plans never materialized. Scandinavia contributed the innovation of grouping student rooms around a communal kitchen and social space. The open-access library was invented and evolved a specialized form. Research laboratories became another highly developed building type and highlighted the enormous problems of functional obsolescence in university buildings in that they, the most expensive, are the least permanent.

Since that period, ideas about the university can be said to have focused on its relationship to the city. There has been regret for the location of new universities in isolated locations. The scale of some new universities is that of a city in itself; others, like those in Cambridge, Massachusetts, have generated cities around them. It is now generally accepted that the dissemination of knowledge, which has become one of the major activities of our age, should not take place in isolation.

Museums

The museum is a Western concept which came into being with the collections of objects of art, antiquity, and natural science begun during the Renaissance. The essential function of a museum is the display of a collection of objects so that it may be viewed in sequence by circulating visitors.

The early private collections

The first collections were housed in the Renaissance palaces of their collectors. Long, wide passages, known as "galleries," were used to display sculptures. Paintings and other objects were accommodated in rooms which opened into each other in a continuous sequence. These types of spatial organization suited the needs of viewing, and the continuity of the route has remained a guiding principle of museum design ever since. It can be seen in designs which are otherwise radically different to the original palace galleries, as for example in the spiral form of the Guggenheim Museum in New York, designed by Frank Lloyd Wright (1869–1959), which was built in 1956–59.

The display of objects was at first determined by the taste of the collector, and was not necessarily arranged systematically. However, with the Age of Enlightenment in the 18th century, collections tended to become more specialized and were arranged into categories of objects. The viewing of these early collections was arranged privately, and usually required letters of introduction. One of the first collections to be publicly owned and open to the public was the Hans Sloane collection, acquired in 1753 by Parliament for the original British Museum. But visiting by the public was restricted to three hours a day, and formal application had to be made for admission.

The development of the civic museum

During the 19th century many of the great private collections of Europe and America were made available to the public, either because of the development of democratic principles as in America, Britain, and France, or through the enlightened despotism of rulers of the German States. Many great institutions were founded and built, such as the Altes Museum in Berlin (1823–30), the British Museum in London (1823–47), and the Metropolitan Museum in New York (1874–80). The Berlin Altes Museum was designed by Karl Friedrich Schinkel (1781–1841), who expressed the new public function of a museum by making the whole of the front facade into an entrance. The importance of the public entrance was a feature of many of the new institutions.

A great expansion of museum collections and buildings took place during the second half of the 19th century. Specialist institutions for science, natural history, fine arts, and applied arts were created. Two notable examples are the British Natural History Museum of 1871–81, designed by Alfred Waterhouse (1830–1905), and the American Natural History Museum in New York designed by J. C. Cody in the 1870s. During the same period a series of international exhibitions, such as in London in 1851, Paris in 1889, and Chicago in 1893, popularized the viewing of collections of objects and affected the design of exhibition spaces and the display of items. (See EXHIBITION BUILDINGS.)

Interior of the Guggenheim Museum, New York (1959), by Frank Lloyd Wright.

The British Museum, London, England (1823–47). The classical facade was designed by Sir Robert Smirke, the portico and the dome of the reading room by Sydney Smirke.

The World War I Gallery in the Smithsonian Institute, Washington D.C. The institute was started in 1847 by James Smithson and James Renwick.

Display in museum design

Methods of displaying objects followed two main trends, often at the same time. The first trend regarded a museum as a storehouse of great treasures: this applied particularly to collections of paintings, sculpture, and the applied arts. In terms of the display the totality of the collection rather than the individual work of art was emphasized. The second trend was concerned with education and placed great emphasis on the label, the written description of the object, and its origins. This trend applied to collections of natural history, science, and antiquity, and such a museum could be described as a series of illustrated labels. The late 19th-century museum, therefore, tended to be composed either of extremely long galleries or of huge spaces with side galleries and top lighting, and was filled with innumerable display stands and cases containing labeled objects.

The great innovatory reaction to this concept of museum design was not fully realized until the reconstruction of many Italian museums after World War II. The museum renovations in Genoa, Florence, Verona, Venice, Milan, Palermo, and elsewhere by Franco Albini (b. 1905), Studio BBPR, and Carlo Scarpa (b. 1906), were carried out mainly between 1950 and 1965, and emphasized the visual uniqueness of the works of art on display. The arrangement of items became less crowded so that objects could be seen separately. Display supports or backgrounds were often designed individually and labels were kept to a minimum. These renovations have generally been carried out within the original museum spaces, which again underlines the appropriateness of the Renaissance palace plan type. This recent Italian innovation has now influenced museum display in most countries and has, in general, helped the ordinary visitor although it has occasionally annoyed the specialist. The danger of the Italian approach is an overemphasis on design at the expense of the object itself.

The Italian contribution to change in museum design was confined to the display of works of art. Its application to museums of science and technology was less direct, and here the main change has been due to the introduction of alternative means of visual communication. In museums of art, ethnography, and natural history it is important to see the original object rather than a facsimile. This is less true when communicating ideas in science and technology. It may, in fact, be helpful to use other methods besides the display of the products of science and technology, such as charts, films, recorded talks, multiscreen presentations, and similar devices, and moreover to allow visitors to manipulate and test machinery specially created for the purpose. The Evoluon at Eind-

hoven in Holland, founded in 1966, is a good example of such a museum, and several temporary exhibitions designed by Charles Eames (b. 1907) make use of similar methods.

All these changes in display and presentation have their own specific design requirements, but may not necessarily affect the design of the museum as a whole. The large-scale architectural considerations have been most affected by the increased range of subject matter of museums. In Lucerne, for example, the Transport Museum houses railroad engines and paddle steamers; at Beaulieu in Hampshire, England, a whole museum complex is devoted to automobiles and motorcycles. In Stockholm, The Waasa—a timber warship raised from the bed of Stockholm harbor—was enclosed within a large structure and had to be continuously sprayed for several years to allow it to dry out gradually. In Washington D.C., the National Air and Space Museum contains aircraft suspended from the roof and space exploration craft. Such material poses architectural problems different in both scale and character from the exhibition of paintings and sculpture in rooms that might at one time have been those of a Renaissance palace.

A further extension of the scale of museum design has occurred with the creation of whole sites as museums. The first example of this, which has since been widely copied, was Skansen in Stockholm. A whole island was layed out in 1891 as a zoo for national animal species and a museum of vernacular building types and national folklore. A more recent example is at Ironbridge in Shropshire, England, where artifacts of the early Industrial Revolution were manufactured and still exist. The design problem of these large-scale thematic museums is to make a coherent display of a dispersed group of elements.

In most cases, however, the crucial problem is the design of the middle- and small-scale parts of a building which form the immediate background to the exhibits. It is the combination of such details with the idea of a route that provides the distinguishing features of a museum. These details involve the wall and its ability to accept fixings, or the design of rails from which objects can be hung, the provision and control of natural and artificial light to deal with a wide variety of conditions depending on the type of display, the floor, and its ability to take supports for screens or to have power points for servicing showcases, and, most important of all, for walls and floors to be clear of elements such as grilles, heaters, pilasters, security points, and light switches, which can cause visual interference with the viewing of objects.

Lighting

Currently a major problem of design concerns

methods of lighting, and is caused by the serious limitations placed on the display of many objects by the needs of preservation. All organic materials containing carbon—textiles, paper, ivory, leather, feathers, wood, and many pigments—are subject to varying rates of deterioration under the action of light. The amount of deterioration is proportional to the intensity of light and the length of exposure. Recommended lighting levels for the preservation of different materials are very low—50 lux for fabrics, watercolors, and similar sensitive exhibits, 150 lux for oil paintings. In comparison, the normal lighting level for offices is 300 lux. In a museum the eye has to be gradually accustomed to the required lower lighting levels through the organization of the space and the control of daylight.

This need has led to the design of elaborate methods of reducing daylight by deep baffles and overhead louvers, sometimes automatically controlled to prevent the light intensity exceeding a given amount. These devices tend to occupy a large proportion of the building volume and are also expensive. The simplest solution to the lighting problem has been to omit daylight altogether and to rely entirely on artificial sources of light. The Hayward Gallery in London (1968), demonstrates both approaches: the upper galleries have natural overhead lighting controlled by a complex louver system, while the lower galleries are windowless, and artificial lighting is reorganized to suit each new exhibition.

There has been a discernible reaction to both these solutions—the museum as a machine controlling daylight and the museum as a black box—and alternatives are being explored in which deep rooms receive some side lighting through windows. This would provide a sense of daylight but much of the actual illumination would come from artificial sources selected for the specific lighting needs of particular objects.

Museum and exhibition design has had to respond to the pressures of an increasing number of visitors. The need to keep museums open through large parts of the day, irrespective of natural lighting conditions, has made it necessary to include artificial lighting in any design. Other ways in which museum design has adapted to the pressures of numbers are by enlarging and reemphasizing the route through the exhibition, and by reorganizing many service functions—from postcard counters to toilet facilities. In recent years many exisiting museums have reorganized their entrance halls and restaurant areas to accommodate the crowds of visitors which may attend during the weekend and peak evening hours.

The organization of space

Besides these public spaces there are other areas which are crucial to the proper functioning of a museum. These comprise storage, conservation, research, and administrative areas. These functions may occupy as much as one-third to one-half of the total floor area of the museum building.

The ways in which we look at objects and the messages we seek from them change in time, as can be seen from an analysis of museum design over the last two centuries. The main architectural problem is, therefore, one of creating a permanent enclosure together with certain service functions, while at the same time allowing for variety and change in settings to fit the content and nature of particular displays.

These requirements have recently been interpreted as an argument for anonymous flexible space, which has therefore been seen as the main innovating requirement of museum architecture. Two major examples of this approach are the Centre Pompidou in Paris (1977), by Piano and Rogers, and the Sainsbury Center For The Visual Arts in England (1978), by Foster Associates. These buildings provide large empty spaces. The spatial organization of exhibitions is seen as a separate task from the design of the building which does, however, provide the technical infrastructure for servicing the display of objects. But no space is genuinely anonymous; in the Paris galleries, for example, the deep beams, the exposed services, and the views of the surrounding rooftops are all visually strong and specific. Also, no single space is flexible over the entire range of possibilities; an artificially lit enclosure does offer a great number of possible ways of arranging a route and screens, but does not allow the arrangement of exhibits so that they are seen in silhouette against natural sunlight and foliage.

Other recent approaches to museum design have also been concerned with variety and change in exhibition settings, but have tended to concentrate on providing the most appropriate settings for particular collections. An interesting example of this is the J.P.Getty Museum in California (1974). Here the building is a reconstruction of an ancient Roman villa. The original spaces of such a villa provide a great variety of settings for objects—open and covered exterior spaces, side-lit, top-lit, and totally windowless interior spaces. The ancient plan and volumes have been ingeniously adapted to accommodate modern museum technology, including visitor parking lots, air conditioning, and security systems. Another museum which seeks to provide a variety of types of space and to include natural settings is the Oakland Museum (1968) by Kevin Roche (b. 1922). A further innovation of this museum is that it is designed to integrate with the surrounding area and community, and can be entered in an

Oakland Museum, California (1969), by Roche, Dinkaloo, and Partners.

Interior view of museum space in the Centre Pompidou, Paris, France (1977), by Piano and Rogers.

informal way at many different points. In conclusion, flexibility in museum design may often most usefully consist in the provision of a variety of spaces covering a wide spectrum of potential museum needs. Each space may be suggestive of certain purposes but also allow some alternative use.

Laboratories

The laboratory was originally conceived simply as a work place, although it has now come to signify a room, building, or complex of buildings devoted to experimental or routine procedures associated with the sciences, or the teaching of sciences. The history of laboratory buildings before World War II reveals little as to the nature of actual or future laboratories. The traditional pattern of laboratory layout was established around 1800, but with the dramatic changes in science research and teaching techniques since the 1940s, laboratory buildings widely disparate in nature have emerged.

In recent years, primarily as a result of massive government support, science has become an establishment. At the forefront is "Big Science"—a label given on the one hand to public enterprise that builds, for example, particle accelerators, and on the other, to private enterprise that constructs vast laboratory installations for endless routine control procedures. Between these extremes fall the college-based teaching and research facilities that range from very basic to highly sophisticated buildings.

Public and private "Big Science," the universities, and hospital-based research laboratories have this in common; they are all threatened by obsolescence at an accelerating rate. The life expectancy of a body of knowledge in small particle physics, for example, has been estimated at no more than four years. The model rate of revision for all sciences is now probably about 15 years, and yet laboratories are probably the most technically sophisticated of all buildings, and are generally expected to last at least 40 years. Acknowledgment of the question of obsolescence is not recent, however. Serge Chermayeff (b. 1900), in his laboratory designs for Imperial Chemicals Industry (ICI) Dyestuffs Division at Blackley in England (1935), produced many now familiar motifs. One such solution was the installation of benches at right angles to external walls for both ease of servicing and segregation of less costly office space from the expensive laboratories themselves. Although many similar solutions have since been devised, none have completely overcome this major problem.

While it is true that some laboratory buildings have been designed by such well-known architects as Louis Kahn (1901–74) and Eero Saarinen (1910–61), in terms of the evolution of solution types, skillful pragmatism has been more of a virtue than the expressive embodiment of science. The buildings are so expensive to construct and maintain that the successful synthesis of different engineering systems and complex planning requirements is more important than the packaging. In fact, most of the key ideas in laboratory design since World War II have been dressed up in a wide variety of architectural fashions.

Early designs

During the 1940s and 1950s the emphasis in laboratory design was on applied building science in the service of the individual laboratory worker or group of workers. Daylighting studies and mechanical engineering predominated to create well-serviced work stations. The Bell Telephone Laboratories in the U.S. (1941) exemplify this period. Here, all the services are distributed in a matrix of horizontal and vertical ducts in the external walls, with branches carried in floor ducts to bench positions. However, the building structure makes few concessions to its complex burden of pipes, wiring, and convectors. Other laboratory buildings ran the services in ducts lining a central corridor. Although such buildings were, by contemporary standards, heavily serviced, no special logic was invoked to relate the working environment to the services provided. "More is better" was the acceptable axiom of servicing demands of scientific research. Nevertheless, the Bell Building exemplified both American and European practice for nearly 20 years.

British rationalism

In Britain in the 1950s, laboratory design was the concern of the Nuffield Foundation for

Richards Medical Research Center, Philadelphia, Pennsylvania (1960), by Louis Kahn.

Architectural Studies. Detailed studies of laboratory bench activities, daylighting, and artificial lighting, and of the ease of installing and maintaining services were carried out. All these studies informed a great boom in academic laboratory building in the late 1950s and 1960s. Laboratory work increasingly demanded a controlled environment, and the need for air handling and distribution prompted theories of rational services distribution. The key notion was of planning based on regular grids within which functions would be defined and parts made interchangeable.

Interest in modular coordination and industrialized building was not confined only to the laboratory design field. In some designs, services were installed above or below floor slabs. In others they were grouped in vertical structural ducts; these were often positioned at the perimeter of the building. Alternatively, a double or "tartan" grid was used with wide bands for corridors and narrow bands for services. The most notable example of this type is Loughborough University of Technology designed by Philip Dowson in 1967.

American rationalism

Architects in the U.S. were less attracted to theories of services distribution. Taking air conditioning for granted, they began in the early 1960s to go directly for solutions which put ease of servicing on the same level of importance as occupant requirements. Two buildings by Eero Saarinen (1910–61)—the International Business Machines (IBM) Research Laboratory at Yorktown, New York (1961) and the Bell Telephone Laboratories at Holmdel (1962)—collect all services into large linear vertical cores feeding laboratories either side. These double bands of laboratories alternate with double bands of offices reached from the laboratories across a corridor—a people duct. The work spaces in these buildings have no windows; rest areas with windows are provided instead, an arrangement which apparently suits the occupants but which won U.S. laboratory design few friends in Europe. The grouped services design most admired by architects was not the essentially horizontal scheme of Saarinen, but that of Louis Kahn (1901–74) in his Richards Memorial Medical Research Center. This building has windows, but the researchers have obscured many of them themselves. Kahn's dislike of services and his overriding obsession with his symbolic intentions also led to needlessly complex distribution patterns.

Probably the most elegant technical designs of the period are the State University College at Cornell (1969), designed by Ulrich Franzen (b. 1921) and the Salk Institute for Biological Sciences at La Jolla, another of Kahn's

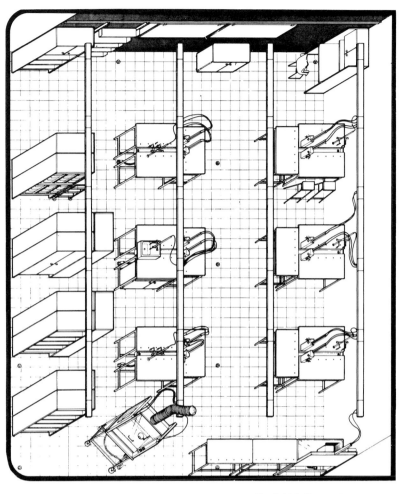

The British Sintacel Metriscope range of movable laboratory benching serviced from above illustrates a general trend away from built-in furniture toward systems that can be adapted by users. In this system liquid waste can be pumped away automatically via the overhead boom.

designs. The former is a six-storey windowless tower of agronomy laboratories with very clear and simple articulation of structure and services—alternate coffers of standard T-planks are occupied by air-distribution or bench services fed from perimeter main ducts. The Salk Institute provides a loft space of 245 × 65 ft. (75 × 20 m) with all services carried in an interstitial space within a Vierendeel truss floor structure; offices are housed in "pods" plugged into one side of the three-storey building.

These two buildings may be seen as the end product of the first major thrust of research into science buildings. They embody the fruits of both British and American experiments. But both were very expensive to construct.

Indeterminacy

Many of the laboratories and experimental facilities of Big Science have become highly particularized, but the bulk demand for laboratories is for teaching and research. The rapid rate of growth and change in teaching and research ensured the obsolescence of older buildings in postwar years and the impossibility of pinning down needs for long

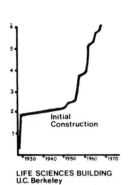

LIFE SCIENCES BUILDING
U.C. Berkeley

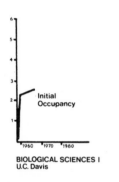

BIOLOGICAL SCIENCES I
U.C. Davis

REVELL COLLEGE BUILDING B
U.C. San Diego

Graph of cumulative capitalization
of laboratory buildings built in
1928, 1958, and 1960 show cost of
coping with changing needs. Most
recent building incurs alteration
costs as soon as it is completed,
whereas 1928 laboratory enjoyed
20 years of stability.

enough to realize new building that could meet them. Poor utilization of space and very costly upgrading and alteration were the consequences.

Laboratory designers reacted in two ways. First, there was an interest in improving space utilization through better understanding of teaching patterns and research practice. This could be described as a management approach to laboratory design, and it coincided with a more general optimism about the value of scientific management in the late 1960s.

Second, rather than simply coping with demands for change, some architects set out to design with future change in mind. Flexibility, indeterminacy, and adaptability became the watchwords in laboratory design. Flexibility lay in the choice of construction technology rather than in modifying users' behavior patterns. Much work in this field was done by the Educational Facilities Laboratory in the U.S. and by the Laboratories Investigation Unit in Britain. They concerned themselves with laboratory furniture, equipment, and services that would be more responsive to short-term changes of use and layout than the traditional fixed benching.

However, flexibility led to increased costs, and was therefore not a complete answer. Furthermore, many designers lost sight of the original aims of flexibility in pursuit of clever technical devices which might never be used or, more often, which tied the future usability of the building to some contemporary gadgetry. Redundant flexibility could be tolerated in prestige buildings (such as the Salk Institute), but the mass market demanded an equation between capital cost and the cost of operating and altering its less sophisticated laboratories.

The new pragmatism

An important study of the criteria for cost effectiveness in laboratory buildings was carried out for the University of California by Building Systems Development (BSD) in 1970. The study focused on bioscience teaching and research laboratories where the problems of rapidly changing requirements were, and still are, acute. An examination of the cost histories of many laboratories built from 1928 onward showed a progressive shortening of the period of routine expense incurred following completion. Buildings started in 1966, for example, incurred continuing costs of nearly half the initial construction spending rate as soon as they were completed. Analysis of the costs indicated that plumbing and electrical work, and the disruption caused to adjacent spaces, were the major cost components.

From these analyses, BSD developed an approach to laboratory design aimed at minimizing the total owning cost over the

building's life. The fundamental premise of the approach is that since plan requirements change faster than basic technology, a sensible approach is first to optimize the construction technology according to its intrinsic properties (the most economic spans for structures, the optimum volumes for air-handling units, for example), and then to test the result against a wide range of possible plan configurations. This approach led to the "space module" concept—large-scale (10,000–12,000 sq. ft./930–1,116 sq. m) building blocks of serviced space which can be arranged in many different configurations but which always obey the logic of efficient use of the construction technology. Some elements such as structure, main ducts, and ceiling grid, are treated as permanent, while lighting, partitions, and secondary services are designed for ease of alteration. A key feature is the strict zoning of the services distribution in the ceiling space.

This approach, in contrast to earlier, more ideological solutions is pragmatic because its aim is to find an optimum balance, based on life-cycle costs, between general requirements and specialized ones; for example, animal houses and cold rooms remain outside the scope of the system. In this way redundancy of flexibility is minimized. The BSD approach has to date found most widespread application in the U.S. Veterans Administration hospitals of Loma Linda (1977) and Martinsburg (1978), a building type with similar problems to laboratories.

A counterpart for this style is also found in the recent work of John Weeks in Britain. An early prophet of indeterminacy, he now advocates a minimum of technologically "special" construction to achieve a long-term chance to fit between the majority of laboratory activities and the building. Special needs may always be met on an ad hoc basis.

Future designs

Increasing automation or encapsulation of routine. Laboratory work continues to diminish the need for special design or building techniques, particularly since standards of environmental and services design for non-laboratory buildings have risen to meet present-day expectations. Many speculative office buildings could ably meet needs which previously required a laboratory environment. This is particularly true since the advent of adaptable and movable laboratory furniture and equipment. In a similar way, present-day school and junior college teaching space amply meets the needs of basic science teaching with its emphasis on prepackaged experiments and similar teaching aids.

The frontiers of laboratory design may be found in research laboratories, particularly for multidisciplinary research and in the housing

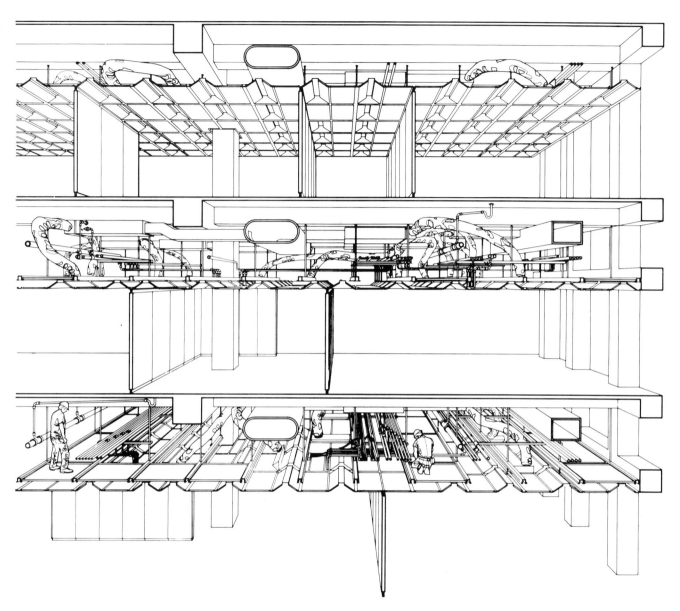

of large-scale testing and experimental equipment. However, the latter, as in the case of particle physics and astronomy, may more accurately be classed as pieces of equipment, (which may or may not require to be enclosed) than as laboratories. They will inevitably continue to throw up the need for buildings where the cost of enclosure will be insignificant alongside the cost of the equipment. Elsewhere, the demand for energy conservation, together with the need for environmental servicing, will probably ensure that the deep-plan, small-window format for research laboratories will persist for some time, although an increasing number of commonplace laboratory functions will probably be adequately housed in quite ordinary and habitual building forms.

The Academic Building System of 1970 for the Universities of California and Indiana was developed for the complete laboratory requirements of science and engineering teaching and research. The system is a set of design rules aimed at reducing the cost penalty of unforeseeable change. All services are pre-coordinated; no two component systems try to share the same space. Contrast with the complex integration of structure and services of earlier design approaches.

Entertainment and recreation

Theaters

Ancient, classical, and medieval theaters

Archaeologists have found evidence of liturgical drama in ancient Egypt but its form can only be surmised. It was the Greeks whose drama first reached a pitch of achievement which persisted as a major influence, even to the present day. Their dramatic performances were part of religious festivals staged in the open air, originally in natural amphitheaters, with the audience sitting on a hillside and a flat circular area (the "orchestra") cleared for dancers and singers. The lead was taken by the Athenians and it was in the theater of Dionysus on the sloping side of the Acropolis at Athens that plays by Aeschylus, Sophocles, Euripides, and Aristophanes were first performed in the 5th century BC.

One of the best preserved and most beautiful of Greek theaters is at Epidaurus (c. 350 BC), where the audience of up to 10,000 sat in an extended semicircle around the circular orchestra. Behind this was the *skene* (hence "scene"), a raised platform with openings in its rear wall for actors' entrances. The main setting was still the open landscape behind. Theaters approximating to this layout are found in all the ancient Greek or Hellenistic cities around the Mediterranean.

Greek drama was the basis of Roman drama, but by the time the Romans began building theaters of their own, the religious content had virtually disappeared; the later Greeks introduced very broad comedies of manners which the Romans translated and developed.

The typical Roman theater was still open to the sky, but enclosed by walls. The audience now sat in an exact semicircle about the semicircular orchestra, and the stage had an elaborate architectural background of columns and portals in two or three tiers. A theater built under Pompeii (c. 55 BC), was the first stone-built theater to which Roman engineering skill and architectural taste were applied, and the pattern was repeated all over the Roman world; there is a well-preserved example at Orange in the south of France (c. AD 50).

After the collapse of the Roman Empire theater building ceased, but drama gradually reemerged, again within the context of religion. This time, Christian religious ritual was accompanied by mystery plays, recounting Bible stories which were enacted in cathedrals, churches, or in the streets leading to them. Sometimes actors performed against tableaux mounted on horsedrawn carts, or on temporary wooden stages.

Renaissance theater—the Italian influence

As secular elements, in the form of comic interludes, began to take an increasingly important part in the Church's drama, this, together with Renaissance interest in the classical past, led to a new kind of theater. In the courts of the late 15th-century Italian city states there were productions of rediscovered Latin comedies and the tragedies of Seneca, but the presentation was medieval rather than classical, with the action on a raised stage.

At the same time, there was a growing interest in the ruins of Greek and Roman theaters and in ancient writers, such as Marcus Pollio Vitruvius (active 46–30 BC). When Sebastiano Serlio (1475–1554) published his *Architettura* in 1551, it contained a plan for a theater in the classical manner; a semicircular theater surrounded a flat open space—corresponding to the ancient orchestra—connected by steps to a raised stage.

The particularly Italian contribution was the use of elaborate perspective painted on canvas-covered wooden frames, and eventually this led to more complex stages. While stage machinery had been used from the earliest times, the Italian perspective settings needed a special system for changing scenes. This was achieved by locating painted wing-flats in grooves along which they could slide back and forth in front of a painted canvas backdrop. Flying scenery above the stage and the use of trapdoors and elevators beneath it were further developments.

The Teatro Olimpico at Vicenza in Italy, built to the design of Andrea Palladio (1508–80), and completed in 1584 four years after his death, stands as a good example of the early Italian theater. Its permanent architectural background in the Roman manner has three "portals," with streets in false perspective behind them. The large central portal developed in later buildings into the familiar proscenium arch, with doors for the performers at the sides. The first permanent proscenium theater, the Teatro Farnese at Parma (1617–28), set the pattern not only for the typical Italian theater, with its ornate proscenium and tiered boxes around shallowly sloping stalls, but also for most European theaters for the next 300 years. This was largely due to the widespread influence of the Italian Baroque, which itself put great stress on the theatrical arts. Opera, which virtually fused all the arts, developed directly out of Italy's choral and musical tradition, and in the theater there was as much emphasis on the art of the scene painter as on that of the dramatist.

Some of the greatest achievements of Baroque architectural thought were to be found not in stone and brick but painted on canvas flats. Some artists worked on both; one remarkable 18th-century dynasty of designers, the Bibienas, not only designed marvelous stage settings, they also built some beautiful Italian theaters one of which, the Margrave's

Roman theater at Orange, southern France, designed for 7,000 spectators.

View of auditorium from the stage of the Teatro Olimpico, Sabbionetta, near Mantua, Italy (1588), by Vincenzo Scamozzi. Scamozzi also completed the Teatro Olimpico started by Andrea Palladio.

Opera House at Bayreuth, survives intact. In England too, Inigo Jones (1573–1652) sowed the seeds of his architectural career as a designer of court masques.

While the Italian emphasis on the picture inside the frame pushed the actors behind the proscenium, the English theater retained the tradition of a forestage until the 19th century. Also, most English theaters were much smaller than continental ones, and there was more emphasis on acting than on spectacle. The theaters that began to be built in America in the 18th century followed this pattern.

English theaters

England's first permanent playhouse as such was built in 1576 by the actor James Burbage, just east of the City of London. Toward the end of the century, Southwark in London became the center of theatrical activity, in response to the extraordinary flowering of dramatic writing culminating in the work of William Shakespeare (1564–1616). We have, however, very little definite knowledge of these theaters, even of the Globe which Shakespeare helped Richard Burbage (c. 1567–1619) found in 1599, for none has survived. What is certain is that a raised thrust stage projected into the middle of a space surrounded by three tiers of shallow balconies. These balconies and the stage were roofed, but the yard in the middle where the "groundlings" stood was open to the sky. The stage was joined to one side of the structure and actors made use of its storeys as part of the permanent setting. Only in the 17th century was there a trend toward properly roofed buildings.

Theaters from the mid-18th century

In Europe, the stylistic lead eventually passed from the Italians to first the Germans, with the full flowering of Rococo decoration in, for example, Munich's Residenz Theater (1753), and then to the French, who were challenging so many established ideas at the time of their Revolution. Exteriors became more grandly classical and the theater itself became a separate monumental element, its foyers and approaches now as important as the auditorium. The Grand Theater at Bordeaux, built between 1773–80, was aptly named, with a grand colonnade leading to a magnificent staircase in a great hall the whole height of the building. This was the style which dominated the European theaters of the 19th century, culminating in the Paris Opéra of 1875, designed by Charles Garnier (1825–98), a palace of fantastic splendor decorated with marble, sculpture, mosaic, and gilt.

Backstage areas increased in size and complexity in order to make possible the use of increasingly more elaborate scenery. Fly

Reconstruction drawing of the Fortune Theatre, London, England (c. 1588).

towers, lateral wings, and undercroft areas below the stage accessed by trapdoor, together with ancillary workshop areas and greenroom facilities, began to constitute a high proportion of the total volume of the theater. Furthermore, the need for rapid scenery changes led to many special devices apart from the use of flying sets; these included the use of mobile sets on tracks, and arrangements of scenery on turntables.

Throughout Europe the erection of an opera house became the symbol of civilization. The rows of private boxes, with the central royal box, and the parterre for lesser folk, suited the social climate; people came as much to be seen as to see, and the lighting in the auditorium was as strong as that on the stage.

This did not suit Richard Wagner (1813–83), who determined that the audience must become immersed in his music dramas without being distracted by their fellows, or even by the sight of the musicians. With the help of a local architect, and the financial backing of King Ludwig of Bavaria, he opened his Festspielhaus at Bayreuth in 1876, a year after the Paris Opéra—and in total contrast to it. Gone are the rows of boxes, the exotic decorations, and the plush seats; instead the audience sits on cane seats in a single, steeply raked, fan-shaped tier, while the large orchestra is hidden mostly under the stage. The acoustics are extremely successful.

In the 19th-century theaters of Britain and the U.S. the boxes had retreated to the sides and the tiers were more open, though interrupted by columns supporting the balconies. It was not until the 1890s that a system of cantilevered balconies allowed the elimination of columns and made possible deep tiers with improved sight lines for large audiences.

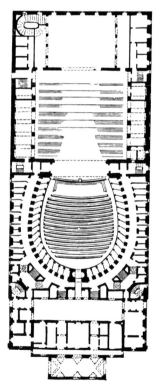

Plan of La Scala, Milan (1778), designed by Giuseppe Piermarini. The auditorium is surrounded by seven tiers of boxes. Behind the proscenium arch there is an area almost the size of the auditorium used for stage sets and flats.

Interior view of foyer, Paris Opéra, France (1875), by Charles Garnier.

New York State Theater in the Lincoln Center complex (1964), by Philip Johnson.

Opera House, Sydney, Australia (1957–73), by Jørn Utzon and engineers Ove Arup. The complex houses two major theaters and other facilities.

Other technical innovations included gas lighting, which allowed much greater control than candles and oil lamps, while electricity later brought still more intensity and flexibility. Dankmar Adler and Louis Sullivan's capacious Auditorium Theater in Chicago (1889) was one of the first to introduce electric lighting, with thousands of carbon filament lamps giving the gilded auditorium a magical atmosphere. The Auditorium Theater also enjoyed a form of air conditioning.

Fire was a constant scourge. The greatest danger was on stage, where there was a constant risk of candles, gas, or electric lights igniting hanging canvas and wooden frames. Safety measures gradually introduced from the end of the 18th century included an iron-framed safety curtain to shut off a stage fire from the auditorium, drenchers and sprinklers over the stage, and smoke ventilators in the roof. Means of escape for the public was also the subject of legislation from the end of the 19th century.

The 20th century

In this century, the traditions of the 19th-century theater allied to new safety regulations became inhibiting to many directors and dramatists. They reacted against over-elaborate scenery and effects, the confines of the picture frame, and the conventions forced on actors in the very large-capacity theaters which, for commercial reasons, were built in the Victorian Age. In 1910, for example, Max Reinhardt staged an in-the-round production of *Oedipus* in an old Berlin circus building, later converted by the architect Hans Poelzig (1869–1936) into the Grosses Schauspielhaus, with the audience on three sides of the stage.

Many experiments in open stage productions were made by Vsevolod Meyerhold (1874–1942) in Russia, Jacques Copeau (1879–1949) in France and New York, and Erwin Piscator (1893–1966) in Germany. It was Piscator who briefed Gropius for the "total theater" project which, though never built, has had a great and not altogether beneficial effect on design theory. The desire to experiment with audience-to-actor relationships has, in a machine-dominated age, inevitably led to attempts to design buildings which can be transformed by machinery. A more successful approach has been to design a complex with two or three auditoriums, as in many projects built since 1945: at the Lincoln Center, New York, designed by Philip Johnson (b. 1906) and others; the National Arts Center, Ottawa; at Mannheim and Düsseldorf in Germany; at the National Theatre in London, designed by Denys Lasdun (b. 1914); and at the Sydney Opera House, designed by Jørn Utzon (b. 1918). These are on the grand scale, but the many more modest theater buildings erected in

Europe and the U.S. since World War II indicate that the theater is still very much alive.

A recent design concept which deserves special mention is the series of thrust-stage theaters inspired by the theater director Sir Tyrone Guthrie (1900–71). He first adapted a non-theater building, the Assembly Hall in Edinburgh, and was then invited to Canada to advise on the Festival Theater in Stratford, Ontario. This was followed by Powell and Moya's Chichester Festival Theatre in England and the Tyrone Guthrie Theater in Minneapolis. They vary in detail but each has a large amphitheater with a thrust stage surrounded on three sides, and a back wall to the stage which can be modified within limits.

(See also SERVICES.)

Movie Theaters

Early forms of cinematographic entertainment

The movie theater is essentially a 20th-century building type, although its early development took place in the 1890s. After inventing the phonograph in 1876, Thomas Edison (1847–1931) was keen to devise its visual equivalent and by 1889 he and George Eastman (1854–1932) had perfected a celluloid film strip whose sequential frames could be projected to give moving pictures. The first commercial use of this invention was in the Kinetoscope, a manually operated slot machine in which short films, usually of a risqué or slapstick nature, were projected for the individual viewer. Starting in 1892 these machines could be found in rows in penny arcades or Kinetoscope parlors.

In the next few years a number of people experimented with the projection of much larger moving images on a sheet or screen which could be viewed by many people at once. This first took place in public in Berlin and Paris in 1895, in London at the Empire Music Hall in 1896, and in New York at Koster and Bial's Music Hall, also in 1896. Because of this early association, the design of the first movie theaters was based on that of contemporary music halls: the auditorium faced a stage (for entertainment between films), with a proscenium arch and orchestra pit, and connecting with the street was an entrance lobby with a box office and candy kiosk.

The dramatic popular success of early movies was, however, more connected with the development of amusement arcades than with this type of theater. By 1900 projection equipment was more easily obtainable and cheaper, and in the Los Angeles area Thomas L. Tally set up a movie show at the back of an amusement arcade. So successful was this venture that he very soon took over the rest of the arcade as an auditorium and named it "The Electric Theater." This arrangement was quickly imitated around the world and was generally known as the "nickelodeon." By 1905 they were found throughout the U.S., and by 1907 there were 300 of them in New York City alone. Each consisted of a simple screen with a curtain, a space for a piano or small orchestra, seating for the audience, and advertising at the street entrance. The English equivalent was the "penny gaff."

The spread of the nickelodeon led directly to a demand for higher standards of public safety in movie theaters generally. In Britain the Cinematograph Act of 1909 specified the number of emergeney exits and the fire extinguishers that must be provided, also laying down as a requirement a fire-resistant wall between the projection room and the audience. Similar controls were adopted in the U.S. but varied from state to state.

Grauman's Chinese Theaters in Hollywood Boulevard, Los Angeles (1927), by Mayer and Holler.

American movie theater design 1910–30

After 1910 the design of movie theaters became more competitive, the general aim being to increase audience sizes, both by attracting those who previously would have nothing to do with nickelodeons and those who frequented music halls. Upholstered tip-up seats replaced simple benches, and lounge furniture was put out in the foyer. At this time, the main elements of a movie theater consisted of a highly decorated street facade with a barnlike hall behind it, accommodating a decorative interior.

Gradually the designers gained confidence. In the U.S., Thomas Lamb was instrumental in providing luxurious, spacious buildings that could be enjoyed by the public at popular prices. His approach was that of the classical or "hard top" school which gave the public the elegance of chandeliers, elaborate staircases, and fine draperies, and by 1921 he had designed over 300 movie theaters. Another designer, John Emberson, developed a different approach with "atmospheric" interiors reflecting the romantic and exotic character of the epic films of the day. The Capitol and Paradise Theaters, Chicago; Lowes Paradise, New York; and the Olympia, Miami, exemplified this approach. In Hollywood, the Egyptian Theater of 1922 and Grauman's Chinese Theater of 1927 are also of this type, but have large forecourts in order to accommodate the crowds on premiere nights. The size of theaters also increased; the Roxy in New York, designed by Walter Ahlschlager, had a seating capacity of 6,000.

European movie theater design 1910–30

The design of movie theaters in Europe, although following many of the same trends, was in some ways very different from that in the U.S.; it also varied from one country to another.

In Germany the traditional importance of the small opera house, with its wide aisle encircling the auditorium, resulted in a strong emphasis on the auditorium as a separate design element. This emphasis was clear in Oscar Kaufman's theaters of 1909 and 1911 in Berlin. German designers also realized that their theaters were primarily used at night and developed facades based on elaborate displays of neon lighting.

In France there was a general tendency to follow the American "atmospheric" pattern for large theaters, but with the audience at several different levels. But the French also developed small, specialist theaters whose design was specifically directed at certain sections of the public.

Finally, in Europe, the Modern Movement was beginning to influence theater design. The Skandia Cinema in Stockholm (1922) designed by Gunnar Asplund (1885–1940) is an early example of this influence, relying as it does on sound planning and simple decor. A later example is the Universum Kino in Berlin (1926–28) designed by Erich Mendelsohn (1887–1953) which has a horseshoe-shaped auditorium derived entirely from functional considerations.

The 1930s

The next major influence on the design of movie theaters was the advent of "talkies" in 1928. Suddenly it became important that the shape and finish of auditoriums should reinforce sound production and eliminate echo, and in these respects many of the older barnlike interiors were seriously deficient. Also, acoustic insulation against extraneous noise from outside the theater became important.

Most of these problems were successfully dealt with within a few years, and during the same period the quality of sound reproduction was improved with multicellular, high-frequency horn speakers replacing one simple loudspeaker. The overall result of these improvements was a significant increase in audience sizes. In London the Gaumont State of 1937 had a seating capacity of 4,000 and waiting space for another 4,000, and in Britain generally at this time the average seating capacity for a local movie theater was 1,400–1,800.

Besides these general developments, two new types of movie theater were developed in the 1930s: the news theater, and the drive-in theater. The news theater had existed previously but not in a specialized form; during this period many were built in urban locations wherever people might have time to spare. Very often the sites chosen were restricted and noisy, and required considerable planning skill to overcome these problems. The De Handelsblad Cineac in Amsterdam, built in 1934, by Johannes Duiker (1890–1935) was a good Modern Movement example of this type.

The drive-in theater was first developed in the U.S., usually in suburban locations, and consisted of a fan-shaped parking lot, with each parking space supplied with a loudspeaker which could be detached from a post and taken into individual automobiles. The screen was supported by a structure that could withstand extreme wind conditions. The screen itself had to be as large as possible in order to be seen from a distance; in a theater at Point Florida, Trinidad, the screen is 110 x 54 ft. (33 x 16 m) and the height of the whole structure is 74 ft. (22 m). Also included in the layout were rest rooms, snack bars, and play areas, and sometimes food was available on trays that could be taken into the automobiles. Obviously this type of theater was particularly suited to countries with a good climate.

Postwar movie theaters

After World War II, there was growing competition between the movies and radio and television which necessitated a fresh drive to attract larger audiences to movie theaters. During the war several technical advances had been made in cinematography, specifically in the development of new types of photographic emulsions, color systems, and camera lenses. These advances were eagerly taken over by the movie industry and as a result it was soon possible to project onto larger screens without any appreciable loss in picture quality.

In 1952 Cinerama was introduced with the aim of creating a powerful visual and aural

One of a large chain of English movie theaters; the Odeon in Burnley, Lancashire, England (1937), by Harry Weedon.

impact by using three synchronized projectors on a wide screen and stereophonic sound from five speakers. The following year 20th Century-Fox brought out the CinemaScope system which used an auxiliary anamorphic lens in front of the normal lens to project images of twice the width that was previously possible. The CinemaScope system with its deep curving screen and single projector proved to be more easily adaptable to most existing theaters than the Cinerama system with its three projectors.

The CinemaScope system is still widely used today; more complex systems have been developed, often to project 360° images, but these have so far had only limited use. Walt Disney's Cinerama of 1958 encircled the audience with a continuous image on a perimeter screen in a circular auditorium. More recently the "New York Experience" and the "London Experience" have combined still and moving images projected simultaneously, together with stereophonic sound, in order to create a complete atmospheric experience. One of the most advanced theaters of this general type is the Space Theater at the San Diego Hall of Science. Here the audience sit in reclining seats to view continuous images projected onto a tilted hemispherical screen which is the underside of the dome enclosing the auditorium. The systems in use here are the Spitz Transit Simulator and the Omnimax Film Projection System.

In recent years commercial pressure on the cinema industry has caused a reduction in the number and size of movie theaters. Promoters have subdivided existing theaters so that several small auditoriums under the same management can provide a diversity of film entertainment with flexibility of auditorium size. Automation is reducing the staff required for projection. Cinemation used by the Rank Organization has the complete program, auditorium lighting sequence, internal

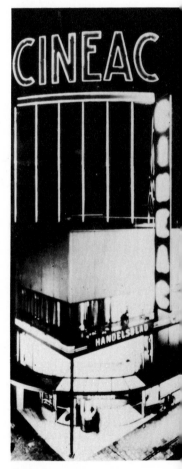

The De Handelsblad Cineac, Amsterdam, Holland (1934), by Johannes Duiker.

music, and curtain control monitored electronically. Few cinemas have been built in Europe in recent years, but by adapting them in this way, they have managed to survive in the face of increasing competition from color television.

Sports buildings

Sport as a formalized demonstration of skill and prowess appears to have existed in almost every civilization. Certainly buildings for sport were known in Ancient Egypt, Mexico, Classical Greece, and Imperial Rome, and major sports festivals were recorded in Imperial China. The basic types of sports and recreation buildings were developed 2,000 to 2,500 years ago in the form of stadiums, palaestrae, gymnasiums, thermae, theaters, hippodromes, basilicas, walled courts, and domed hall structures.

Sports buildings can be divided into two distinct categories; those whose primary function is the promotion of sport as a mass spectator event, and those designed for participation. In their modern forms, these two categories, although based around the same core activities, bear little relation to each other as building types. They appear to have distinctly different roots and differ greatly in both physical and institutional scale.

Mass spectator sports buildings

The basic geometric requirements for sports arenas have remained unchanged throughout their 2,500-year history and are the same for athletics and ice hockey in the 20th century as they were for gladiator events in the 1st century. The plan forms of all mass spectator stadiums are controlled by the ergonomic requirements of good views of the field of play for standing or seated crowds. The variations of seating arrangements—horseshoe, oval, round, rectangular, etc.—are generated by optimizing the relationship between players, the field of play, and the audience, within given topography and orientation.

The division of seating tiers into segments and blocks develops both from the practical considerations of access and exit and also from the desire to denote social rank or price in relation to the best views. The addition of ancillary accomodation for both audience and performers is related to the primary movement routes. The development of all stadiums can be seen as variations on these four factors using available technology.

Classical Greece By far the most influential center for sports in recorded history is Olympia, home of the Olympic Games. It is also one of the earliest sites where evidence of building remains. The Games developed over a period of 1,000 years from 776 BC to AD 394 when they were banned as a pagan ritual by Theodosius 1 (346–395). The site at Olympia had always been a place of cultural sig-

nificance comprising many places of worship, and this significantly influenced the form of the stadiums.

The first Olympic stadium was designed as an embankment stadium with a capacity of 20,000. Spectators were seated on banks cut into the Cronus Hill on one side and on banks built up artificially on the other. A small grandstand for honored guests was situated near the finish of the racecourse which was 100 ft. (30 m) wide and 660 ft. (200 m) long. The athletes entered from the east through a special archway from the Altris or sacred grove—the original site of places of worship. Adjoining the stadium was the hippodrome, with a field of play some 1,050 x 2,525 ft. (320 x 770 m) for horse racing and with spectator banks in a U-shape on the surrounding slopes and hills; again the open end faces east toward the Altris. The Gate of Triumph for the horsemen, however, was out to the west, keeping the horses away from the sacred grove. The layout of the site at Olympia, with the main stadiums to the west of the Altris and the training buildings to the east, reflects the cultural and religious significance of the Games as the temples are en route between training and performing areas. This allowed athletes to invoke assistance of the gods before they raced and to give thanks afterward.

Other major festivals of sport and demonstrations of physical skill were held in Greece in association with festivals of the arts and theater. The circular theater forms of Classical Greece established seating tier patterns which are still present in stadiums today. The theater at Epidaurus, the best preserved of the Greek theaters, has 32 rows of seats forming the lower tier separated from the top 20 rows by a wide gallery. The seating tiers are segmented by 24 radial stairways giving access to all the seats. The seating tiers were constructed by cutting into existing land forms and setting marble slabs in the cut ground. There was a theater of this kind, with a circular area for the orchestra and a raised stage, in every major town.

Imperial Rome The Romans developed more urban forms of stadiums and arenas explicitly for mass entertainment rather than cultural celebration. Sport as a mass entertainment was seen as one way of subduing a potentially restless populace. It was also seen as an essential part of educating the young for military training. As the maintenance of the Empire was a primary concern of the ruling classes, sporting prowess was a matter of general social concern.

The Romans continued the plan form of the Greek hippodrome in the development of their circuses. The largest circus built (in AD 311) was the Circus of Maxentius which had the same field of play as the Hippodrome at Olympia but the seating tiers were built up in

J.M. Kirov Stadium, Krestovsky Island, Leningrad, USSR (1937). A massive open-air arena built for spectator sports.

masonry and concrete to seat, according to Pliny, 250,000 people. The seating tiers were surrounded by colonnaded galleries and stairways giving access to the ground, a scale of construction unknown to the Greeks.

Amphitheaters were also a form unknown to the Greeks but were found in every important Roman settlement. They are good exponents of the character of life of the Romans, who preferred displays of mortal combat (considered to be good training for a nation of soldiers), to the demonstrations of skills favored by the Greeks. The largest Roman amphitheater was built in AD 80. Initially known as Amphitheater Flavium, it was later renamed the Colosseum after the enormous statue of Emperor Nero (37–68) which was erected nearby. In plan it is a vast ellipse, like two Greek theaters face-to-face, 620 x 513 ft. (190 x 156 m) with an internal arena 287 x 180 ft. (87 x 55 m) surrounded by a wall behind which was the Imperial podium. Above the podium level were seats in four diminishing levels for 50,000 people with galleries and access stairs behind. The main structure for the seating tiers and galleries was concrete, with 80 stone, arched openings on each of the four levels, the ones on the second floor forming entrances to the staircases. The performers, both gladiators and animals, were housed under the lowest seating tier at arena floor level. The arena floor could also be flooded for naval displays. The positioning of the Imperial podium just behind the arena wall allowed the Emperor to compete for the crowd's attention with the performers.

The organization and form of Spanish bullrings continued the form of the Roman amphitheater long after the end of the Roman Empire. They adhered to the elliptical or circular layout with the same strictly hierarchical relationships between performers, important personnel, and audience. With the exception of the bullrings, there is little evidence of a revival of mass spectator sports stadiums until the 19th century, when sport once again became a matter for general social concern with the rising importance of physical education in Europe.

Modern times Attempts to revive the Olympic Games were made as early as the 18th century in Greece and Germany. Their true revival, however, was due to the French educationalist Pierre de Courbertin in 1894. His vision gave a new impetus to building for sport. The first modern Olympics were held in Greece in 1896 using the reconstructed Panathenian Stadium in Athens with a 1,093 ft. (333.3 m) track within the U-shaped tiers of seats. The second and third Games in Paris and St Louis were staged alongside world exhibitions in temporary accommodation. In 1906 Pierre de Courbertin tried to introduce the arts competitions into the Games; five competitions were introduced, of which

architecture was to have pride of place.

The 1908 Games in London provided the first permanent Olympic Stadium at the White City London, designed by James B. Fulton. Although a very plain functional elliptical stadium for 80,000 people, it gave impetus to the revival of Olympic architecture. However, the architecture competition within the Games did not generate the excitement that de Courbertin expected, and it was not until standardized rules and dimensions were established that the challenge of building stadiums was taken up. It is worth noting that the most exciting sports stadium developments of this period lay outside the Olympic competitions, notably the racetrack stadium of Zarzuela near Madrid, Spain (1935), designed by Eduardo Torroja (1899–1961). Here the main stand, in the form of a fluted deadweight concrete cantilever roof, was counterbalanced by vertical tie rods behind the stanchions.

It was not until 1928 that fixed dimensions for athletics tracks were settled, and the period up to 1948 was one of rationalization. Moving the Games to a new location every four years made it essential to have universally accepted dimensions for all aspects of sports, and their equipment. Sports arenas became one of the only building types with unified functional requirements. In addition, increasingly sophisticated measuring and timing devices introduced during this period created complexities which demanded the development of new forms and skills beyond those envisaged by the builders of the original Olympic stadiums.

The acceptance of international regulations for sport influenced all sports stadiums not just those for the Olympics themselves. The regulations not only introduced set sizes but also considerations of wind speed and direction. With the added influences of staging sports events in countries with more variable climates than that of Greece, coupled with demands for increasing spectator comfort, stadium forms began to develop around the problem of building roofs.

The range of structures developed for the partial roofing of large field stadiums and the complete roofing of many smaller arenas have developed a new range of forms through the application of reinforced concrete, steel, and lightweight materials to long spans. The Olympic buildings of the last 30 years provide useful examples of these building forms.

The first Olympic stadium at White City, London, had a simply supported steel truss roof spanning over the seats on the straights from the top of the seating tiers to columns in front of the stands. This form of roof had the serious disadvantage of columns cutting across everyone's view. The introduction of this form of roof did not change the basic amphitheater form of the stadium. The Amsterdam Games stadium for 1928 designed

Interior view of the Colosseum, Rome (AD 80); the largest Roman amphitheater.

Exterior view of the Plaza de Toros Monumental, Barcelona, Spain (1913-14), by Raspall y Mas. Spanish bullrings continued the tradition of Roman amphitheaters.

by J. Wils had a more sophisticated roof of a similar form using a counterbalanced steel truss cantilevered out beyond columns one-third of the way down the seating tier. Therefore only one-third of the seats had a partially obstructed view.

The first completely covered spectator arena for the Olympics was the swimming arena in Melbourne, Australia (1956), designed by Borland, McIntyre, and Murphy. The steel trusses for the roof are supported on raking lattice beams carrying the seating tiers and giving the building a trapezoidal form which reflects the layout of seats around a rectangular pool. The Palazzetto sports palace designed by Pier Nervi (b. 1891) and A. Vitellozi for the 1960 Games in Rome is a circular, domed structure which spans clear over the central arena and surrounding seating tiers. The dome is constructed from precast concrete members forming a self-supporting, lattice-grid shell held up on fine, raking, reinforced-concrete columns which are free of the seating tiers.

The Aztec stadium designed by P. R. Vasquez and R. Mijares and built primarily for football for the 1968 Mexico City Olympics, is an example of the new forms generated by completely roofing a multiple tiered stadium. The basic form of the Aztec Stadium is similar to that of the Colosseum in Rome, however the introduction of an in situ reinforced-concrete frame with the cantilever roof beams contiguous with the supports for the seating tiers produces an extraordinarily light appearance—a dematerialized Colosseum. The Swimming and Sports Arena for the same Games, designed by Rosen, Recamier, Gutierrez, and Valderde, introduced another new form of structure again generated by the roof spans and independent of the seating tiers. Using a structural form derived from suspension bridge construction, the two rectangular halls are lined at their ends by tall pylons which carry the curved steel suspension ropes for the roofs, with externally exposed guys stabilizing the pylons.

The Munich Stadium for the 1972 Games, designed by Behnisch and Partners with the engineer Frei Otto (b. 1925), is the first stadium where the roof structure becomes the dominant form for the stadium as a whole. The structure of the seating tiers becomes a secondary element. The roof consists of a net of cables covered with plexiglass sheets suspended by steel ropes from masts which are guyed down to concrete anchor foundations buried in the surrounding landscape. This produces an undulating form in sympathy with the landscape but independent of the rigorous geometric requirements of the seating tiers and track. The use of this translucent roofing material is the first example of the direct influence of television on the construction of

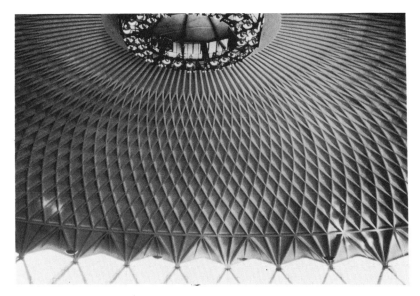

stadiums. It was feared that the shadows cast by a solid roof would have made color filming of the Games impossible.

The Montreal Stadium (1974) designed by R. Taillibert and A. Daoust was the first athletics stadium to have completely covered seating areas. The immense reinforced-concrete cantiliver beams are tied back into the structure for the seating tiers as in the case of the Aztec Stadium. The arched form of the cantilevers with their outer tips restrained by a concrete ring beam dominates the overall form of the stadium. The stadium is still incomplete as it was proposed to build a raking concrete tower rising above the cantilever roof that would carry a rectractable fabric roof to cover the central ellipse and so allow the stadium to be used throughout the year.

The Sports City designed by John Roberts and Partners, which is proposed as the basis for a bid for the Olympics in the late 1980s in Riyadh, is a further development of roofing. It comprises an immense steelwork lattice arch spanning 1,200 ft. (366 m) across the center of the arena carrying cable nets on either side to cover the whole stadium. Sunbreakers within the cable-net roof are proposed to give a controlled environment while maintaining conditions for "natural turf" to survive. When completed, this stadium will produce a form where the roof does not simply dominate but will dwarf the rest of the structure.

The development of large field arenas has now reached a decisive turning point whereby the technology exists to cover them completely and may result in "outdoor stadiums" becoming simply large-scale versions of "indoor stadiums."

Future developments The range of examples described above shows not only the increasing technical feasibility of covering spectator sports stadiums, but also that as the dimen-

Palazzetto dello Sport, Rome, Italy. One of two sports halls built for the 1960 Olympic Games by Pier Luigi Nervi and A. Vitellozi.

Interior of main Olympic Stadium, Munich, Germany (1968–72), by G. Behnisch and Partners, and Frei Otto.

Astrodome (Harris County Stadium) in Houston, Texas (1965), by Lloyd and Morgan. The enclosed stadium can accommodate 66,000.

Exterior of the Astrodome (Harris County Stadium), Texas.

sional and environmental requirements for each sport have been codified and fixed so the technical solutions for the building envelope have become divorced from the strict geometry of the sports. Many building forms have evolved in the last 30 years which are quite distinct from earliest stadiums although the core activities remain the same. However, the immense capital and running cost of these buildings is causing great concern as the level of use of the facilities does not appear to justify the level of investment. There appear to be three areas of development in response to this concern.

First, there are attempts to find cheaper solutions to roofing large spaces. Experiments such as those carried out at the Pontiac Metropolitan Stadium at Michigan and at the University of Iowa using cable-retained, air-supported structures spanning 722 ft. (219 m) and 424 ft. (129 m) respectively, claim considerable cost savings and may develop a further form of stadium building.

Second, there are attempts to increase the range of events held in any one stadium. There have been some attempts to develop banks of seats which can be moved to accommodate different field sizes, and hydraulic floors have been installed capable of converting swimming stadiums into dry sports arenas. The development of these mechanical aids has been restricted to the U.S. and their economic viability has yet to be proved.

Third, there are attempts to find new forms of mass entertainment events which will attract sufficiently large audiences, both live and on television, to offset the costs of the buildings. The sports extravaganzas and sporting superstar competitions have undoubtedly introduced an unprecedented level of finance into the staging of sports events and, along with professional sport, have introduced stringent technical requirements into stadium design in order to exploit television revenues to the full. The Houston Astrodome (1965) designed by Lloyd and Morgan, and the New Orleans Unidome are examples of this. The continued development of stadiums may well reflect the need to maintain this revenue income.

Participant sports buildings

Unlike stadiums, participant sports buildings do not have a consistent basic controlling geometry and have developed more from the use of existing buildings or building types. For this reason there appears to be little direct connection between early and modern forms, except in the case of specific games such as walled court ball games where the court configuration is itself derived from an earlier building. Games such as squash rackets, fives, real (royal) tennis, and peleste all have their origins in games played in courtyards of

existing buildings. Fives courts are directly based on the form of a courtyard at Eton College, England, and real tennis courts are thought to be based on the court at Hampton Court, England, where the game was played in the reign of Henry VIII (1491–1547). Buildings in and around courtyards was the accepted form of early sports buildings.

Classical Greece The Gymnasium at Olympia was built in the 3rd century BC and was intended for practice of events requiring a great deal of space. The central open courtyard was surrounded by pillared archways and halls, the largest of which was the racing hall 690 x 38 ft. (210.5 x 11.5 m)—slightly longer than the length of the main stadium racecourse. The palaestrae of the same period, built for gymnastics, wrestling, boxing, and jumping, were constructed in the form of a central open square surrounded by restrooms, classrooms, dressing rooms, and baths. The biggest sports building at Olympia, the Leonidaeum, was also of this form. Originally designed for gymnastics, this building typifies the pure courtyard form with a central square surrounded completely by a shallow pitched-roofed building colonnaded both externally and internally.

Imperial Rome The Romans built extensively for participation in sport as a leisure activity. Every major town had thermae, which are normally thought of as simply baths buildings but usually contained many other facilities as well. The thermae building form also developed from the courtyard form, although the outer wall is no longer made up of the roofed buildings.

The thermae at Caracalla was the greatest sports and recreation building of antiquity. Started in AD 211, it consisted of a surrounding colonnaded wall housing small apartments, lecture theaters, shops, accommodation for slaves, and the water reservoirs for the baths. Within the walls the main baths building stands in a park with avenues lined with trees and with sculptures and fountains; part of the park was set out as a small arena for wrestling and gymnastics. The central building was an immense basilica structure covering 285,000 sq. ft. (26,473 sq. m) with a central hall 183 x 79 ft. (55.5 x 24 m) roofed with a three-bayed intersecting vault built in concrete and held up on eight masonry piers faced with granite columns. The surrounding halls, all of vaulted and domed construction, were highly decorated with mosaics and sculpture.

The courtyard form continued through from Greece and Rome to the Spanish plaza, such as Plaza Madrid (1619), and to the Zwinger Dresden, East Germany (1728), but until the real revival of interest in mass participation in sport in Germany and Britain in the 19th century, there appears to be little reference or interest in these traditional forms despite archaeological information being published at the time.

Modern times The revival of interest in sport around the 1890s coincided with both the height of Imperial effort and with the concern for public health and fitness resulting from the appalling conditions in cities. Sport was conceived in a strenuous abnegatory spirit not as leisure to be enjoyed. The forms of Roman and Greek origin may have appeared inappropriate for this reason. The first generation of public baths in Britain had more to do with public cleanliness than leisure or entertainment, and the development of gymnasiums associated mainly with education were seen more as centers for endurance and fitness training than for elegant demonstrations of prowess. These buildings were of a basically utilitarian form normally built in masonry with cast-iron roof trusses with lantern lights mounted above them. The spans involved posed no particular challenge, and the building forms remain within the range developing for other hall structures of the period such as industrial premises.

The rationalization of dimensions and control requirements for individual sports is clearly reflected in the forms of public sports centers in Europe. Those centers which group together facilities for a variety of sports, such as Billingham Forum, Carlton Forum, and Picketts Lock sports centers in England, are planned as a series of courts and halls, each with its correct height, length, and breadth grouped around centralized common facilities such as changing rooms. By minimizing circulation and placing the main spaces around a central core this generation of sports centers tend to have a massive and unrelenting exterior with none of the leisured generosity of the colonnaded form.

Similarly, the change in form of swimming pools from public health pools to measured competition pools generated a group of buildings during the 1950s and early 1960s of a slightly more open form. But it was not until the early 1970s that participant sport reached a level where it became a marketable commodity. Sports buildings have now started to include bars and restaurants as standard accommodation, and free-form swimming pools with wave machines and palm trees are being introduced more akin the facilities of luxury hotels than the "public health" pools of 50 years earlier.

Future developments The assimilation of leisure into sports centers otherwise dominated by the dimensional and control requirements of particular sports has not yet been resolved in terms of the building form. Some hybrids have developed, such as the sports center at Swindon, England, designed by Gillinson and Barnett, where a free-form pool is housed in an elegant domed building and adjoins the sports halls which are still massive unrelieved

sheds. The future form of sports buildings will depend on the relative pressures of demand between the leisure side and the competition training side to produce economically viable facilities.

Increasing sophistication in timing and measuring equipment alongside the international acceptance of dimensional criteria has produced ever-increasing demands for buildings producing optimum controlled environments for each sport. At the same time, concern over the economic and social viability of institutions generated by these factors is growing. The future developments in the form of sports buildings must lie in the resolution of these demands in order to produce solutions capable of answering society's sporting needs at an appropriate social and economic cost.

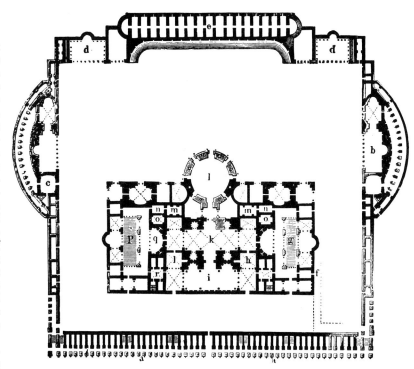

Ground plan of the Thermae of Caracalla, Rome.

Empire Pool, Wembley, London (1934), by Sir Owen Williams. A large enclosed swimming pool with a movable floor.

Interior of the multi-purpose leisure center at Swindon, England (1976), by Gillinson and Barnett.

Institutional

Hospitals

Medical knowledge in early societies was nearly always confined to the priesthood. Medical schools, therefore, arose in the guise of temple complexes such as those in Edfu in Egypt, Benares in India, and Epidaurus in the Peloponnese. At the close of the Roman Empire a few slave hospitals existed, but in general wealthy landowners could afford their own doctors; peasants or slaves were too disposable to warrant the expensive hospitalization. Valetudinaria were, however, built for the Legions, and the example at Inchtuthil shows a rectangular building about 200 x 300 ft. (60 x 90 m) around a court with single or two-bed rooms down either side of a wide corridor. Hospital buildings are known to have existed in the civilizations of India, Ceylon, Persia, and Arabia. But they were rare because a hospital is the product of the developed society and is dependent upon the moral and economic support of the community.

The Middle Ages

Medieval man lived in a highly structured feudal society where change was slow. His buildings reflect this and were constructed slowly and used over very long periods of time. Innovation was concentrated in ecclesiastical structures, and only insofar as these were adapted for hospital purposes could technical invention be credited to hospitals. The basic medieval hospital unit is the church nave of the early Christian church. This came to be converted into a ward by placing beds down either side. The simple bed-nave then developed ancillary accommodation, sometimes around a court, and in time several bed-naves would be linked together in a haphazard form. There were exceptions, such as the specially built infirmaries attached to the great abbeys, and the isolation hospitals.

By the 12th century infectious patients, usually lepers, were separated from the non-infectious and were placed in a lazar house or *Maladerie*. Ordinary hospitals were known as *Hôtel Dieu*. The Maladerie de Tortoi, built in the 14th century, demonstrated the clear division of function thought necessary in the treatment of infectious sick. The entire building, sited outside the town wall, was itself surrounded by a high wall through which the patients were fed. They could be supervised from a high-level gallery in the bed ward, and a corridor linked the ward to the chapel, where patients were separated by a barrier from outsiders. At Tonerre the Church of Notre Dame de Fontenilles, founded 1293–95, is a fine example of a simple Hôtel Dieu. It had a massive wooden barrel vault with openings for ventilation. One window per cubicle was provided for the 40 beds,

Church of Notre Dame de Fontenilles, Tonnerre, France, founded c. 1293. An early example of a Hôtel Dieu type of hospital with an open-plan bed-nave.

arranged down either side of the great bed-nave, with its altar at the far end. The Hôtel Dieu at Beaune, founded in 1443 by Nicholas Rolin and completed around 1450, was a further development. Its ancillary accommodation was arranged around a court, and the great hall 148 x 45 ft. (45 x 14 m) had 30 beds divided by timber partitioning. It is thought to be the first example of a hospital where rich patients, accommodated on the upper floors, were thereby segregated from the poor patients.

As the Church became wealthier and better organized, great abbeys were built and the monastery hospital, perhaps the nearest thing to a medieval teaching hospital, emerged. As size increased, so did the problem of sanitation. At Fountains Abbey, founded in England in 1132, many buildings were sited along or over the river, which was widened for sewage. The infirmary itself rested on great piles and spanned the multiple drains that ran beneath it.

Medieval hospitals continued to be used and often expanded during the Renaissance.

In the 15th and 16th centuries hospitals averaged 300-500 beds, but by the 17th and 18th centuries these same hospitals would have increased to take between 1,000 and 2,000 beds. There was tremendous over-crowding, particularly of the urban hospitals. In France commerce developed and towns expanded rapidly. The Hôtel Dieu in Paris, before it was burned down in 1772, was perhaps the greatest medieval example. Its original bed-nave became interlocked with other bed-naves over the centuries, each a complete hospital in itself, until this enormous medieval agglomeration bridged the Seine and extended along the south bank. In 1515 up to eight people shared a bed and 7,000 patients inhabited the warren, with a mortality rate of one in four. There were few stairs for escape and these acted as airborne infection routes due to stack effect. In 1748, the ventilation was examined by Duhamel, who proposed that fresh air be admitted through high windows and warmed by stoves.

The Renaissance

In the early decades of the 15th century Filippo Brunelleschi (1377–1446) and his colleagues deliberately set out to break with their medieval inheritance and create a new art. In 1419 he built the Foundling Hospital in Florence, with its famous arcade and terra-cotta medallions by Luca della Robbia (1400–82). Buildings now began to reflect the geometric harmony that Renaissance man perceived in the universe. There was a mathematical exactitude in the balance and symmetry of design, and inspiration was drawn from classical Greece and Rome. Buildings were seen as self-contained and finite, often conceived on an axis and composed for maximum proportional harmony. Hospitals began to have formalized master plans in the shape of stars, squares, or tees. Sometimes a master plan would take several centuries to complete, as did the Ospedale Maggiore built by Filarete (1400–69), but the concept was adhered to doggedly through the years, which was quite different to the haphazard growth of the medieval period.

The Ospedale Maggiore is a transitional building with a Renaissance plan and Gothic facades. The austere geometry is monumental in scale, with medieval bed-naves retained but set out in the form of a double cross with altars at their intersections. He placed men on one side of the great court and women on the other. The building was sited near the town moat and canals were channeled through undercrofts which were flushed by rainwater in times of storm. The stormwater pipes doubled as ventilation pipes to bring air to the undercrofts, which included store rooms for bread, wine, and live cattle. Laundry, too, was undertaken here. Originally designed for 300–350 patients, the building has in recent times accommodated up to 2,000. Each patient had access through a door between the beds to the sanitary undercroft below and in this aspect it was far in advance of other great buildings of the time, such as San Spirito in Rome, or the great hospital of the Order of St John in Valletta, Malta. Both these fine buildings were T-shaped in plan and the latter, built by Grand Master La Cassière in 1575, and enlarged by Cottoner in 1662, became one of the major medical centers of Europe in the 16th and 17th centuries. It was badly damaged in the last war, but the great ward, 520 x 35 ft. (158 x 11 m), can still be seen today, sited above the harbor where the galleys landed to offload the wounded. Beneath the great ward were galley slaves' quarters, and the iron-pillared beds were backed up by a chapel, library, linen store, laundry, and chief physician's quarters.

In Spain Filarete's cruciform plan reached its highest perfection. In 1504 Enrique Egas built the hospital of Santa Cruz in Toledo and here two-storey bed-naves with an altar at the richly vaulted crossing enabled eight wards to celebrate mass simultaneously. In 1591 Juan de Telosa introduced the double-sided loggia at Medina del Campo; previously bed-naves would be either without loggias or provided with a single-sided one, being part of an abbey cloister. He now provided both inner and outer loggias down both long sides of the bed-nave, not only to give shaded access, but to provide secluded ambulatory space for patients. His bed-naves were subdivided by central altars and had deep alcoves for additional privacy, while the whole building was arranged around a central courtyard.

The 18th century

In England leprosy had declined in the 15th century and the *Maladerie* fell into decay. The dissolution of the monasteries between 1530 and 1540 put an end to almost all the remaining hospitals for 200 years. As the Industrial Revolution gathered pace and the closure of common land occurred, the sick congregated in the industrial centers. The population of England increased from 5.5 to 9 million during the 18th century. The threat of epidemic diseases became imminent, but the importance of good ventilation was now recognized. With the advent of mechanization, people began to experiment with hospital ventilation. In about 1743 Hale's ventilators were installed at Winchester, while Sir John Pringle, an army surgeon, advocated certain minimum distances between beds in naturally ventilated wards.

The mid-18th century saw the beginning of the corridor and pavilion plan form, specifically designed to reduce airborne infection. For the first time a deliberate and methodical approach was applied to health buildings;

Ospedale Maggiore, Milan, Italy, begun in 1461 by Antonio Filarete. Only one court of a much larger plan (centered on a chapel) was completed.

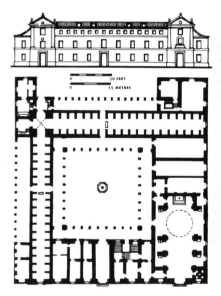

Medina del Campo (1591) by Juan de Telosa : elevation and plan. This introduced a bed-nave with loggias for shaded access and with ambulatory space for patients.

wards, at right angles to the communication system instead of alongside it, became a standard feature of hospitals from then on. The Royal Naval Hospital at Plymouth, England, designed by Rovehead in 1756–64, was possibly the first hospital to be designed on the pavilion system; single and two-storey ward pavilions were connected by colonnades, the roofs of which served as terraces. The whole building housed 1,250 beds and probably formed the basis for the great French hospital of Lariboisière, since members of a French Royal Commission visited Plymouth with a view to adopting its design in the rebuilding of the Hôtel Dieu in Paris.

In Paris the Académie des Sciences initiated a competition for hospital design. One of the greatest submissions was by Jean Baptiste Le Roy in 1773. He believed ''a hospital ward was really a machine for the treatment of the sick'' and had his building been built, it would probably have been the most splendid example of hospital architecture since Filarete. Bernard Poyet's design advocated a radial building to increase natural ventilation to the bed areas and had gigantic dimensions. It was the ultimate in formalism and was rejected by the Académie. Lavoisier and his colleagues, however, designed a hospital that was not built until 1846–53. This was Lariboisière, designed specifically to reduce the spread of airborne infection. It was classical in layout, with colonnades at either side of the central courtyard, adapted to the climate of northern France. The 612 beds were in self-contained, three-storey wards, each of 32 beds, with a sister's room, office, toilet, washroom, and sluice. Lariboisière was artificially ventilated and warmed, but it produced a high mortality rate because the ventilation system was a negative-pressure one. The contaminated air, warmed by water stoves, was sucked out by central, vertical flues at a 1,000 cu. ft./hr. (28 cu. m/hr.) and contaminated the patients it passed. The hot water was provided by a high-pressure, hot-water system. To counteract this airborne infection, Thomas Laurens later installed a positive-pressure system using a steam engine to fan clean air inward and extract it through vertical shafts. Nevertheless, the block planning was superb and Florence Nightingale (1820–1910) in her book, *Notes on Hospitals*, advocated it as the standard for future hospital designs. The wards formed the basis of "Nightingale wards," themselves a development of the medieval bed-nave, central work tables replacing the altars.

Heating, lighting, natural and artificial ventilation were now the most important topics of hospital design. In the late 1870s there was a further development which took the form of a brief vogue for circular wards, in the belief that this would aid natural ventilation still further.

The 19th century—Renkioi

One of the products of the Industrial Revolution was the ability of nations to make war on one another more effectively. As man became more obsessed by mechanization, so he developed prefabrication techniques and began to apply them to hospitals. In 1855 the British Army, at war in the Crimea, commissioned a hospital from Isambard K. Brunel (1806–59). At Renkioi he produced the cheapest, lightest, and fastest building yet seen. The ward units weighed 200 lb. (90 kg) per bed, were timber-framed with attachable, winter-insulated linings, and had polished tin roofs with external whitewashed walls to reflect the heat in summer. An iron kitchen per 1,000 meals and an iron washhouse per 2,000 beds were provided with other ancillary buildings. The pavilions were placed either side of the central spine—22 ft. (7 m) wide—to give extensibility to ground levels, and his organization diagram remains valid today. Down the center of this spine ran a railroad from which the troops would be unloaded directly from the ships. The sewers were constructed with interlocking wooden trunking mains, and a positive-pressure, decentralized ventilation system was installed by means of a rotary fan at the corridor end of each 50-bed pavilion. This gave 1,500 cu. ft./min. (42 cu. m/min.), piped through an underground duct, and prevented bad air from the toilets penetrating the wards. The air itself was humidified by being passed over water.

A few years later America benefited from Brunel's designs at Renkioi. The American genius for mass production was seen in the vast amount of building that was undertaken during the Civil War. In four years the Union Army built 204 hospitals with nearly 137,000 beds, mostly prefabricated, lightweight, and disposable. The Satterlee General Hospital in West Philadelphia, built in 1862, was the largest hospital north of the Mason-Dixon line Its total capacity was 3,519 beds. John McArthur Jnr in 1862 built the Mower General Hospital at Chestnut Hill, Philadelphia. Its 51 timber pavilions provided 3,100 beds around an enormous elliptical corridor 2,400 ft. (730 m) long, with administration and service buildings in the center. The average mortality rate of Union hospitals was 8%, which was lower than in many civilian hospitals and considerably lower than that in Europe during the Franco-Prussian Wars.

In the latter half of the 19th century, Britain was at the height of the Victorian empire and was influencing hospital design in Europe and America. Hospitals were considered to be civil monuments. The Civil Hospital at Antwerp (1880) by Baekelmans and Bilmeyer is a good example; it shows the brief vogue for circular wards and it supplied its 380 patients with water from an iron-piped water

Mower General Hospital, Chestnut Hill, Philadelphia (1863), built by John McArthur Jr, with 51 timber pavilions providing some 3,100 beds.

system. There were two different drainage systems: one for storm and one for sewage, the latter drained into two huge brick cesspits with road access for removal. Innovation in the Victorian era can primarily be seen in the field of engineering rather than hospitals, though the new technology was applied in hospital sewer systems, gas installations, and numerous mechanical devices.

In medical practice specialization developed, while in hospitals the dominance of the ward gave way to the diagnostic and treatment areas. Anesthesia, discovered in 1846, made surgery an exploratory science, while Joseph Lister and Von Bergman found ways of preventing infection. In 1895 Wilhelm Konrad Roentgen (1845–1923) invented the x-ray, and by the 1920s diagnostic laboratory services were fairly advanced.

The 20th century

Compared to the Victorian building boom, few significant advances in hospital planning occurred in the first half of the 20th century. Reinforced concrete and steel technology enabled tall buildings to be built and at the end of World War II, the U.S., Switzerland, and Sweden were building vertical hospitals, where ward towers sat upon a podium of diagnostic and treatment departments. All but Britain had abandoned the "Nightingale ward" for smaller rooms and each of these vertical hospitals relied upon a central elevator stack to reduce walking distances and optimize volume to reduce air-conditioning costs. The most significant postwar hospital was St Lo in northern France by Paul Nelson, and this was closely followed by Gordon Friesen's work at the United Mineworkers' Hospital in Pennsylvania. He fundamentally reorganized the supplies problem by harnessing existing technology, such as elevators, trayveyors, pneumatic tubes, and ejection devices, to further reduce staff numbers and revenue costs.

During the early 1950s, work study and organization and methods study, were applied to hospital planning, with the general intention of replacing the generally empirical contemporary design methods with industrial study techniques. However, as hospital briefing processes were developed, and as the first postwar hospitals were reassessed in use, it became clear that the functions around which they were so carefully constructed always changed, often before the buildings were complete, and continued to change. In response to this, a number of new planning approaches and technical solutions were developed. In principle they offered alternative strategies for the definition and relationship of constructionally "hard" and "soft" elements, and the degree and kind of flexibility required in detailed planning.

One solution was represented by the design

of Northwick Park Hospital and Clinical Research Centre in England, begun in 1963 (architects Llewelyn-Davies Weeks). This was the first deliberate attempt at an unfinished, wholly flexible, horizontal hospital. Here the overall complex is loose, the structure and main communication routes are "hard" and the buildings served by the communication routes are "soft," and independently variable. They are modular, and have regular services reticulation. This pattern, categorized as "indeterminate," is designed to allow for episodic growth.

An alternative solution focuses attention on the importance of services engineering, which now accounts for over one-third of the cost of any hospital. The integration of ducts and pipework has become nearly as important as the medical planning. To free the latter from engineering constraints, the "interstitial space" hospital has emerged. Greenwich Hospital, London, and the Veterans' Administration Hospital, San Diego, are among the first such hospitals. They have been closely followed by McMaster Medical Center, Ontario, and, more recently, by the Woodhull Hospital in Brooklyn, New York. In these hospitals, every other floor is devoted to engineering services, thereby segregating pipes, ducts, and mechanical plant from occupied spaces. In this type, the structure and horizontal service voids are "hard" and everything else is "soft." This type offers a very high level of internal flexibility, but may be subject to "catastrophic" pressures. There may come a time when growth overtakes the capacity for adaptability, in which case the whole building becomes obsolete.

In planning hospital provision within the community, the emphasis is now placed more

Northwick Park Hospital and Clinical Research Center, London, England (1963), by Llewelyn-Davies Weeks.

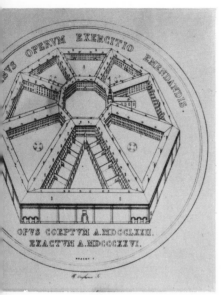

The Maison de Force at Ghent, Belgium, begun in 1772. This shows the original plan for the building, in fact only five of the eight sides were built during the 18th century.

on small additions to existing plant rather than total replacement. So long as the pressures for growth and change can be contained by gradual incremental alteration, an organization can avoid the enormous capital expenditure and organization shock which catastrophic growth entails. When new hospitals are being considered as part of a regional plan for hospital bed provision, large new hospitals are now less frequently planned. Greater emphasis is placed on smaller bed and medical service packages, more easily accessible to local communities. Nevertheless, the balance between the requirement for wide-spectrum availability of sophisticated specialties in a large central hospital, and the more limited availability in smaller hospitals, is not easily attained.

The British Department of Health and Social Services has developed a system which is likely to become a prototype for hospital construction in the 1980s. It is based on central, extensible, linear communications which serve standard departmental "shells" on either side. This system is highly flexible and adaptive, and allows small or middle-sized hospitals to be extended and integrated with new construction. It therefore responds positively to the two points made above: the need to avoid catastrophic growth pressures by allowing episodic growth and change, and the trend toward gradual development of small peripheral health facilities.
(See also SERVICES.)

Prisons

Although prisons of one sort or another existed in Greek and Roman times and were common in medieval Europe, they were generally incorporated into other types of building or were little more than impromptu cages. The prison as an institution in its own right emerged in the Low Countries and England during the 16th century when bridewells and houses of correction were set up to take care of the indigent, disorderly, and immoral elements of the population. (Criminals were disposed of in other ways.) These new institutions did not give rise to an architectural type, most being housed in existing buildings such as disused palaces, monasteries, or convents.

Thus, in the 1770s, on the eve of the great penal reforms, Europe's prisons were rarely buildings specifically constructed for the purpose, and even when this was the case, as with London's 15th-century Newgate, the building was modeled on the traditional city gate and bore no specific relation to its use as a prison. John Howard (1726–91), who installed momentous prison doors at Newgate (1774), showed that most prisoners, whether debtors, misdemeanants, or felons, were maintained in buildings hardly different from cottages, lodging houses, and tenements; their

only distinguishing feature being some kind of surrounding wall.

Meanwhile, there had been a number of proposals in which the architecture of the prison was becoming more specific to its purpose. The House of Correction in Rome (1702–04) designed by Carlo Fontana (1634–1714) and built under the auspices of Pope Clement XI; Bugniet's project for a prison at Lyons (1762); and the Maison de Force at Ghent (c. 1772), built for Count Vilain XIII, are the most notable examples. Fontana's *silentium* was for undisciplined children, and its solitary sleeping cells and common central aisle reflect the monastic basis of its penitential discipline. Bugniet's prison and the Maison de Force both adopted rigid centripetal plan geometries to consolidate enormously extended institutional layouts.

During the 1780s, the purpose of the prison was, in any case, to become redefined. The blatant coercion of bridewell and house of correction association with forced labor was overlaid with a new motive when the reforming power of solitude on the mind was added to the unanswerable power of compulsory toil on the body.

The late 18th-century penal reforms were mainly the work of Cesare Beccaria (1738–94) and John Howard. Beccaria was primarily concerned with the criminal law and his work had no effect on the form of prison building or administration. Howard, on the other hand, was concerned almost exclusively with prisons. His journeys had convinced him that practically every prison in the civilized world was the breeding ground of disease and vice, and that reform could only be effected by a total rebuilding.

The first of Howard's suggestions was that prisons should be thoroughly ventilated so as to eradicate endemic jail fever (typhus). Although his faith in ventilation was due to the mistaken idea that the disease spread only in confined air adulterated with effluvia and animal steams, it remained the foundation of his campaign. Second, he maintained that prisons should also be architecturally subdivided to prevent "evil communication," a Pauline phrase much in vogue with 18th- and 19th-century penal reformers. Divisions were to be made between male and female prisoners; the innocent and the convicted; debtors and felons. In the same cause, they would be provided with cells for solitary confinement during some or all of the day to further stem the advance of corruption by communication between prisoners and to allow solitude to breed virtuous thoughts—another central reformist belief which, although rooted in the monastic tradition, was given new meaning in rationalist psychology.

These requirements were drawn together into an architectural form almost single-handed by an obscure English architect William Blackburn (1750–91). He is known to have been involved in the design of at least 19 prisons, and was regarded as Howard's architectural amanuensis at the time. Most of his prisons were courtyard plans in which the need for subdivision to control communication was at odds with the need for wholesale perforation of the building fabric to effect ventilation; however, the plans are recognizably of a specific building type with a program peculiar to itself. The external appearance of the prison was also made recognizable with a high boundary wall, away from other buildings, punctuated only with an entrance portal of suitably ponderous proportions to which the insignia of confinement were then applied—chains, manacles, fasces, exaggerated rustication, and mottoes such as "solitude."

The death of both Howard and Blackburn in 1791 put an end to this first phase of prison reform which had been restricted to England. A second, international phase began in the same year with the setting up of Walnut Street prison in Philadelphia, U.S., and, more importantly, with the publication of plans for a model prison, *Panopticon,* by the English philosopher and reformer Jeremy Bentham (1748–1832), also in 1791. In early radial prison plans, such as the Maison de Force, the geometric center was also the center of the highest authority—chapel and government offices. Nevertheless, these plans were no more than pictures of a power hierarchy; they did not help enforce authority through the medium of architecture. The Panopticon did. Here the governor sat in the hub of a rotunda, his officers perambulated around "annular galleries" suspended in the space of the drum, and the prisoners were locked in cells around the outer surface. The purpose of this arrangement was, via the manipulation of light, image, and even sound, to make all information flow inward toward the center, and none outward. Two, less radical, and more immediately practicable versions of this idea were a detached wing radial plan and a polygonal plan, and there were, during the second decade of the 19th century, various disputes about the superiority of the one over the other. These were resolved in favor of the radial.

Meanwhile, the basis of reformed penal discipline had changed little from Howard's time, except that the prison itself was now considered the vessel of authority. Its function was to impose and maintain order under the direction of the governor. One problem, apparently administrative but in fact of a more general character, was emerging in the 1820s. Divisions between one type of prisoner and another were increasing. In the 1780s a good reformed prison might have six classifications of prisoner, and six segregated wards. In 1818 it would have at least 12, and by 1830 there

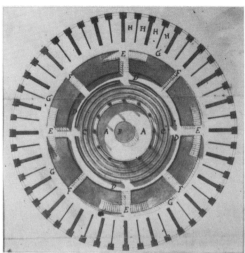

The Penitentiary Panopticon (1791) by Bentham and Reveley.

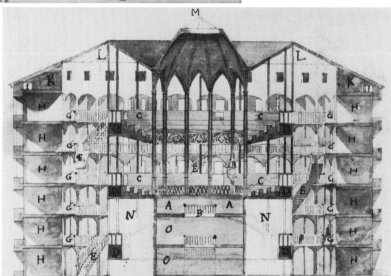

Sectional view of Bentham's Panopticon. The design of the building with the governor's office in the center, helped to enforce authority.

were prisons, like Tothill Fields, Westminster, with 24 classifications from female juvenile misdemeanants to male felons awaiting transportation to penal colonies; each corresponded to a rung on the ladder of moral corruption from the most innocent to the most depraved. The dense, segmental geometry of prison plans during the 1820s and 1830s was therefore a result of the increasing sensitivity to the different shades of criminal malevolence. It was not easily compatible, however, with the desire to unify the whole prison under one sovereign eye, that of the governor, and so the centers of power had to proliferate in order that each class could be properly overseen. A certain type of programatic architecture had thus been developed to a point of impasse where either classification or centralization had to be compromised.

Just at this critical point, the existing type of penal discipline, based on solitary night cells and daytime classification, was given up. In its place was introduced a kind of solitary confinement far more thoroughgoing than that of the 1780s. Cherry Hill Prison in Philadel-

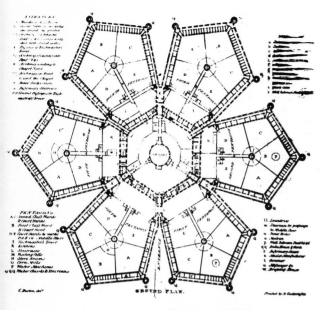

Ground plan of Millbank
Penitentiary (1812–18) by William
Williams and Thomas Hardwick.
Six pentagonal prisons are located
around a hexagonal administration
and services core.

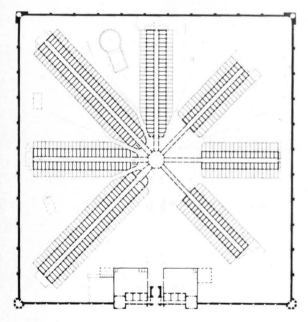

Cherry Hill Prison in Philadelphia
(1821-29) by John Haviland.

phia (1821–29), was designed by John Havi-land (1792–1852), was the first of this new regime. It was a radial of the conventional type except that the wings joined the center and the corridors fed directly off the central rotunda. The accommodation comprised cells only (plus adjoining solitary exercise yards). Internal surveillance space was again unified, and the number of possible classifications was infinite, each prisoner being his own class.

The technological and architectural prob-lems of creating total solitude had seemed insoluble in the 18th century, but between the 1820s and 1840s enormous efforts were made to devise suitable aids. These studies of the engineering of human separation within a densely occupied building type culminated in the construction in London of Pentonville Model Prison (1840–42) by Joshua Jebb (1793–1863), an engineer by profession. With its galleried radiating halls, lined with several storeys of solitary cells, Pentonville remained the genotype of the prison in 19th-century Europe. In the U.S., the back-to-back, barred cell block, surrounded by surveillance gal-leries, was the preferred type. This derived from another form of penal discipline—the Auburn or Silent system—in which prisoners worked together during the day. No sophis-ticated technology of separation was employed, but policing was accordingly more intimate, and surveillance more pervasive.

By the 1850s the matter of prison design became an increasingly hermetic area of expertise over which the prison authorities had firm control. Only in the late 19th and 20th centuries did the layout, organization, and servicing of prison buildings change rad-ically, and then it was in an effort to rid the prison of the stigma attached to its past. This amelioration of the power of the prison over its inmates commenced early on with the foundation of the Colonie Agricole for young offenders at Melbray (1839–50) designed by Abel Blouet. This was organized into family houses and looked like a model alpine village. In terms of overall plan, the most significant change was the introduction of the telegraph pole layout with parallel cell blocks attached to either side of a central service corridor. This was first seen at Fresnes, France, in 1898, in a design by F. H. Poussin. It is a type that has been much used in the U.S. during the 20th century. A good example is the Louisiana State Penitentiary, completed in 1955. The Metray pavilion layout, which is increasingly informal in appearance, was employed at the Illinois Women's Prison, Dwight, in 1930, and continues to be used in prisons for first offenders, juveniles, petty criminals, and those with social problems.

Certainly in the 20th century the prison as a special and particular type of building is tending to disappear as the morality of sep-aration is brought into question.

The planning of this modern
French prison, with its five
star-shaped cell blocks enclosed
from the outside by a high wall
formed by outer buildings, owes a
great deal to 19th-century
precedents.

Defense, emergency, and portable

Fortifications

Since man ceased being a nomad and began to live in groups he has sought to protect himself and his possessions against aggressors. Few human activities have absorbed so much effort as the construction of fortifications. The effect that these have had on the form of human settlements cannot be over-estimated—it is only in the last century that the need to secure defense has ceased to be the major element in urban morphology.

The means adopted for defense has always depended on the weapons used for attack—an innovation by the attacker provoking a response in the design of the defenses. At any one time, the effectiveness of particular fortifications depends on the relative strengths of the means of attack and the means of defense. Sometimes the balance favors the defenders, as in the late Middle Ages, at other times it favors the attackers, as in the 16th century.

In achieving protection, the first necessity is to exploit whatever defensive opportunities are offered by natural features. An enemy attacking uphill is at a disadvantage and thus hilltop sites have always been a first choice. Water is also a hindrance to attack and this explains the construction of Iron Age villages on lakes, or the building of Venice: a medieval city on a lagoon. Caves also provided refuge for primitive man but were still proving useful as late as the 16th century, when a castle was built in the mouth of the Predjama Cave in Slovenia. The choice of a site is always the most important decision in defensive building and the art of military architecture lies in using the manmade work to exploit the potential of the terrain. Where the configuration of the land offers no natural opportunities then artificial defenses must be created.

Early times

The simplest artificial defensive system is exemplified by the Iron Age fort where a ditch is backed by an earth rampart. The rampart offered a height advantage not only in hand-to-hand fighting and hurling stones but also a vantage point for observing the enemy's approach. These type of forts are found in Britain and Central Europe and vary considerably in size and complexity. Maiden Castle in Dorset, England, covers 15 acres (6 hectares) and accommodated 4,000 inhabitants within its triple line of ditches and ramparts. The weakest point in the defensive system was the entrance, and this was protected by additional lines of ramparts so that an attacker who breached the gate would find himself in a maze of passages and subjected to attack from above. These forts clearly demonstrate the two elements which were to be adopted over millennia in the quest to counter improvements in offensive tech-

niques: a defensive perimeter or enceinte with special arrangements at its weakest point—the entrance. The earth rampart, with perhaps a timber palisade, evolved in Europe into a stone-faced wall wherever stone was available; a variation is the vitrified wall where the stones are welded together by fire. Masonry fortifications were, however, used much earlier in the Middle East; the remains of a 23 ft. (7 m) high wall dating from about 7500 BC have been found at Jericho. This is the earliest example of a wall enclosing a settled area.

A different system was used to defend Anatolian cities of around 6000 BC. Catal Hüyük and Hacilar consisted of one-storey, mud-brick houses built contiguously with no ground-level street, and accessible from their roofs. The outside walls of these houses formed a defensive perimeter. An enemy who successfully breached the wall would find himself trapped in a single room and subjected to attack from above. Although suited to a situation where all access was from the rooftops, houses built against a city wall hindered rapid movement of its defenders in a city with ground-level streets.

The Hittites, who ruled Anatolia around 1900 BC, were adept at taking advantage of the defensive possibilities of land forms. Bogaz-koy, their capital, is located with a deep ravine on three sides and is subdivided into a number of independent quarters, each capable of separate defense in the event of other parts of the city being taken. This is an important concept that emerges in different epochs of military architecture. Hittite walls were built of stone on brick and varied between 6 ft. 6

The Cliff Palace, Mesa Verde National Park, Colorado, is the largest known prehistoric cliff dwelling. The circular structures are the remains of ceremonial chambers, called *kivas*. The square buildings were the living quarters. The site had the advantage of being easily defended against attack.

in. (2 m) and 13 ft. (4 m) in thickness. Gateways were protected by towers, with the entrance passageway turning through 90° to hinder enemy penetration. Another notable Hittite innovation of the 2nd millenium is the caisson wall—two parallel walls connected by short cross walls.

Ancient Egyptian hieroglyphs for castles indicate a rectangular walled space with a gateway in one corner, and often show a tower in the furthest corner from the entrance—perhaps a forerunner of the medieval keep. Enclosures with double walls and protected gateways dating from the 29th century BC have been found at Abydos and Hierakonpolis. Other hieroglyphs show isolated circular towers with what appear to be battered walls and a crenelated balcony, and during the civil wars of the 21st century small brick forts were built which also had crenellated walls. At the same period the newly conquered territories of Nubia were secured by a chain of fortresses built on the banks or islands of the Nile. Where the terrain is irregular the defense works follow the contours but where the ground is flat the fortress is rectangular and axial in plan, establishing a formal precedent that was to endure for 3,000 years. Buhen is a fine example of the latter, with a rudimentary form of concentric defense. A low rampart about 656 ft. (200 m) square overlooked a rock-cut moat and was backed by a massive brick wall 30 ft. (9 m) high and nearly 16 ft. (5 m) thick, reinforced with square bastions. The gates in the middle of each wall were guarded by projecting towers and other towers protected each corner.

Double walls also surrounded the citadels of the Assyrians; they were 100 ft. (30 m) high and wide enough to accommodate a chariot pulled by four horses abreast, while at Babylon they were reinforced by projecting towers along the perimeter. The enceinte defending the city was usually rectangular although at Zincirli, built around 1000 BC, it forms a nearly perfect circle 1,968 ft. (600 m) in diameter.

The most important Mesopotamian innovation was the Medean Wall, which was built between the Tigris and the Euphrates rivers by the Samarians during the 4th millennium BC. It was the prototype of a strategic concept—building a frontier where no natural barrier existed—which was to be emulated by many civilizations widely separated in space and time. Later examples were the Great Wall of China which was started in 246 BC and eventually stretched 3,720 mi. (6,000 km); Hadrian's Wall which was started in AD 127 and stretched the 73 mi. (117 km) from the Tyne to the Solway in order to stop the Caledonian tribes from invading the Roman province of Britain; and the Limes Germanicus, the wall which the Romans built on their northern frontier between the Rhine and the Danube. Examples from this century are the French Maginot Line and the wall which now divides east and west Berlin.

If the crenellated, towered walls of the fortresses of the Ancient Near and Middle East resemble those of medieval Europe it is scarcely surprising, since their techniques of siege warfare as depicted in Assyrian reliefs were very similar to those employed in the 14th century. Battering rams, siege towers, and scaling ladders were all used to assault walls, which were also subjected to undermining operations.

Greece and Rome

On mainland Europe the Myceneans built citadels which were neither purely military installations like Egyptian forts, nor shelter for the surrounding population, but rather heavily defended palaces. The best preserved are those at Mycenae and Tiryns, which date from around 1500 BC. The former consists of a hilltop surrounded by a continuous, towerless wall. It was entered by a ramp through double gateways the first of which, the Gate of the Lions, is notable for its massive monolithic jambs. At Tiryns, the hilltop is crowned with even more massive masonry—in places the walls are over 27 ft. (8 m) thick. To enter the citadel an attacker had to move along a ramp which exposed his right side, unprotected by a shield, to the wall. He then had to pass through three gateways set at right angles to one another before reaching the royal residence at the center.

The offensive armory was extended around 400 BC by the Greeks who invented catapults and ballistae which could shoot arrows and hurl stones over an effective range of 1,312 ft.

Section of the Great Wall of China (c. 246 BC) which extends along the former northern boundary.

(400 m). In a siege these weapons could be used to neutralize the defense, so that battering rams could be brought up to a city's walls. Defenders could also use the new weapons to frustrate an assault as well as bombarding the besieging batteries. Greek architects responded to the new threat with a number of important innovations. Towers were built higher to outrange the enemy, and they were also enlarged to accommodate heavier weapons.

To protect catapults from wet weather, to which they are particularly susceptible, the enclosed chamber, or casemate, was devised. Missiles were fired through slits in the masonry walls, and to obtain a wide field of fire circular or pentagonal towers were built. These were arranged to give flanking fire along a curtain wall. Another feature of classical Greek defensive systems is the idea of defense in depth. Concentric rings of defense were established to force the enemy artillery out of range of the center of the defensive position.

Castel Eurialo at Syracuse is the most perfect example of a Greek fortress which incorporates these innovations. An attacker would have to negotiate three rock-cut ditches protected by outworks before reaching the main enclosure. The outworks were connected by a system of underground passages which enabled large numbers of concealed troops to be rushed to any critical point. This type of fortress defended cities such as Miletus and Priene, both of which consisted of an orthogonal grid of streets loosely surrounded by a city wall which cut across the contours and surrounded a cliff top to prevent an enemy seizing a dominant position. The walls follow a sawtooth plan which exposes the attacker's unprotected right side. At Miletus, each sector of wall is protected by a tower with a small sally port every 197 ft. (60 m). Although the Romans never achieved the subtlety and sophistication of Greek fortifications the standardized designs of their defense works were to be emulated in Europe for 1,000 years after the fall of the Empire. The legionary fortress, or *castrum*, as exemplified by Chester in Britain or Timgad in North Africa, was based on two main streets crossing at right angles, and was surrounded by a rectangular enceinte. This consisted of a ditch and earth rampart about 16 ft. (5 m) high, which was made permanent by an outer skin of stone. It was sometimes punctuated by towers with a twin-towered gatehouse guarding each entrance. This type of Roman structure was to form the foundation of many future cities.

Chains of smaller forts were built to provide shelter for marching armies at intervals of one day's march, and to protect vulnerable provinces from sudden attack. The forts built in the 4th century AD, as Britain's first defense

system, were an example of the latter. Detached watchtowers, or *burgi*, were built in the middle of the 2nd century. Unable to accommodate more than a few troops, these were for defensive purposes only and are important as the forerunners of the medieval keep. With the increasing threat to the Empire from the barbarians, the need to protect the hitherto undefended Roman cities became imperative. The 12 mi. (19 km) Aurelian wall was built around Rome, and in about 400 BC Theodosius II provided the eastern capital Constantinople with a defensive system which resisted invaders for 1,000 years until it was breached by the Turks in 1453. This consisted of three walls and a ditch, each line of wall overlooking that in front, and with the two inner walls protected by battlemented square or octagonal towers.

Permanent fortifications of the Roman type. Towers projected from the crenelated walls forcing the attackers to expose their flanks to the defenders, in any assault on the walls.

The Middle Ages

The castle is as much a symbol of feudal Europe as the Gothic cathedral, and thousands were built to dominate the fragmented territories that succeeded the Roman Empire. In some countries, particularly those on the edges of Christendom, their abundance has given the name to whole provinces—Castile in Spain and Burgenland in Austria. Before the castle emerged in its final form in the 13th century, builders had to relearn the skills of the classical civilization of Greece and Rome; for European fortresses of the Dark Ages were simple variations on the Iron Age hillfort. A good example is the Viking encampment at Trelleborg in Denmark which, in spite of its careful geometry of squares contained within circles, is still defended by a ditch and earthen rampart.

The Viking descendants, the Normans,

City walls of Avila, Spain, which completely surround the town. Begun in AD 1090.

evolved the motte and bailey castle. The motte was an artificial earth mound topped by a wooden tower. A palisade enclosed ground at the foot of the motte to form a courtyard or bailey in which people and livestock could find shelter. The tower on the motte was developed into a massive stone keep or donjon (the earliest has been dated to 992 at Langeris on the Loire) with a square ground plan as in the White Tower in London. The main living quarters of the feudal lord were on the upper floors of these towers which were entered at second-floor level.

Throughout the feudal era the keep remained a feature of castle building. During the 13th century it was sometimes incorporated into massive gate towers, as at Harlech, or simply became the largest among a number of towers, as at Caernarvon. With the breakup of feudalism, greater use was made of mercenary troops whose loyalty was always doubtful, and again the keep emerges as capable of defense from an internal as much as an external threat. Tattershall Castle (1434–46) is an example of a late type of keep.

The first change to the square plan tower was made necessary because of the vulnerability of its corners to attack by mining. Round towers were to prove more effective in this respect and also in deflecting missiles. At Château Gaillard, built by Richard the Lionheart in 1196, a cylindrical keep presents a massive masonry prow to the only possible line of attack. In England, shell keeps consisting of a circular wall with buildings inside enclosing a circular courtyard were built at Restormel and, although altered later, at Windsor.

In Germany, towers or "bergfrieds," derived from Roman *burgi*, remained tall and slender with only limited accommodation and were often attached to a surrounding wall which enclosed the main living quarters at ground level. This combination produced such romantic silhouettes as the Rhineland fortresses of Eltz or Pfalzgrafenstein, where the island tower is surrounded by a boat-shaped enceinte. In the 19th century Ludwig of Bavaria used these buildings as inspiration for follies like Neuschwanstein.

The palisade surrounding the Norman bailey was also replaced by a stone curtain wall. The danger of an attacker breaching a wall by a battering ram or by scaling it was countered by arranging projecting towers higher than the wall so as to provide flanking fire on the entire length of the perimeter—a technique used by the Greeks. Framlingham is an early English example of a curtain strengthened by towers. Walls were also strengthened by buttresses, as in Connisborough keep. Sometimes the two were combined—the Gravensteen fortress at Ghent owes its unique appearance to the towers built on top of the buttresses of the outer curtain.

Frederick II's Castel del Monte (1240) in Puglia, Italy, is a variation of the shell keep. It is a regular octagon enclosing an octagonal courtyard and with projecting octagonal towers on each corner. The symmetry of its planning, and the detail of its entrance, point to oriental and classical antecedents, for Frederick II had taken part in a crusade to the Holy Land in 1227. These wars had already given a tremendous impetus to castle design by transferring the ideas of Byzantium to western Europe. These concepts were being implemented as early as 1185 when the keep at Dover Castle was surrounded by inner and outer curtains to provide a system of concentric defense based on the walls of Constantinople.

In Palestine the Crusaders established a system of rule by a military aristocracy based in castles. They sought to secure the territory by a chain of massive fortresses which were so well sited that even today one of them, Beaufort in south Lebanon, still fulfills an important military function. The hilltop Krak des Chevaliers is perhaps the finest of these castles. Consisting of two concentric curtain walls with circular towers strengthened by massive battered plinths, its only approachable flank is defended by a moat. The keep has been replaced by three connected towers in the center of the system. The approach used the old techniques of exposing an enemy to attack on his right side, while machicolations projecting from the walls and towers enabled missiles and boiling oil to be hurled on the heads of those approaching the base of the defense enceinte.

Edward I was another Western monarch who had taken the Cross, and the chain of castles which he built to subdue the Welsh mark the culmination of a technique of fortress building that started with the Greeks. Caernarvon Castle, with its octagonal towers and its decorative use of masonry, must have been derived from the walls of Constantinople. Like Conway, the space inside the curtain is divided into two wards. At both these castles the main towers are crowned by smaller turrets and unlike Edward's other works the walls follow an irregular outline. At Harlech and Beaumaris the same designer, Master James of St George, produced carefully composed symmetrical concentric designs with massive keep-gatehouses attached to the inner curtain.

At Caernarvon and Conway, the castles are part of the defenses of two bastide cities built as part of a colonization program. Like the French bastides at Montpazier and Aigues Mortes, or the Italian Montagnana, these cities have an orthogonal grid plan reminiscent of the Roman *castrum* and are surrounded by a rectangular curtain wall with projecting towers. Saxon colonies planted in Transylvania in the 12th century to defend the

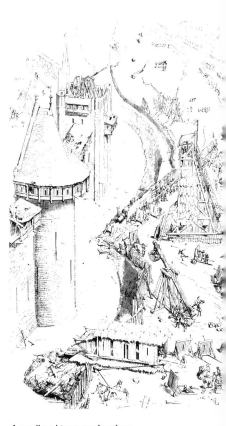

A medieval town under siege. Catapults, scaling towers, movable shields, and other siege engines are being used in an attempt to breach the walls and bridge the moat.

Krak des Chevaliers, Syria (1210–1252). A crusader's castle, built with concentric fortifications.

Early 15th-century siege cannon.

Low round bastion containing embrasures with cannons (early 15th-century).

Late 16th-century triangular bastion being abandoned during an attack. The assailants have bridged the moat and have built earthworks to protect themselves from the fire of the defenders.

eastern border of Hungary evolved an original form in their village fortresses, which consisted of a circular wall 40 ft. (12 m) high and 10 ft. (3 m) thick enclosing a circular space with a large church in the middle. Living quarters for the civilian population were on four floors of a timber structure supported by the outer wall.

City walls were the symbol of a free citizenry and the right to fortify was jealously guarded. Carcassonne, the best preserved of medieval fortified cities, has two concentric walls flanked with towers and separated by a walkway. The two gates are protected by barbicans and the whole system is strengthened by a citadel located on the western wall. In Italian cities in particular, the Middle Ages were a period of internecine strife between the nobles. This led to the construction of a high tower of refuge as part of the town house—the bristling silhouette of San Gimignano stands as a monument to this practice.

The age of the gun

Just at the time when the design of the castle reached its zenith, when a well-victualed fortress was virtually impregnable and it seemed that the balance had tipped irretrievably in favor of defense, a weapon was being developed which was to render obsolete the high stone walls which had proved effective for thousands of years. Although the earliest illustration of a gun dates from 1326 it was in the 15th century that firearms had a significant effect on warfare. The English defeat in the Hundred Years War had as much to do with the efficiency of French artillery as the inspiration of Joan of Arc.

The first reaction to the new weapons was to thicken existing walls since the high, relatively thin walls of the medieval castle not only presented a good target but were incapable of providing the wide platform needed to accommodate defending cannon. Existing fortresses had new works added which were capable of resisting the impact of cast-iron shot either through sheer thickness of stone or by backing a stone wall with an earth rampart. Transitional fortresses such as Senigallia or Ostia in Italy have low towers and battered walls but relatively few gunports, and they still use machicolations. Salses in France, built in 1498 with large gunports and curved parapets, represents a further step toward meeting the new threat.

The close of the Middle Ages and the eclipse of feudalism by the nation state also saw an increasing functional division between the fortress and the residence. Although some large houses, particularly French châteaux like Chenonceaux or Azay-le-Rideau, retained drum towers and machicolations, the symbolic and decorative nature of these defensive devices is emphasized by their juxtaposition with large windows. In Scotland, where there was for much longer a threat of sudden raids, a unique type of tower house was evolved which was capable of resisting attack by small firearms. They were built on an L plan as at Balbengo or, like Claypotts and Castle Fraser, on a Z plan.

The most spectacular demonstration of the inadequacy of medieval fortresses had occurred in 1494 when the artillery train of Charles VIII had swept through the Italian peninsula. It was not surprising therefore that the polymaths of the Italian Renaissance should have applied their talents to the problems of defense and the late 15th and early 16th centuries are notable for the number of published works discussing this topic. Leone Battista Alberti (1404–72) advocated the star-shaped citadel and Francesco di Giorgio and Antonio Filarete (1400–69) proposed a serrated plan for the defense perimeter. The former introduced the idea of the *caponier*; a protected position from which fire could be directed along the floor of a surrounding ditch. Leonardo da Vinci made many studies of fortresses with casemated gun positions and with carefully profiled embrasures to deflect shot.

The medieval round or square tower produced an area in front of it which could not be covered by fire from neighboring towers and it also presented difficulties in concentrating defensive fire. The Italians therefore replaced the tower by the triangular bastion, both faces of which could be swept by fire from the flanks of adjoining bastions. Antonio da Sangallo the Elder (1455–1534) built a fort at Nettuno in 1520 which had a square plan with four triangular corner bastions and introduced a defensive system that was to be developed over the next 250 years. Earthwork bastions were used at Ravenna in 1512 and had become common practice by 1527 when Michele Sanmicheli (c. 1484–1559) used them at Verona; consequently, during the next century, Italian engineers were in demand throughout Europe for their knowledge of the bastioned system of defense.

While the Italians were developing the triangular bastion the English, under Henry VIII, were building a dozen coastal forts which exploited the planning possibilities of the circle in a typically Renaissance concern with geometry. Built around 1540 they were intended to protect the south coast from a continental invasion and were probably derived from the published work of Albrecht Dürer. The largest, Deal, has a central circular tower surrounded by two lower rings, each of six semicircular bastions mounting guns in the open and in casemates.

These forts represent an interesting aberration because the future lay with the bastion which was already being exported from Europe to the New World. The Spanish during their domination of Central and South America protected their ports by elaborate systems of bastioned fortresses. At Havana and Cartagena the usually rigid geometry used in Europe is carefully adapted to the local topography while the Castillo de San Marcos in Florida, built a century after Nettuno, represents the ultimate development of the symmetrical quadrilateral fort.

The bastion, however, was not new to the Americas. Around AD 1200 the Chimu had built a fortress at Paramonga in central Peru which, in addition to surrounding a hill with three lines of walls, extended a platform at each corner to form bastions, presumably to remove the core of the citadel out of bow or slingshot range of an attacker. The Chimu capital, Chan Chan, covered an area of 8 sq. mi. (20 sq. m) and consisted of ten or more rectangular walled citadels which, it is suggested, accommodated different clans or tribes. The greatest pre-Columbian fortress was Sacsahuaman, built by the Incas to defend their capital Cuzco. This was constructed of massive granite blocks up to 20 ft. (6 m) long and 8 ft. (2.5 m) wide, with mortarless joints. It has three lines of ramparts arranged in a sawtoothed plan which exposed an attacker to flanking fire in the same way as a projecting tower.

It was more difficult to adapt city walls than to build single forts according to the new bastioned system. Existing buildings had to be respected in laying out fields of fire and narrow streets made it difficult to shift cannon around quickly so as to bring maximum firepower on the point of attack. One of the most comprehensive examples of the new system applied to an existing city is Lucca, in Tuscany. Here by 1561 low, wide bastions with blunt corners were encircled by a ditch with a sloping glacis, extending into the surrounding countryside and forcing enemy siege batteries beyond range of the city center.

The new defensive requirements could only be fully met in the new cities which were built to defend the boundaries of the emerging nation states. These took up the ideas of the Renaissance theorists. Francesco di Giorgio's 1480 plan for a city, based on an octagonal perimeter with streets radiating from a central space, became the prototype for projects such as Philipville (1555) on the Franco-Belgian border or Palmanova (1593) on the eastern frontier of the Venetian republic. The original plan for the latter had nine bastions on the corners of a regular polygon connected by radial streets to a central square where troops and artillery could be held in reserve and rushed to any point of attack. A century later when Sébastien Vauban (1633–1707) built a garrison town at Neuf Brisach in Alsace, he abandoned the radial plan for an orthogonal grid contained within a fortified octagon.

The French school of military engineering, and in particular Vauban, dominated fortification building during most of the 17th and 18th centuries. The simple bastion and curtain of the Italians was developed into "three systems." The first of these protected the curtain by a low outwork beyond the ditch while the second detached the bastion from the curtain so that in the event of its being captured the city could still be defended. In Vauban's third system, used at Neuf Brisach (1698), the diminutive bastions are protected by counterguards with a third line of outerworks, ravelins between the counterguards. The whole is protected by a system of covered passageways, or caponiers, covering the ditches. Vauban's great rival in both defense and attack was the Dutchman Coehoorn who had developed an even more complex "three systems," which relied on wet ditches and hollow construction in several layers. His first method was used at Bergen-op-Zoom but his second and third methods would have required such vast areas of land for their lines of defense that the small area remaining within the fort could not have accommodated an adequate garrison.

Although Vauban's fame today rests on his defensive building, his contemporaries respected him for his success in attack. By a method of digging a series of parallel trenches and bombarding walls at close range from breaching batteries, he took every stronghold that he attacked. With the advantage apparently in favor of the attacker, by the end of the 18th century engineers were proposing radical changes to the bastion system. One of these was the sawtoothed or tenaille enceinte which eliminated the weak curtain wall and replaced it by an arrangement of contiguous triangular outworks. Montalembert advocated backing the tenaille with two-storey gun towers and using three-storey casemated caponiers; two ideas which were to be developed during the 19th century.

Circular towers came back into use; martello gun towers of various sizes were used for coastal defense by the British during the Napoleonic wars, and the Austrians used them in the defenses they constructed at Linz and Verona in the 1830s. Three-decker forts had repulsed the Allies at Sebastopol during the Crimean War but during the American Civil War increasingly effective shells demonstrated the limitations of masonry in resisting an explosive impact. Steel and concrete, which were being developed for structural purposes, especially in France, offered an effective alternative.

The increasing power of 19th-century rifled guns forced defenders to extend their perimeters even further than the 2 mi. (3 km) of Vauban's day, in order to keep the urban center out of the besiegers' range. It therefore no longer became possible to surround a city

Naarden, Holland: one of the few settlements retaining a complete 17th-century fortification network.

Part of the Maginot Line, a defensive system built along the eastern boundary of France in the mid-1930s.

Submarine pen built by German forces in occupied Trondheim, Norway (c. 1941).

A fort in the Thames estuary; one of several surrounding the English coastline.

with a continuous enceinte and it was replaced by a series of small fortresses several miles from the center, covering one another with interlocking fields of fire. The abandonment of continuous walls was of great consequence, for it permitted the suburban expansion of the 19th century.

Detached forts had proved very effective in the Peninsular War when the British lines of the Torres Vedras, a series of 59 earthworks, successfully resisted the French advance on Lisbon. By the end of the 19th century the detached fort was being constructed in the new materials, often comprising a reinforced-concrete block buried in earth. A limited number of artillery pieces was mounted in armor-plated cupolas and infantry attacks were repulsed by machine guns, sometimes mounted in retractable turrets. Antwerp and Liège were surrounded by rings of forts of this type and around Verdun the same engineer, Brialmont, built examples with protective concrete roofs 3 ft. (1 m) thick, separated from the main structure by 3 ft. (1 m) of sand.

The 20th century

The new underground forts were tested in World War I and stood up to German attack with varied success. The Belgian forts were quickly overwhelmed by enemy artillery barrages but the more heavily fortified works at Verdun proved a more formidable obstacle. With the establishment of a line of field fortification stretching 496 mi. (800 km) from the Channel to Switzerland, and the high cost in lives of attempting any attack, the pendulum had once again swung in favor of defense. The Germans in particular developed the single slit trench into deep systems of interconnected dugouts, trenches, and pillboxes up to 1.25 mi. (2 km) deep. Initially the dugouts could be as much as 40 ft. (12 m) underground, but in the later stages of the war shallower concrete structures were used to house command posts, barracks, telephone exchanges, and field hospitals. Where ground conditions did not permit excavation both sides used concrete pillboxes with walls over 3 ft. (1 m) thick.

After the war, the French developed this apparent defensive superiority into the fortifications of the Maginot Line. The line, which was started in 1930, was compared to a battle fleet in that it deployed units of various sizes ranging from border minefields, fortified houses, and barbed wire entanglements through concrete "avant-postes" manned by 25 men and two-storey artillery casemates, to the famous underground forts or "ouvrages." These were sited every 3–5 mi. (5–8 km) and held garrisons of up to 1,200 men in five storeys of underground accommodation with only armored cupolas protruding above the surface. They were capable of resisting the

heaviest bombardment so, in the event of their being reached by an enemy, could call down fire on themselves from neighboring batteries. The Maginot Line forts were never taken by assault but they could not save France from defeat in 1940 by the Germans who exploited two offensive weapons developed during World War I—the aircraft and the tank.

The threat of aerial bombardment had led to defensive measures being taken during World War I. German submarine pens at Bruges used a combination of precast beams and in situ reinforced-concrete slab roofs which became the basis of structures built on the French coast during World War II.

In World War II the Germans tried to protect their armament industry against Allied bombing. A submarine assembly plant near Bremen had a roof of reinforced concrete 23 ft. (7 m) thick and fighter-plane factories were built with concrete shell roofs, using gravel mounds as framework. An underground factory was constructed during 1943 by slave labor at Nordhausen to produce V1 and V2 weapons and jet engines. It consisted of two parallel tunnels 1.25 mi. (2 km) long, connected by 61 ft. (200 m) of parallel galleries. By the end of the war work had started on other underground weapon factories as well as subterranean liquid oxygen and synthetic oil plants.

As well as protecting their installations, the Germans adopted a policy of dispersal away from concentrations of population. This strategic policy in Britain contributed to the success of the New Town movement which was intended to accommodate people from London and the provincial conurbations in new towns in the countryside. The idea was that by reducing the population density in existing cities, any future bomb attack would be less effective.

Aerial bombing created a completely new type of building: the air-raid shelter. The most interesting British innovation was the mass production of small steel "Anderson" and "Morrison" shelters. Germany made shelters available for a much larger proportion of its population: some of these concrete bunkers were designed to accommodate as many as 18,000 people. They were built above ground and their massive concrete walls were camouflaged with painted windows or as burnt-out buildings. Their construction could resist direct bomb hits but they could offer no protection against the fire storms caused by the heaviest Allied raids.

Another building type unique to World War II was the anti-aircraft or flak tower. German towers, as many as ten storeys high, gave an unobstructed field of fire from their roofs while accommodating guncrews, air-raid shelters, and hospitals in their lower floors. Similar towers housed radar equipment and, whether square or circular in plan, are reminiscent of medieval keeps complete with projecting gun platforms which look like huge machicolations. The high observation towers built by the Germans in the Channel Islands as part of the Atlantic Wall used rough shuttered concrete in a way which anticipated the postwar work of Le Corbusier.

The British built forts in the sea approaches to provide a defense against enemy surface vessels in addition to aircraft. Navy forts, 9 mi. (15 km) offshore, consisted of two concrete towers connected by a steel superstructure. They were constructed onshore and then towed out and sunk in position. Army forts were also floated out after fabrication. These were sited about 4 mi. (6 km) offshore and consisted of several three-storey accommodation units, with gun platforms sitting on top of steel and concrete frames.

The threat of nuclear war hanging over the world during the last half of the 20th century has provoked a number of responses in defense building. The logic of nuclear deterrence means that a defender must have knowledge of an incoming attack in order to launch his own planes and missiles: hence the building of the DEW (Distant Early Warning) line across the Arctic wastes of North America. These are not conventional forts but radar and communication complexes housed in plastic geodesic domes. This electronic shield extends into the Atlantic where "Texas Towers" accommodate similar equipment on platforms derived from oil dwelling rigs. Electronics in the form of "people sniffers," heat detectors, and laser beams have also invaded the conventional battlefield although more orthodox defenses still have their uses—the underground forts of the Bar lev line played a very important part in the Yom Kippur war and the Vietnamese built elaborate dugout systems against American saturation bombing.

In a planning response to the atom bomb a number of decentralization proposals were made in the 1950s on a scale far exceeding that previously proposed in Britain. A suggestion was made to divide the whole of the U.S. into 25 mi. (40 km) squares, each with a factory in the middle and housing arranged in linear strips of 160 dwellings to the mile. Similar Soviet decentralization proposals were based on the experience of the industrial dispersion of World War II.

The height of the Cold War saw a brief period of shelter building in the U.S. There was argument between the advocates of large blast shelters and those of light fallout shelters which, it was claimed, could mean a difference between 20 and 80 million casualties. The Federal Civil Defense Administration prepared a program for the strengthening of existing buildings and the provision of blast-proof concrete cores for all new shelters.

Emergency, portable, and temporary buildings

The concept of built structures generally assumes the idea of permanence. However, there has always been a significant and influential minority of structures which have not been permanent and whose very essence has been their temporary nature. These structures fall into three broad categories. Firstly, there are structures designed for easy initial transportation; a good example is the World War II Nissen and Quonset huts, a further example is the factory-finished, road-transportable mobile home. Secondly, there are structures designed for a continuous cycle of transit, erection, use, dismantling, packing, transit, and so on. Here the best example is the circus tent. Finally, some have been designed for a short-term specific function; the Crystal Palace in London, England, was an example of this type of structure. Some structures combine characteristics of two categories.

The architectural interest in temporary structures stems from an extremely rapid time scale for erection and demolition and the unusual freedom from aesthetic, legal, and financial constraints which have characterized the erection of such structures. As a result, temporary mobile structures often display a range of remarkable innovative skills in solving formidable problems. The resultant design freedom acts as a persistent challenge to the well-worn conventions of traditional architecture.

Structures designed for easy initial transportation

Houses Prior to industrialization and the consequent development of transportation, it was rare for all the components of a building to be moved any distance. There are exceptions, as in the 18th century when East Anglian bricks were shipped to St Petersburg in Russia for the building of the Czars' palaces. The need to move entire buildings in prefabricated form occurred with the immigrant movement to the various colonies in the period before 1820. In 1624 the English brought a prefabricated, panelized house to their fishing settlement at Cape Anne in the U.S. In 1727 two timber houses were manufactured in New Orleans for use in the West Indies. Later buildings of this type formed ready markets in the California Gold Rush and in the development of the Prairie States of the U.S.

One major innovation, dating from 1829, was the development of corrugated-iron sheeting. An English firm called "Richard Walker—Carpenter and Builder" carried out the initial development of the material (which was invented by Henry Robinson Palmer)

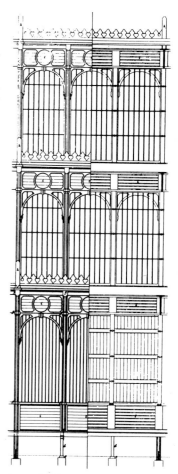

Internal and external elevations and section of one 24 ft. (7 m) structural bay of the Crystal Palace, London, England (1851), by Sir Joseph Paxton.

Interior view of the Crystal Palace after its reerection at Sydenham, south London, England in 1853.

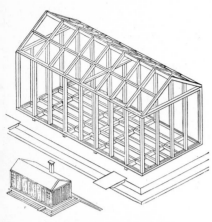

Portable cottage for emigrants designed by Mr Manning of High Holborn, London, England, in 1830. Built out of accurately cut standardized timber elements.

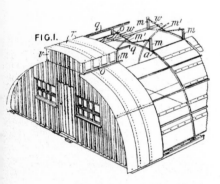

Patent drawing of a Nissen hut by Captain Nissen (1916). These steel-framed and corrugated-iron buildings were used extensively in both World Wars.

The Tabernacle (c. 1491 BC), a massive portable tent structure used by the Israelites on their wanderings through the Sinai Desert following the exodus from Egypt.

recognizing its potential for use in roofing, doors, shutters, partitions, and external walls. In 1833 Walker advertised his product: "(corrugated iron) is particularly commended for portable buildings for exportation. The small space occupied in storing them, when the respective parts are separated, renders their convenience cheap and easy. For new settlements, the facility with which they may be erected or removed from place to place is a desideration of great consideration."

In addition to Walker's advertisement, Loudon's Encyclopedia of 1833 described two major wooden prefabricated houses both built in the late 1820s. They were manufactured by a Mr Manning of Holborn, working with a Mr Richard, a carpenter from the East End of London. The first of these houses was built to accommodate the exiled Napoleon living on the island of St Helena. It was 118 x 49 x 23 ft. (36 x 15 x 7 m). The second was a rather more ambitious house designed for the Governor of New Zealand. It included an enclosed promenade 328 ft. (100 m) long, weighed over 250 tons, and cost £2,000.

Thus, from the 1820s there were firms producing prefabricated houses. Walker's corrugated-iron structures were less developed in detail than Manning's timber structures, and their acceptance was probably limited due to the corrosion problem. Manning's designs included a wide range of "mobile homes." He had already solved many of the inherent difficulties of prefabrication and standardization—assembly, erection, and portability.

Military structures Prefabricated military buildings of one kind or another have existed from the earliest times. During the 19th century many different systems of hutting for temporary barracks, hospitals, and stores were developed. During the Crimean War in the 1850s, timber and corrugated-iron shelters of many designs were used. Isambard Kingdom Brunel (1806–59) designed his famous timber Renkioi hospital in this campaign, which was erected and brought into operation in a very short period of time (see HOSPITALS). The American Civil War provided a large market for camp buildings of various kinds. Skillings and Flint lumber dealers of Boston and New York began producing standard panelized components for buildings before 1861 and found a large market in the Union Army.

In 1916 Captain Nissen invented the "Nissen hut," a prefabricated frame structure covered in bolted sections of corrugated iron. In addition to this, there were prefabricated homes built in Britain and Europe after World War I.

However, it was not until World War II that prefabrication techniques, coupled with the development of road transportation became an everyday activity and these "tem-

porary" homes are still with us. During the war, prefabrication techniques had widespread military and civilian use. In 1943 for example, German authorities built 1,625 prefabricated homes in Hamburg following the bombing of the city. In the approach to the "D Day" landings the South of England became a vast camp with over 1.4 million troops accommodated in "hutments." A wide variety of prefabrication techniques were used, one of the most common being the corrugated-iron C'testiphon system. The American Quonset system was also extensively used.

Structures designed for continuous movement

The Tabernacle in the wilderness In chapters 25–31 of the biblical book of Exodus there is a highly detailed account of the building of the Tabernacle. This was a portable sanctuary in the form of a tent structure built to symbolize God's presence with the Israelites on their desert wanderings. It has been calculated that the Tabernacle was built c. 1491 BC, thus making this one of the earliest fully documented accounts of a building. The structure was used for 40 years in the desert wanderings of the Israelites and then for 343 years in Shiloh, giving the structure a life span of 383 years.

The Tabernacle was approximately 157 x 79 ft. (48 x 24 m). It was made up of standardized units of curtains on timber framing (19 x 9 units). Within this open-roofed courtyard the Tabernacle (which means the "Tent of Congregation" in the original Hebrew) was positioned. This structure was the most sacred part of the edifice; it contained the Ark of the Covenant, which contained the stone tablets with the Ten Commandments inscribed upon them.

Portable prefabricated tent shrines similar in form to the Tabernacle had been constructed in Egypt prior to this structure, but precise details are not known.

Membrane structures On the upper wall surface of the Colosseum in Rome there is a rhythm of projecting stone brackets. Originally these supported timber masts that held up a vast "velaria." This was a linen fabric canopy to provide protection from rain or sun and it was slung across the structure, supported by natural fiber ropes. This membrane and cable structure was widely used throughout the Roman Empire to support roofing to amphitheaters and theaters.

During the intervening centuries the development of membrane structures was limited to military tents and similar structures. Probably the next major development occurred with the circus tent. The development of the circus began about 1800 with performances being held in large halls which were often specially built for the purpose. In the 1860s in the U.S., circus companies began to use the railroads to transport their equipment to towns which did not possess such large halls. This required the development of structures that could be quickly assembled and dismantled. In 1867 an American circus, complete with its own tent and technical apparatus, visited Paris. This created much excitement, and resulted in the use of similar structures by European circus companies. These tents, such as the classic "Chapiteau," were up to 164 ft. (50 m) in diameter, supported on four primary masts made from machine-woven linen or hemp canvas. By 1872 the Stromeyer Company, which manufactured large tent structures, was established in Germany and is still in existence today.

In 1917 an English engineer, F. W. Lanchester (1868–1946), took out a patent for the design of large tent structures, which had no need for poles or supports. His patent was for an air-supported structure, with entry confined to various air locks. However, the patent was not developed until the 1940s. Nowadays, the inflatable structure has widespread use, particularly for temporary warehousing. It has also been used in combination with the geodesic dome. Here, the rate of inflation of the membrane support structure is related to the speed at which operatives can bolt together the metal sections of the dome structure. In this way the membrane forms a type of scaffolding that raises the structure as it is continually added to at ground level. This removes the need for complex scaffolding, and the membrane structure can serve as the weather protection.

Geodesic domes Dome structures, the innovation of Buckminster Fuller (b. 1895), have been progressively developed since the early 1950s. In 1957, for example, the Hawaiian Symphony Orchestra was able to give a concert inside a 157 ft. (48 m) diameter dome that 24 hours earlier had been no more than a collection of subunits unloaded from an aircraft. The U.S. government sponsored the design of a dome-shaped Trade Pavilion which was used in various sites in different countries from 1956 onward. However, despite the qualities of these dome structures, there are very basic inherent problems such as the difficulty in dividing up the internal space. This is not a particular problem with exhibitions, but for many other functions it may rule out this form of structure.

Structures design for short-term functions

Pavilions–the Field of the Cloth of Gold Perhaps the most sumptuous temporary structures ever built were erected in a field outside Calais in the year 1520. This was for a 13-day "summit meeting" between King Henry VIII of England and Francis I, King of France, and their courts. The meeting place became known as the Field of the Cloth of Gold; an apt description of the finery that was liberally applied to the vast array of temporary structures. The French relied upon a series of spectacular tent structures, while the English built a vast temporary palace as well as a series of banqueting marquees.

The French marquee was an astonishing 120 ft. (37 m) high, supported by a central mast of wooden posts lashed together. The British were more ambitious, building a temporary palace 180 ft (5.5 m) square. The building was a curious mixture of materials. The roof was of canvas and below this was a timber-framed structure. The walls were pierced by vast windows of over 5,000 sq. ft. (464 sq. m) of glass that gave great delight to the occupants. Then below the frames was a brick base which in turn sat on a stone foundation.

Masques The first opera was staged in 1597 and complex art form required elaborate scenery with complex machinery to control its movement. Masques were initially confined to indoor theaters but they soon outgrew the confines of the four walls. Large-scale structures were devised for use in the open air; many of them were movable and were operated by complex pulleys and winches. Similar structures existed in medieval times for pageants and religious plays.

Exhibition halls Ever since 1851 there has been a succession of Great Exhibitions. The Crystal Palace (1851), designed by Joseph Paxton (1803–65), was made up of standardized, mass-produced, cast-iron, and timber details and a vast acreage of glass. It was designed on a module of 8 ft. (2.4 m) and covered an area of over 91,507 sq. ft. (8,500 sq. m).

From this date, there has been a long succession of "landmark buildings" built for various exhibitions and world fairs. They range from the Eiffel Tower of the Paris Exposition of 1889 up to the air structures designed for the Osaka World Fair in Japan of 1970. (see EXHIBITION BUILDINGS.)

Disaster housing Since the early 1960s relief agencies have attempted to design disaster shelters that can be flown to areas of need in great haste. One such innovation is a polyurethane dome structure 16 ft. (5 m) in diameter which is sprayed onto an inflatable mold. This is then deflated and the completed dome removed. However, with a greater understanding of the survivors' needs it is now apparent that there is very rarely any need for such provision. Surviving families can rebuild their homes very rapidly, or move in with relatives. In terms of time, the critical factor is always the acquisition of land for the structures, not the time to fabricate them.

A demountable tent dome devised by students of the Architectural Association in London, England, and used for an exhibition in 1969.

Painting depicting the "Field of the Cloth of Gold"—the meeting place between King Henry VIII of England and Francis I of France in the year 1520.

Polyurethane igloos donated by the Red Cross after the 1972 earthquake in Managua, Nicaragua. They were however unoccupied until 138 days after the earthquake.

Section 3 Structural design, elements, systems and processes

Structural design and its bases

Structural theory

Structural theory, as it now exists, is a growing body of mathematical models of the ways in which structures carry loads and deform under them or, at their limit, collapse through instability or local overstressing of the materials. Most of these models relate to particular types of structural element or system, and particular types of loading and response to the load. They are based on a few much more fundamental and widely applicable models of the conditions of static and dynamic equilibrium and the deformation of materials under load. They have to be supplemented in use by data on, and theories about, the loads likely to be experienced.

Even the fundamental models—the basic theories of statics, dynamics, deformation, and failure—are mostly of recent origin. Prior to their formulation, there was a different kind of codification of the experience gained in actual construction to serve as a guide to design. But we will consider here just the development of these basic modern theories and the other models or theories derived from them.

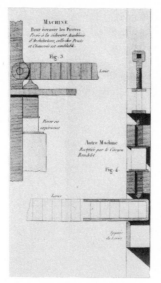

The practical application of structural theory usually calls for a knowledge of material strengths. These French testing machines (late 18th century) were used to determine the crushing strengths of building stone.

A view of the right-hand machine shown above in use. The loading lever, pivoting on knife edges, allowed the applied load to be magnified up to 63 times.

Subsequent development has been in the practical application of these theories. Until recently, it has been limited to statics, initially to the determination of forces in structures that were "statically determinate," but increasingly for the past century also to similar determinations in structures that are "statically indeterminate." The distinction here is between structures whose geometry, support, and internal jointure allow only one internal distribution of forces for a particular pattern of loading, and structures in which they would allow numerous different distributions, so that the actual distribution depends also on the material properties and the resulting deformations. Only the former (or, more strictly, the idealizations of them that are considered for the purpose of analysis) are statically determinate. The first structural forms to be analyzed were the hanging flexible chain and the arch envisaged as an inverted chain. The next major application was to pin-jointed trusses in the mid-19th century. Since applications to statically indeterminate structures must take deformations into account, they have involved also the basic theories of deformation and will, therefore, be considered below. Applications of dynamics to problems of major structural interest, like the response of buildings to fluctuating wind loads or earthquake shock, involve the deformational characteristics in yet another way and have become practical only in the last few decades.

Statics and dynamics

The basic theories of statics and dynamics can at least be traced back to Greek beginnings. Only Archimedean statics were of practical relevance though; and they were limited to the determination of centers of gravity and the laws of balance for the simple case where the weights on the balance arm act vertically. This limitation was a grave shortcoming because it precluded any valid application to the balance of forces in any case, like that of an arch or dome, in which major forces were inclined to the vertical.

A precise concept of a force acting in any direction was arrived at only slowly between the 13th and 16th centuries by men like Jordanus, Leonardo, and Stevin. They envisaged an inclined force in terms of the effective weight of a body partly supported by an inclined plane, or in terms of the pull on a balance arm of an inclined cord passing over a pulley and then carrying a weight vertically at its free end. On this basis, they were able to form some idea of the conditions of balance or static equilibrium of forces acting in different directions. A more powerful concept was arrived at in the following century in terms of dynamics; that is of the movements that different forces would produce. This concept received its classic statement in Newton's laws of motion. It was more powerful because it was completely unrestricted by particular physical models like weighted cords, and because it included the situation where movement does take place as a result of the forces acting.

Deformation and strength

Complementary theories of deformation and strength did not really begin to appear until the publication of Galileo's *Two new sciences* in the early 17th century. Though Galileo's conclusions had been foreshadowed to some extent by Leonardo, it was Galileo who first clearly formulated them in a coherent set of general propositions. The main subject of interest was the bending strength of a beam, which observation showed was not simply proportional to the cross-sectional area for a given material, as was the tensile strength. Galileo correctly deduced that the bending strength of a beam of rectangular cross section was proportional to its width and the square of its depth. However, he arrived at too high a ratio of the bending strength to the strength in direct tension because he assumed that the entire cross section would fail simultaneously in tension, pivoting about the bottom edge.

Galileo's error may have stemmed from the observation that a stone beam does, in fact, fail throughout its depth in tension. But it is not a simultaneous failure. It starts at the top and rapidly progresses down as the remaining unbroken section is reduced in depth. To understand this it was necessary to take into account the deformations associated with bending, and the relationship between these deformations and the development of internal resistance to the applied load. The brittle fracture of a rectangular stone beam is, moreover, a simple phenomenon compared with the deflection and failure of a beam of non-rectangular section made of a material like wrought iron, steel, or reinforced concrete. Fuller understanding was only gained slowly and was dependent on basic theories of elastic and plastic deformation.

The basic theory of elasticity postulates a direct proportionality of stress (internal resistance) and strain (local deformation) and is a restatement of Hooke's Law, published in the late 17th century, that the force required to produce a given extension is proportional to the extension. On this basis, the stress in a beam varies uniformly from a maximum tension on one face to a maximum compression on the opposite. Coulomb first applied similar ideas to the problem of bending in the late 18th century, and they were developed further by Navier in the early 19th century. On the foundation laid by C.L.M.H. Navier (1785–1836) there was then an ever-widening range of application to fresh problems of statically indeterminate behavior. In the 19th century these remained predominantly problems of bending—including the bending of columns, of beams continuous over several supports, and of continuous frames—and of interactions between members in structural systems like statically indeterminate plane trusses and stiffened suspension bridges. In the 20th century, there have

The simple theory of the bending of beams was largely worked out between the mid-17th and early 19th centuries. But considerable reliance was still placed on direct testing when Peter Barlow made this test in about 1810.

been many further applications to members of composite cross section—notably of reinforced and prestressed concrete—and to complex structural elements and systems that have had to be considered three-dimensionally.

The postulate of a direct proportionality of stress and strain is, however, an idealization of varying validity for different materials. For some, it is valid almost up to failure. For others, it is valid only for relatively small deformations compared with those they are capable of sustaining without failure. And, for some of each, the stress at which failure occurs is not constant but varies with the type of stress. In particular, it may be much lower in tension than in compression. Throughout the 19th century and well into the 20th, the consequent limitations in the validity of predictions based on this postulate tended to be a source of confusion. There was not usually much trouble when they were used only to give indications of stiffness or safety under loads well within the ultimate capacity; but they could give very misleading indications of ultimate strength. An early hint of this arose in the course of the initial search for the ideal cross section for a cast-iron beam. Elastic theory suggested an I–section with equal top and bottom flanges, which is indeed the stiffest section. But tests, which were usually designed only to measure strength, indicated that a very different I–section, with a much larger bottom flange, was considerably stronger for the same cross-sectional area. In this case, the theoretical predictions could be reconciled with the experimental observation simply by allowing for the different strengths of the material in tension and compression. Less easily resolved discrepancies between predicted and true strengths became apparent later when materials like steel and concrete were used, particularly when they were used in systems which were statically highly indeterminate. Here they were the result of considerable capacities possessed by these materials for deforming further without failure after the stress had reached a value near its maximum.

The simplest alternative postulate to cover this situation was the simple theory of plasticity. According to this, the material does not fracture but continues to deform at a constant yield stress once this stress is reached. Under lower stresses, its deformation is relatively so small that it can be ignored. On this basis, the stress in a beam made of a single material increases at failure to a uniform tension over one part of the cross section and a uniform compression over the remaining part—not quite Galileo's assumption because the compression must balance the tension and will usually have to act over a comparable depth. Hitherto the main applications of this theory have been in predictions of the ultimate strengths of reinforced-concrete slabs and continuous frames, and of steel rigid-jointed frames; but, since they are again based on idealizations, these predictions usually only indicate possible limits or bounds to the real strengths.

The present trend is toward more realistic basic postulates recognizing both an initial elastic deformation and a subsequent plastic one, and toward more comprehensive predictions of the response of complete structural systems to increasing loads. In particular, more attention is being paid to stability which usually deteriorates as deflections increase, even when an ample margin of strength might seem to remain. Considerable attention is also now being given to dynamic behavior. All these developments have involved calculations on a scale which involves the use of computers.

Design criteria

When the basic theory of static equilibrium for forces acting in any direction was first applied in structural design in the second half of the 18th century and the early 19th, the criterion of a safe design seemed obvious enough. The structure would be safe if it could support its own weight, and perhaps the weight of a wagon passing over it, or of machinery on a floor, without overloading any crucial element—arch rib, beam, column, masonry pier, or tie rod. The strengths of these elements could be assessed by loading specimens to failure, or by similarly loading specimens of the material if the strength of the element could then be estimated by simple proportion. For greater safety, some factor would be allowed on the measured or estimated strengths.

During the 19th century, this approach had to be supplemented in two ways. Loads other than the weight of the structure itself became more important, particularly for railroad bridges, and the development of elastic theories of the behavior of the main structural elements and some complete structural systems called for further criteria to bypass the reliance on strength tests of these elements and systems. Tests were made to determine both wind loads and the effective loads imposed by moving locomotives, but the data obtained remained of limited and somewhat questionable validity for want of adequate understanding of the nature of these dynamic loads. To ensure the safety of each element of the structure, it became usual to estimate the maximum stresses that would arise under the maximum loads that would normally be expected, and to compare these "working stresses" with "allowable stresses" for the materials used. These latter stresses were substantially below the stresses at which tests had shown that the material would fail, the margin being a matter of judgment, initially by the individual designer and later by groups of the more experienced designers.

In the first half of the 20th century, much more data was acquired on the loads to be expected, and design criteria for particular classes of structure—like steel frames and reinforced-concrete frames—were progressively codified for normal design in terms of design loads and allowable stresses. Since it was increasingly realized that the estimated maximum stresses would be subject to varying margins of error, according to the sort of approximations or idealizations implicit in the estimations, the allowable stresses were usually stated in relation to particular specified means of estimation. They also took some account of the nature of the loading (steady or fluctuating, closely predictable or highly unpredictable) and the likely mode of failure (sudden or gradual; catastrophic or purely local) as well as of the variability of the material. However, they remained, to a large extent, a matter of professional judgment and were ultimately justified or found lacking only by the test of the safety of structures designed to them. Some aspects of design remained outside these criteria and were still covered, as they had long been, by direct use of proven practice.

In the last few decades, far more again has been learned about likely loads, particularly wind loads and earthquake shocks, and about their variability, and that of the strengths of materials and of the structural elements made from them. In the light of this knowledge, and of the newly available theories of ultimate strength based on the theory of plasticity, there have been two new trends. The first has been toward recognizing the variabilities explicitly in the design criteria, coupled with the use of probability theory to arrive at the precise criteria.

The second has been a move away from an almost exclusive concern with allowable stresses at normal loads, toward a parallel or primary concern with ultimate strengths. The criterion of safety here reverts almost to the original criterion, but with a precisely specified margin between the estimated strength and the expected load. Additional criteria are then introduced to guard against possibilities of excessive deflections and other undesirable behavior under normal loads. In principle, each eventuality considered is guarded against by ensuring that it has a sufficiently low specified probability of occurrence.

Structural design

Design must always be distinguished from analysis—structural design as much as any other kind. In the limited sense of conceiving and drawing or modeling a form to be built, it should be based on one sort of analysis—an appraisal of the relevant requirements—and must precede another, the testing of the design against appropriate criteria to obtain some assurance before it is built that it will meet the requirements. In practice though, there tends to be a certain amount of alternation between conception and these two sorts of analysis, so that it is probably better to consider design as embracing all three activities. In these more embracing terms at least, structural design has clearly changed considerably with the growth of modern theory and of the associated possibilities of analysis. On the other hand, much structural innovation occurred without the benefits of this theory, so it is worth considering how this was possible.

Before doing so it will be helpful to consider the general course of structural design. This has always been far less innovative, but has been the soil from which innovation has sprung. Usually it has gone no further than developing small variations on what has been built previously. Heights or spans may have been changed slightly, a different material may have been substituted somewhere, or the plan may have been adapted to fit a different site or meet a different functional requirement. Today, the safety of such variations is checked by routine analyses of stresses, strengths, and deflections in accordance with some "code of practice" whose rules, though framed in terms of the kinds of criteria discussed above, are essentially rules for the safe, but limited, development of past experience. The chief difference in the past, before such rules were available, was that past experience had to be drawn on more directly, and with less understanding of the likely limits of its validity. The corresponding rules, whether explicitly formulated or just part of the builder's know-how, mostly related directly to the safe proportions of elements used in particular ways and constructed of particular materials. Because they were not framed in terms of the strengths of the materials and the loads to be carried, they had to be learned afresh by trial and error whenever there was a significant change in material or scale or loading if the strength of the element was the main consideration. They were generally useful because in many structures the stresses were almost entirely compressive and well within the strength limits of the material if they were sufficiently uniformly distributed. Safe design was largely a matter of ensuring this reasonable uniformity by suitable geometric proportioning. A further important factor was that the loading was nearly always predominantly self-weight, which was also determined by the proportions, though scale was also important here.

In these circumstances, innovation was for a long time a very slow evolutionary process punctuated not infrequently by collapses or partial failures when the designer went a little too far in some direction. Many such collapses are on record, as are the remedial works which often followed. There must have been many more of which we now know nothing. As the height of Gothic churches was increased, for instance, there was fairly frequent trouble from the high vaults pushing out the supporting piers, and one origin of the flying buttress was the addition of external props to some churches to halt such movements before the vaults collapsed. Elsewhere, ties were added across the springings of the vaults for the same purpose.

Occasionally, however, more radical innovation was attempted. Two instances were in the rebuilding of the Church of Hagia Sophia in Constantinople in the 6th century, and in the construction without centering of the great dome of Florence Cathedral in the 15th century. In both these cases, the designers must have had a considerable intuitive grasp of the relevant conditions of static equilibrium, though they must have seen them primarily in geometric terms and fallen back, like their contemporaries, directly on past experience when they had to decide on the dimensions of a buttress or a tie. In Constantinople, Anthemius and the elder Isidorus underestimated the necessary buttressing, with the result that more had to be added as construction proceeded. It was also necessary, some years later, to rebuild the dome to a raised profile. But this does little to diminish an achievement that was almost without parallel. In Florence, Filippo Brunelleschi (1377–1446) built only the dome, and was fortunate in the work of his predecessors who had provided a very secure base for it. His design for constructing it was based throughout on a very clear recognition of the stability of a circular dome with a central circular opening of any size by virtue of the continuous horizontal arch at this opening, and he seems to have very imaginatively applied lessons learned from a study of flat Roman concrete arches in overcoming the problem presented by the octagonal

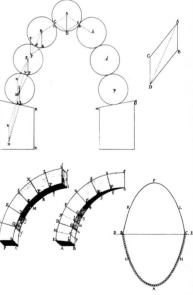

By a theory of statics made in the early 1750s, the cracked dome of St Peter's, Rome, was envisaged as composed of a ring of orange-slice arches (bottom center), and each arch was then considered as a mirror inversion of a similarly weighted hanging chain.

plan of his dome. It is notable that he made one large brick model before starting construction of the dome itself, presumably to test the basic concept.

Though it is possible that both Leonardo da Vinci (1452–1519) (in projects for the dome of Milan Cathedral) and Sir Christopher Wren (1632–1723) (in designs for the dome of St Paul's, London) were guided by a fuller understanding of the conditions of static equilibrium, it is not until the mid-18th century that there was clear evidence of practical application of statical theory. It was used then in several analyses of the stability of the cracked dome of St Peter's, Rome, built more than a century earlier. A little later it was used again in analyses by Gauthey of the stability of a projected dome for Ste Geneviève (now the Panthéon) in Paris; this being probably the first time that it played a direct part in innovative design.

Ste Geneviève was, however, still a masonry structure in which the stresses were predominantly compressive and self-weight was the major load. Analyses based on statical theory and the theory of elasticity, closely coupled with tests on prototype elements and models, really began to play a major part in design in the first half of the 19th century. Without them, the unprecedented rapid development of new types of beam, truss, and suspension system in cast and wrought iron, often to carry heavy imposed loads, would have been impossible. Notable examples were Sir William Fairbairn's (1789-1874) and Eaton Hodgkinson's (1789–1861) search for efficient forms of beam, initially for mill buildings and then, in association with Robert Stephenson (1803–59), for the 400 ft. (122 m) and 450 ft. (137 m) spans of the Conway and Britannia Tubular Bridges, and the contributions by Thomas Telford (1757–1834), and C.L.M.H. Navier (1785–1836) to the development of wide-span

suspension bridges (see BRIDGES). Long series of carefully planned tests played the major part in guiding the beam designs. Telford also relied largely on tests to establish the profiles and cross sections of the suspension chains both for his projected 980 ft. (300 m) central span over the Mersey at Runcorn and for the slightly later executed bridges over the Menai Straits and the Conway River. But Navier, having studied Telford's and other designs, then provided a theoretical basis for future designs. Truss design finally emerged from an early, very confused phase only with the introduction of simple and accurate ways of calculating the forces in the members in mid-century; and further development was greatly facilitated a little later by the introduction of purely graphic methods of analysis.

In comparison with the innovative design of the first half of the 19th century, design in the latter part of the century and the first half of the 20th century was mostly a series of more direct extrapolations of previous designs. Developments in theory and related developments in design criteria made it possible to vary forms, increase their scale, and even introduce new materials like reinforced concrete, with much greater freedom and assurance than would have been possible previously. But, in so doing, they must have reduced the incentives to innovate more radically, and they probably led to a frequent overemphasis on the analytical aspects of design. Hardy Cross (1885–1959), a great American teacher, was among those who recognized this overemphasis and helped to counter it by introducing a new way of analyzing complex building frames that enabled the designer to focus throughout on the physical reality rather than a mathematical abstraction. A few designers in reinforced concrete, from Robert Maillart (1872–1940) and Eugène Freyssinet (1879–1962), to men like Eduardo Torroja (1899–1961) and Ove Arup (b. 1895), also saw that, as a guide and stimulus to creative design, structural theory was valuable mainly as a source of insight into the ways in which structures behave. Where they could not justify the safety of their designs by calculation, they resorted again to testing, and were able to learn considerably more from the tests than their predecessors could have done, thanks to their deeper theoretical insights.

Recent further developments in theory and design criteria, together with improvements in model-testing techniques and the revolution in calculation brought about by the computer, have made a new wave of innovative designs possible. Typically, these designs have been developed much like some of those of a century or more previously through both tests and analyses. But these have explored types of interactive structural behavior far beyond the reach of the earlier tests and analyses. Analysis by computer can, moreover, be arranged to explore very rapidly a range of variations on a

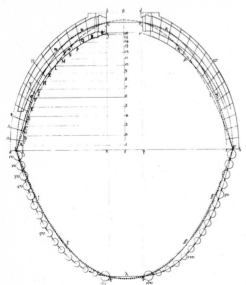

The detailed application of the hanging-chain analogy (see previous picture).

particular design to enable the most appropriate configuration or proportions to be selected. The selection can even be made automatically if the designer feeds into the computer the necessary basis of selection; but there is no need for him to hand over responsibility to the computer to this extent. In the design of buildings at least, the structure is only a part of the whole. There is therefore a strong feeling among the best designers that, unless they have to push a form to the very limits of practicality, they should remain in more direct control.

Structural elements

Arches

Throughout most of architectural history, the arch has been the chief means of overcoming the spanning limitations of single blocks of stone or lengths of timber. It carries its loads in simple compression, acting along a line which curves downward from the crown toward each abutment. Because of this simple compressive action, the individual units of a brick or stone arch need merely be butted against one another, either with or without an interposed bed of mortar to give more uniform contact. Provided that the joints are aligned roughly at right angles to the compression, the precise curve of the arch is not very critical until its depth becomes a small fraction of its span. The relative immobility of the abutments (which tend to be forced apart by the arch thrust) is more important. The flatter the arch, the less critical is its precise curve, but the greater the outward thrust becomes and the more important the immobility of the abutments. In the extreme it may have a completely flat soffit, but very strong abutments are then necessary except for the smallest spans.

One obstacle to the early realization of this true arch form would have been the need for some kind of centering to give temporary support to two halves of the incomplete arch during construction. The earliest forms were therefore probably merely approximations to it that eliminated this need. One, which survives today in Mycenean cyclopean construction, consisted of only three rough blocks of stone, the central one somewhat larger than the gap between the other two and wedged between them. A second, of which monumental examples survive in Egypt from the 3rd millennium BC, consisted of only two long blocks inclined toward one another as an inverted V-shape. This form was probably constructed even earlier in timber. The third, of which surviving examples are very widespread, was what is commonly known as the false or corbeled arch. This consisted, in fully developed Egyptian and Greek examples, of four large blocks laid on horizontal beds with the upper two projecting forward over the lower two to close the gap and

with the undersides of all four cut to a continuous semicircular or segmental soffit. In other examples, larger numbers of small blocks were similarly laid, usually to form an inverted V-shaped opening in a wall.

None of these early forms was very efficient. Spans rarely exceeded 6 ft. 6 in. (2 m). The spanning of substantially wider gaps called for true arches constructed on centering from large numbers of bricks or stone voussoirs. Small true brick arches appeared first in Mesopotamia and Egypt, to be followed, by the 5th or 4th century BC, by arches of accurately cut stone voussoirs. Initial caution was reflected in the use of voussoirs of a depth exceeding the radius of the arch. But, with increasing confidence, the Romans reduced this depth to about one-tenth of the radius on spans up to 82 ft. (25 m) by the 1st century BC. This, however, was mainly for bridges. In buildings, concrete was then beginning to be used in place of cut stone, and was used in Rome for all the longer spans from the 1st century AD. Superficially, Roman concrete arches might be mistaken for brick arches, because they were invariably faced with brick. But only about one brick in four penetrated the thickness of the arch to divide the concrete core into voussoir-shaped sections. Elsewhere, in the later Empire, this concrete form was often transposed back into pure brick or stone. A semicircular or segmental profile was almost universal; partly, perhaps, because it was considered to be strongest, but also, no doubt, because it made construction easier. On short spans the Romans did, however, introduce the flat arch, and then sometimes joggled the voussoirs (i.e. made them interlock).

Later brick and stone arches departed from Roman precedents mainly in the adoption of other profiles. Of these, the most important were the pointed profiles characteristic both of most Islamic and of Gothic arches. The Islamic form appeared first and was preceded by a Sassanian form of roughly parabolic profile,

Roman concrete arches c. AD 200. The brick facing has gone, showing the concrete mass behind divided into voussoir-like sections by other bricks penetrating the thickness.

Timber arches are the principal elements of the fine hammer-beam roof of Westminster Hall, London.

where the main objective was probably the same—a reduction in the amount of temporary support required during construction and hence in the quantity of timber required for centering. To economize further in this scarce resource, brick arches were constructed of superimposed rings in such a way that only part of the first ring had to be carried, during construction, by the centering alone. Gothic pointed arches were usually built in stone and economy of centering was a less important consideration. The chief merit of the pointed profile was probably the ease with which it could be used in ribbed vaults of any plan shape and, without aesthetic inconsistency, throughout structures that vaulted in this way. Many other profiles were also adopted, sometimes highly fanciful. But they were of no structural significance because they involved no more than the cutting away of part of the soffit of a deep arch of a more normal profile. A final development was the 18th-century French flat arch which was heavily reinforced with iron to make it function more like a beam.

Arches of other materials—timber, iron, steel, and reinforced concrete—have usually been, structurally speaking, somewhat impure forms; in that their action has not been purely compressive. They have also been capable of some bending resistance to the loads to be carried and could therefore be made correspondingly more slender. In later developments of the timber inverted V (or cruck) form, it was usual to peg or strap together three or more parallel arch rings with the joints in one ring offset from those in adjacent rings. A fine example may be seen in the late 14th-century roof of Westminster Hall, London. This led to the timber arch with multiple lamination that became common in the 19th century and has been given a further lease of life with the introduction of more durable glues in the last few decades. Additional stiffness has usually been provided by some form of bracing, either to a second parallel arch or to other structural elements. Early cast-iron arches of the late 18th and early 19th centuries all closely resembled braced timber arches. Later steel and reinforced-concrete arches have usually been given the necessary stiffness simply by the adoption of an I-shaped, boxlike, or tubular cross section.

Beams and slabs

Already in the early 3rd millennium BC, in Zozer's tomb complex at Saqqara, blocks of stone were being deliberately shaped for use as ceiling beams instead of just used as found. The fact that their undersides were cut to a rounded form suggests an even earlier use of cut timber, probably palm logs. Indeed, the prior use of cut timber is generally suggested by the detailing of prototypical stone column and beam forms down to the Doric temple. This is to be

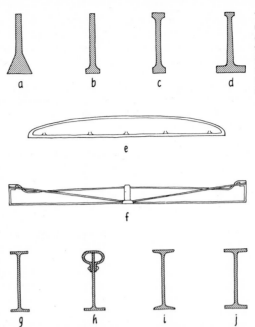

The development of cast- and rolled-iron and steel beams: a-e, cast-iron cross sections and a typical profile, 1796-1824; f, a cast-iron beam trussed with wrought-iron rods, c. 1840; g-j, wrought-iron and steel cross sections, 1850-1955.

expected, since timber is easier to cut and transport. Apart from its inferior durability, it is also a more appropriate material because the structural action of a beam involves internal tension as well as compression, and timber, unlike stone, has a tensile strength along the grain to match its compressive strength. Simple cut timber beams spanned about 26 ft. (8 m) between the columns of the great audience hall at Persepolis, and were probably transported from Lebanon. At the expense of a much greater effort, certainly of cutting, the far more massive stone slabs and lintels of the Hypostyle Hall at Karnak have clear spans of only about 20 ft. (6 m). For most types of stone, the limit was substantially lower, though greater spans have been attainable from certain types of timber.

To overcome these spanning limitations, various other means have been tried. With stone there was really only one possibility—the use of metal reinforcement at the bottom to improve the tensile weakness. Though there do seem to have been earlier attempts, it is unlikely that anything significant was achieved by such reinforcement until the 18th-century development of the reinforced flat arch. With timber, there was a wider range of possibilities including, in addition to the use of metal reinforcement, the building up of beams of greater length or greater effective depth than could be cut from available timber in one piece. Mostly this involved, until very recently, the use of techniques very similar to those used in building up wide-spanning timber arches, except that, instead of simply butting successive lengths of timber at the bottom of the beam against one another, they were joined in ways that permitted some direct transmission of tension. Recently, the introduction of much more efficient glued joints has vastly extended the useful range of possible built-up forms and permitted the construction of wide-spanning beams of I or box section. A further possibility was a hybrid trussed form, best exemplified by early 19th-century beams with wrought-iron rods constituting the ties of a simple shallow truss.

Iron beams have a long history too; but they became structurally important elements only in the late 18th century with the widespread introduction of cast iron. The most efficient form of cast-iron beam was found to be one of I-shaped cross section with a larger bottom flange, on account of the lower strength of the material in tension than in compression. In the 19th century, spans were extended, as in the case of timber beams, by trussing with wrought-iron rods. In this way, it was also possible to support the heavier loads that were then becoming more common, but the application of the rods was often badly conceived (with the end attachments too high in relation to the compression flange), and this led to some serious collapses. By the middle of the century, wrought iron began to oust cast iron and was then, in turn, ousted by steel (see BRIDGES). They did away with the disparity of tensile and compressive strengths and, hence, with the need for trussing or unequal flanges. The associated change from casting to rolling as the method of fabrication also led to the adoption of constant depths and cross sections for the standard beams, instead of depths increasing to midspan, as had been adopted for many cast-iron beams. For the largest loads and spans, variable cross sections could however be built up by riveting on (or more recently by welding on) additional plates as required. For lighter loads, lightweight beams have been made with open webs.

Reinforced-concrete beams may be regarded as the modern counterpart of the reinforced flat stone arches. Early development in the late 19th century was mainly empirical, as had been the development of most earlier forms. But the right choice and placing of the reinforcement to develop the full potential strength of the concrete section called for at least some quantitative analysis of the internal stresses to be expected. Design was placed on a much more certain basis in the first decades of the 20th century, and further improvements resulted from a better understanding of the factors that determined concrete strengths. Reinforced concrete then became a highly versatile structural material, since the strength of the concrete, the overall geometry of the element, and the quantity and placement of the reinforcement rods were all under the designer's control. This versatility also permitted, for the first time, the construction of monolithic slabs of any desired width transverse to the main span. These could be constructed simply as laterally extended beams. But it was soon realized (notably by

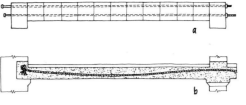

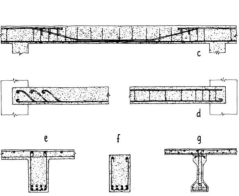

The development of reinforced- and prestressed-concrete beams: a, an early American proposal for a prestressed-stone beam, 1811; b-f, reinforced-concrete beams, 1854-c. 1920; g, prestressed-concrete, c. 1950.

Turner in the U.S. and by Maillart in Switzerland) that such slabs could be reinforced to span in any direction and could, for instance, span directly between the four corners of a rectangular bay. Such slabs, of a uniform depth less than half that of the beams that would otherwise be required between the corner supports, are sometimes referred to as flat plates.

Prestressing of the reinforcement has permitted a more efficient use of modern high-strength concrete. The reinforcement is initially tensioned to a high enough stress to ensure that all the concrete will remain in compression even when the beam is subjected to its maximum load. Because some of the initial tension is inevitably lost, higher strength reinforcement than is normally used is essential, but this became more readily available from the 1930s onward. Beams for use in buildings are most economically precast and such beams, usually of I section, have been widely used in the last few decades. Similar larger beams have also been used for bridges with spans up to about 60 ft. (18 m). For longer spans, box sections are preferred and, as an alternative to precasting these in one piece, in situ prestressing has permitted a return to the reinforced flat arch form with the complete span assembled from short voussoir-like sections.

Columns, piers, and walls

The average compressive stress in a brick or stone wall has, until very recently, rarely been more than a small fraction of the compressive strength of the brick or stone. Development was therefore seen largely in terms of a repeated search for economy in construction, without loss of cohesion through the thickness or excessive nonuniformity of stress leading to

The welded-steel box girder deck of the Severn Bridge, England, under erection in 1964. The closed box form is especially suited to carrying loads that may cause twisting as well as symmetrical bending, and the trapezoidal cross section adopted here also had considerable aerodynamic advantages.

instability. Much the same could be said of freestanding columns and piers which can collapse in any direction and not just at right angles to their length, and in which a local weakness cannot be supported by an adjacent margin of strength.

Brick walls and piers have, from the beginning, usually been of fairly uniform construction throughout their thicknesses. Different bonds, or methods of laying the bricks, have been devised to avoid the weaknesses that would be introduced by continuous vertical joints. Stone walls, on the contrary, were rarely as uniformly built. To economize in the effort of cutting, they were built with two skins enclosing a filling of loose rubble—bound more or less effectively, by earth or mortar. The blocks of the skins were dressed to fit fairly closely on the surface in many different bonding patterns. But behind the surface they were usually left much rougher and less closely fitting, so that much of the total compression was concentrated on the skins. Stone piers were usually built in the same way, with one continuous outer skin. Only in relatively slender, freestanding columns was the construction uniform, although the superimposed drums were usually dressed to fit only around the circumference when the whole column was not monolithic.

Within the broad pattern of the many variations on this theme, it is possible to distinguish only one significantly new form—the Roman concrete wall or pier. This may have come into being through the accident of the existence of a natural pozzolanic earth which made excellent concrete when used, in the usual way, to bond the rubble core. Once the strength of this concrete was realized, more and more reliance was placed on it, and the skins became little more than a facing to it—a kind of permanent shuttering. Wall and pier thicknesses remained, nevertheless, almost as great as had been common previously, giving a great margin of extra strength.

Only in the 20th century have further new, but related, forms appeared. The much more slender "calculated" brick wall is a direct development from earlier brick walls, exploiting both the availability of a wider range of bricks of different guaranteed strengths, and a clearer understanding of actual and desirable strengths. The modern concrete or reinforced concrete wall and its counterpart, the reinforced-concrete column, are, on the other hand, more closely associated with the development of reinforced-concrete beams and slabs than with Roman concrete walls and piers. The column developed at the same time as the beam. The wall, characteristically even more slender than the "calculated" brick wall, has been largely a development of the last few decades.

Timber is as immediately adaptable to the role of a column as to that of a beam. Indeed the upright growth of the tree must have suggested this role in manmade construction at a very

Cast-iron columns supporting primary and secondary cast-iron beams in a boat store at Portsmouth, England, built in 1845. The principal beams are trussed with wrought-iron rods.

early date. But it was not a role that was capable of much continued development, and timber is not a naturally appropriate material for the construction of load-bearing walls other than framed walls in which columns (or studs) are the main load-bearing elements. Further development in this direction came about through the substitution of cast iron, and later of steel, for timber.

Cast-iron columns, of single-storey height, were introduced at about the same time as cast-iron beams, and continued being used throughout the 19th century. They were usually tubular in cross section, this being the most efficient section to avoid buckling under an axial vertical load. Steel columns of many different cross sections, both rolled and built up, have been used since the latter part of the 19th century. The usual rolled section has been an I or a broader flanged H, but tubular sections have now become available again. Sections built up by welding from huge plates can now be obtained to carry loads through 100 or more storeys. Even larger columns are built up, partly in situ, for long-span bridges.

Domes and related elements

The dome may be regarded as the three-dimensional counterpart of the arch. In its true circular form, a vertical arch is rotated around a vertical axis and sweeps out, at every level, a continuous circular horizontal ring. Loads can be transmitted both along the meridian lines of the vertical arches and around the horizontal rings. This dual mechanism allows, in principle, a much freer choice of profile than for an arch of similar thickness, though the full freedom is realized only with appropriate support conditions plus a capacity to resist circumference tensions in the lower rings when the meridians become steeply inclined. In masonry and unreinforced concrete domes this was never fully realized. But one related possibility was widely exploited—that of leaving an open eye in the center, with the uppermost horizontal rings acting, in effect, as the keystone to all the incomplete vertical arches.

There are good reasons for believing that the simple dome form, set directly on the ground, was the first completely manmade spatial enclosure. Simple domed huts, constructed from a wide variety of materials, can still be found throughout the world. It certainly seems probable that the masonry form preceded that of the two-dimensional masonry arch because it can be built entirely without centering, taking advantage of the keying action of each successive horizontal ring. In fact, it would have been seen as a circular wall gradually closing in on itself, rather than in terms of a ring of vertical arches. The dome, constructed with horizontally bedded rings and sharply pointed profile, had already achieved monumental proportions by about the 14th century BC in the great tombs at Mycenae. But these tombs were not completely freestanding. They depended partly for their stability on the earth piled against them outside.

The full development of the potential of the truly freestanding dome owed much to Roman concrete. Casting the concrete in horizontal layers and varying the constituents toward the top to lighten it, Roman builders constructed over the Pantheon in the early 2nd century a dome that has twice since been equaled but never really surpassed. As was their usual practice, they made it immensely thick at the base, but stepped it back externally as it ascended and left an open eye at the top. Taking advantage of the ease with which the wet concrete could be made to take on any shape that was first given to the timber formwork, they also experimented with different variations on the basic circular geometry of the inner surface. The inside of the Pantheon dome was deeply coffered. Elsewhere the surface was scalloped or lobed; but always the circular plan form was retained outside. Yet rarely, if ever, does the thick monolithic base seem to have had enough tensile strength to contain the outward thrusts developed higher by the meridian arch actions.

With these Roman precedents the development of similar forms in other materials was fairly rapid, much of it occurring initially in parts of the Empire which lacked a natural pozzolanic earth and the skill of making good concrete. Brick was the usual alternative material, but timber and stone were also used, as were specially made, interlocking, open-ended earthenware tubes set in a continuous spiral. Two associated developments were of considerable importance later. One arose from the need for more clearly defined, transitional elements where the substructure from which the dome rose was not circular in plan. The corners somehow had to be bridged, and the two basic means devised were a diagonal

Roman concrete dome of the Pantheon. A circular dome, unlike an arch, is self-supporting with such an opening at the crown.

The octagonal dome of Florence Cathedral, whose construction between 1420 and 1432 gave great impetus to subsequent Renaissance developments.

archlike form known as a squinch and a spherical triangular form known as a pendentive. The latter was, in effect, a small part of a larger dome springing lower down. The other development was the double dome, in which the outer dome served primarily to protect the inner one from the weather but also offered the possibility of achieving a more impressive external silhouette—a possibility that was widely exploited both by later Islamic architects and by Western architects from the High Renaissance onward.

The later Western development was initiated by an achievement that, in the circumstances of the time, probably exceeded that of Hadrian's architect of the Pantheon dome. This was Brunelleschi's construction of the dome of Florence Cathedral in the early 15th century. A major difficulty here was the octagonal plan form which Brunelleschi was constrained to follow throughout the height of the dome itself. His central idea was to construct it, nevertheless, as if it were a circular dome of the same internal diameter as the diagonals of the octagon—a diameter that slightly exceeded that of the Pantheon dome. In this way, and by means of numerous related devices, he succeeded in completing it without any centering, as the first Renaissance double dome. Nearly all later major domes returned to the circular plan and were built partly on centering. There was also, after the construction of the dome of St Peter's, an increasing separation between the inner and outer domes, even to the extent of calling for one or more intermediate domes and reducing the outermost to a timber-framed roof, as it had already become in most bulbous Islamic and Byzantine domes.

In these later domes some attempt was made to contain the outward thrusts developed above by means of circumferential ties in the lower parts where radial cracking would previously have been observed. But it is probably not until the 19th century that the ties were fully effective. Toward the end of that century very thin brick or tile domes were being built (see SHELLS). There was, however, an earlier form that could be made quite thin without calling for circumferential ties if it had a firm base. This was the tall spire, often octagonal in plan, which is a special category of dome that rises to the crown without ever curving inward. In it, the circumferential stresses are compressive throughout the height if the outward thrust developed is fully resisted at the base. Tension did, nevertheless, develop and led to cracking when this resistance occurred only after some outward spread had occurred.

Floor systems

The continuous slab constitutes a self-contained floor system, though it may be desirable for nonstructural reasons to add a separate top surface and a separate ceiling

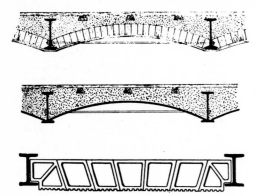

Some 19th-century fireproof floors: top, an early form with shallow brick arches spanning between iron beams and carrying a concrete topping; center, a variant form with corrugated iron in place of the brick arches; bottom, a lighter form with special hollow terra-cotta blocks introduced in Chicago in 1872.

below. Before the development of the reinforced-concrete slab, the nearest equivalents were the floor composed of beams of timber or stone set immediately alongside one another, and the floor provided by a more or less solid fill above a brick or concrete vault. The first of these involved a very extravagant use of material and hence expenditure of effort, so it usually gave way to a more differentiated form with increasing skill in construction. The second was more efficient, inherently strong, and fireproof, and continued to be used for these reasons until supplanted by the reinforced-concrete slab. But it had the drawbacks of greater overall depth than alternative forms, and of greater weight plus the generation of outward thrusts, so that stronger walls were called for. These drawbacks were minimized in the 18th and 19th centuries by the development of appropriately light and shallow tile vaults, but they could not be wholly eliminated.

The alternative to these forms was always some composite system, with beams as the principal spanning and load-bearing elements. In the commonest of these systems, still widely used, light timber beams span at short intervals between opposite walls and are covered by boards or twigs and rammed earth or something similar spanning across. In a slightly more elaborate variant, often adopted in 18th-century mills to give large minimally obstructed floor areas, the light beams spanned between heavier timber beams which were carried by isolated columns. None of these forms is fireproof and that defect led, toward the end of the 18th century, to successive changes in the latter form that finally culminated in the reinforced-concrete slab, either spanning between steel or reinforced-concrete beams or spanning directly between columns. The first change was the substitution of shallow brick vaults for the secondary timber beams and boarding, with iron tie rods between the heavy main timber beams to neutralize the thrusts of these vaults. Later, iron beams were substituted for the main timber beams; and finally the heavy brick vaults were replaced by lighter forms, usually flatter vaults made from special hollow blocks or from some early form of reinforced concrete.

Today the usual floor system, apart from intermediate floors within single dwellings, is the reinforced-concrete slab with or without projecting beams. For very heavy loadings and wide spans, a grid of beams within a bay may be used to stiffen and strengthen the slab without requiring it to be of great thickness throughout. In all cases, the slab has a great advantage over most earlier systems because it is a good horizontal diaphragm, binding the walls or columns together and distributing any side loads between them, as well as serving its primary purpose; though some of this advantage may be lost if it is not continuous over the whole floor area.

Foundations

The loads that a structure imposes on the ground normally reach the ground (or the level of the lowest floor if that is below the outside ground level) through walls, piers, or columns. Ideally, if the ground surface is a firm stratum of natural rock, able to take the loads directly without noticeable settlement, the walls, piers, or columns can simply be ended when they reach it—or rather can be built up directly from it after some preliminary leveling. Unfortunately, such strata have rarely been found in the places where men have wanted to build, and some means have had to be provided to spread the loads more widely or carry them down to rock or firmer ground at a lower level.

Apart from shallow excavation to reach rock close to the surface, there were three means that were already widely practiced in Roman times, and the first two at least were much older. These were the spread footing, piling, and the continuous raft. The first and last spread the load fairly near the surface, simply by providing each wall, pier, or column with a substantially wider base or providing a more continuous and still wider base for a number of piers or columns. The same materials were used as for the superstructure, though the Romans generally preferred concrete, especially for continuous rafts and foundations below water level. The second, piling, carried the load further down without necessitating deep excavation. The piles were almost always of timber. Once hammered into the ground, they acted as columns, usually transmitting part of the load to firmer ground at the foot and spreading part of it through the intermediate strata by surface friction. A group of piles might be capped by a timber grillage, to provide a level platform on which masonry could be set.

With the exception of the continuous concrete raft, which was not used again until the later 18th century, these methods continued in use well into the 19th century, with little change except in such matters as methods of pile driving and of working below water level. By the middle of that century, however, further development was stimulated: first by the need for deeper underwater foundations for large bridges; then for foundations capable of supporting buildings of 20 and more storeys in places like Chicago, where rock was far below the surface.

For deep underwater foundations the answer was the pneumatic caisson. This was a development of the earlier cofferdam—a wall within which, after pumping out the water, it was possible to excavate and then build the base of the pier in the dry (see BRIDGES). The caisson was a prefabricated continuous wall furnished at the foot with a cutting edge. It was towed to the site and then sunk through the ground to a sufficiently firm stratum under its own weight as the upper strata were excavated within it. In the pneumatic caisson, the top was closed and provided with airlocks, so that inside enough pressure could be maintained to keep out water even at a considerable depth, thereby still allowing the excavation to proceed in the dry. By making the wall of iron, or later of steel or reinforced concrete, it could also serve, when excavation was completed, as the outer wall of the pier itself.

The caisson could also be used on land when the requirements were similar, and was used in this way toward the end of the 19th century. However, the new requirements for tall buildings were mainly met by the substitution of grillages of steel beams for the less efficient, earlier spread footings. These have since given way to footings and rafts of reinforced concrete, while there have been parallel developments in piling with the substitution of steel and reinforced-concrete piles for the previously universal timber pile. The heaviest reinforced-concrete piles are nowadays cast in situ in a pre-bored hole.

Equally significant has been the increasing exploitation of the buoyancy principle—that of creating open basements below ground level of sufficient volume to displace a weight of earth comparable with the total weight of the building, so that there is only a small net change in pressure at foundation level when construction is completed. In terms of structural form, this calls for a rigid, boxlike form below ground level, usually achieved with a heavy reinforced-concrete base slab and reinforced-concrete side walls braced by the floor slabs and framing at the intermediate basement levels; but it has also led to further developments in construction processes that are referred to under that heading.

Rigid frames

In a structural system composed of columns and beams, the beams may simply rest freely on the supporting columns. Both then remain independent elements. The system will remain stable under a small sideways disturbance along the line of the columns if these have broad bases

or are otherwise restrained against overturning; but otherwise it will, in itself, be unstable. It can be made inherently stable in its own plane in one of two ways: by diagonal bracing or by connecting the beams to the columns in such a way as to prevent any relative rotations at the joints. In the latter case, the beams and columns will respond as one to any load and lose much of their independence. From this point of view the whole assembly—or at least the typical basic assembly of a beam carried by two columns—may be regarded as the structural element. Looked at in another way, it is the limiting case of an arch whose profile departs so far from a possible purely compressive line of resistance to the imposed loads that it has to resist them largely in bending. Since the columns must participate in the bending, the form cannot be realized in unreinforced masonry.

The true rigid-jointed portal frame (as the basic assembly is known) was first realized in iron in the mid-19th century. It was, for instance, the basic structural element of the Crystal Palace and of several slightly earlier structures in the English docks, though it was introduced with caution as indicated by the additional provision of diagonal braces in some frames. Before this, approximations to the form in timber had been fairly common. Usually a partial rigidity of the joints was achieved mainly by the use of short diagonal braces across the corners. Toward the end of the 19th century complete rigidity of the joints was obtained in heavy steel construction by a somewhat similar means—the use of rounded fillets between the underside of the beam and the column. In modern frames of steel, reinforced concrete, or timber, the necessary rigidity of the joints is achieved more directly and elegantly. In reinforced concrete it is achieved primarily by continuity of some of the reinforcement between column and beam; in steel by welding or high-strength bolting that prevents slip by creating large frictional resistances; and in timber by closely comparable means, including the use of high-strength glues and special bolted connectors. In single-storey frames the beam may be hipped, giving a more archlike form with reduced bending action.

Shells

The term shell is used here to denote a spanning and space-enclosing element of domed or other vaultlike form, but with a thickness and order of magnitude less than was usual for these masonry and mass-concrete forms. Like the latter, a shell may be curved in two directions or in one only; but the two curvatures of the doubly curved form may be of opposite sense, like those of a saddle—a possibility almost restricted to the fan vault in masonry—and the singly curved form may be taken to include barrel-shaped and folded or corrugated forms that span along the length of the barrel or the

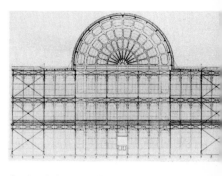

Sectional elevation of the Crystal Palace, London, (1850-51). Rigidity was obtained by the use of deep trussed beams, but was supplemented by means of some auxiliary diagonal bracing.

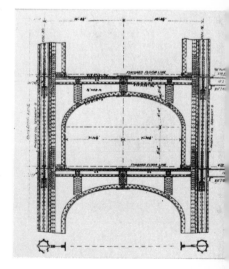

Heavy steel portal frames with deep rounded filets between columns and beams as used throughout the 17 storeys of the Old Colony Building, Chicago (1893-94).

very easy to calculate, and the whole surface could be generated by sweeping one straight line (or set of straight planks) along two others spaced apart at opposite extremities and inclined in opposite directions. The barrel-shaped and folded forms appeared a little later. None of these forms is as suitable for really large spans as the domed form, and for shorter spans other types of roof call for less labor in construction and are now, therefore, usually cheaper. But the shell, together with the doubly curved tensile membrane or cable net, has so enlarged the formal vocabulary of architecture that it will continue to play an important role where economy is not the overriding consideration. The Saarinen/Ammann and Whitney roof of the TWA Terminal Building at Kennedy Airport demonstrates its versatility at the limits of practicality; Jørn Utzon's (b. 1918) original impracticable proposal of sharply ridged shells for the Sydney Opera House went beyond these limits and called for a different arched type of construction.

Trusses and space frames

Trusses and space frames are assemblies of linear members that act primarily in axial tension or compression as ties or struts. The term truss denotes an assembly in one plane, and the term space frame describes a three-dimensional assembly in which the interconnections are such that a load at any point is distributed in all directions through the assembly. The joints need not be rigid and, ideally, should allow free relative rotations of the members. But they must be capable of transmitting tension as well as compression. The usual role in a building is of carrying a roof—in place of the arch, dome, vault, beam, or slab.

Like the arch, dome, and vault, they overcome the spanning limitations of the individual members. But the parts they have played in architectural history have been briefer, partly on account of the greater difficulties of understanding sufficiently clearly the structural behavior, and partly because of the difficulties of making suitable joints.

The simplest roof truss consists of a pair of rafters joined at the foot, and thereby prevented from spreading, by a horizontal tie. It is just possible that the Greeks arrived at this form. But there is no clear evidence of it until Roman times, and the oldest surviving examples are in the roof of the Monastery of St Catherine on Mount Sinai, constructed in the 6th century. The major French Gothic cathedrals were roofed with more steeply pitched trusses of essentially similar kind, and trusses more closely resembling those of St Catherine's seem to have remained the normal means of roofing the basilican church in Italy and further east. But in England, for instance, most surviving roofs constructed prior to the 18th century show a very poor understanding of the

The freely modeled shell roofs of the TWA Terminal at Kennedy Airport, whose support at only a few isolated points necessitated considerable thickening of the concrete to withstand large bending actions.

folds, and act as deep beams. To achieve the reduction in thickness, tensile strength must be provided in the shell itself, or at the level of support, or in both places, in accordance with the requirements of the surface geometry, the pattern of loading, and the type of support. Both because an adequate understanding of these requirements has been gained only in the course of the past 100 years, and because the means of providing the tensile strengths have largely been developed over the same period, the true shell is a recent innovation.

The first shell-like domes were built around the turn of the century in the eastern U.S. The outstanding example was the dome (intended as only a temporary closure) over the crossing of the Cathedral of St John the Divine in New York City. This was constructed of several layers of flat tiles set tangentially to the surface to give an average thickness of only about 1/250 of the span. It was reinforced near the base with circumferential steel rods. Since the early 1920s, reinforced concrete has been used, with prestressed circumferential reinforcement for the larger domes. In very large flat domes, where all the tensile strength is required at the level of support, the surface has sometimes been corrugated radially to reduce the risk of buckling under the meridional compression.

The other forms have all been introduced in the 20th century, and reinforced concrete has been the natural choice of material, though timber has also been used on a small scale—giving a less durable structure but eliminating the wastefulness of first constructing the form in timber, then casting the concrete shell on it and dismantling the timber form. The anticlastic or saddle-shaped forms were introduced first, chiefly because the ideal state of stress in one of them—the hyperbolic paraboloid—was truss principle and must have acted more as beams or (as in the case of the hammer-beam roof) as arches. The term "tie beam" for the bottom horizontal member is symptomatic of this confusion.

In the early 19th century, the true timber truss, necessarily somewhat elaborated, and with the bottom tie made from shorter lengths of timber with lapped joints, was stretched to span about 150 ft. (45 m); but the first wide-span iron roofs (of basically arched form) had then been built, and future development was in iron and steel. With the introduction of wrought iron for the ties, there was a clearer differentiation between these and the struts that was carried over into steel construction. Because there was no risk of the ties buckling, they were made appreciably more slender.

Alongside the 19th-century development of the roof truss, there was a much more intensive development of the bridge truss, stimulated particularly by the great demand for bridges on the new railroads. In the early trusses, whether constructed wholly of timber, of timber with iron ties, or wholly of iron, there was usually a fairly close lattice of diagonal members with or without additional uprights between a horizontal bottom chord and a horizontal or arched top chord. By the middle of the century, a clearer understanding of the structural behavior permitted the elimination of unnecessary members and a virtual standardization of a few simpler and more efficient forms. These could easily be adapted later to the requirements of longer spans, including spans in which the entire truss was arched, or in which it was cantilevered out, with diminishing overall depth, on both sides of each support. They could also be adapted for use in buildings when the span or the load would have called for an excessively heavy, solid-webbed beam.

Architecturally, the most important space frames are lighter framed equivalents of domes and vaults, or of slabs spanning in two or more directions simultaneously. The framed dome is a very early form, particularly if we include primitive dome-shaped huts. But even in fully developed timber-framing systems, the ribs were invariably aligned radially and circumferentially, and the system was then braced by additional diagonals or by the outer covering. Early iron-framed domes merely reproduced this timber form, and it was only in the second half of the 19th century that an inherently stiff, triangulated pattern of framing was substituted. This might be regarded as the first true space frame.

Further development of the framed dome or vault has taken place almost entirely in the 20th century and has lagged somewhat behind parallel developments in airframe structures, where there was a greater incentive to seek the most efficient use of material to save weight. There have been three objectives: triangulation of the members to give inherent stiffness and

distribute applied loads; a uniform distribution of the members to match the usually fairly uniform distribution of maximum stresses in the equivalent shell; and the minimum variation in the length of members. In a barrel vault or other form curved in one direction only, these objectives were easily met by aligning most of the members in two directions, equally inclined to the principal curvature to form a diagonal grid, with others parallel to the axis. In a dome it was impossible to achieve them fully with the numbers of members that were desirable because there is a definite limit to the number of ways in which a spherical surface can be divided into identical equal-sided parts. It can, at most, be divided into 20 triangular parts or 12 pentagonal parts. Buckminster Fuller (b.1895), in his geodesic domes, has taken the triangular division as the starting point and has further divided the basic equilateral triangles into smaller, and inevitably slightly irregular, triangles. These triangles have then sometimes been grouped into hexagons as the basic units for fabrication and erection.

The space-frame equivalent of the slab is entirely a recent 20th-century innovation, partly on account of the computational difficulties of carrying out an adequate analysis of its structural behavior. It usually takes the form of a two-way or three-way grid of intersecting plane trusses that are interconnected at each intersection by having a member in common. Because it is much lighter than the equivalent solid slab, it can economically span much further at the expense of some increase in overall depth.

The St Louis Climatron – a Buckminster Fuller geodesic dome constructed in 1960. The predominantly hexagonal structural grid is double to give more resistance to buckling.

Suspension elements, tensile membranes, and cable nets

Suspension is, in itself, easier than support from below. There is no risk of a loaded suspender overturning or buckling. Provided that the point of attachment remains firm, failure can occur only through inadequate tensile strength, though problems may arise through lack of stiffness. The use of suspension elements in both buildings and bridges has therefore been inhibited through most of architectural history largely by the near impossibility of making large rods or cables capable of carrying loads comparable with those that could be supported by a large pier or column, or by an arch. From at least the 6th century (in Constantinople) iron rods were forged large enough to serve as ties across arches and vaults. But this, and a similar use as circumferential ties around domes, remained the only major use until the end of the 18th century. Then, with iron more widely available, eye-bar chains and wire cables began to be made for use in suspension bridges. In the 20th century, with the advent of much higher strength steel cables and bars, these have found use also in wide-span and tall buildings.

Tensile membranes, in the form of simple tents, have a much longer history. The development of new forms, capable of spanning greater distances and enclosing large areas with minimal obstruction, has similarly been made possible in the last few decades, partly by the availability of cables and fabrics of higher strength than previously available. In this case however, an equally important and complementary role has been played by a newly acquired ability to analyze the stress distributions and determine surface geometries that will result in fairly uniform stresses, and thus minimize the required strengths. For a given loading per unit area, stress in general depends on curvature. The flatter this is, the greater the stress becomes. But fabric stresses can be kept within feasible limits over large spans that necessitate flattish overall curvatures by using a network of cables to carry the main loads, and allowing the fabric merely to span between them. The main design problem then becomes one of determining the cable geometry, including the locations of the cable intersections. A further possibility has been opened up by developments in mechanical engineering. By means of air pumps, the entire weight of a membrane can be supported by air pressure. To prevent it billowing in the wind, the pressure must be kept slightly higher than would be necessary for this purpose alone, and the tensile strength of the membrane and of any auxiliary cables is called upon only to resist this slight excess pressure.

Large inflated membranes were proposed as fixed spatial enclosures early in this century. Today they are still under development and viewed with some suspicion if long life is

Tension roofs for the 1972 Munich Olympics. The cable networks (visible above the more prominent joints in the transparent weather shields suspended from them) are tensioned through the edge cables and guy ropes.

wanted, or large numbers of people are to be accommodated. Among the forms that have been built are simple single skins, single skins strengthened for longer spans by auxiliary cables, and double skins. The last are inflated between the skins, so that there is no need for airlocks to maintain a complete enclosure, and rigidity is more easily obtained. A double skin also allows more freedom in the choice of geometric form. With a single skin, this is restricted to smoothly curved forms that are essentially bulbous.

In a tent, the fabric is kept in shape by tightening the guy ropes. This has the effect of tensioning the fabric directly in one direction, and indirectly in a direction at right angles, where the manner of cutting leads to it assuming an opposite curvature. Tentlike membranes or cable networks are similarly tailored to have opposite curvatures in different directions. Overall tension is then imposed by tensioning the fabric or cables in one of these directions against pulls induced in the other. Within the general requirements of opposite curvatures everywhere in the two principal directions, and a surface geometry that will give a reasonable uniformity of stress, a wide choice of form is possible with different shapes of boundary and different main points of support. There are also different ways of imposing the necessary overall tension. Recent large roofs have been of very varied form and have usually had the necessary tensions imposed either by guy ropes or by means of continuous arched edge members. The latter method was used in the first large roof, that of the Dorton Arena constructed in the mid-1950s. The former method was used in all the large roofs constructed for the Munich Olympics in the early 1970s.

Vaults

The simplest type of masonry or mass concrete vault is the barrel vault. It is just an arch extended so that its width is comparable with, or more usually exceeds, its span. The extension removes the risk of lateral instability and collapse and reduces the risk of collapse as a result of a local weakness, in much the same way as does the lateral extension of a column to make a wall. If, as was often the case, the vault was closed at one end by a wall, construction without centering was also possible if thin flat bricks were used and set in successive flat rings, each leaning back toward the wall.

For these reasons the barrel vault was, almost certainly, the first type of vault, after the dome, to be devised as an alternative roof to the simple beam or slab. And, while its development has been closely associated with that of the arch, it was probably in the lead in the earlier stages. Thus the extensive remains of brick barrel vaults over the storerooms of the Ramesseum at Thebes, built without centering as described in the previous paragraph, antedate by about six centuries the nearby earliest surviving, freestanding brick arch of similar span. Likewise the 82 ft. (25 m) span of the Taq-i Kisra of the Palace of Ctesiphon near Baghdad, a Sassanian brick barrel vault, was matched only many centuries later in the similar arches of some Iranian bridges. Both the storeroom vaults and that of the Taq-i Kisra were roughly parabolic in profile.

This pattern was repeated over a shorter period in the development of the Roman concrete barrel vault and arch from the early 2nd century BC to the 1st century AD. Indeed, the pure concrete arch never really appeared. What has been described above as a concrete arch might perhaps be better regarded as a short length of concrete barrel vault, faced on both sides by a brick arch. The use of wet concrete did necessitate centering to support the timber formwork on which the concrete was placed. Probably because it was easier to construct this to a semicircular profile, that was the profile nearly always adopted. The use of centering and formwork did, however, facilitate a more adventurous choice of vault form, as also in the case of the dome. The most important innovation was the groined vault, formed by the intersection of two barrel vaults aligned at right angles to one another. This permitted the vaulting of a square bay open, or partly open, on all sides. It was the usual Roman way of doing this: the dome being used only over a circular base, or an octagonal or other near circular one, from which the transition to the circle could easily be made.

Later vaults in the West have mostly been developments of the transposition into stone of these Roman forms; those in what was the Eastern Empire were developments of a similar transposition into brick; and those in the Islamic world were developments of the earlier brick forms, built without centering or with a minimum use of it. The least change occurred in what was the Eastern Empire and later became the Byzantine (see EARLY CHRISTIAN AND BYZANTINE ARCHITECTURE). It went little further than a blurring of the distinction between a groin vault and a dome on pendentives. This came about through an adaptation of the centerless method of barrel vault construction to the intersecting barrels of the groin vault, leading to the upward arching of the crowns of these barrels toward the center of the vault and the disappearance of clearly defined groins in the upper half of the vault.

In the Islamic world, the roughly parabolic profile of the barrel vault gave way to a much more clearly defined, pointed one. Without the groined vault as the basis for further developments over a square base, the new forms that did emerge here were related, instead, to the dome set on squinches, and to the opposite of the groined vault—the pavilion or cross vault, in which sections of barrel vaults rise directly from the four sides of the base to meet one another on the diagonals. Ribs, composed of bricks set on edge, were incorporated in these forms with the apparent primary objective of serving as permanent integral centering, but with an obvious secondary delight in the surface patterns that they created. The most characteristic pattern was that of a geometric interlace, very similar to the starlike interlaces used on a smaller scale as a purely decorative motif. Its close relationship to the squinch form is also made very clear by the existence—side by side in the Friday Mosque in Isfahan—of vaults ribbed in this way, and others formed by multiple tiers of squinches.

In the late Romanesque and Gothic West, the semicircular barrel also gave way to one of pointed profile and, much more importantly, the groined vault gave way to a different kind of ribbed vault. In this, the ribs followed the groins, as embedded brick ribs had done in the earlier Roman concrete vaults. But there was a fundamental difference in that the ribs seem usually, if not always, to have been built first as a skeleton that defined the form and simplified the rest of the construction, whereas the Roman ribs were constructed integrally with the rest of the vault at the same level. Usually, also, the webs between the ribs were considerably thinner and lighter, and were arched up slightly between the ribs and the boundary arches. The use of pointed profiles of different rise–to–span ratio resolved the geometric problem of keeping the crowns of the diagonal ribs at the same height as those of the boundary arches, though this was not always attempted; some Gothic ribbed vaults were of more dome-like form, like Byzantine groined vaults. Structurally, the manner of construction would usually have resulted in much of the load passing through the ribs to the fairly solid (and horizontally coursed) *tas de charge,* from which each group of ribs and boundary arches sprang.

Early Gothic ribbed vaults had only diagonal ribs and boundary arches, except for some of

The earliest surviving barrel vaults of true arch form, built of mud brick at the Ramesseum, Thebes, in the 13th century BC.

A typical Islamic brick ribbed vault in the Friday Mosque in Isfahan, probably 11th century. The interlacing of the ribs may be compared with the usual Gothic preference for diagonal ribs meeting at the crown.

those that covered two adjacent rectangular bays of the high central nave of a church (see GOTHIC ARCHITECTURE). In these, there was an additional rib parallel to the transverse boundary arches and connecting the two intermediate points of support along the sides of the nave. The vault was then known as sexpartite. In later vaults, additional ribs were usually added, either to further simplify the construction of the webs or for their decorative value. Two extreme results of this trend were the fan vault and the net vault. In the first, bundles of ribs of identical profile radiated fanwise from each point of support. In the second, a diagonal network of ribs covered the whole surface and largely obliterated the division of a sequence of vaults into individual bays. Closely related to this latter form was a rarer type of vault over a square bay that looked almost identical to the Islamic vault, with a starlike interlace of ribs. This had one brief sequel in the 17th century when Guarino Guarini (1624–83) and Bernardo Vittone (1704–70) largely eliminated the webs to create diaphanous inner domes.

Before the masonry and brick vault were largely superseded by reinforced-concrete shell forms, there were three further developments of varied structural significance. The first was a revival of the stone groined vault, constructed of precisely dressed voussoirs—but now deliberately constructed to complex surface geometries to demonstrate virtuosity in stone cutting. The scale was always small, and so was the structural significance. Most important architecturally were the larger scale, ribbed vaults of Johann Balthasar Neumann (1687–1753) and others that were given similarly complex surface geometries and tended to call for some hidden reinforcement. Finally, there were the first thin lightweight vaults, constructed of flat tiles laid to follow the surface, usually over fairly small spans. These provided the first modern fireproof floors and were also used as ceilings. Mostly they were barrel or cross vaults of low rise–to–span ratio.

Structural systems

Early forms

The combination of structural elements such as walls, columns, beams, piers, arches, and domes to create complete space-enclosing systems involves the provision of adequate support for each element and the ensuring of overall stability, sometimes under adverse conditions of wind or earthquake. It also involves devising appropriate connections. It is therefore inherently simpler to conceive and construct what might be called a unitary structure, composed of just a single element. This is the main reason for believing that the first completely manmade spatial enclosures

were simple dome-shaped huts. Another reason is the unitary form of other structures like birds' nests that are similarly built, rather than the product of natural growth or an adaptation of an existing form. The first such huts may well have been built about 10,000 years ago. Almost 3,500 years ago the great Tholos tombs of Mycenae recalled the lineal descendants of some of these prototypes on a vastly increased scale, with diameters up to 47 ft. 6 in. (14. 5 m).

The more complex pre-Roman systems included both domes and barrel vaults set on low walls. However, by far the commonest system was a combination of walls and/or columns as the vertical supports, with beams for horizontal spanning. It may have originated with the laying of fallen branches over existing clefts in the rock or in some similar manner; or it may have been a development of the simple domed form. Archaeological evidence shows that it had probably appeared by the end of the 8th millennium BC. Fairly thick walls of mud brick or stone enclosed roughly rectangular rooms and must have been roofed by closely spaced timber beams.

The fully developed form of this wall and beam system, with internal columns to permit wider rooms, is well portrayed in a wall painting in a New Kingdom tomb at Thebes. This represents a three-storey Egyptian house with few openings in the external walls and at least

A wall painting of a three-storey house (shown in cross section with somewhat distorted proportions) from the Egyptian New Kingdom tomb at Thebes, probably 12th century BC.

one column or row of columns in each room to assist in carrying the timber floor, or the flat timber roof. The proportions shown are distorted but excavations show that the mud brick walls would have been about 3 ft. (1 m) thick. With walls also running at right angles to one another, this would have given an ample margin of lateral stability; the internal columns being kept upright by being connected, through the floor beams bearing on them, to the walls. Three storeys were probably the limit to which this structural system was then used.

It was reproduced, more substantially and durably, in a variety of public buildings and in the column and beam temple. Usually these were single-storey only, though the Greeks built several two-storey stoas and used double tiers of columns to reduce the internal roof spans of some of their larger temples. The temple was usually very substantially built. In Egypt, from the middle of the 3rd millennium BC, it was built entirely in stone, including the flat roof of heavy slabs and lintels. Partly for this reason, the columns were of such massive proportions that they needed little stabilizing by the surrounding walls, as did the timber columns in the house. This is amply demonstrated in those instances where the collapse of the roofing slabs has long left a row of freestanding columns, or where the roof was never completed (as in the case of the colonnades of Amenophis III at Luxor). The Doric column was also of substantial proportions and has in a few cases, (as at Segesta), demonstrated a similar inherent stability. On the other hand, there are instances of Doric colonnades being overthrown by earthquake. If their collapse had not been brought about partly by previous cutting at the base, connection to the walls of the cella through the roof beams would have originally provided an additional margin of stability. Where the columns were slenderer, the stability of the structural system as a whole would have been assured primarily by the continuous walls running in two directions, and by the interconnections provided by the roof beams.

Other unitary and near-unitary forms were the walls and towers built for defense, similar towers built occasionally for other purposes, the pyramid, and ziggurat. The walls were usually much more substantially built than domestic walls and had ample margins of stability under normal conditions. The defense tower was, typically, a section of wall built to a greater height on a closed square, polygonal, or circular plan. Floors or galleries and internal stairs were usually constructed of timber beams or stone slabs, sometimes cantilevered out from the walls for a short distance rather than spanning right across the tower. The closed-plan form conferred great inherent strength, so that it was not necessary to make the walls quite as thick as the adjacent sections of the main defense wall. One other purpose was to carry a

The Temple of Concord at Agrigento, about 440 BC. One of the best preserved examples of the fully developed Doric form.

warning light for shipping. The famous lighthouse of Alexandria, built in the mid-3rd century BC, rose in medieval times to a height of 430 ft. (131 m) from a base 99 ft. (30 m) square, and must have been constructed in much the same way. On account of its greater height, though, it was constructed in three superimposed sections, each of which tapered somewhat as it rose. At the base, the walls were probably immensely thick.

The ziggurat was simply a high temple platform that acquired a stepped profile as it was progressively enlarged and heightened by successive rebuilding. It was therefore of far from homogeneous construction and is of less structural interest than the pyramid which is both a much earlier form, and one which was the product of deliberate design. If we ignore the small internal chambers and access passages, the geometric form is as simple as that of the defense wall. Nothing would have been simpler, in principle, than to construct it throughout of uniform courses of well-fitted blocks of stone; but this was beyond the resources of even the early Egyptian kings. Most of the mass was a loose fill of uncut stone and sand, and the problem was to stabilize this. In the later pyramids of the Old Kingdom, this was achieved by facing the fill at intervals through the thickness with inwardly sloping walls of cut stone. The technique was learned in a very revealing series of experiments that began with the Step Pyramid at Saqqara, an enlarged flat-topped mastaba, and continued at Meidum and Dahshur.

Later wide-span buildings

Timber-, concrete-, and masonry-roofed systems. Three Roman, or largely Roman, innovations vastly increased the structural possibilities of creating large, unobstructed, spatial enclosures. They were the timber roof truss, the concrete dome, and the concrete groined vault. They appeared, moreover, at times when architects and their patrons were very ready to exploit the possibilities offered and, thereby, to stimulate further development.

The greatest architectural impact was made by the dome: the groined vault being of only slightly less importance at the time, and of no less importance in relation to later innovations. Both were employed in association with the closely related element, the arch. Since both the groined vault and the arch are essentially outward thrusting elements, and since the dome also exerted outward thrusts in practice, on account of the relative weakness in tension of the concrete, the chief structural problem was that of thrust containment without excessive structural movements. Typically, it was solved partly by the provision of massive supports that were not easily overturned, and partly by an increasingly skillful opposition of thrust and counterthrust.

The main initial exploitation of the dome took place in a period of about 60 years, extending from the reign of Nero in the mid-1st century to that of Hadrian in the early 2nd century. Thereafter, architects were largely content to adapt the forms then established or introduce variations on a smaller scale. It began with the construction of the domed octagonal room of Nero's Golden House. This has an unusually large central eye and is steep-sided externally right to the level of this eye, so that its outward thrusts can never have been large. They are buttressed by walls running outward from each supporting corner pier. It culminated on the one hand with Hadrian's Pantheon, and on the other with some of the spatially more exciting domed rooms of his Villa near Tivoli.

The Pantheon dome is notable, above all, for its unprecedented size. Since the diameter of its central eye relative to its span is also less than half as great, it also thrusts outward much more than the dome of the Golden House. Understandably, therefore, it was given a more continuous support which appears externally as an almost unbroken, tall, circular drum. Closer examination reveals, however, that this drum consists of three superimposed rings of semicircular or part-circular arches, mostly filled on the exterior, but either open to the interior or open to voids within the total thickness of the drum. The thrusts of these arches largely balance one another circumferentially, but do have components that act outward, and add to the outward thrusts of the dome. The resultant total outward thrusts are resisted by the buttressing action of the overall thickness of the

The octagonal room of Nero's Golden House in Rome (AD 58-64). The dome still shows clear impressions of the boards on which the wet concrete was tipped.

The Church of Hagia Sophia, Constantinople (Istanbul), looking westward from the apse. This was the outstanding achievement, both structurally and architecturally, of the Byzantine Empire.

drum, where it is solid from face to face.

In the octagonal room of the small baths at the Villa, a dome that, in its lower part, curved boldly inward on alternate sides of the octagon, was interrupted at the same level by large window openings on the remaining sides. In the nearby vestibule of the Piazza d'Oro, an internally scalloped dome was carried over tall open arches on four sides, and by outwardly projecting semicircular walls on the other four sides. In the latter case, the sinuous and interrupted support was clearly expressed externally, since the vestibule was freestanding. In both cases, arched forms were used to deflect the weight of the dome over the openings, with outwardly projecting walls to resist the outward thrusts. If the larger pavilion at the opposite end of the Piazza ever carried a vault like that in the small baths, the further step of carrying the main weight vertically down through columnar screens across the openings, while the outward thrust was resisted as in the other structures, must also be credited to Hadrian's architect. If not, this was the most important subsequent Roman innovation in the use of the dome.

There were, in principle, three possible ways of supporting the groined vault while leaving the sides of the square either completely open or partially open with clerestory windows. All, of course, included adequate vertical support for the downward weight at the corners and some means of resisting the thrusts that acted diagonally outward. One way of resisting the thrusts was with iron ties at about the springing level, either on both diagonals or across the sides. Another was to rely on the sheer bulk and weight of very massive and broadly based corner piers. A third was to take advantage of the partial neutralization of the thrusts of adjacent vaults, and to resist the remaining thrusts by means of walls projecting from the corners at which the vaults met, and at right angles to their open sides. The first was impractical at the time, and the second is inefficient except for fairly small, single vaults. Almost from the start, when using such vaults in sequence over long rectangular rooms, the Romans adopted the third. To extend the space enclosed further, barrel-vaulted bays were created between the projecting walls, also aligned at right angles to the open sides, and thus contributing to the buttressing action. In Trajan's Market, in Rome, this was done on two levels, but the side bays created were, as shops, not fully open to the central space. In the typical large *frigidaria* of the Imperial baths, the side bays were fully open to the central space, and open to one another through arches cut in the projecting walls. Here already, in essence, was the structural system of the Gothic cathedral.

Provided that the roof truss had a fully effective bottom tie, it called merely for the same purely vertical support as a beam, except when subjected to a side load by the wind. While it offered the possibility of wider spans, it did not therefore need, like the dome and groined vault, new types of support. The most important Roman use was in the *basilica*; a large rectangular hall serving as a place of general assembly. In Trajan's Basilica Ulpia, not far from the Market, the main rectangle was flanked on both long sides by two parallel colonnades, each of two storeys. The columns were probably connected longitudinally by stone beams or architraves, and spanned transversely at the intermediate level by timber beams to create galleries over the aisles below. The structural system would be still essentially that of earlier columned halls with continuous outer walls. It was taken over with very little change by the early Christian church.

In post-Roman times it was the New Rome—Constantinople—and the Byzantine Empire that kept closest to Roman precedents. The chief innovations were made in a single outstanding structure. Here, in the 6th-century church of Hagia Sophia in Constantinople itself, an earlier basilica was rebuilt in a form that fused some of the possibilities inherent in Roman domed structures with the fully developed buttressing system of the large *frigidaria*. The chief innovations were the support of the central dome on giant pendentives springing from only four piers, and the buttressing of the outward thrusts of this dome on two sides by half-domes of the same diameter rising to its base. A single, completely open space was thus enclosed, measuring double the length covered by the dome alone. It was flanked by aisles and galleries on both sides, as the nave of the earlier basilica had been. But these were now given the strong buttressing characteristics of the side bays of the *frigidaria*, though vaulted with domed groin vaults over the aisles and small domes over the galleries. To increase the buttressing, the walls dividing the bays (actually pairs of parallel walls) were extended above the gallery roof level to the base of the main dome. Smaller half-domes opened off the large ones to swell the open central space further and contribute to the buttressing in the longitudinal direction.

With the Ottoman conquest of Constantinople in the 15th century, there was some further development of this form in the Imperial mosques. Typically, the support system of the central dome was simplified and openly expressed, and the disparity between two different types of buttressing either reduced or eliminated. In the nearby Mosque of Ahmet, for instance, the buttressing half-dome, as used only at the east and west in Hagia Sophia, was used on all four sides in precisely the same way. This was consistent with an earlier Islamic preference for simple structural forms and, in particular, for a simple, fully centralized support system for the dome, differing from the Roman chiefly in the use of the squinch as the usual transition element from a square or polygonal base, and in the avoidance of sinuous curvilinear support boundaries.

In the truss-roofed basilica, as taken over by the Western Christian Church, there was usually one difference from the structural system of the Basilica Ulpia. Side aisles, sometimes double, were only single-storey and were roofed at a lower level than the central nave in order to permit clerestory lighting. This tended to leave the long clerestory walls, carried on relatively slender colonnades, with inadequate lateral bracing. In Italy, this weakness, when it became apparent, was made good by the addition of transverse arches at intervals, sometimes in combination with transverse timber struts and ties. In France, Germany, Spain, and England, it was avoided in the typical Romanesque, timber-roofed church by the adoption of more sturdy proportions throughout, including the replacement of colonnades by substantial piers.

It was already becoming commoner in these countries, however, to construct stone vaults over the nave and aisles. Initially, the nave vault was a continuous barrel, then a sequence of groin vaults. Outward thrusts again had to be resisted. Most significantly, this was done by means of lean-to arches over the aisle vaults and under a steeply pitched lean-to roof on each side, or continuous half-barrels of similar profile. In the Gothic structural system, as seen in the great French cathedrals of the early 13th century, the ribbed vault replaced the groined vault, the supporting piers became more slender, and the lean-to arches over the aisle vaults rose above the aisle roofs to counter directly the thrusts from the main nave vaults. Often these arches were doubled; the upper arches similarly resisting wind loads on the steeply pitched, timber roofs above the main vaults. Outer buttress piers rose above the aisle roofs to receive the bases of the arches, themselves commonly referred to as flying buttresses.

Amiens Cathedral showing the Gothic structural system of the great 13th-century cathedrals of the Ile de France.

Flying buttresses around the exterior of the apse of Notre Dame, Paris – the external hallmark of the Gothic structural system.

Although the manner of transfer of loads from the high vaults to the ground was essentially similar to that in *frigidaria* of the Roman Imperial baths, this Gothic structural system differed greatly in other ways. It used the pointed arch throughout in place of the semicircular. It had a slenderness of proportion, almost a linearity, that had no precedent in masonry-vaulted construction and was in complete contrast to the massiveness of the Roman system. This slenderness (which was matched by a much reduced average thickness of the vaults) called for a more precise placement of the elements in relation to one another to ensure that all loads could be carried, as far as possible, in simple axial compression. Badly eccentric loads could cause excessive bowing over a period of years, and thereby lead to collapse. Numerous such collapses are on record, and others were forestalled only by the addition of new flying buttresses or, later and particularly in Italy, of iron ties across the main vaults.

The use of iron ties to contain thrusts became more important from the Renaissance onward. Also the dome, the semicircular arch and barrel vault, and the flat entablature in the form of a flat arch came back into favor in the West. Iron ties became normal around the base of the dome, and the unprecedentedly massive piers that supported the dome of St Peter's in Rome gave way to much lighter and more daring supports, even allowing for the reduced spans, in churches like Guarino Guarini's San Lorenzo in Turin and Jacques Germain Soufflot's Ste Geneviève (now the Panthéon) in Paris. Though here the outward forms were no longer Gothic, the lessons of the Gothic structural achievements had been learned and were reapplied to other spatial and formal concepts.

Iron-, steel-, and reinforced-concrete roofed systems. Jacques Soufflot (1713–80) also made wide use of iron reinforcements in the architraves of the portico of Ste Geneviève, and, on the basis of this experience, designed the first completely iron-framed roof for the Louvre. This was followed a few years later by a larger iron roof over the new Palais-Royal Theater. From the early 19th century onward, iron roofs, then roofs of steel and reinforced concrete, increasingly became the usual choices for buildings ranging from market halls, conservatories, and train sheds to exhibition halls, concert halls, and aircraft hangars. They were lighter in weight than mass concrete or masonry roofs, not combustible as timber roofs were, and were capable of spanning further without necessarily calling for very heavy supports. The basic support requirements of the dome, the vault, the beam, and the truss remained as they always had been. However, they were reduced in magnitude for a given span, and where there was, in principle, a choice between

the use of a tie or a buttress to resist an outward thrust, the tie became an even more attractive choice with the wider availability of wrought-iron rods, and subsequently of high-strength steel wire, coupled with a better ability to estimate the required strength. In addition, the new architectural requirements were usually more flexible than the earlier requirements of the church or the mosque, and could often be met by a simpler overall structural system more akin in this respect to the original, simple dome-shaped hut.

These simple, almost unitary, forms were employed first for structures like conservatories and train sheds, though this use was heralded at the start of the 19th century by the ribbed iron dome of 128 ft. (39 m) diameter of the Paris Corn Exchange supported on a two-tier masonry arcade. In the Palm House at Kew Gardens, London, two long wings consisted simply of semicircular, iron-ribbed, glazed barrel vaults bearing on low masonry strip foundations. In the center there was a slightly more complex cross section consisting of similar barrels raised on iron columns to the level of the top of the wings and flanked by half barrels of the same form. For the train shed roofs, timber arches and trusses were widely

The Panthéon, Paris – a Neo-Classical reinterpretation of the Gothic structural system in which the central dome (upper right) was carried on proportionately much lighter piers than had been usual in Renaissance churches like St Peter's. Outward thrusts were resisted partly by circumferential ties and partly by hidden flying buttresses.

used at first, and these were then simply copied in iron, in many cases without any major innovation. Rivalry between competing companies then led to more novel designs like that at St Pancras Station, London. Here, eight tracks were bridged by a single iron-ribbed and partly glazed barrel vault of 230 ft. (70 m) span. The ribs rose directly from platform level and were carried below on massive brick piers. The piers were relieved of outward thrusts by means of tie rods beneath the tracks, and the whole roof was given additional longitudinal rigidity by means of diagonal bracing between the ribs and the main purlins. Twenty years later, toward the end of the 19th century, the steel-arched roof of the Galérie des Machines in Paris introduced the only major variation on this form—the insertion of pins at the feet and crown of each arch, making the structural action fully determinate and easing the construction process. The arch thrusts here were taken by large concrete foundations.

In the 20th century, a number of closely comparable barrel vaults springing at or near ground level have been built in reinforced concrete. Notable examples are Eugène Freyssinet's airship hangars at Orly Airport, and exhibition halls in Italy constructed by Pier Luigi Nervi (b. 1891). In these, the ribs were made contiguous and given trough-shaped cross sections for greater stiffness. There was no need of diagonal bracing for longitudinal stiffness. Nervi adopted a similar form for his large Sports Palace in Rome, varying it chiefly by rotating the basic rib around a vertical axis to give a shallow dome rather than translating it along a horizontal axis. The thrusts were taken to the ground by legs that continued the line of the vault and were, in effect, a part of it.

Reinforced-concrete shells have usually been used in similar ways or as straight alternatives to trussed roofs. On account of their slenderness, and the consequent need to avoid major deformations of the surface or bending actions, it has been necessary, however, to give more precise attention to the conditions of support. In particular, it has been necessary in the case of large roofs to avoid the situation in which the support, in taking the load, shortens, extends, or otherwise moves in a way that is incompatible with the natural deformation of the shell itself. Therefore, it has frequently been necessary, rather than just helpful, to prestress ties or reinforcement. Further difficulties have arisen where, for architectural reasons, it has been desired to provide only local supports where continuous support would have been preferable, or to use one shell to stabilize another, either continuously along a surface discontinuity or locally. All these difficulties had to be faced at the TWA Terminal Building at Kennedy Airport, New York, and called for very considerable thickening along all the shell boundaries and ridges.

A new use, to which the space-frame

equivalent of the slab also lends itself, is in the large cantilevered roof or canopy. The cantilever can either be balanced around the points of support or extend largely to one side so that it is unbalanced, but even where it is balanced under normal conditions some imbalance must be allowed for under wind or snow load, for instance, and stability under side loads must be ensured. To ensure lateral stability, stiff supports well anchored in the ground are needed. An even firmer anchorage will also withstand overturning in the direction of overhang; but it has been usual, in grandstand roofs for instance, to reduce the anchorage requirement by having a secondary tie pulling down at the rear to reduce the effective imbalance. Sometimes, as in the Alitalia Hangar at Leonardo da Vinci Airport, Rome, this tie has been attached to a secondary structure at the rear, whose own weight has thereby served to balance that of the cantilever overhang.

A counterpart to the cantilevered roof, that avoids the problems of unbalanced load, is the beam or slab roof supported on both or all sides, and either with or without some symmetrical overhangs. This first became capable of covering wide spans in about the middle of the 19th century with the appearance of large open-webbed, wrought-iron beams. Beams of this type with a span of 72 ft. (22 m) were the main roofing element of the original Crystal Palace, London. They were rigidly joined to cast-iron columns as early portal frames, and lateral rigidity was largely ensured in the plane of these frames by this jointing and, throughout the building, by the use of similar frames of one third of the span in two directions in flanking aisles and galleries. It was a modern equivalent of Trajan's Basilica Ulpia, freed of dependence

St Pancras Station roof, London, England (1866-68). This was the widest spanning of the mid-19th-century iron roofs, its arches being almost four times the span of the timber arches of Westminster Hall.

on outer masonry walls for its stability, but incorporating some diagonal bracing as an extra safeguard. A more recent equivalent, although of simpler form, since it lacks the aisles and galleries, is the welded steel portal framing of Mies van der Rohe's Crown Hall at the Illinois Institute of Technology, Chicago (1952). Pure slab roofs in reinforced concrete cannot achieve comparable spans. The two-way spanning roof is therefore one with an associated grid of beams or is the actual space frame equivalent of this. Such roofs have been used recently where a large area is to be covered with at least four widely spaced supports, rather like the legs of a table. Some cantilevering beyond them on all sides is not only possible but desirable since it reduces the maximum bending actions, and thus the necessary depth of the roof. Lateral stability is ensured by rigid connections at the tops of the supports to give an omnidirectional portal frame action. A common method has been to splay out the column heads or even to use several legs splayed out from the foot at each support position.

A further possibility with roofs of beam or slab type, used for instance by Nervi in the Burgo Paper Mill, Mantua, is to suspend the roof from supports rising above it. In this way, very large spans can be achieved without the depth of roof that would otherwise be necessary, because much more closely spaced points of support can be provided by the suspension system (which resembles that of a suspension bridge or a stayed girder bridge). In effect, the whole roof including the suspension system becomes a much deeper beam, of which the roof proper is the bottom member. This means that it is subject to some overall compression that would not otherwise arise.

Systems with lightweight roofs. In lightweight roofs the whole roof is in tension. In the type that is closely related to the suspended roof just described, the suspension system becomes a fairly close mesh of cables in two directions. As described elsewhere (see CABLE NETS), the cables in one direction must everywhere have an opposite curvature to those in the other direction and one set must be pretensioned against the other set to give enough stiffness to the whole system. This is done through edge members to which families of cables running in a particular direction are attached at regular intervals. For roofs of relatively simple shape, stiff reinforced-concrete arches have been used, as for the Dorton Arena, Raleigh, and the Yale Hockey Rink. For roofs of more complex shape, like those erected for the West German Pavilion at Expo '67 in Montreal, and for the 1972 Munich Olympics, heavier cables stretched between the tops of tall sloping legs and ground anchorages, or between one ground anchorage and another have been substituted—almost a return to the structural system of the early tent. A light weather skin is

hung from the cable nets to complete each roof.

In the second type of lightweight roof, a continuous membrane is the roof. The support in this inflated type of roof consists, in part, of a simple continuous tying down at ground level to a foundation beam or slab which is able to resist uplift. Besides this, it takes the completely novel form of slightly pressurized air supplied by air pumps, the pump requirement being minimized by the provision of airlocks or a double skin. For longer spans it has, however, been found desirable to strengthen the membrane by the addition of an external cable system that is also supported, through the membrane, by the internal pressure.

Later multistorey buildings

Bearing-wall systems from ancient times. The structural form consisting of timber floors spanning between bearing walls, with or without intermediate column supports, has continued to be built, with only minor variations on much earlier examples like the Egyptian house referred to previously. Still with mud-brick walls, fairly closely spaced behind narrow frontages, it served for the tenements of Republican and early Imperial Rome. Elsewhere, sometimes with the substitution of fired brick or stone for mud brick, it has served for structures rising to about ten storeys. But, where the height has been sought partly for reasons of defense, it has assumed more of the character of the earlier, purely defensive tower. Such towers also continued to be built for simple defense until superseded by squatter forms with changes in methods of warfare, and

A detail of the space-frame roof of the McCormick Place Exhibition Center, Chicago (1970). This roof spans in two directions between isolated columns 148 ft. (45m) apart and cantilevers 74 ft. (22.5m) beyond the outermost.

The cable roof used to great architectural effect in the Yale Hockey Rink, Newhaven, Connecticut (1958). The cables span longitudinally and transversely between a vertically arched spine beam and two horizontally arched side beams.

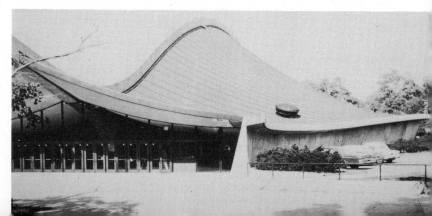

they served as prototypes for other types of tower like the church bell tower and the slender minaret.

The Roman tenements were far from ideal structures and far from being the best examples of the form; but their very defects did lead, particularly after the disastrous fire of AD 64, to the first major improvements. These were the substitution of concrete for mud brick in the walls, and of concrete vaults for at least the floors that separated one dwelling from another. This latter substitution was probably the first systematic attempt to achieve a fireproof structure for an everyday, non-religious use. The resulting form is best seen today in the partial remains in the Roman port of Ostia (where the upper storeys are largely gone) and in the more fully preserved structure of the Market of Trajan in Rome itself (where the main hall referred to previously was flanked by three storeys of similarly constructed shops). Concrete barrel vaults, with flat topping to provide the floors above, spanned between party walls about 16 ft. (5 m) apart and there were intermediate timber mezzanine floors. Such structures were always constructed in larger blocks or *insulae*, so that there would have been no real problems of lateral stability. The modest thrusts of the barrel vaults would largely have neutralized one another and, where they were unbalanced at the end of a block, the end wall and its returns were strong enough to contain them.

Though there continued to be an intermittent use of stone vaults of various kinds, chiefly over basements or to carry a *piano nobile* in post-Roman times, there was no further major advance until the 18th century. Then, initially in France, lighter kinds of fireproof floors were introduced and, since they were lighter, they could readily be used again throughout the height of the building. Usually they were still of vaulted form, using either flat tiles or hollow pots, until the introduction of the reinforced-concrete floor in the late 19th century. Only with this was the whole structural system significantly changed by virtue of the ability of such a floor to act as a much more efficient horizontal diaphragm than any earlier floor of comparable depth and weight. This ability was not properly exploited until well into the 20th century when there were complementary developments in walling and in structural understanding.

The full exploitation of the reinforced-concrete floor began with its upended counterpart, the modern thin concrete wall, as the vertical structural element. Walls of uniform thickness throughout their height were arranged, internally and externally, in two directions at right angles. Together with continuous floor slabs, they gave a system like a single cellular vertical cantilever rooted in the ground, with excellent natural lateral stiffness and resistance to wind loads. At a height of only eight storeys there

Three-storey shops flanking Trajan's Market in Rome. These are the best surviving example of the later substantial Roman brick-faced concrete multistorey structural forms.

A French 18th-century counterpart to the Roman form seen above, with lighter incombustible floors substituted for the heavy Roman concrete vaults. Buildings of very similar construction were erected at Versailles toward the end of the century.

was, in fact, stiffness to spare. In subsequent structures this height has been more than doubled, not only with concrete walls but also with brick walls that are little thicker and only a quarter of the thickness that would have been adopted for such heights toward the end of the 19th century.

Two variations of this system have been a simplification in which the main walls all run in one direction, and a version of the original form in which walls and floors are both precast as large panels of storey height or room width. The first is inherently much less stiff in the direction at right angles to the walls, and is suitable only for buildings that are fairly long in this direction and of only moderate height. The second can be just as stiff as its prototype, but only if the joints between panels are equivalent to those in concrete cast continuously which has been difficult to achieve in practice at reasonable cost. All these forms—prototypes and variants—call for a fairly close spacing of the main walls and for their continuity throughout the main height of the building, so that they have been found suitable only for apartment buildings.

Hybrid systems that utilize partial timber or iron framing. Any of the bearing-wall systems referred to above that has intermediate timber column supports for the floors might, to that extent, be described as a hybrid system. The term is used here only for those in which there is a complete internal framing system of columns and beams with external brick or masonry walls or in which a complete timber frame is infilled to create a type of bearing-wall system.

Structures of the latter type range all the way from ones in which the timber framing is little more than a series of uprights and horizontals embedded in brickwork or rubble masonry, to others in which closely spaced timbers virtually

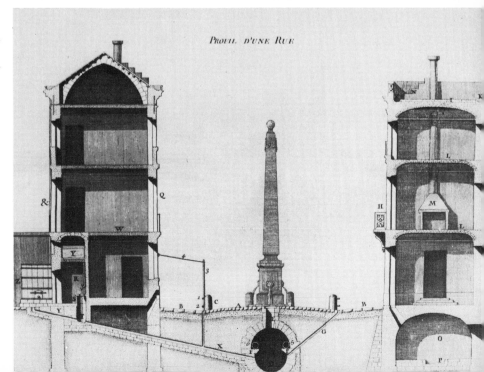

PROFIL D'UNE RUE

constitute a wall. They have been built, as a rule, where the strength and stability of the brickwork or masonry alone was inadequate, where there was a risk of earthquake or differential settlement leading to distortions that simple masonry was less able to accept without collapse, or where timber was the most readily available material. Only where the columns and beams of the frame were fairly widely spaced and braced across their joints with short diagonals, as, for instance, in later English timber-framed houses, was there much similarity with later fully framed structures.

The complete internal framing of structures of the former type was the more immediate ancestor of later fully framed structures. Complete internal framing was used in warehouse and mill buildings to give wide expanses of open floor, and was distinguished from the other type of frame by the absence of either infilling or bracing between the columns and beams to give lateral stability. All lateral stability was provided by the enclosing box of substantial outer walls. Up to the end of the 18th century, and well into the 19th in the case of warehouses, both columns and beams were of timber, and the floors were also of timber boards. Then, after numerous disastrous fires, there was a progressive substitution of cast iron, first for the columns and then for the beams also, coupled with the substitution of the early brick-vault type of fireproof floor for the timber-boarded floor. With this substitution, the frame itself usually became two-dimensional with beams running only across the shorter width of the structure, typically over two intermediate columns. This frame was, therefore, further dependent for its stability on its connections, through the floor system, with the end walls.

Gardner's Store, Glasgow (1856). Internal cast-iron framing has here spread to the facades, but stability is still ensured by tying the beams back, on alternate floors, to masonry walls at the rear.

Fully framed systems of iron, steel, and concrete. The complete multistorey frame of columns and beams needed another source of lateral stability which could, in principle, be provided in three ways: by means of rigid joints and the inherent stiffness of the frame itself when so jointed; by means of diagonal braces; or by means of infills of masonry that acted as braces without being called upon to carry floor loads, as were the unframed outer walls of the warehouses and mill buildings described previously. All three means were adopted in a number of convergent developments in the mid-19th century. These developments included the use of rigid portal framing two or three storeys in height with some supplementary diagonal bracing in the aisles and galleries of the Crystal Palace and in roughly contemporary English dockyard buildings; the use of four-storey rigid framing of surprisingly modern detailing and without supplementary bracing in the aisles and galleries of a slightly later boathouse at Sheerness Dockyard in England; and the expansion of the complete internal warehouse frame to the facades of similarly framed commercial buildings near the city centers in London, Liverpool, Glasgow, and New York. In the large six-storey warehouse built a few years later around three sides of the St Ouen Docks in Paris, the iron frame was carried into all the outer walls except the relatively short principal facade, and all the floors were fireproof shallow vaults of hollow brick topped with concrete. Partial brick infilling of the framed facades contributed to the lateral stability, but was carried, storey by storey, by the frame.

From these beginnings the frame was developed rapidly in the later stages of the rebuilding that followed the Chicago fire of 1871. Notable achievements were the nine-storey Home Insurance Building, the first building in which the frame was fully protected by fireproof casing, and the 14-storey Reliance Building, probably the best of the steel-framed structures built toward the end of the century with all lateral stiffness provided by the frame itself, no party walls, and only light external cladding of terra-cotta and glass. Reinforced-

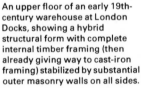

An upper floor of an early 19th-century warehouse at London Docks, showing a hybrid structural form with complete internal timber framing (then already giving way to cast-iron framing) stabilized by substantial outer masonry walls on all sides.

concrete frames of similar height followed early in the 20th century. These also had beams between the columns. At the same time, reinforced-concrete frames of fewer storeys, designed for heavier floor loadings, were built with beamless floors; splayed column heads being provided to collect the floor loads and achieve overall rigidity.

While the reinforced-concrete frame was being developed, the steel frame was pushed to its limits, if not beyond, in a series of ever taller skyscrapers that culminated in the Empire State Building, 85 storeys high without the observation tower, which was built later. Not only the need for lateral stability, but also the need to avoid unacceptable movements of the upper storeys in the wind, necessitated greatly increased amounts of steel in these buildings for bracing purposes, either in the form of diagonals or as deep portals, notwithstanding the bonus of stiffness contributed by heavy masonry claddings and additional internal walls. At heights above about 20 storeys, new hybrid forms that are structurally more efficient have therefore been introduced. At lower heights recent development has concentrated on overall simplification of the frame and its design as a single entity rather than an aggregation of individual elements. This has been facilitated by the possibility, for instance, of varying concrete strengths to achieve the required strengths of column at different heights with little or no change in cross section, and by an architectural preference for simple, repetitive, open-floor plans. Welding and bolting, too, have now made it easier to achieve the desirable rigid joints in steelwork. At heights of only two or three storeys, new light forms of steel frame have been introduced using hollow section columns and open-web beams, both often fabricated as standard components.

Recent hybrid systems for tall buildings. The inefficiency of the tall column and beam frame of the type developed in the late 19th century, stems from the fact that its stiffness as a vertical cantilever had to be provided by the bending stiffness of the individual columns or by very extensive diagonal bracing. In the first case, the columns had to be much heavier than they need have been, solely to carry the extra weight of additional storeys. In the second case, there was the additional cost and inconvenience of the diagonals. If the structure is considered as a single entity, a more efficient form is that of a single vertical tube or something similar. This is, of course, the form of the tower with a continuous outer masonry wall—a form that has recently been built successfully in reinforced concrete to heights up to 1,640 ft. (500 m). One modern hybrid system has relied, as the main source of stiffness, on one or more such towers or cores occupying part of the total area and accommodating elevators, stairways, and other ser-

vices. A second has similarly relied on a few transverse walls, continuous through the height of the building but isolated from one another. A third has transformed the whole frame, apart from a central core, into the equivalent of a giant outer tube.

The first two of these systems were much less radical innovations than the third because they may be said to have done no more than deliberately exploit sources of stiffness that were, to some extent, already present in most earlier framed skyscrapers but largely ignored in the design. However, by taking them explicitly into account and making the most of them, considerable economies have been achieved in buildings of up to about 40 or 50 storeys. The first system is the most widely applicable, though it is most appropriate to buildings of squarish or circular plan where a single central service core is called for. In buildings of long rectangular plan, this core may have to be split in two for both structural and functional reasons, and this will give less stiffness at the expense of more wall. For such buildings, the second system is likely to be more efficient, with continuous gable walls and perhaps one or two transverse walls or equivalent continuous systems of diagonal bracing.

For buildings above about 50 storeys, the third system has proved to be the only one capable of limiting the penalty for height to not much more than that which inevitably arises from the greater vertical loads to be carried at the base. Without its introduction it is unlikely that any more such buildings would have been built this century. It calls, even more than the first system, for a roughly square or circular plan, preferably identical from top to bottom or slightly reduced in area on successive floors. It is most efficient where the architectural requirement is for the maximum openness of the floors between the central service core and the perimeter, to allow the maximum freedom of use. Ideally, all wind loads are resisted not by bending of the individual lengths of column, but by overall tension on the windward side and overall compression on the leeward side, superimposed on the compression on both sides due to gravity. One way of achieving this, first adopted on a fairly small scale for the 13-storey IBM Building in Pittsburgh, but since used for the 100-storey John Hancock Center in Chicago, is to depart from a pure column and beam grid in favor of a more trusslike form. In the IBM Building, an overall diagonal lattice was used, of constant mesh and cross section from top to bottom, but with different steel members according to the load to be carried. In the Hancock Center, normal beams and columns were retained, but a few very large diagonals traversing the whole facade were added. The other way (of varying efficiency according to the proportions adopted) is to make the beams which interconnect with the columns far stiffer than in a normal frame. This has been

The Reliance Building, Chicago (1895). Fully framed both internally and externally and a prototype for the lightly clad steel-framed buildings of 50 years later.

Twentieth-century developments of tall multistorey framed buildings in Chicago: Hancock, De Witt, and Lakeshore Drive Buildings.

done in a number of buildings of both steel and reinforced concrete by both spacing the columns more closely and increasing the depth of the shorter beams that result, up to the limit of the full spandrel depth between windows. In the twin towers of the 110-storey New York World Trade Center, the columns were set only 3 ft. 3in. (1 m) apart and the beams made 4 ft. 3 in. (1.3 m) deep to come very close to the ideal. Response to the wind is not, however, static since the wind force itself fluctuates. Further means may therefore be desirable to limit movements in such tall structures. In the case of the World Trade Center, dampers were incorporated in the connections of the wide-spanning floor beams to the outer columns.

Two variants of the first system with a stiff core may be more briefly mentioned. In the first, adopted by Frank Lloyd Wright (1869–1959) in the Laboratory Tower for the Johnson Wax Company at Racine, Wisconsin (1949), the core carries all vertical loads and provides all lateral stiffness, the floors being individually cantilevered out from it. In the second, there are just a few larger cantilevers, usually of storey depth and accommodating service plant, or one only, and the floors are either suspended from these by peripheral vertical suspenders or carried by peripheral columns. These systems cannot be justified in the same way. Their chief merit, in certain circumstances, is in freeing a large area at ground level, either for later use or according to the proportions adopted) is to make to facilitate construction on a constricted site.

Protection against fire and other hazards. The risk of fire has been present at all times and in all places. It has most obviously stimulated innovation when it has led to the replacement of combustible timber systems or elements by incombustible ones—notably the replacement of timber roofs by stone or concrete vaults (or the construction of such vaults as ceilings below the roof proper, as in Romanesque and Gothic churches), the replacement of timber floors by a variety of "fireproof" floors, and the replacement of timber column and beam framing by iron framing. But fire can also weaken a structure by raising its temperature excessively. Iron and steel, though incombustible, suffer fairly rapidly by losing strength and distorting badly. Reinforced concrete can also suffer if its reinforcement is overheated. When these possibilities were recognized, they were first guarded against by means of protective layers of insulation. In the case of iron and steel this has taken various forms—lightweight, specially fabricated hollow blocks, casings of dense or lightweight concrete, and, most recently, sprayed-on insulants. In the case of reinforced concrete (and prestressed concrete) it has naturally taken the form of an adequate concrete cover to the essential reinforcement, though this cover has had to be prevented from spalling off as its temperature rises by sec-

The World Trade Center, New York, nearing completion in 1972. The twin 1,345 ft. (410m) towers are also made to act as very stiff tubular cantilevers, here by means of the close spacing of the peripheral columns and the use of very deep beams to connect them at each floor.

ondary binding reinforcement in beams and columns. The most novel approach has been to protect steel columns by keeping them filled with water to keep their temperature down. This approach has been adopted for several recent tall buildings and has entailed a return to the tubular cross section of many early cast-iron columns (which sometimes doubled as internal rainwater drains). In other cases they have again been left bare and set some distance outside the building to reduce their exposure to the heat of a fire. Further indirect influences on structural form have arisen through provisions to limit the spread of fire and ensure means of escape in large buildings.

The main earthquake risk is fortunately limited to certain parts of the world. In some of these—in parts of Greece, Turkey, and Iran for instance—there is evidence of long traditions of forms of construction that are better fitted than some other forms to survive an earthquake. The use of embedded timber framing in walls, as mentioned earlier, is one of the best examples. It is difficult to say, however, to what extent these forms were developed or subsequently used on this account. There are other possible reasons for their adoption, they are also found elsewhere, and they were not used to the exclusion of much less suitable forms and of undesirable practices like that of making very heavy roofs. It is possible to see a more deliberate response to the earthquake threat in the extensive and systematic use of both timber and iron ties across arches in Byzantine churches, perhaps because architects were very conscious indeed of the threat during the most innovative period in 6th-century Constantinople. It is only very recently, however, that the nature of earthquake movements and of the response of a structure to them has been understood well enough to enable proper consideration to be given in design. On the whole, this consideration has not so much led to new forms as to a greater selectivity and the provision of greater margins of lateral strength. It is recognized that certain things are undesirable, like asymmetrical, broken, or sprawling plans, weak columns, weak joints, any top-heaviness, especially in tall structures, and any marked variation in foundation support. The main difficulty arises with tall multistorey buildings. Various means of partly isolating these from the ground shaking have been proposed, such as making the bottom storey more flexible than the others. But the problem of absorbing a very large amount of energy without the collapse of this storey remains. The more general and fundamental problem of predicting the type of shaking to which a particular structure is likely to be subjected also calls for further research. It is known that the most appropriate design depends a good deal on the dominant frequency of this shaking, so the next advance may have to await better predictive techniques.

It is considerably easier to design structures to survive other types of ground movement that occur more slowly. The slowly advancing wave of surface settlement that follows underground mining is one such type of movement. The obvious structural answer to this, if the settlement is not likely to be too great to preclude building on the site, is to make the building and its foundation stiff enough to ride the wave as it passes; at least it is today, now that this is perfectly feasible without devising any new system. In the past, the building that survived would more likely have been one able to undergo considerable deformation without collapse. One recent innovation has been a new type of flexible steel-framed system for use initially in schools of up to three storeys—the CLASP system, used in Britain. This sits on a flexibly jointed, reinforced-concrete raft and incorporates telescopic-sprung, diagonal braces which act as normal stiff braces except when accommodating movements of the raft.

Finally, extreme local loads have usually been deliberately provided for only in defense walls and towers and, more recently, certain types of shelter that differ from other contemporary structures mainly in being more substantially built, and in their plan form. The chief exception to this rule now is a deliberate provision of a measure of "fail safe" strength, in tall buildings particularly, so that local

accidental damage does not lead to more extensive collapse. Although this has not led to any radically new structural systems, it has led to a different approach to the structural detailing of some bearing-wall systems and some others without a large natural measure of structural continuity.

Means and processes

Construction plant

All construction involves carrying, lifting, and placing materials, either in an unformed state, or partly or entirely preformed. Usually it has also involved some cutting and positive joining (as distinct from just laying units on one another). In recent years it has increasingly involved operations of stressing or jacking up to obtain the desired distributions of internal stresses, or otherwise assist in the construction process. Foundation works have, in addition, involved operations like excavation, pile driving, and pumping away unwanted water.

By early dynastic times in Egypt (3rd millennium BC), if not earlier, basic tools like the chisel and axe, the saw and the bow drill, were already in use, and wedges were used to assist in freeing blocks of stone in the quarry; but the only motivating power was that of human hands. Direct human effort similarly served to move, lift, and place even the largest blocks of stone. Large ramps were built in mud brick, and the blocks were hauled up the ramps on sleds or rollers. The ramps were later dismantled on completion of construction. Such methods were probably adopted by all megalithic builders up to and including the builders of the earliest classical Greek temples. Immense expenditures of labor compensated for the as yet undeveloped skills of building durable, large-scale structures from small units, though examination of the details of stone cutting and fitting shows clear evidence of attempts to reduce the labor required.

In classical Greek and Roman times, labor requirements were greatly reduced; partly by progressive improvement of the basic hand tools and, more importantly, by the introduction of simple cranes with guyed masts and ropes manipulated by windlasses and passing through pulley blocks to give a mechanical advantage. The motivating power might have been either animal or human; large treadmills were sometimes used in the latter case. Similar arrangements, perhaps with men hauling directly on the ropes, must have been used for pile driving. There is also evidence of the existence of fairly efficient pumps, and of a variety of wheeled vehicles for transportation. Some of the earlier constraints on design were thereby removed and replaced mainly by limits on the sizes and weights of individual structural

A Roman guyed mast crane powered by a human treadmill as shown in a contemporary relief now in the Vatican Museum. Note the use of pulley blocks to gain a mechanical advantage.

Construction of the Eddystone Lighthouse in 1757-59. Lifting gear is shown mounted on the structure itself at successive stages of construction.

A French late 18th-century crane still powered like the Roman one but better adapted to traversing a load as well as lifting it. Similar cranes were coming into use in the 15th century.

units of construction.

From the Renaissance onward, surviving evidence of the plant used is much more plentiful. Cranes capable of moving loads horizontally, as well as lifting them, became common, there was a widespread use of gears to obtain a mechanical advantage (though usually at the expense of a great deal of friction), and devices like jackscrews were available. However, apart from the occasional harnessing of wind or waterpower, the work was still done by men or animals until well into the 19th century.

The main improvements on Roman plant began with the substitution of more durable and efficient iron gears and transmission systems for ones made predominately of wood. They continued, in the latter part of the 19th and early 20th centuries, with the introduction of steam engines and then of electric and diesel motors as the sources of power, and numerous further detailed improvements and innovations in the design of the plant itself. These included new types of cranes and hoists, mechanical pile drivers and excavators, mechanical concrete mixers, hydraulic jacks capable of exerting large controlled forces, a wide variety of hand-held tools operated by compressed air or electricity and, at the opposite end of the scale, new large static plant for rolling steel and fabricating or casting complete structural and other components. Even more than the Graeco-Roman innovations, they have contributed immensely to the economy of construction and—together with new materials and methods of structural analysis—to widening the designer's range of choice. Structurally, the most significant innovation has been the hydraulic jack as used in most prestressing operations. Modern electric welding equipment comes a close second in relation to steelwork.

Measuring equipment

Only the most primitive types of structure could be built by eye alone. Measurement soon became necessary to control widths and heights, and to keep floors horizontal and walls vertical. Very simple means seem, nevertheless, to have sufficed for a remarkably long time. The plumb line served to establish verticals. Attached to an A-frame with a mark on the crosspiece, or to some similar device, it also served as a level. Alternatively, water levels were used, or the whole site might be temporarily flooded. Bubble levels seem hardly to have been used until the 19th century. Setting a building out was usually done geometrically with only a few base measurements. For this purpose cords would be used, or square and compass. For larger scale surveying, table-mounted direct optical sights took the place of cords. Divided scales, either linear or angular, seem to have had very limited use in building work up to the Renaissance. The chief consequence of reliance on these simple means was

that a high premium was placed on the adoption of geometrically simple forms.

The merits of forms that could be simply set out and checked did not disappear with the widespread introduction of divided scales and measuring chains and tapes, nor with the later introduction of more versatile and precise counterparts to the table-mounted direct optical sight such as the theodolite; but a freer choice of form did become possible, provided that the form could be adequately defined. In addition, a far more important structural freedom of choice has been gained, largely in the last 100 years, through a growing ability to measure and monitor forces and resultant movements in the course of construction, and to prooftest structural materials and elements. This ability has been gained through the development of instruments to measure pressures and strains (small dimensional changes) in particular, and thereby determine forces applied by hydraulic jacks and internal forces and stresses.

Scaffolding, centering, and formwork

The ramps of mud brick or earth constructed up to the working level by early Egyptian and other builders to enable large blocks of stone to be manhandled into place, also served as the only necessary working platforms. Once construction was complete, they could be used for final dressing of the stone and decoration of its surface as they were progressively dismantled. Other methods of lifting and placing materials have usually called for separate working platforms supported by temporary scaffolds. Until well into the 20th century, when tubular steel scaffolding was introduced, these scaffolds were always made of timber as they still are in less industrialized countries today. In the commonest form, horizontal putlogs were set, at one end, in holes left for the purpose in the ascending structure, and they were lashed at the other to freestanding uprights. Other horizontal timbers ran parallel to the working face to give stability and assist in supporting the working surface of hurdles or planks or something similar. In one variant, used mainly at higher levels, and in vault and dome construction, there were no uprights. The putlogs were more deeply anchored in the ascending structure and further secured by raking struts or ties attached a little lower or higher.

A third type of scaffold provided a much more substantial and extensive working platform at an upper level, such as the springing level of main arches or vaults, in a tall vaulted structure without intermediate floors. This platform probably supported a crane as well as men and materials, and sometimes centering for the arches or vaults as well. Surviving evidence suggests that it was carried, as far as possible, off the ground by a combination of heavy horizontal timbers and raking struts built into or supported by the structure already built. Provisions for its support would have called for the greatest amount of forethought. However, even so, they can never have influenced design as much as requirements for temporary support of the structure itself. Framed structures in general have called for much less scaffolding since, to a greater extent, they have provided their own working platforms.

The need for temporary support of the structure itself during construction has arisen mainly in the case of spanning elements like arches and vaults, and more recently reinforced-concrete floor slabs, put together or cast in situ. The former have not been self-supporting until completed, and the latter not until the concrete has hardened sufficiently. The usual support for an arch or vault was a timber center of the desired profile, either strutted up at intervals from directly below, or carried by fans of raking struts anchored near the springings, or spanning as a timber arch between the springings. A ribbed vault usually required centers only under the ribs. Reinforced-concrete spanning elements cast in situ were, until recently, usually supported by timber beams and props. Now, in multistorey buildings, these timber beams and props have mostly been superseded by easily adjusted telescopic beams and props of steel.

In addition to centers and other props for spanning elements, continuous local support has always been necessary for any element cast in a material like wet concrete that initially has no strength and gains strength only slowly. This local support is provided by what is known as formwork. Until the recent introduction of forms of sheet steel and other materials, timber must have been the usual material, and the impressions of timber boards can still be clearly seen on the exposed faces and undersurfaces of early Roman concrete walls and vaults. The use of timber in this way would, however, have been extravagant until mechanically sawn boards became available, and it would have restricted the choice of surface geometry for a vault. Ways of economizing and at the same time gaining greater freedom in the choice of surface geometry included the later Roman practice of casting behind a veneer of brick or tile which was, in effect, permanent formwork, and a less easily substantiated but probably widespread practice of molding the actual form of a vault in earth or loosely bound rubble above much more roughly executed centers and formwork. The modern use of steel formwork imposes a different restriction on design, since it must be reused repeatedly to justify its initial cost. In multistorey buildings, this tends to call for repetitive floor plans so that the same forms can be used on each floor. On the other hand, with walls continuing unbroken over a considerable height, it has been possible to make self-contained forms that can be progressively slid up the wall itself and anchored to it as casting proceeds.

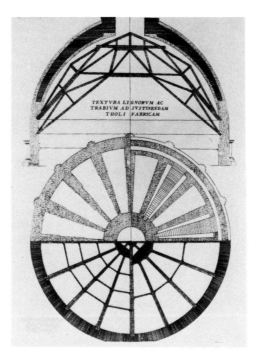

Centering for the construction of the dome of St Peter's, Rome, in 1588-90. The dome was of double shell form with 16 radial ribs between the two shells (top and upper half below). The centering spanned between the springings to provide support beneath each rib (top and lower half below).

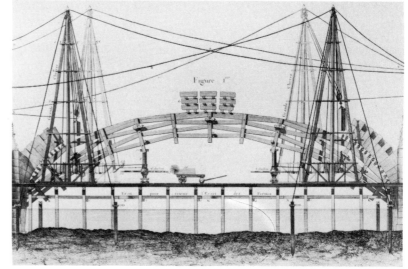

Construction of a bridge at Mantes, France, in 1764, showing centering, temporary access bridge, and lifting and traversing arrangements.

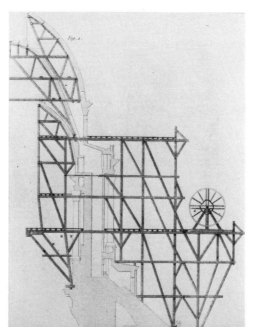

Working platforms, centering, and lifting wheel used in constructing the triple shell of the dome of the Panthéon, Paris, in 1787-90.

Construction processes

A new building or bridge may be envisaged by its designer in its entirety before construction starts, and it may even finally assume this intended form, though changes in plan during construction often occur. The resulting different and incomplete structure must, nevertheless, be stable throughout the building process and it may, for instance, need temporary weather protection. When, as in the past, the construction of large buildings and bridges was often a very lengthy process, and it was much more difficult than today to foresee how the completed structure would behave, more thought was probably given to ensuring the stability of successive incomplete forms than to the final state. Today, careful consideration of intermediate states has again assumed great importance, both where prestressing is carried out to achieve the desired final state, and where the stresses and scale of construction are great enough to lead to significant dimensional changes or changes in geometric configuration as construction proceeds. Considerations of overall economy must have always been of some importance too, and they have tended at different times and places to favor different distributions of effort between the building site and the sources of supply of materials. Here we shall give a few examples of the influences of the adoption of different construction processes on structural forms, and then briefly consider some aspects of the prefabrication of components and elements before they reach the building site.

Problems associated with the incompleteness of the structure during construction. Up to the 19th century, the chief problems of stability were those experienced with arched and vaulted structures. In multibay buildings and multispan bridges the thrusts of adjacent arches or vaults along the axis neutralized or largely neutralized one another in the completed structure, but did not do so to the same extent during construction if the arches or vaults were constructed one at a time, as was usually desirable from other points of view. Each arch or vault normally became self-supporting only when completed (see also ARCHES AND VAULTS).

In both arched and vaulted structures, the problems could, in principle, be overcome by using temporary supports; but, while arches and barrel, groined, and ribbed vaults were constructed on centering when they needed such support, analogous temporary inclined shores seemed rarely, if ever, to have been used to resist temporarily unbalanced thrusts. Where temporary support was provided it took the form of ties across the arch or vault. Use was sometimes made of such ties, later unhooked or sawn off close to the piers, in Gothic construction. Similar ties (of iron) were placed across the main arches that carry the dome of Florence Cathedral and were sawn off when the dome was almost complete. Nevertheless, sufficient width was usually given to the piers to resist the temporarily unbalanced thrusts during construction. This was invariably done in bridges until, in the late 18th century, J.R. Perronet (1708–94) bridged the Seine at Neuilly and was able to halve the proportionate thickness of the piers by constructing all five arches simultaneously. With this reduction in thickness he eased the flow of water and thereby reduced the risk of scouring at the feet of the piers.

In the later 19th and 20th centuries, more diverse structural systems have called for a greater variety of temporary support without completely outmoding these earlier types. In wide-span structures of arched or cantilevered form, the chief innovation has been the use of tensile stays, either radiating from raised supports above the two ends of the span, as in the construction of the Eads Bridge in St Louis, Missouri, and other iron and steel bridges, or incorporated within the depth of the structure, as in some modern prestressed concrete bridges. In tall buildings the usual need has been for additional lateral bracing of tall frames before all the walls are added. Temporary diagonal cable braces have served this purpose.

Dimensional changes have always occurred in the course of construction as the inevitable accompaniment of taking up loads and developing resistance to them. Since they are usually only small changes, they are not a direct constraint on design; as the need to ensure stability of the incomplete structure has been. However, they can influence the practicality of a construction process and have a considerable effect on the final structural behavior, so they have indirectly influenced designers' choices.

It has, for instance, always been necessary to allow for the progressive deformation of centering—particularly timber centering—as an arch or vault is built over it, and for the further deformation of the arch or vault when the centering is removed. To minimize the first deformation, it was realized long ago (as Leone Battista Alberti (1404–72) noted in the 15th century) that construction on centering should be undertaken as rapidly as possible. Where a lengthy construction process had to be envisaged, it was preferable to adopt a form that did not require centering.

Examples of important dimensional changes in modern buildings and bridges are the shortening of the usual concrete core (containing elevators, services, etc) of a tall building, and the progressive deformations in numerous highly interactive structural systems. When the core of the tall building is built ahead of the surrounding structure, as it frequently is, the subsequent shortening of the outer columns must be properly allowed for if the floors are to

The National Westminster Building, London, under construction in 1977 showing the reinforced-concrete core that provides much of the lateral stability rising above the outer steel framing that is being erected around it.

finish horizontal. In the highly interactive system, the whole way in which the building's own weight is carried may depend on the sequence and manner of construction, particularly when a material like concrete is used. It is this dependence that is deliberately exploited in modern prestressing.

Prestressing. The pretensioning of guy ropes must have been practiced for several millennia. Iron ties also had to be pretensioned if they were to be fully effective in restraining undesired movements, and the practice of tensioning them by driving wedges into eyes formed in the ends of the bars dates back at least to medieval times. It was carried over into the pretensioning of the wrought-iron bars used in the mid-19th century to truss cast-iron beams. But it became more usual at this time to form screw threads on the ends of such bars and on the ends of the ties in roof and bridge trusses and of the diagonal braces in column and beam frames. This would have allowed better control over the tensioning. Another procedure introduced then for modifying the stress distribution in a composite iron structure was employed in the construction of the Britannia Tubular Bridge over the Menai Straits. Each of the four spans was fabricated separately at a nearby site and then lifted into position by hydraulic jacks. Had all four been set on their final bearings before they were connected together, each would have independently had to support its own weight: by the raising of the far end of each span after the first above its final bearing at the time of making the connection, all four spans were made to support themselves as one continuous beam.

Modern applications of prestressing include examples similar to all these mid-19th century examples, but the commonest application has been to reinforced concrete, where a redistribution between the steel and concrete of the stresses that carry the self-weight can be particularly beneficial on account of the relative weakness of the concrete in tension. If, on completion of construction, the concrete is under adequate compressive stress everywhere, it is able to play a full part in resisting subsequently applied loads. The usual technique has been to tension the reinforcement by means of hydraulic jacks at one stage of construction and then, at an appropriate later stage when the reinforcement has been anchored or bonded to the concrete, to remove the jacks. Even more than new joining techniques, like the welding of steel and gluing of timber, such prestressing has contributed immensely to the development of structural forms in which material is used to the greatest advantage. Most of the pioneering development was undertaken by Eugène Freyssinet (1879–1962) in France in the second quarter of the 20th century. Subsequent development was unusually rapid, thanks partly to the availability of the analytical techniques necessary for a full use of the possibilities.

From prefabrication to industrialization and systems building. For a surprisingly long time there has been a measure of prefabrication in building—for instance in the Roman mass production of marble columns and in precutting of members of timber frames and trusses before they were taken to the building site. Such prefabrication became significant in relation to design when industrialization of the fabrication process called for the standardization of major structural components and elements. It became most significant with the introduction of columns and beams of cast iron. Its major, though untypical, achievement toward the end of this phase was in permitting the completion of the Crystal Palace for the 1851 London Exhibition in less than six months. Prefabrication of reinforced concrete followed, rather tentatively, toward the end of the 19th century; its main initial merit being that it allowed pretesting in a situation of still inadequate knowledge of the strengths to be expected.

In the 20th century, industrialized prefabrication has become much more common, both for reasons of economy and to gain the benefits of the better working conditions and possibilities of quantity control in a factory. There has been a great production of prestressed concrete beams, concrete wall and floor panels for use in housing, and a somewhat smaller production of a variety of structural components for use in schools, offices, and factories. However, in most of these cases, the components have been designed to fit together with others from the same source to make complete structures, rather than being incorporated in structures otherwise fabricated in situ, or from components from completely independent sources.

Coordinated production of this kind is one facet of what is known as systems building. Outstanding examples have been various systems of housing construction employing the large panels, mostly designed in the decades immediately after World War II to relieve an acute housing shortage. The CLASP system of school construction, employing a fairly light steel frame, was also designed at this time. The heavy concrete large-panel systems have called for great ingenuity in solving the problems of making joints that are adequate both structurally and in keeping the weather out. They have also called for heavy investment in plant at the casting factory, and have entailed the maximum restriction of the designer's freedom in planning individual buildings. Less restrictive or more "open" systems are now preferred, and there has been a parallel development toward different concept systems, more in terms of a basic idea and a package of relevant design skills, and construction and management expertise.

Modern structural prefabrication – a precast storey-height concrete wall panel of the Balency system being lifted from its mold.

Section 4
Services, mechanical and environmental systems

Services, mechanical and environmental systems

The development of services in buildings has a lengthy and complex history. It is possible, however, to introduce some order into the subject by drawing a series of dividing lines to define relatively distinct historical periods. The first of these lines can be drawn at the beginning of the Industrial Revolution. Much of what is nowadays referred to as building services depends upon the availability of sources of mechanical energy. Just as the Industrial Revolution itself depended upon the invention of the steam engine, so the use of steam power, and slightly later coal gas, made possible fundamental developments in the way in which buildings were conceived and used.

The next dividing line occurs with another major development in power sources—electricity—and the enormous steps forward which followed, first from the availability of a safe, controllable, and brilliant source of artificial light, and then from the rapid application of electrical power to other uses. The final line is only recently drawn, but the recognition of the finite limits of the earth's energy resources has provoked new thinking about building services.

Another set of lines can also be drawn and these distinguish between building types: assembly buildings, hospitals, offices, schools, and domestic buildings. Each of these types, at various stages in its development, has been the vehicle for a number of important innovations in services systems. Through these it is possible to examine the question of the relationship between the growth and development of building services, and developments in architectural form itself.

A final distinction which can be made in the organization of such a subject is that between practices in Europe and in the U.S. For reasons of culture and climate, building services, and particularly those for environmental control, have developed along different paths on opposite sides of the Atlantic.

Before the Industrial Revolution

Before controllable sources of energy were to hand, the supply of heat, light, and ventilation to buildings, and to some extent the removal of unwanted waste products, was an unreliable and laborious process. The open fire was only capable of warming the very smallest of spaces and required constant attention. Similarly, when artificial lighting depended upon oil lamps or candles, the intensity of illumination was extremely low and its source was short lived and erratic. Furthermore, by virtue of their method of construction, buildings were inherently difficult to seal, and because smoke from the fire demanded the construction of an aperture—later of chimneys—ventilation was usually abundant, often to the point of discomfort.

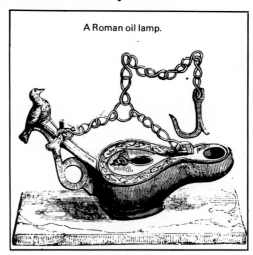

A Roman oil lamp.

There is evidence that as early as 1700BC houses in Crete enjoyed the amenities of bathrooms, water closets, and their associated piped water and drainage services. In the Middle East, the center of city evolution, ditches and drainage channels were dug to serve as sewers; remains of these have been found at Khorsabad (800 BC), some being constructed of brick and protected by a vault brick. The Romans aspired to high standards of domestic and urban comfort, applying their engineering skills to the construction of public toilets and public baths and, most spectacularly, in bringing water supplies to the cities. (They also invented the candle in the 1st century AD, but the evidence suggests that this was regarded as an inferior substitute for the oil lamp and it is clear that their standards of domestic lighting were never high.) In the Middle Ages in Europe, most of this earlier knowledge was lost, although, as at Canterbury in England in the 12th century, there is documented evidence of its survival in isolated instances. In general, in towns and cities of this period the disposal of sewage was generally to a cesspit provided either in the garden or, in certain cases, under the floor of the building. These pits were often constructed of stone with vaulted roofs and examples have survived in many old towns such as Winchelsea in Sussex and Winchester in Hampshire, England. Although usually provided with a ventilation shaft these cesspits became offensive in use and overflowed, contaminating the surrounding soil. This contamination affected water supplies and led to repeated outbreaks of serious epidemics including bubonic plague which, in 1348 and 1349 caused the death of one-third of the population of England. It was only when the urban populations began to increase in numbers with industrialization that attention was again given to these questions.

Hamburg is thought to have been the first major city in modern times to construct a sewer system. This was built in the reconstruction of the city following a major fire in 1843. Even so, as late as 1854 there were cesspools in Red Lion Square and Bedford Row in the center of London. One of the earliest sewers in the city was constructed in the Strand in 1802 but it was not until 1865 that there was a comprehensive main drainage system in London. Within buildings, attention was given to proper sanitation as early as 1596 when Sir John Harrington (1561–1612) designed what is thought to have been the first water closet. By the 18th century the idea had been conceived of using a water sealed trap to protect the interior of a building from the odors of the drainpipe.

Even before the Industrial Revolution the needs of the growing urban populations led to the construction of water supply systems in a number of European cities. The Germans were the leaders in this and by 1558 Augsburg had an extensive and plentiful supply. In London in the 16th century water was pumped from the River Thames to a reservoir from which it was supplied to nearby houses through lead pipes. From contemporary accounts of the state of the river water this must have been a dubious amenity. In 1613, water drawn from pure springs near Hertford, 20 mi. (32 km) to the north of London, was piped to a reservoir at Islington and from there to the city. In France, in about 1608, a "lifting pump" was constructed beneath the Pont Neuf over the River Seine in Paris in order to supply water to the Louvre and the Tuilleries, and a similar machine was constructed in 1669 at the Notre Dame. The most elaborate waterworks of all at this period was that completed in 1669 by the Dutch engineer Rannequin, to supply the gardens at Versailles. This transported water a distance of three-quarters of a mile (0.9 km) from the river and lifted it through a height of 533 ft. (160 m). It must be recognized, however, that these supplies were enjoyed by very few houses. The majority, at best, would have access to a standpipe in the street from which they could draw water for perhaps an hour each day.

In terms of heating, apart from the central fire with logs burning on a stone hearth, the earliest heating apparatus was probably the bronze tripod brazier of the classical Greek household. Readily portable and utilizing carbonized wood (charcoal) as a fuel, it survives today in remote areas of the Middle East. A new system of heating—the hypocaust—was developed by the Romans, who incorporated it into their houses. This construction comprised a stokehold in the external wall from which a large flue passed to a brick chamber formed under the tesselated concrete floor of the principal room. Radiating flues carried smoke and hot gases to flues contained within the thickness of the walls, discharging at eaves level.

Following the extinction of Roman influence in Europe, heating methods reverted to primitive central fires for hundreds of years until simple recesses in the thickness of medieval walls connected to the exterior by a short flue began to make their appearance. However, the wasting away of natural forests and the growing use of coal as fuel for heating emphasized the wasteful design of large open fireplaces. Also, coal burned badly in them and a great number of experiments were conducted to increase efficiency. In 1624 Louis Savot provided a novel fireplace in the Louvre in Paris which incorporated passages under the hearth from which warmed air was delivered from grilles in the mantlepiece. A further notable advance in fireplace design was the introduction in 18th-century France of the canopy incorporating a restricted throat complete with a movable damper. The principles of modern firegrate design were laid down by Count Rumford (1753–1814) who discovered the importance of the distance between the fire opening and the flue throat: if excessive, the flue temperature dropped and the fire refused to draw, and if too small, excessive combustion ensued.

Enclosed fires or stoves were developed in Europe in the 15th century. They were usually of brick and placed centrally in the house to promote maximum warmth. From this developed the Scandinavian stove with its tall iron flue incorporating baffles to extend the travel of hot gases from the fire. In 1744 in the U.S. Benjamin Franklin (1706–90) developed the improved cast-iron, wood-burning stove which bears his name (and is still available in Britain today). An important development of the freestanding iron stove came in 1792 when an Englishman, William Strutt (1756–1830), produced an iron stove in which air passing over a heated surface was distributed by natural convection. From this was developed in 1806 the Belper stove incorporating a greatly increased output for commercial premises.

All these inventions and developments were for single apartment heating. The heating of a number of rooms and floors from a single source developed from the 18th-century use in France of hot-water heating for horticultural purposes. Using large bore pipes and a simple boiler, the first commercial installation was for the new Bank of England, London, in 1792. Early systems operated by gravity; cold water being denser fell back to the boiler through pipes, forcing the lighter warm water to rise to the radiators. Boilers were placed in basements or stokeholds until the introduction of motor-driven pumps or circulators made boiler location a matter of convenience rather than simple physics.

Steam heating was invented by James Watt (1756–1819) in 1784 using waste steam from boiler testing to heat his workshop by intro-

Diagram of a Roman bath building with a hypocaust heating system.

Cast-iron stove with an oven, early 19th century.

ducing it into high level pipes running around his workshop. He did not, surprisingly, see the commercial possibilities and it was an English inventor—Hoyle from Halifax in Yorkshire—who in 1791 patented a method of steam heating. One of the earliest commercial installations was the heating of a silk mill in Watford, England, designed by Thomas Tredgold in 1824.

It is evident from these two examples that even at the end of the 18th century most buildings had very little service equipment. Consequently, the utilitarian role of buildings was to provide shelter from the elements and to offer a degree of modification of the external climate. For the majority of people, in both the cities and the country, the pattern of domestic and productive life was fundamentally conditioned by the external climate, with its daily and seasonal variations. This meant that useful light was only available for an average of 12 hours a day and that a comfortable temperature could only be achieved in the winter months in a restricted area of a building. These limitations were only barely ameliorated by the rudimentary sources of artificial light and heat that were available.

Contemporary references illustrate conditions in pre-industrial buildings. In the early 17th century John Aubrey reported that the wife of William Oughtred, the English mathematician, "would now allow him to burne candle after supper, by which means many a good notion is lost." As a further example, the records at the Jacobean mansion at Audley End in Essex, England, show that, in the winter of 1765, the daily candle consumption was a mere 3 lb. (1.4 kg) in weight. On the subject of the thermal environment in pre-industrial buildings, there is a record at Trinity College, Cambridge, dated February 1739, in which the librarian is instructed to prepare a new catalog in the magnificent college library designed by Sir Christopher Wren (1632–1723). The work was to be undertaken "as soon as the weather permits."

An important by-product of the Industrial Revolution was the proliferation of institutions and of the buildings and plant necessary to sustain them. In this situation a fundamental change was wrought in the expectations of buildings. No longer were the limitations of the natural climate allowed to determine how people should live and work. A building whose use was restricted to certain periods of the year, or which was inconvenient to use after dark, was no longer acceptable. Similarly, a building which required its occupants to devote a good deal of their time to its operation was equally undesirable. What was needed was a building in which the environment was appropriate to the activities at any time of day or night and any season of the year, and which achieved this as automatically as possible.

During the 19th century all of these desires came to be satisfied in buildings of all types. Although the benefits were not enjoyed by all members of society, the fundamentals of the technology of building services and, perhaps of even greater significance, a set of new expectations about the services which a building could accommodate, were well established. From this platform the subsequent developments of the 20th century could easily proceed.

The 19th century

Assembly buildings

From the point of view of services design, assembly buildings are interesting in three ways. The first of these is the problem of ventilation, namely the maintenance of an acceptable thermal environment in densely occupied spaces. Second, the importance of good vision focuses attention upon the technology of lighting systems. Third, the importance of good sound for both music and drama introduces the subject of architectural acoustics.

To outline the main stages in the development of ventilation systems up to this period, the old and new buildings of the Palace of Westminster, London, clearly illustrate the relevant technological changes which occurred as society moved into the Industrial Age.

It is recorded that Sir Christopher Wren (1632–1723) attempted to provide ventilation in the original building by placing truncated pyramids, each with an openable lid, above the ceiling at the four corners of the House of Commons. It was hoped that these pyramids would conduct the heated air from the chamber. However, when the air in the roof space was cooler than that in the House, undesirable downdrafts were created. In 1723 a Dr Desaguliers (1683–1744) was called to attend to the problem. He adapted Wren's pyramids by running a trunking from them to a firegrate which then guaranteed a consistent flow of air from the chamber by assisted convection. He later installed a "centrifugal, or blowing wheel," which was to be turned by a man, to be called a "ventilator." This device remained in use until the building was destroyed by fire in 1834.

The associated House of Lords also suffered from ventilation difficulties and here Sir Humphry Davy (1778–1829) added brick flues alongside its earlier, hypocaust-like heating system, and led fresh air through these to numerous small holes in the floor. Extraction was provided by two metal tubes in the ceiling, each of which passed through a furnace to assist convection.

Following the fire of 1834, the question of

environmental control was a central issue in the design of the new building. A whole series of authorities was involved in the work. The first was a Dr John Reid, who was also responsible for the apparently successful ventilation arrangements in the Commons' temporary home in the reroofed shell of the House of Lords. In this he made provision for cooling the air in summer, in addition to winter heating, by filling the water pipes over which the air passed with cold water. A similar system was installed in the new Commons' chamber but after a dispute the architect Charles Barry (1795–1860) took charge of the job in the House of Lords using a system of supply and extract through the ceiling. In the event neither of the original systems proved to be very successful and other specialists were called in. In 1865 a Dr John Percy installed a system which was sufficiently successful to survive into the 20th century. This was, to all intents and purposes, a full air-conditioning system in the sense that it heated the space in winter, and cooled it in summer by passing the air over blocks of ice. Control in the chamber was in charge of an attendant who watched a thermometer and covered or uncovered areas of the pipes.

Environmental control in theater buildings also inspired great ingenuity. For example, early in the 19th century the Marquis of Chabannes—"an earnest and successful candidate for smoke-doctor distinction"—had installed a ventilation system in the Covent Garden Theater, London, based upon a large gas chandelier supplemented by "caloriferes." These were positioned in one of the galleries, another over the stage, and others at every entrance and staircase so that the incoming air was warmed before it entered the auditorium. A steam-heating system was placed under the stage. Extract ventilation was through three ducts which passed through the roof, each placed over a calorifere.

In the U.S., control of the thermal environment in theaters had become very sophisticated by the 1880s. The theater at New York's Madison Square Garden (1880), designed by McKim, Mead, and White, was described as "the first theater in New York at least, to be efficiently ventilated as well as properly heated and satisfactorily cooled in summer." The system delivered air to openings under the seats. In winter, the air was warmed by being passed over steam radiators, and summertime cooling was achieved by passing the incoming air over "enormous blocks of ice." Extraction was through openings in the ceiling and under the galleries. The success of this installation, which required a high degree of integration of the services with the structure of the building, was such that similar installations rapidly became commonplace in most theaters in New York. Adler and Sullivan's McVickers Theater in

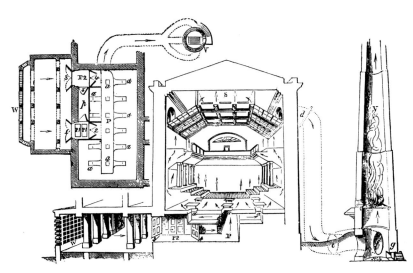

Dr Reid's ventilation system for the temporary House of Commons, London (1830s)

Chicago (1885) had a very similar system except that there the input was through the ceiling and extract through the floors. This system achieved a ventilation rate of four air changes an hour.

By 1895 the utility of "the fan system" was acknowledged as the most suitable arrangement for theaters. It was considered that a dual duct system—one for warm air, the other for cool—would be best. In one proposal the mixture of warm to cool air was to be controlled by attendants, each operating a damper between the two ducts, who were seated in the various parts of the theater. In effect these attendants were human thermostats.

In Europe, the basis for heating and ventilating theaters remained relatively unscientific throughout the middle decades of the century. Nevertheless, a good example of that subsequently achieved was the Hofburg Theater in Vienna (1874) designed by Gottfried Semper (1803–79). This had a combination of collecting, heating, and mixing chambers from which air at an appropriate temperature was conducted throughout the building. Delivery to the auditorium was beneath the seats. During cold weather extract ventilation relied upon the buoyancy of the warmer air inside the building, but when it was warm outdoors exhaust fans were brought into action. Cooling was achieved by passing the air over a water surface of 1,440 sq. yd. (1,203 sq. m) supplied with water from a deep well. This was capable of reducing the temperature of the air by 3–5°C on a summer's day. The center of the whole installation was the lavish "engineer's room" from which the system was controlled. However, even this advanced installation relied upon the fact that warm air rises to achieve adequate ventilation in cold weather. This is an important contrast with 20th-century ventilation practice which has an almost exclusive reliance upon mechanical extract from such buildings.

Mechanical forced ventilation using fans of various types provided a reliable method of moving large volumes of air. Their use increased rapidly toward the end of the 19th century.

Theater lighting progressed during the 19th century from a state similar to that which had prevailed since the evolution of the enclosed theater in the 16th century to a level which contained the essence of present-day practice. Until the advent of limelight in 1794, 18th-century theater relied upon oil lamps and candles for illumination. The difficulty in managing these meant that the houselights generally remained lit during the performance. Gas lighting was first demonstrated in 1788 at the then English Opera House (later the Lyceum Theater), but this was as a display itself and not as a means of lighting drama. (Municipal gas lighting and power distribution commenced in 1813 with the formation of the London and Westminister Gas Company.) In 1817, both the stage and auditorium of the Lyceum were gaslit, and around 1856 Sir Henry Irving (1838–1905) began to insist upon the houselights being dimmed during a performance.

Electric light made its appearance as early as 1846 when an arc lamp powered by batteries was used to simulate the rising sun at a performance of Meyerbeer's Le Prophète at the Grand Opéra in Paris. In 1879 the Bellacour Theater in Lyons was lit by Jablochkoff candles—an improved arc lamp powered by a generator. In 1881 the Paris Opéra and the Savoy Theater in London were both lit by incandescent lamps. At the Savoy the system was designed and installed by the Siemens Company and power was produced by six generators. There were nearly 1,200 lamps in the building; 114 in the auditorium, 715 clear lamps and 100 tinted blue over the stage—the latter for night scenes—with the remainder in the corridors and dressing rooms. Five years later the Vienna Opera House was lit by as many as 5,000 lamps.

In addition to the advantages of artistic effect and control which electric light brought to the theater, it also eased the problems of ventilation by producing light without fumes, and, as an even greater advantage, it considerably reduced the fire risk.

In the design of theaters and other assembly buildings in the 19th century, acoustics—the third criterion of performance—was largely a hit-and-miss affair. Charles Garnier (1825–98) sought an understanding when he was designing the Paris Opéra, and was forced in the end to declare his inability to come to terms with "this bizarre science." The Opéra did have good acoustics, but these were achieved by adopting the standard Italian "horseshoe plan," tried and tested over a century or more.

Modern architectural acoustics was born in the U.S. in the closing years of the 19th century. Wallace Clement Sabine (1868–1919), a young physics professor at Harvard, was asked in 1895 by the president of the university to investigate the cause of poor acoustics in the large lecture room at the newly constructed Fogg Art Museum. Following a remarkable series of experiments he not only solved this particular problem but also discovered the precise mathematical relationship between the dimensions and construction of an auditorium and its acoustical quality. Armed with this knowledge he was able to act as a consultant of McKim, Mead, and White in the design of the New Boston Hall (1900), now Symphony Hall, the home of the Boston Symphony Orchestra. This building, although relatively conservative in its form, may thus be recognized as the first "scientifically" designed auditorium since antiquity.

Sectional elevation of New Boston Music Hall (1900). W.C. Sabine acted as acoustics consultant for the designers McKim, Mead and White.

Hospitals

In the 18th century most hospitals owed their form and arrangement as much to the canons of architectural composition as to any consideration of the special requirements of their function. During this period it is easy to recognize the similarity in plan between hospitals and country houses. On the continent of Europe, and particularly in France, however, attention was beginning to be paid to the question of the form of the "hygenic" hospital. This was in reaction to the difficulties experienced in unsuitable buildings inherited from the Middle Ages—in particular the Hôtel Dieu in Paris. Out of this situation was born the pavilion plan, which rapidly became the standard basis for hospital design. Nevertheless, scientific modes of, for example, ventilation, were rejected as inadequate, and the importance of tall windows to promote good daylighting, in addition to ventilation, was stressed. Hospital heating too, even in a building based on wards for more than 20 patients, was often by open fireplaces or stoves "because heat both promotes the rapid decomposition of foul excretions and also increases the discomfort of the feverish sick by surrounding them with an unnaturally dry, hot atmosphere."

Even after the 1850s authoritative opinion in Europe was reluctant to accept the convenience of central heating and scientific ventilation for hospitals. The predominant reliance upon simple natural ventilation in the hospitals of the second half of the 19th century, however, was not due to a shortage of ideas about mechanical aids; it must be concluded, therefore, that these proved to be unreliable in practice, and that the natural conservatism of the new bureaucracies prevailed in insisting upon the use of tried and tested methods. Invariably, it was only when the circumstances of the construction of a building were in some way unusual that experiment proved more acceptable. This was beautifully demonstrated by Isambard Kingdom Brunel (1806–59) in his design for a hospital at Renkioi in the Dardanelles during the Crimean war. This was remarkable in many ways and not only for its ventilation system and other services. Ventilation was achieved by a system of man-powered fans, one to each ward pavilion. The fans propelled air along a system of floor ducts into the wards. Opening windows were also provided, and these were sheltered from the heat outside by the eaves overhang. The aim was thus to avoid bad air entering the wards by this process of forcing air into them rather than drawing it out. There were no stoves or fireplaces, but hot water was supplied by a small boiler, heated by candles. Lighting was also by candles in specially designed lanterns. Drainage was through a system of tarred wooden sewers.

In the U.S., characteristic enterprise was applied to the problem of hospital design, and this is best exemplified in the design by John S. Billings (1838–1913) of the Johns Hopkins Hospital in Baltimore (1877). In plan the pavilion arrangement was unremarkable, but a cross section through a typical ward block shows that considerable ingenuity went into the design of the ventilation system. Each block consisted of two storeys, with the ward proper on the upper level and the sole function of the lower being to serve as an air intake. The air, having passed across the lawns, entered the long windows in the walls of the lower floor. The walls of this floor were lined with water-filled heating coils and the air circulated from these into the ward. Foul air was extracted by two routes. There was first a duct system linked to grilles beneath every bed. This was connected to the main vertical ventilator shaft. In addition, a further series of outlets were located in the ceiling, and these led through the roof space to meet the ventilator shaft at a high point just below a steam-heated coil. This added further motive force to the extraction process. In cold weather the ceiling outlets were not opened, being only brought into use on warm days.

At the end of the 19th century the question of mechanical ventilation for hospitals was revived in Europe. In England, for example, William Henman declared in 1894 that "in our constantly varying climate, with the smoke-

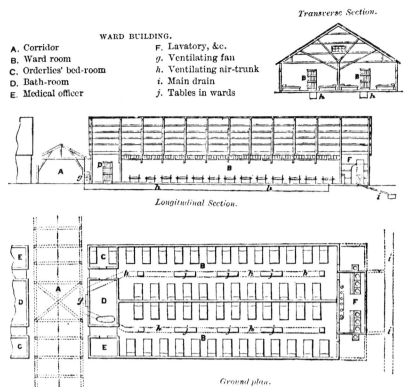

WARD BUILDING.

A. Corridor
B. Ward room
C. Orderlies' bed-room
D. Bath-room
E. Medical officer

F. Lavatory, &c.
g. Ventilating fan
h. Ventilating air-trunk
i. Main drain
j. Tables in wards

Transverse Section.

Longitudinal Section.

Ground plan.

Plan and section of a ward pavilion in I.K. Brunel's Renkioi Hospital showing underfloor ventilation trunking.

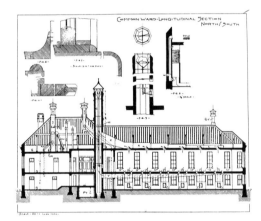

Section of a ward in J.S. Billing's Johns Hopkins Hospital in Baltimore (1877) showing air ducts and ventilation towers.

laden, and often impure atmosphere of our cities and towns, . . . a natural system (of ventilation in hospitals) is impracticable." He illustrated his own solution to the problem by his designs for the New General Hospital in Birmingham. Here he installed four plant rooms in the basement, each with air intakes, filters, heaters, and humidifiers. These connected to a duct system which supplied the wards and other rooms on the three floors above. The building had no opening windows and very few open fires. Extract was through flues to the roof level. The motive power for the system was from electric motors driving eight fans.

The sophistication of this building was not,

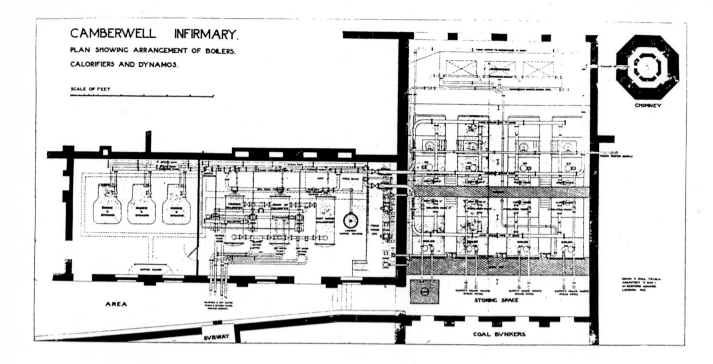

Plan of the plant room of the Camberwell Infirmary by E.T. Hall showing the disposition of boilers, calorifiers, and dynamos. Hall, who opposed artificial ventilation for hospitals, nevertheless employed other advanced servicing systems for the time.

however, typical of everyday practice, and at the turn of the century Edwin T. Hall (1851–1923) presented an extensive discussion of the principles of hospital design. In his opinion, artificial ventilation was "a mistake in any hospital." In his design for Camberwell Infirmary, London, Hall was ambiguous about ventilation. While emphasizing that all wards had openable windows on three sides, he also mentioned that they had electric extract ventilation. His point was that the air for wards should not be delivered through ducts because they were likely eventually to become foul. However, in spite of his opposition to artificial ventilation, Hall was not reactionary in all things since the designs of all the hospitals he described were advanced in their environmental services. Camberwell, for example, had a very substantial plant room containing four boilers, three calorifiers, and three dynamos. Exhaust steam from the dynamos supplemented the heating. The building was electrically lit and had electric elevators serving the ward blocks. The heating system, consisting of hot-water radiators, was arranged on a dual circuit principle to ensure that the building would be heated even if one circuit required maintenance.

Hall's building could be said to represent the best of conventional practice at the time, but William Henman was simultaneously laying down his beliefs once more through his design for the Royal Victoria Hospital in Belfast (1903). In this the implications of the system which he had used at Birmingham were fully realized and applied. The plan of the building was compressed so that the wards were now only separated by a party wall and

in addition they had to rely on rooflighting. The advantages of this arrangement in terms of movement about the building are clear, yet the density of the building was no greater than at Birmingham because the insistence on natural lighting restricted the building to a single storey.

This comparison between Henman's and Hall's buildings neatly encapsulates the basic alternative approaches to hospital design around the 1900s; hospitals were fundamentally determined by the nature of their service installations.

Offices

The office building is essentially a product of the 19th century. However, examined from the viewpoint of services design, and particularly environmental services, even this radical invention, in its early days, owed much to buildings of the past, especially to domestic building. The very first office buildings in the U.S., and many built even during the decades up to and into the 20th century, were aggregations of quite small rooms which were lit by daylight and ventilated by openable windows. Some were heated by open fires and only later were the attractions of some centralized, labor-saving heating system recognized. The New York Life Assurance Company's office designed by Griffiths Thomas, for example, was heated by steam forced into it on the fan principle. This development did not, however, fundamentally alter the basis of the design of the building. In Europe, the domestic influence was more apparent, both visually and technologically, and this persisted for much longer. For example, even as late as 1899 the Parr's Bank in Liverpool designed by Richard Norman Shaw (1831–1912), still relied upon the open fire to heat the smaller offices.

By the end of the 19th century the use of central heating for offices was well established in the U.S., and it was possible for *Architectural Record* to publish in 1895 quite comprehensive guidance on the nature of effective heating systems for these and other building types. In this, a basic distinction was made between the modes of heating for the "corporation offices which occupy the lower floors" and the upper offices, which were for rent. It was recommended that the former should be heated by the indirect, fan-system-warmed air, while the latter, because of the greater subdivision of the upper floors for rental purposes, should be heated directly by a one-pipe system supplying small radiators located in front of each window. Typical speculative plans for offices in New York at this time illustrate how this arrangement worked. This same article also discussed the merits of various types of thermostat, so indicating that automatic control was already part of everyday practice.

At the turn of the century, therefore, particularly in the U.S., there was no shortage of advanced ventilation technologies. But these were only applied where the nature of the activity performed in the building was distinctly non-domestic. Most office activity could be satisfactorily undertaken by small numbers of people in small rooms and the domestic analogy therefore held good. When, however, the nature of a business could be best served by a different building arrangement, the rules of the game were redefined and other technologies were brought to bear. This is clearly what later occurred with the collaboration between the Larkin Company of Buffalo and Frank Lloyd Wright (1869–1959) in 1906. Here the whole building was for a single organization whose functional efficiency could best be promoted by an open-plan arrangement. This allowed Wright to achieve this new synthesis between form and environmental technology. Nevertheless, systems of this kind were already in existence and in widespread use on the corporation offices of conventional office buildings and in other building types. Wright's achievement was to see the potential in these systems to allow a new arrangement of office space.

In terms of lighting in the late 1800s, even the most advanced office designs were still conceived as being predominantly daylit. Yet their lighting had come a long way from the unsatisfactory reliance upon candles at the beginning of the 19th century. In both Europe and the U.S., electric lighting was absolutely standard for office buildings by the 1890s. Often, particularly in the U.S., the power supply was generated in the building itself in

conjunction with the plant of the heating system. By the end of the century architectural journals in the U.S. could publish firm guidelines for deciding upon the number and location of light fittings in a room. A distinction was made between small and large rooms. In the former, "ceiling lights are not as useful as wall brackets." In the latter, "the most ideal light is one which is diffused from small clusters of two or three lights each, distributed uniformly on the ceiling." Also, where a large space is broken up by columns "... a very good illumination is often obtained by rings of lights arranged about the columns and carefully worked into the ornamentation." The precise number of lamps to be provided seemed to be an open question, but "50 to 60 (sq.) ft. (4.7 or 5.6 sq. m) per light may be considered an average" in open spaces. Small offices had two or three such lamps which "makes the average lighting about one to 40–45 sq. ft. (3.7–4.2 sq. m)."

A discussion of the office buildings must inevitably include an appraisal of the role of the elevator. Elevators for goods and passengers were in existence (in a form known as "teagles") in British textile mills as early as 1830. These were operated by cables from the main engine of the factory; they were inherently unsafe. One of the earliest hydraulically operated systems consisted of a cage directly mounted on a long plunger which descended into a cylinder located in a deep pit. Elisha Otis (1811–61) solved the safety problem and thus allowed the passenger elevator to emerge, by using a ratchet and pawl system at the sides of the shaft. He demonstrated his prototype at the New York Fair in 1854 and took out a patent in 1857, followed quickly by Sir William Armstrong (1810–1900) in England. Elevators around this time were invariably powered by steam. Later in the century more highly sophisticated hydraulically and electrically operated systems were developed. By 1870 the term "elevator building" was in common use, and the potential for the construction of convenient high buildings which it

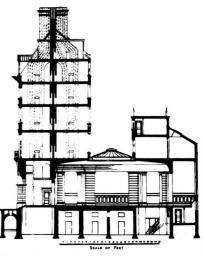

Richard Norman Shaw's Parr's Bank, Liverpool (1899).

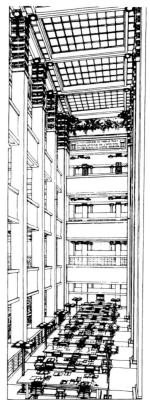

Interior view of Frank Lloyd Wright's Larkin Building in Buffalo (1906).

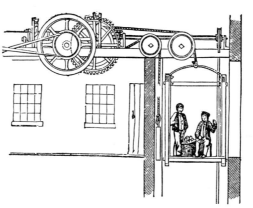

A teagle in a multistorey British textile mill of the 1830s.

offered, was recognized and vigorously exploited.

The problem experienced by designers of tall office buildings in the late 19th century was that other service systems could not all be simply extended vertically. It was not possible, for example, to rely upon natural convection to circulate hot water around a tall building; pumped circulation became necessary. Similarly, the water services could not operate on the normal mains pressure. This was overcome by pumping water up into storage tanks at the top of the building and distributing it downward by gravity. This solution itself led to problems as buildings became taller since the pressure on the lower floors became excessive. This was solved by the introduction of pressure-relieving tanks on intermediate floors.

In the U.S. the fire hazards in tall buildings were solved by technology. In the early discussions of the problem the emphasis was upon fireproof construction and upon the provision of fire escapes which were quickly recognized to be essential features of any tall building. Service systems had their part to play too, and there is a record as early as 1881 of the Chicago architects Burnham and Root receiving an instruction from their client, Peter Brooks, to run a fire fighters' standpipe all the way up the southeast corner of the eight-storey Montauk Building. In Europe, in contrast, the problem was solved for many years by the simple device of legislative prohibition!

An additional problem of early office blocks was that heated air achieved very high velocities in ventilation and elevator shafts, thus causing cold air to be sucked through openings at the lower floors. This "stack" effect rapidly produced a response in the form of the revolving door. This was the brainchild of Theophilus van Kannel who was, in 1889, honored by the Franklin Institute with the presentation of their John Scott Medal in recognition of the value of this innovation.

The growth in the size of the institutions housed in tall buildings also produced the need for improved systems of communication. The potential of the telephone, patented by Alexander Graham Bell (1847–1922) in 1876 as a means of internal communication, was quickly recognized, and the use of elevators and chutes for the conveyance of goods and documents was soon standard practice.

By the end of the 19th century a high level of servicing was regarded as essential to the commercial success of any office building, particularly in the U.S. The common practice was to expose all pipes and fittings. As Peter Brooks wrote to Burnham and Root in connection with the Montauk Building, "This covering up of pipes is all a mistake, they should be exposed everywhere, if necessary painted well and handsomely." Further

developments in services technology in the 20th century, however, led to an almost elaborate concern for their concealment.

Floor plan of Burnham and Root's Reliance Building (1890-94).

Schools

One of the major social developments during the 19th century was the acceptance of universal education. Throughout Europe and in the U.S. the school building emerged as an independent building type with its own special functional requirements carefully analyzed and the nature of appropriate designs widely discussed. By the 1870s it was the established view that the design of a school should follow a logical sequence from a consideration of the teaching method, through to the "elements which control the shape and size of the schoolrooms and classrooms composing the building," and on then to details of their construction and equipment. Such considerations also incorporated the question of "warming and ventilation" and of toilets and washrooms. It was assumed that lighting meant daylight. (The installation of electric lighting in schools was essentially a 20th-century innovation.)

In Europe, reference to research in Germany into building design led many architects to adopt a scientific approach to design of services in schools. In England, for example, lighting was based on a relationship of 30 sq. in. (192 sq. cm) of glass to every sq. ft. (0.09 sq. m) of floor area. The sills of the windows were to be at least 4 ft. (1.2 m) from the floor. It was also stated "that lighting from the side—especially the left side—is of such great importance as properly to have a material influence over our plans." The English architect E.R. Robson (1835–1912) advocated that "warming and ventilation must be treated as inseparable or, at least, in treating of one the other must be always present to our

The United Telephone Company's Exchange, London (1883).

mind.'' Following on from this he outlined the scientific basis for design by pointing to German studies of the effects of occupancy upon the freshness of the air in a classroom. He also discussed the relationship between the size of windows and the heating requirements of a room. This offered a basis for calculating the heat requirement of a room, and established a clear, if conflicting, relationship between lighting and heating.

After discussing such methods of heating as gas stoves and systems using high-pressure hot-water pipes. Robson had to confess that ''It is much easier . . . to determine what we ought not to do, than to draw final conclusions as to the best course for adoption in each case.'' At this time the open fire was still considered to be acceptable, partly because it had ''in its favor a strong prejudice in the mind of the English people.'' Also, the fact that the building would invariably have a caretaker meant that tending the fire would be the latter's responsibility and not the teacher's. Robson therefore recommended that the open fire was suitable for schools of three departments containing 500 children, and that for schools for 750 or more children an artificial system was preferable. In between these numbers the choice must be decided ''by the peculiarity of the plan.''

In the late 1800s technology of heating was wide ranging in both the U.S. and Europe. Open fires could be set in grates which supplemented the radiant heat they produced with convected air. Freestanding stoves, such as the Gurney stove, were not approved unless they were placed in a basement chamber, although even here there were objections on the basis that the caretaker could not be relied upon to operate the apparatus efficiently. Nevertheless, this idea was developed further in a system installed at the Luisen Schule in Berlin. This had a separate extraction system, with its own small boiler which was used to assist the flow of heated air from the heating chamber through the building. This could also be used independently in the summer months to provide ventilation when the heating was not in use.

London Board Schools and Jonson Street School in Stepney (c. 1890) designed by T.R. Smith, built to accommodate 1,500 children, exemplifies the fusion of these design principles and technologies into a very satisfying whole. Such innovations in 19th-century school building illustrates the contrast with the contemporary attitudes to hospital building; there was a much readier acceptance of quite sophisticated services technology in schools. In the 20th century services in hospitals are now equally sophisticated.

Domestic buildings

To illustrate the development of services in

Cast-iron radiator suitable for large rooms, late 19th century.

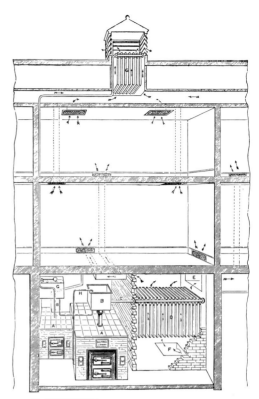

Integrated heating and ventilation system for a school from Robson's *School Architecture* (1874). A central boiler heats air which is distributed to the various rooms by convection.

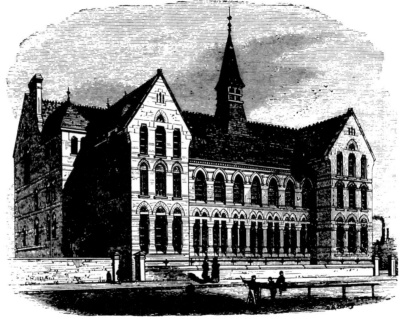

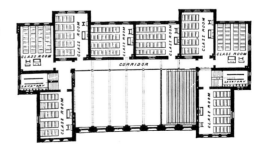

T.R. Smith's Jonson Street School, Stepney, London (1890); perspective and plan of upper floor.

domestic buildings during the 19th century, it is useful to draw parallels between customs and methods in Europe and in the U.S. At the beginning of the century a typical house in each continent could be described, environmentally, in very similar terms. Heating was by open fires, ventilation was by openable windows, generally of the sliding sash variety, and artificial lighting was by lamp or candle. Although during the early years of the 19th century there was no lack of experiment and development in appliances for heating and ventilating houses—Sir John Soane (1753–1837) in England, for example, had steam-heating systems installed at his own houses at Tyrringham (1797) and Lincolns Inn Fields, London (1831)—these practices were not widespread at this period.

By the mid-1800s it was still standard practice in Europe to consider the open fire or stove to be the most appropriate means of heating a house, and ventilation was still by means of openable windows. In many homes, candles and lamps still provided the night-time lighting, but, in the cities, coal gas was widely used.

In the U.S. the constraint of tradition was less strong, and by the middle of the century, American manuals on house design accepted some form of central heating, and in some instances controlled ventilation, as being absolutely standard and demonstrably desirable. Although the use of the open fire was not dismissed completely, a system of heating every room in the house was preferable to a single furnace. Andrew Jackson Downing (1815–52), in his *Architecture of Country Houses* published in 1850, proposed a system with two components; a hot-air furnace—preferably that invented by a Mr Chilson of Boston—and a ventilation system based on that devised by a Frederick Emerson, also of Boston. Chilson's furnace consisted of a cast-iron casing lined with fire brick, and with an elaborate arrangement of flues above it. The air to be heated passed over these flues on its way to the warm-air pipes leading to the various rooms of the house. The furnace was placed in a brick chamber in the basement. But Downing's recognition of the need for good ventilation, and his enthusiasm for central heating as a means of achieving it, did not completely blind him to the traditional attractions of the open fire. In a footnote he declared, "We have a great love of the cheerful, open fireplace with its genial expression of *soul* in its ruddy blaze, and the wealth of home associations that surround its time-honored hearth."

Very soon after Downing's publication, whole-house heating became standard practice in the U.S. whereas the Europeans, particularly the English, remained faithful to the open fire for many years after. In 1880, for example, the English architect J.J. Stevenson

discussed the subject of heating and other services in his two-volume work *House Architecture*, and even though he described central heating systems of the kind advocated by Downing, he concluded that "for heating English houses, the best system, on the whole, is the old one of open fires . . . it has the advantage that we are used to it, and that everyone understands it . . . " Stevenson's book also offers valuable insights into other aspects of English (and hence other European) domestic services toward the end of the 19th century. In a chapter on ventilation he wrote "In good ventilation there must be security that the changing of the air will be carried on in all circumstances. The fresh air must come in without cold drafts, and there must be a power of regulating the quantity of air supplied according to the number of people who have to breathe it. The air let in should be pure, and free from dirt and dust." As to his views on the subject of artificial lighting, although recognizing the greater efficiency of gas over lamps or candles, Stevenson felt that this advantage was outweighed by the undesirable products of its combustion and the necessity to have the light in a fixed position in the room. He concluded that gas was preferable for illuminating stairs, passages, and the "offices" of a house, but that lamps and candles were better in the main rooms. Finally, Stevenson also included a chapter on hot-water service in which he described a straightforward, direct system heated by the kitchen fire, with flow and return pipes and branches to the various taps for baths, sinks, and housemaids' closets.

In the year in which Stevenson's book was published, electric lighting was installed for the first time in a house in England. This followed the almost simultaneous development of a reliable incandescent bulb by Edison (1847–1931) in the U.S., and by Swan (1828–1914) in England. The house was "Cragside," the residence of the inventor Sir William Armstrong, which was designed by Richard Norman Shaw. The power for the system came from a water turbine located 4,500 ft. (1,370 m) from the house, and the installation consisted of 45 Swan lamps. By 1882 several reliable incandescent lamps were available in both the U.S. and in Europe. In August of that year the Electric Lighting Act, 1882, passed into English law "to facilitate and regulate the supply of electricity for lighting and other purposes . . . " In the same year an "Electric Exhibition" was held at the Crystal Palace, London, where 38 English and 13 foreign exhibitors displayed their products.

Gas lighting was not defeated by these developments and the invention in Germany in 1886 of the incandescent mantle by Carl von Welsbach (1858–1929) restored it to the lead in domestic lighting. The inverted burner

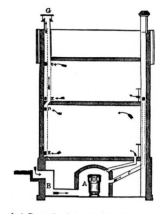

A.J. Downing's central heating system using a Chilson furnace (1850).

which was developed in 1903 by Ahrend, also in Germany, allowed gas to continue even longer as a competitor to electricity by throwing most of its light downward rather than onto the ceiling. It was not until around 1913, when electricity supplies were generally available and costs had become truly competitive with gas, that the issue was finally decided.

Access to the amenity of adequate lighting was limited for many years to city dwellers, therefore the introduction of the kerosene lamp in the late 1860s was of great importance in the impact it had upon rural life. This cheap, safe, odorless, and bright light offered the countryman a quality of illumination almost comparable to that which gas provided in the cities.

The pressures for urbanization in the 19th century also focused attention upon the need for effective drainage within buildings. Early in the 19th century, ironworks, finding their markets for cast-iron guns diminishing, turned their attention to casting pipes. These, together with the development of steam-pumping engines, revolutionized the installation of plumbing in buildings. Water could now, under normal mains pressure, be provided to a reasonable height in buildings enabling apartment houses several storeys high to be built, each storey complete with its own water supply and internal plumbing services. In addition, the water closet, still in its infancy and generally working on a valve principle, could be located in compartments throughout the building connected to soil stacks of cast iron discharging into public sewers in the streets.

The Public Health movement in both Europe and the U.S. led to the construction of public sewers in the cities. Many diseases, such as cholera and typhoid, are directly spread by polluted water supplies. The earliest essays in water filtration were carried out by the Glasgow Water Company in 1806 but the first full purification plant was that installed by James Simpson (1779–1869) for the Chelsea Water Works Company in 1829. It was, however, not until 1854 that John Snow (1815–58) conclusively proved that the London cholera epidemic was directly traceable to foul water drawn from Broad Street well, and the need for filtration and treatment of all drinking water supplies was proved beyond doubt. Nevertheless, it was not until 1907 that the chlorination of water supplies was introduced at Maidstone, England, followed in the next year by the installation of the first U.S. treatment plant at Jersey City, New Jersey.

The advent of the Industrial Revolution also stimulated the desire for personal cleanliness, leading to the development of the public bath house to compensate for the almost total lack of domestic plumbing. These institutions consisted mainly of small indi-

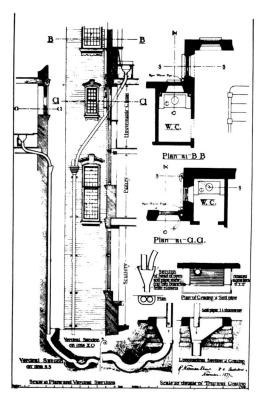

Victorian domestic drainage. Diagram of open soil-pipe system used by Richard Norman Shaw at 6 Elverdale Road, London (1877).

vidual bathrooms controlled by an attendant and served from a central boiler.

By the end of the 19th century the divergent approaches to domestic environmental services in Europe and the U.S. had become confirmed in everyday practice. Although quite sophisticated drainage, hot water, and lighting systems were known in Europe, and were extensively used in large buildings, it was exceptional for a European house to be centrally heated. In the U.S., on the other hand, it was almost inconceivable for a house not to have some form of heating system, usually a warm-air type. These differences in attitude and in technology of environmental control had, by this time, an important influence upon approach to domestic architecture on either side of the Atlantic. This is borne out by a comparison of houses of around the 1900s which may be taken to be representative of practice in each country.

In the U.S., the achievement of Frank Lloyd Wright (1867–1959) in forging a domestic architecture in which the elements of environmental services were fully integrated into a coherent whole, had a major influence. Examination of almost any of Wright's plans for the domestic buildings that he designed in the first decade of the 20th century shows how doors and windows were reconsidered and reorganized within a new conception of domestic space. In the living room, a cold-air intake register is located near the french windows, making it clear that wintertime ventilation was carefully considered but strictly controlled. The fireplace which dominates the plan has more a symbolic than practical role in environmental control. The plan also shows that every window is an opening light and allows one to form the impression of the house in summer as an inward extension of the covered porches, open on all sides with ventilating breezes blowing through.

Also in 1908, Barry Parker (1867–1947) and Raymond Unwin (1863–1940), in their plans for a house of similar size, made very clear reference to the English vernacular, although there is a level of sophistication in the way in which they dispose of the traditional elements to produce a house of surprising modernity. Within the severe rectangle of the plan the major space of the living room is articulated to form a series of areas, each with specific environmental attributes. The inglenook is the obvious source of winter comfort, with its utility enhanced by the carefully positioned windows to supply light for reading in the warmth of the fire. Fuel is immediately to hand through a hatch giving access to the fuel store. Dining takes place at the center of the room opposite the fire and in the light from the bay window. The bay itself is, in effect, a small room in which advantage can be taken of the sun's warmth at all times of the year, even in the winter. On occasions when the space needed additional warmth, a fire could be lit in the hearth in the entry area, helping to extend the effect of the ingle fire. The summers in this northern part of England are both cooler and less reliable than in the midwest of the U.S. so the interior of the house remains relatively isolated from outdoors with the only direct connection being made onto a small protected veranda through doors positioned off the principal axis of the living room.

From the characteristics of these houses it is evident that that by the use of central heating Wright achieves a uniform environment throughout the house in the winter, and this allows him to cut free from traditional constraints and gives him the freedom to pursue his unique architectural vision. Parker and Unwin, on the other hand, accept the European preference for the open fire and its corollary, but go on from this to develop an environmental scheme of great subtlety and utility.

The 20th century

The development of building services during the 19th century reflects how the conceptual and material fruits of the Industrial Revolution were brought to bear upon this aspect of building design. By the end of this period there was no technical reason why any building could not be made to offer a precisely and

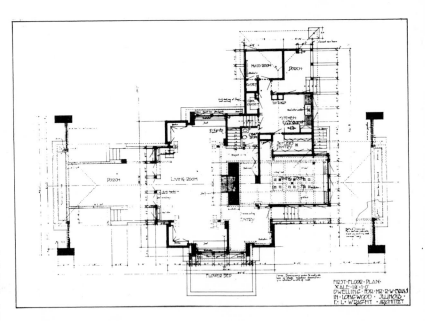

Frank Lloyd Wright's plan of Evans House, Chicago (1908).

constantly controlled internal environment, though this possibility does not seem to have occurred to or attracted designers. In some areas, however, the new technology had been successfully exploited. It was commonplace, for example, for undesirable waste products to be safely and efficiently carried away, and both people and goods could be transported about by mechanical systems.

While it is generally accepted that the Industrial Revolution necessarily depended upon the conceptual advance of earlier scientific revolution, it has been argued that this science directly contributed very little to the technology of the 19th century. The evidence of building services seems to support this view; general application of either theoretical analysis or controlled experimentation in design was clearly lacking. Although many technical authors included data and formulas on aspects of heating and ventilation, most

Plan of Parker and Unwin's Whirriestone House at Rochdale, England (1908).

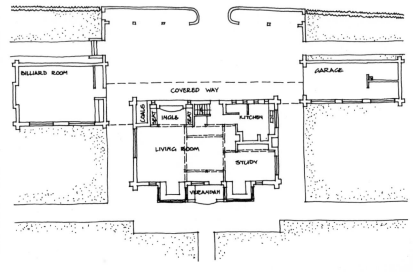

contemporary discussion concerned itself with practical empiricism.

The emergence of a more truly scientific basis for design is the first of the four features which distinguishes the 19th from the 20th century in terms of building services. The second is the availability of relatively cheap and abundant electrical power and the profound influence which this had in extending the scope and application of the systems developed during the preceding century. The third, which to some extent is dependent upon the previous two, is the gradual emergence of the idea of the building which offers total control of its internal environment. Finally, one of the major developments in building design in the 20th century is the emergence of architectural languages which readily allow services systems to be fully integrated into the building fabric. The developments of the 20th century in design of services in the various building types illustrates these features.

Assembly buildings

The technology of environmental control in assembly buildings was already highly developed by the end of the 19th century. The elements of full air conditioning were in existence, and their integration into the fabric of the building was comprehensively achieved. Similarly, in the theater, the attractions of electric lighting had been eagerly exploited. During the 20th century, therefore, the principal developments have been ones of refinement rather than of fundamental innovation. There are, however, two important exceptions to this, first in the field of acoustics and, second, in the lighting of buildings made possible by the emergence of a new sub-type: the movie theater.

The foundations of the modern science of architectural acoustics were laid down by Sabine at the end of the 19th century. Initially, as in its first application in the design of Symphony Hall at Boston (1900), the new science was used to allow the designer to reproduce the qualities of known and admired precedents. It was only later that a secure basis for innovative design was established.

In 1924, in England, Hope Bagenal presented the first paper on the subject of acoustics to be given before the RIBA in London since 1895. In this he outlined the results of new work carried out by the Building Research Board into the sound absorption of materials, and was able to discuss some basic relationships between the shape of spaces and their acoustic performance. The emphasis was still, however, upon precedent as the basis for new design. Nevertheless, there was an indication of the potential of the scientific approach in form making, although tentative, in his recommendation of a fan shape as the basis for the design of large concert rooms.

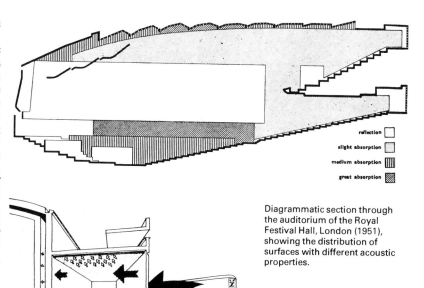

reflection ☐
slight absorption ▨
medium absorption ▥
great absorption ▦

Diagrammatic section through the auditorium of the Royal Festival Hall, London (1951), showing the distribution of surfaces with different acoustic properties.

Royal Festival Hall, London. The auditorium is protected from external noise by subsidiary spaces and by having an independent structure.

It is clear that the formal possibilities of the new science were known to some of the leaders of the movement for a new architecture around this time. The auditorium of the Centrosoyus Building in Moscow (1929) designed by Le Corbusier (1887–1966) had its cross section determined by an analysis of reflected sound, and he made even more explicit reference to these principles in his project for the Palais des Soviets in 1931. In both cases the plan was fan shaped. At the same time, Alvar Aalto (1898–1976) in Finland was generating the marvelous undulating timber ceiling of the lecture room at the Municipal Library at Viipuri (1935) from an extensive analysis of reflected sound.

After World War II, the construction of London's Royal Festival Hall (1951) offered Bagenal an opportunity to apply his ideas to the design of a major building. The nature of the building was fundamentally influenced by the design response to acoustical questions. The rectangular shape and construction of the auditorium clearly predominate. But even in this building the plan shape of the auditorium was selected by reference to precedent. Allen and Crompton, two members of the Building Research Station's team, wrote in 1951 that "no decisive guidance on this point could be obtained from acoustical theory and it was decided therefore to fall back upon tradition. The evidence, even so, was necessarily slim, but it seemed to point toward halls with parallel sides as having generally better reputation for musical acoustics than 'fan' or 'horseshoe' plans.'' Another acoustical determinant was the response to the problem of excluding noise from trains which rattled over a bridge along the edge of the site. This was achieved by both planning and constructional means. First, the auditorium was surrounded by foyers and other ancillary spaces and, second, its structure was made totally independent of that of these other spaces.

With this building the direct and positive influence of acoustical science upon the nature of architecture was first comprehensively demonstrated. In the years which followed, most major auditoriums throughout the world reflected its influence in both method and form. That architectural acoustics is, however, a hazardous science was dramatically and expensively demonstrated at New York's Philharmonic Hall at the Lincoln Center (1956–1962). Here the acoustic consultant, Leo Beranek, carried out the most extensive survey of auditoriums ever undertaken, covering 54 concert halls and theaters throughout the world. But still the combined weight of all of this empirical evidence, plus the application of the most up-to-date theory, failed to guaranteee satisfactory conditions. The hall—which was horseshoe shaped—has subsequently been completely rebuilt.

Almost contemporary with the Philharmonic Hall, but representing a fundamental break from the mainstream of 19th– and 20th–century tradition was Hans Scharoun's Philharmonie in Berlin (1956–63). Scharoun (1893–1972) was determined to create a completely new relationship between audience and orchestra, and between the various sections of the audience itself. He proposed a plan in which the audience is broken up into small groups seated in what he called vineyards. The first reaction to this by the acoustics consultant, Lothar Cremer, was that the idea was too risky. Scharoun prevailed, however, and the completed hall has proved to be a great success. His profound intuition about the processes by which sound is reflected helped to open up completely new horizons for both the design and theory of auditoriums.

An intriguing possibility for the future is the use of electronic devices to modify the acoustics of an auditorium. In 1964 a system known as assisted resonance was installed as an experiment to improve the low-frequency sound of the Royal Festival Hall in London. This used a number of microphone-amplifier-loudspeaker units called channels, positioned in the auditorium so that each channel increased the reverberation time over its own narrow band of frequencies. A total of 172 channels were used, the experiment was successful, and the installation is now in permanent use. This technique has since been

Plan of Hans Scharoun's Philharmonie, Berlin (1956-63) showing seating arrangements.

Section of Philharmonie, Berlin. The audience is arranged on various levels around the orchestra.

developed to allow the acoustics of a hall to be changed to suit the needs of a particular event. Such systems have been installed in halls in the U.S., Africa, and in England, for example. Advanced research in the field is now part of the program of Pierre Boulez' Institut de Recherche et Coordination Acoustique/Musique (IRCAM) at the Centre Pompidou in Paris. Its work suggests fascinating possibilities for redefining the conventional relationships between architectural space and acoustics.

In the period between the two World Wars the most popular form of mass entertainment throughout the western world was the motion picture. To meet this demand an enormous number of movie theaters were constructed. While there are obvious similarities between a movie theater and a theater, there were at that time differences in the social context within which the two building types operated, and these were sufficiently powerful to allow some aspects of movie theater design to develop along altogether new lines. Among these was the extreme originality which was displayed in the use of electric lighting as a decorative medium rather than as a merely utilitarian service.

From a very early date, the creative possibilities of electric exterior lighting were explored in theater design and, later, in movie theaters. In the U.S., by 1905, carbon dioxide-filled Moore tubes, invented ten years previously, were available in lengths up to 200 ft. (61 m) and could be twisted into the shapes of words. Their use was restricted, however, by the fact that they required a current of around 16,000 volts. They produced a white light, and soon the decorative possibilities of the red light from neon-filled tubes were recognized by George Claude (1870–1960) in France. These were used at the Grand Palais in Paris in 1913 while in London, in the same year, the facade of the West End Cinema combined an arch of white Moore tubes with the theater name in red neon.

Much interesting work in lighting systems was done in Germany during the 1920s; the interior of the "Capitol" in Berlin (1925), designed by Hans Poelzig (1869–1936); the "Universum" (1926–31) also in Berlin by Erich Mendelsohn (1887–1953); the exterior of the "Capitol" in Breslau by Friedrich Lipp; and Schoffler, Schlonbach, and Jacobi's "Titania-Palast" in Berlin (c. 1930).

Movie theater designers at this time were also developing "atmospheric," or "scenic" lighting. In many of the auditoriums the decor represented exotic locations, such as Moorish palaces, and these were enhanced by lighting systems which could simulate sunrise or sunset or the star-spangled night sky. These effects required the extensive use of elaborate dimmer systems and very quickly the potential of these, and of color mixing systems, was recognized. In the New Victoria Cinema in London, for example, which was opened in 1930, the color of the interior was more the product of colored lights reflected from neutral-colored surfaces than of applied color. By the development of these techniques it became possible for a movie theater to change its color scheme from performance to performance, and the interaction between light and the folds of elaborate curtain drapes became almost an art form in itself.

One of the leaders in the technology of movie theater lighting was the Holophane Company. In 1930 they installed a three-color reflector system to illuminate the ceiling of the Richmond Cinema in England, designed by Leathart and Granger. This allowed 672 sequences of lighting effects to be arranged automatically. The possibilities of this system provoked Rollo Gillespie Williams of Holophane to liken it to a piano: the circuits were the strings, the controls the keys, and the lighting effects were the music. From this vision it was but a short step to the collaboration between Holophane and the Wurlitzer Organ Company that led to a lighting control system operated directly by the organ keys. The ultimate achievement in this was probably at the Capitol Cinema in Manchester (1930), where the system was capable of producing 5,044 color combinations.

After World War II the film industry concentrated its technological resources more on improvements in the films themselves and, as a result, the cinema building was relatively neglected. The social and economic circumstances which produced the architectural extravaganzas of the interwar years were clearly never to be repeated. All of that inventiveness and technical ingenuity in lighting design was subsequently expended upon a new building type—the "super cinema"—but this was to be short-lived and had virtually no influence upon the design of other buildings.

Hospitals

At the end of the 19th century, hospital building was at an interesting state of development. The technology of environmental control had advanced to the point where very large buildings could be ventilated entirely by mechanical means, but established opinion continued to argue the virtues of natural ventilation. During the 20th century innovations in hospital design can still be classified under these two categories.

More than most other building types, hospitals are influenced in both their form and equipment by extra-architectural considerations. Changes in medical theory and practice can transform the priorities of design overnight. This is clearly illustrated by the example of the repercussions of the attitude of the medical profession to the importance of sunlight in hospitals. The value of fresh air and sunlight in the treatment of tuberculosis was recognized in the 19th century and influenced the layout and orientation of many sanatoriums. This influence continued into the 20th century. In 1912, for example, the American architect William Atkinson published a design for a hospital that was based on an ingenious organization of wards relative to service spaces in order to maximize the exposure of the former to light and air. This influence reached its zenith during the 1920s and 1930s, as illustrated by the design of Alvar Aalto (1898–1976) for the Sanatorium at Paimio, Finland (1928–33). Here the entire form of the building is determined by the desire to achieve maximum exposure to the sun's rays. The main ward block becomes a seven-storey-high slender slab facing south, terminating at each level in an open terrace.

As with other building types, hospital design in Europe and the U.S. has moved in different directions during the 20th century, largely becuase of the effects of attitudes to environmental controls and services. By the 1920s hospitals throughout Europe were generally designed on the pavilion model. In the U.S., however, the experience of using quite sophisticated services in office buildings and apartments led to similar principles being applied to hospital design. The emphasis was therefore upon vertical planning, with the elevator, laundry chute, and various other labor-saving devices playing a leading part in the construction. This approach simplified the problem of the heating, lighting, plumbing, and other services by the use of vertical stacks. The kitchen was most frequently placed in the basement and mechanically ventilated. It was equipped with every sort of device, with refrigerators being in general use.

During the interwar years hospitals in the U.S. came to rely more substantially on artificial control of the environment. In operating blocks, for example, artificial light was becoming the norm, and patients' bathrooms and toilets opened directly from the wards and were frequently without windows, depending upon mechanical ventilation. Another novel service was the almost universal signal system which allowed a patient to call a nurse by a push-button. By and large, European hospital designers rejected these developments from across the Atlantic by reviving the traditional arguments about "light and air." They were equally conservative in their choice of heating systems. Hot-water or steam systems using either radiators or pipes in the wards were favored, and while the theoretical attractions of the warm-air planum system and of a panel system with pipes embedded in walls and ceilings were recognized, these were rejected because of "many objections in practice." One innovation that was approved in some designs was the introduction of a "sunroom," glazed with "Vita-glass" which, unlike ordinary glass, was transparent to ultraviolet rays. Operating theaters were naturally lit, and it was recommended that they should have a large, vertical, north-facing window with a 45° sloping light above this; the latter was to be fitted with a water-spray pipe above it for cooling in hot weather.

Following World War II, the possibility of having a fully artificially controlled environment in a hospital was soon recognized in the U.S. The combination of air conditioning and artificial lighting permitted construction of "fatter" buildings since the traditional compromise between the planning of patients' rooms with outside light and efficient departmental planning could be avoided. A much greater freedom of planning could therefore be achieved.

As environmental service systems used in buildings became more advanced and sophisticated so the traditional differences between the form of buildings of distinct functional types were progressively eliminated. With reference to hospitals, in the late 1950s Skidmore, Owings, and Merrill designed the Northwest Community Hospital at Arlington Heights, Illinois, with full air conditioning of all of its core areas and service spaces and with provision for air conditioning the wards at a later date. The appearance of the building is predominantly the product of its technologies and it shares these substantially with another building type, namely the office.

The potential of this manner of building was fully realized in 1964 in the design of the enormous Bellevue Hospital in New York by Pomerance and Breines. Here is a flexible cube 264 x 240 ft. (80.5 x 73 m) in plan and 25 storeys high, fully air conditioned with natural light restricted entirely to patients' rooms. This design was based upon the argument that "supporting functions can be carried on as well if not better in artificial light, patients' rooms are arranged along outside walls, and supporting services are located in the central areas of each floor. This deployment sets the need for complete air conditioning, and efficiency of the arrangement justifies the expense."

In Europe generally, developments and advances in hospital design since World War II have closely followed U.S. trends. In countries such as Britain, however, it has seldom been possible to justify the budget limits of their National Health Services. The only major exceptions in Britain, for example, are Greenwich Hospital, London (1963–69), designed by the Health Ministry's own architects, and Yorke, Rosenberg, and Mardall's new block for St. Thomas' Hospital in Central London (1966). With their obvious references to U.S. practice, these are clearly good designs in the technical sense. The restrictions of both capital and running costs have, however, kept alive the tradition of the pavilion plan. Powell and Moya's Wexham Park Hospital at Slough (1957–60) exploits its suburban location and generous site to produce a traditional form in which high-level environmental technology is applied to treatment areas, but where the ward blocks are

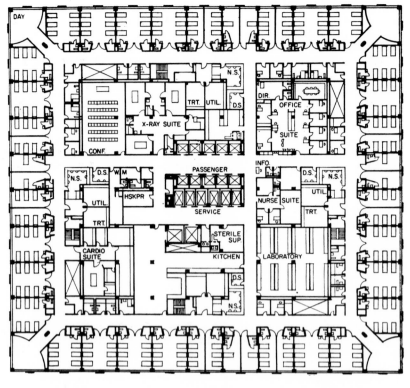

Plan of a typical floor of the Bellevue Hospital, New York (1964) by Pomerance and Breines. This fully air-conditioned multistorey hospital relies entirely on artificial environmental controls. Only the patients' rooms along the perimeter receive natural light.

modeled more on ideas from domestic building. These are L-shaped units of side-lit and top-lit naturally ventilated areas enclosing open spaces into which patients may move when the weather and their health permit. In view of present economic conditions, it is probable that designs for future hospital buildings will also consist of a mixture of 20th-century and traditional ideas.

Office buildings

At the beginning of the 20th century the office was well established as a building type in both Europe and the U.S. Because of the enormous expansion of commercial and administrative activity in the preceding 50 years it had been possible to lay down extensive and reliable ground rules for its design. These embraced all aspects of form, layout, construction, and services. The majority of these buildings, on both sides of the Atlantic, were constructed for rental, and their design very quickly became stereotyped within the framework of constraints imposed by legislation, the calculations of real estate, and the available technology. Furthermore, their origin in domestic building continued to dominate design in the first decades of the new century, although the conscious precedent was now the office building itself.

A good example of "up-to-date" office design in Europe after World War I is Sir John Burnet and Partners' Adelaide House next to London Bridge (1922). As the plan shows, this was conceived as a daylit building. Because of its relationship to the river the building was allowed to rise to 11 storeys, higher than the building regulations normally allowed, and it had an installation of four elevators. Heating was provided by a low-pressure, hot-water circulating system with radiators beneath the windows. Ventilation was provided through fresh-air inlets behind the radiators.Other services were an electric vacuum-cleaning plant and a mailshoot linking all parts of the building directly to the postal room in the basement.

Contemporary U.S. practice can be illustrated by McKenzie, Voorhees, and Gmelin's Barclay-Vesey Telephone Building in New York (1923–26). This had a 16-car elevator system which was seen to be fundamental to the whole conception of the 32-storey building. The building's original electrical equipment included direction sign, light, ventilating fan, pump, vacuum cleaning, heat control, and communication systems, plus office machinery and a photographic plant. All the electrical outlets were located off-center on columns to allow flexibility in locating office partitions. The drainage system was located in the service core, and the toilets were mechanically ventilated into the plumbing shafts, with outlets at the 32nd storey. The general prac-

tice for heating New York office buildings at this date was a two-pipe, low-pressure vacuum system with radiators under the windows. Ventilation was still by natural means, except for toilets, basements, and those cases where whole floors were in single occupancy. In these mechanical extraction was used. Although air-conditioning systems were contemplated for structures for large corporations, it had not become common practice and was not installed.

The benefits of air conditioning were first realized in 1902 by Willis Carrier, by general consent the father of this innovation. He discovered that by controlling the temperature of air it was also possible to control its humidity. This allowed every aspect of the air in the building to be controlled; temperature, rate of movement, velocity, and cleanliness. Probably the first fully air-conditioned office block was the Milam Building in San Antonio, Texas (1928) designed by George Willis. This used a simple system consisting of one plant serving the principal lower floors and a series of smaller systems to serve the standard office floors above. Input of conditioned air was through a duct above the central corridor, and the corridor itself was used as the return duct. Contemporary textbooks dealing with air conditioning, such as that by Moyer and Firtz published in 1933, argued that the commercial attractions of the air-conditioned office building were in terms of the greater efficiency of the employees who would enjoy constant comfort and would be released from the distractions of manual adjustments of radiators, windows, and desk fans. (It is interesting to note that, even at this date, ice was still considered to be a viable cooling medium, particularly for systems for small buildings.)

In these developments of heating and ventilation systems were the beginnings of one of the most significant influences upon office building design in the 20th century, but, as with many innovations, the full implications of the advances went unrecognized. In other respects, such as layout and lighting systems, these buildings were absolutely conventional.

It was more than a decade later that the concept of a totally artificial environment—thermal, visual, and acoustical—became the logical extension of the application of air conditioning in office buildings. A vital additional component of this equation was the development in 1938 of the fluorescent lamp—marketed simultaneously by GEC and Westinghouse in the U.S. This offered greatly increased efficiency over the incandescent lamp and made it possible to achieve high levels of lighting without overheating the building. By 1942 technology was sufficiently advanced to allow a building of the size of the Pentagon, Washington D.C., to be fully air conditioned. Its requirements were supplied

Adelaide House, London (1922) by Sir John Burnet and Partners.

from a separate powerhouse building through an underground tunnel 1,500 ft. (458 m) long. A separate distribution system supplied the requirements of the interior areas. This huge installation had five massive boilers supplying the heating, and the air distribution was handled by 570 fans located in service areas around the building. Other services in the building included an underfloor duct network for signal systems, including a 12,000-extension telephone system, a pneumatic inter-office communication network, a centrally controlled synchronous clock system, and fire alarm systems. It is clear that a building on this scale would have been inconceivable if it were not for the availability of these mechanical services.

An important secondary theme in the development of the technology of air conditioning was the introduction of the unit air conditioner. This could be installed in a room and provide full environmental control without the need to install an elaborate network of ducts throughout the building. A prototype room cooler devised by General Electric was installed experimentally in Willis Carrier's own house in 1929, and the Carrier Company themselves marketed an ''Atmospheric Cabinet'' room cooler in 1932.

In the years after World War II office buildings in the U.S. and in Europe developed along fundamentally different lines. This divergence was influenced by such factors as climate, economics, scale, and tradition. In the U.S. an office building without air conditioning would be almost inconceivable. In Europe circumstances have meant that a much wider range of options has been explored.

Concerning events in the U.S., development was very rapid after the war. Kahn and Jacobs' Universal Pictures Building in New York (1947), for example, no longer relied upon either daylighting or natural ventilation, with a full 11 storeys of deep-plan space. By the early 1950s, with the completion of the United Nations Headquarters in New York under the executive control of Wallace Harrison (b. 1895), and of Skidmore, Owings, and Merrill's Lever House (1952), the vocabulary of a lightweight, sealed, heat-resistant, glass envelope wrapped around a fully air-conditioned, artificially lit environment was established, although paradoxically in both these cases the buildings, with their slender cross sections, appear to be naturally lit. A vernacular fundamentally founded upon this approach to environmental control soon emerged, and was applied with both consistency and flexibility to numerous projects in a wide variety of situations. Skidmore, Owings, and Merrill's own work typifies this phase through such examples as the Connecticut General Life Insurance Building (1954–57)—a low-rise suburban design—the

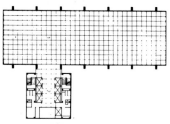

Plan of the Inland Steel Building (1956-58). Elevators and other services are linked to an area of free office space at each level.

Inland Steel Building, Chicago (1956–58) representing the urban solution, and perhaps the most typical of all, the Union Carbide Building in New York (1957–60).

The utility of this vernacular as a solution to the problem of the office building in the U.S. and, by the 1970s, throughout the world is demonstrated by recent designs. *Architectural Record*, in 1976, published Hugh Stubbins and Associates' designs for the Citicorp Center, New York; Marani, Rounthwaite, and Dick and Arthur Erickson's Bank of Canada, Ottawa; and Philip Johnson and John Burgee's Pennzoil Place, Houston. While each of these represents a degree of reinterpretation of the vernacular and new sophistication in the mechanical systems, they nevertheless rest fundamentally on the earlier environmental philosophy of Lever House.

In Europe, postwar designs for office buildings clearly owe a considerable debt to Lever House, but at first, in the late 1940s and in the 1950s, environmental conception was foremost about daylighting. Many buildings with large areas of clear glazing and narrow cross section were constructed. Ventilation was invariably by natural means and heating was supplied by hot-water systems. However, this approach soon fell from favor. The reasons for its downfall were almost entirely environmental. First, the combination of large areas of glazing and lightweight structures led to serious problems from solar overheating. The second reason, which was closely related to the first, was that with the growth of prosperity in the 1960s, building owners and designers became attracted to the benefits of air conditioning, particularly since they had by then acquired firsthand experience of buildings which failed environmentally in the summer months. An early example of the use of air conditioning in such an office building was Robert Matthew, Johnson-Marshall, and Partners' New Zealand House in London (1958–63). This is, in form, in the center of the mainstream, with its tower above a podium, and it was conceived as a daylit building. The air conditioning was therefore primarily installed to overcome the environmental shortcomings of the type, but did not influence the nature of the design in any fundamental sense.

Even though this development overcame the difficulties of the earlier type by the application of cooling, it was to enjoy an even shorter life. The combination of the higher

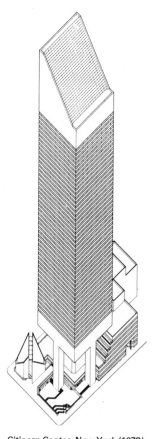

Citicorp Center, New York (1978). This recently completed office building continues the tradition of building with lightweight, sealed, heat-resistant glass envelopes enclosing a fully air-conditioned artificial environment.

running costs that were inevitably involved, the growth of interest in new ideas on office space planning, in particular the German "Bürolandschaft" approach and of new research into the problem of environmental control, led quickly to the adoption of a deep-plan design. In some instances, especially in city centers, this moved much closer to the American model of a deep tower with solar control glass helping to reduce the influence of ambient energy upon the functioning of the environmental control systems. In other cases, a very deep, low-rise form emerged. In Britain, buildings such as Foster Associates' Willis Faber Building at Ipswich, and Yorke, Rosenberg, and Mardall's Central Electricity Generating Board Building at Gloucester, owe much to American practice.

By the mid-1970s a new development could be detected in European office design, particularly in suburban and rural locations. Here the emphasis is upon the reconsideration of the attractions of natural lighting and, in some cases, of natural ventilation. The inspiration for this move was the result of a combination of interrelated circumstances. First, the effects of the enormous increase in energy costs of the early 1970s. Second, a shift of emphasis in the "sociology" of office organization away from the management science image of the 1960s toward a less formal view, with more emphasis being placed upon the individual. Finally, an almost inbuilt resistance to artificial environments, as demonstrated principally by the English, made this approach self-evidently attractive.

In all such approaches to office building design the total integration of the services systems and the building fabric is fundamental. In the past 30 years there has been a shift from the idea of the suspended ceiling as a cosmetic layer applied beneath the structural floor to conceal unsightly services, to the acceptance of the idea of the "service zone" as a major element in any highly serviced building. The ceiling plane itself has become a complex element of environmental control containing lighting fixtures, air input and extract points, smoke detection devices, and water sprinklers. The ceiling also has an

Section of CEGB Building, Bristol (1970s) by Arup Associates.

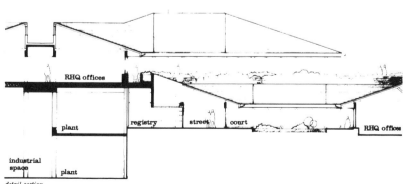

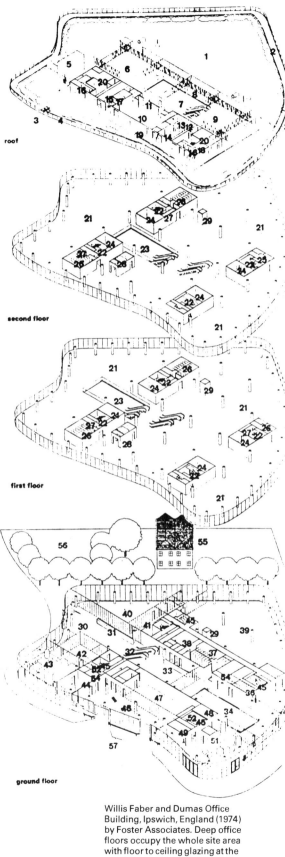

Willis Faber and Dumas Office Building, Ipswich, England (1974) by Foster Associates. Deep office floors occupy the whole site area with floor to ceiling glazing at the curvilinear perimeter. Access to the various levels is by means of escalators in the central wall.

acoustic control function, primarily by offering an absorbent surface to reduce sound levels in the space below. In some buildings even acoustics are the province of systems, with loudspeakers delivering a continuous background noise as part of a "sound-conditioning" installation.

Schools

Any discussion of environmental systems in schools of the 19th century is inevitably concerned primarily with urban schools located on small sites and in hostile urban environments. These constraints prompted considerable ingenuity of environmental and service systems design and, from this point of view, schools had reached an advanced state of technological development by the end of the 1800s.

In the 20th century the sociological context within which schools are designed has changed considerably from that of the previous century. The emergence of a suburban life-style for city workers necessitated the construction of many schools in low-density locations and this has allowed other ways of creating an internal environment to be adopted. Whereas the 19th-century school used advanced environmental technologies, in the 20th century, at least in the first part, the emphasis has been upon a naturalistic view of the environment. The attractions of daylight, sunlight, and fresh air seized the imagination of both educationalists and architects and resulted in the concept of the "open-air" school.

In the period between the two World Wars these ideas about the priorities of environmental design in schools and the technologies necessary to satisfy them remained unchanged on both sides of the Atlantic, although mechanical ventilation was more readily accepted and used in the U.S. than in Europe. Toward the end of this period the influence of the analytical methods of the Modern Movement in architecture began to be evident in designs, and the Corona Avenue School at Bell, California (1935), designed by Richard Neutra (1892–1970), for example, represents a sophisticated response to the local climate. Each classroom has an outdoor teaching area and the connection between indoors and outdoors can be controlled by a double system of adjustable blinds and sliding windows.

Following World War II, daylighting requirements became a dominant environmental influence on the design of European schools. Quantitatively expressed design standards and design tools to ensure that these were achieved were introduced. In consequence, the standard pattern for school design was of a dispersed plan with large areas of glazing in the teaching spaces. This pattern was to survive with only minor

amendments for almost a quarter of a century.

In the U.S., where school design has tended to be less influenced by centralized control, experimentation has been more common. As an example, in 1947 Laurel Creek School in California, by architects/engineers Franklin, Kump, and Falk, had a classroom cross section into which all of the environmental control systems were fully integrated. The deep-plan classroom—32 ft. (10 m) square—was lit by floor-to-ceiling windows on one side and a clerestorey at the other, both protected by a wide roof overhang. Working light at the center of the room came from a continuous skylight at the apex of the roof. Ventilation air was supplied through a system of ducts along the window walls and exhaust was through a vent in the skylight. The principal heating of the building came from heating coils embedded in the floor slab, and preheating, when necessary, could be supplied through the ventilation air system. Each classroom had its own thermostatic control system. Although the design as published did not incorporate hot-weather cooling, its provision was anticipated and was claimed to require only minor modifications. This building represented a significant development in school design since it both abandoned the "responsible" mode of environmental control represented by Neutra's 1935 design and took a major step toward a mechanically controlled, largely artificial environment. This direction of development progressed further in 1950 when Alonzo J. Harriman published a study of school operating costs. In this he calculated heating costs for four school forms. The argument was extensive and moved inexorably to the conclusion that "daylight proves expensive." From these studies Harriman developed his "K-8" prototype school. In this the inner half of each classroom was designed to be permanently artificially lit and the outer half to have predominantly natural lighting, but even here with some artificial supplement.

A combination of the technologies of Laurel Creek School and of K-8 provided all the elements of a totally artificial environment, with automatically controlled heating, cooling, and ventilation, and with permanent artificial lighting. Such a scheme was used in 1956 in the North Hagerstown High School in Maryland, designed by McLeod and Ferrara. Here sophisticated heating, cooling, and ventilating systems were installed but, more important, a major step toward open planning was made. An additional innovation was the incorporation of an experimental educational television installation, fully integrated into its services system. This functioned not only to receive commercial television but also to provide facilities for closed circuit work throughout the building.

By the early 1960s the idea of a fully

artificial environment was quite commonplace in U.S. school design. The fundamental shift in the attitude of architects to environmental control is well illustrated by John Lyon Reid who wrote in 1964, "we know that young people learn better when environmental conditions are right for them. Air conditioning must be considered as one of the components. Architecture cannot be considered as merely a matter of light, shade, form, and texture."

Against this background, the School Construction Systems Development program in California (1962) can be seen to be a logical step in a process of evolution which had been in operation for two decades. Its achievement was the complete degree of integration which it achieved between all the elements of the program, both functional and technological. At the outset, four criteria for the design of schools were established—long-span structures; varied mobility of partitions; full thermal environmental control, with the ability to adopt to changing plan configurations; and an efficient and attractive low-brightness lighting system which adapts to changing plan configurations. The design solution finally adopted consisted of an "umbrella" roof covering a deep space. Beneath this a flexible partition system allowed enormous flexibility of internal planning. The roof itself carried unit air conditioners with flexible ducting and a variable system of modular lighting fittings. From this basic kit a wide range of schools could be fashioned and, within them, education could proceed unimpeded by the vagaries of the natural climate.

In Europe daylighting standards remained a dominant influence upon school design until the very end of the 1960s. Then, under the combined influence of ideas from the U.S. and developments in educational theory, experiments were made with deep-plan forms and with an increasing reliance on mechanical modes of environmental control. With this relaxation of daylighting requirements in schools a further development followed. The theory of Integrated Environmental Design, which was an important influence on office building design, was applied to schools with the intention of reducing energy consumption and, at the same time, providing a highly controlled environment. The designs that resulted from this clearly shared the essential features of the IED office buildings; deep rectangular plans with small areas of glazing. In this they were more evidently the product of a theory of environmental design than of the earlier traditions of European school building.

As a postscript to this account of school design it is essential to mention St. George's School in Wallasey, England, designed by Emslie Morgan and completed in 1961. The fundamental conception of this building was far from modern but by applying the prin-

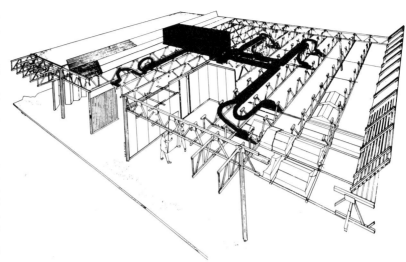

Diagrammatic perspective of SCSD system showing long-span roof trusses, integrated ceilings, roof-mounted packaged air-conditioning system, and movable partitions.

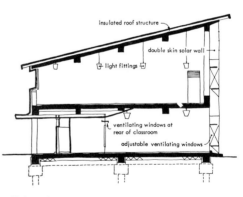

St George's School, Wallasey, England (1961), designed by Emslie Morgan, uses a solar wall with controllable panels to achieve a satisfactory internal environment without the use of complex, mechanical, energy-consuming systems.

ciples which underlie traditional conservatories to the design of a school, Morgan was able to make one of the very few radical innovations in environmental control of the whole century. The main building of the

Reflected ceiling plan diagrams of duct installations in an SCSD School.

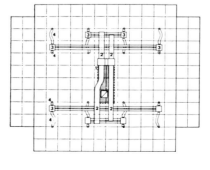

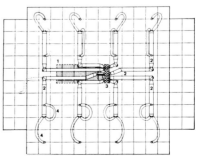

school consists of a two-storey classroom block with a solar wall along the long south face, with the remainder of the enclosure of heavy, highly insulated construction. The solar wall has two glazed skins, the outer one with clear glass, the inner mainly of obscured glass. Certain areas of the inner skin have reversible panels, with one face painted black to act as a heat absorber and the other of reflecting polished aluminum. These offer such a high degree of heating control that the building has been able to maintain a satisfactory thermal environment without recourse to any other heat source.

Unfortunately such truly innovatory designs in architecture take many years before they become models for general practice. The lessons of St. George's School have yet to find their way into the repertoire of present-day school designers.

Domestic buildings

The main theme of servicing of domestic buildings in the 20th century is the installation of labor-saving systems. The burden of all domestic chores may now be lightened by mechanical aids. The storage, preparation, and cooking of food has been transformed by technology to a point where all earlier constraints have been abolished. Cleaning the house is assisted by mechanical tools, as is the maintenance or modification of a house or its contents. In terms of the more traditional concerns of building services design, it is now almost inconceivable that a house in the western world should not have full internal sanitation, mechanical ventilation of kitchens and bathrooms, an electrical lighting installation, and extensive central heating.

For most western countries this state has been reached by gradual development and improvement, but right at the outset of the century there were visionaries who, perhaps by a mixture of ingenuity and eccentricity, offered a glimpse of the future. A remarkable example of this was the Villa Feria Electra near Troyes in France. This was the invention and home of Georgia Knap, an automobile manufacturer. Among its features the Villa Feria Electra boasted electrically operated entrance gates, floodlit at night, with an intercommunication system to allow the occupant of the house to speak to the visitor at the gate. The house was electrically heated and lit, and had an automatic fire alarm. But it was in the preparation and presentation of food that Knap excelled himself. The kitchen had an electric range of immense proportions with automatic controls. There was also an array of gadgets including a dishwasher, a mincer, a miniature butter churn, a mayonnaise maker, a coffee grinder, and a knife polisher. An American visitor to the house in 1907, the architect Frederic Lees, later offered Knap's extraordinary house as a pointer of "the way toward progress."

With reference to the mainstream of architectural theory and practice, ideas about the servicing of the house were central to what was to become known as the Modern Movement. Le Corbusier (1887–1966), in his book *Vers Une Architecture* (1923), pointed out that the technologies of heating and lighting had developed to a point which allowed the functional performance of traditional materials and arrangements to be abandoned. He also rejoiced in the effect that they would have upon life-style: "A house is a machine for living in. Baths, sun, hot water, cold water, warmth at will, conservation of food, hygiene, beauty in the sense of good proportion." These statements were reflected in his designs, with their compact planning, absence of fireplaces, and celebration of kitchens and bathrooms.

The influence of Le Corbusier's ideas rapidly spread and in 1934 Raymond McGrath introduced his survey *Twentieth Century Houses,* published in England, in which he drew attention to the importance of automatic systems of heating, lighting, etc. in achieving the smooth operation of the complete house. McGrath stressed the importance of electrical power for this scheme. He also emphasized that these new systems would be visually unobtrusive but would, at the same time, have a profound effect upon house design. It is apparent, therefore, that the Modern Movement incorporated an approach to environmental control and services that was both extravagant in its use of resources and perhaps over-enthusiastic in its faith in

Villa Savoie at Poissy, France (1929-31) by Le Corbusier and Jeanneret before its restoration.

technology.

In the U.S. in the period between the two World Wars, one of the major contributions to both the theory and practice of domestic architecture came from Frank Lloyd Wright (1869–1959). His project of 1934 for a "zoned house" showed three arrangements of a similar series of spaces for rural, suburban, and urban sites. Common to each design was a clear expression of a centralized service core. This incorporated all the systems: oil-burning boiler and fuel tanks; air compressors; oil and gasoline supply for an automobile; heating, and air-conditioning units; electric wiring and plumbing; and vent and smoke flues. Each bathroom was a one-piece, standardized fixture directly connected to the stack—as were the kitchen sink, ranges, and refrigerator. By these means what Wright called "the wasteful tangled web of wires and piping (at present) involved in the construction of the ordinary dwelling," was replaced by a factory-produced, standardized unit. This was a significant development insofar as by collecting together all the service systems into a compact unit they became more economical and efficient in their own terms and also allowed freedom of planning elsewhere in the house.

Wright's "Usonian" houses of the late 1930s have a clear relationship to his earlier scheme. They also incorporate a further innovation—a means of heating described by Wright as "gravity heat." This was an under-floor system with steam or hot-water pipes buried in a rock-ballast bed beneath the concrete floor. This system, while restricted to a single-storey house, freed the architect from the problems of integrating heating devices, such as radiators, neatly into his designs although, as his earlier work demonstrated, this was a problem which Wright had already mastered.

In designs such as Wright's Usonian houses, the vision of the fully serviced, labor-saving house was substantially realized. Since World War II there have been many stylistic developments in house design, but in most respects these make use of the technologies which were already in existence by 1940. In the U.S., houses equipped with packaged air-conditioning units were not uncommon by the mid-1950s and were commonplace a decade later. Nevertheless, as with other building types, and following the earlier history of domestic architecture, European practice has followed a somewhat different path. In England, for example, it was only in 1961 that it was recommended that government-financed housing should, in future, be centrally heated. Nowadays, in Europe generally, central heating is standard in new buildings in both public and private sectors and most households are equipped with labor-saving services. But there is little evidence of these developments having the

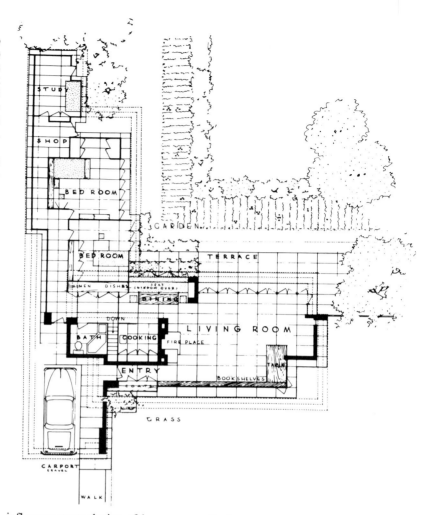

influence upon design of houses as radical as that predicted by the pioneers of 20th-century design, and which, to a large extent, has happened in the U.S.—European conservatism still prevails.

No account of domestic services in the 20th century can be complete without mention of the special servicing problems which arise in high-rise, high-density forms of housing. As in the case of late 19th-century office building, such a radical change in the scale of building inevitably led to new problems. In housing, waste disposal is the major problem when large numbers of people live in apartments above ground level. By the early 1930s this had been recognized and a number of solutions had been tried. Arguably the most sophisticated of these was the French "Garchey" system. In this, all domestic waste is collected in a hopper under the kitchen sink and is carried away through a mains system by waste water from the sink. The waste is collected at a central disposal station where liquids and solids are separated and the latter burned in an incinerator. The system is frequently used in conjunction with a district heating scheme—the heat is derived from the incinerator—from which all the apartments

Frank Lloyd Wright's Herbert Jacobs House, Westmorland, near Madison, Wisconsin (1937). Compact services occupy a pivotal position in the plan.

receive their space heating and domestic hot water. Waste disposal apart, however, service systems in high-rise apartment houses presented few problems that essentially had not already been overcome in dealing with office buildings.

Postscript—the Responsible Age

Innovation in building services since the Industrial Revolution has clearly been characterized by a process by which the activities of man have been liberated from the restrictions of climate and the rhythm of the seasons by the increased control over his environment which his buildings have been able to offer. This progress has been almost entirely dependent upon the massive use of energy-consuming devices of environmental modification and control. The realization that the earth's resources of energy are finite, and already largely consumed, marks the beginning of a new era in the latter part of the 20th century—the "Responsible Age."

Architects, and others involved in the design of buildings, were among the first to recognize and respond to the question of better energy use. In this discussion a distinction may be drawn between "alternative" technologies, which predominantly seek to make use of ambient, renewable energy sources such as the sun and wind, and other approaches which seek to reduce the use of conventional types of energy within buildings. In the sense that they introduce radically different parameters into the equation of building design, the former are more obviously exciting but, in the short and medium term at least, the latter are demonstrably more necessary and they too have a fundamental influence upon the nature of buildings and their services.

The principal technical problem of the "alternative" approach is to have ambient energy available at the time when it is required. This new correlation between climate and architecture has directed considerable attention toward problems of the storage of energy, either from day to night or from summer to winter. While such problems may be solved quite easily in some locations, it is likely that in major centers of population the problem will prove to be technically or economically intractable.

Interest in the use of solar energy in buildings goes back many years. One of the most advanced early designs was the Massachusetts Institute of Technology (MIT) Solar House No. 1 (1939) built at Cambridge, Massachusetts. Here are many of the elements found in solutions to the problem of long-term thermal storage. Water heated in a solar collector, which is an integral part of the building fabric, is passed to a large insulated

storage vessel under the house. Heat is drawn off from this and delivered, in this case as warmed air, to the habitable rooms.

A fundamentally different approach, based upon short- rather than long-term storage was adopted in the design of the Baer House in New Mexico (1972). Here collection and storage of energy are combined and consist of a series of water-filled metal drums arranged to form the south wall of the house behind a single sheet of glass. At night and in cloudy weather the collectors are covered by insulated shutters. The house is heated by the transfer of heat from the drums to the interior.

As a third variation of this theme, the design by Integrated Life-Support Systems Laboratories, also in New Mexico (1972), seeks to optimize the relationship between the habitable space of the house and its enclosing envelope by the use of a dome structure. Collection and storage of solar heat takes place outside the dome allowing the designer to avoid the problem of balancing the relationship between the form of the building and the optimal design of its energy collectors. This house also has a wind-powered generator to supply its electrical requirements.

Two designs which attempt to achieve total self-sufficiency are Project Ouroboros at the University of Minnesota in the U.S., and the Autarkic House Project at Cambridge University in England. In both designs waste products are used to produce methane gas as an additional renewable source of energy. Also, each plan distinguishes between a primary habitable zone and a secondary glass-house space for food production and as an extension of the living space. In both cases the systems and the fabric of the building are

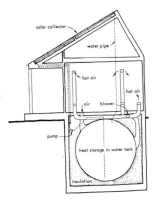

Diagrammatic section of MIT Solar House No. 1 (1939). Water heated by the solar collectors on the roof is stored in a large insulated tank under the house. The hot water heats air which is ducted to the habitable spaces in the house.

Project Ouroboros House, University of Minnesota. The guiding principle in the design of this house is energy conservation and the exploitation of ambient energy sources in an attempt to be totally self-sufficient.

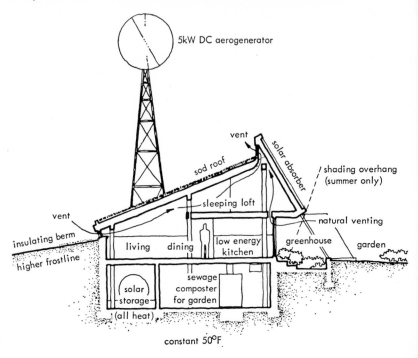

highly integrated.

In France, Professor F. Trombe, in association with architect Jacques Michel, has developed a device using the thermal storage capacity of a thick concrete wall behind a glazed skin to set up warming convention currents through the house in winter. In summer the operation of simple controls allows the system to induce ventilation. In locations such as the Pyrenees, where there is a good deal of winter sunshine, between 50–75% of the total heat requirement can be supplied in this way.

There are currently many experiments seeking to improve the performance of building elements of this alternative kind, and it is clear that aspects of this technology will have an impact on more conventional designs and upon the design of non-domestic buildings in the years ahead. It is prophetic that a new computer building in New Mexico, designed by Emilio Ambasz and due to be completed in 1979, will use a solar wall to provide the power for the air-conditioning plant which is vital to the operation of environmentally sensitive computers.

In 20th-century houses, offices, and schools, the high cost of sophisticated environmental services has led, particularly in Europe, to a number of designs which, within the framework of conventional design aims and technology, are seeking to reduce energy consumption. It is clear from such designs that, by paying attention to the form and construction of the building's envelope and to the design of its systems, considerable economies can be made. In England, for example, collaboration between the London architects MacCormac and Jameson and researchers at Cambridge University has resulted in a proposal for a low-energy building form which achieves a compromise between the advantages of a daylit building and of a deep-plan, air-conditioned design without the environmental shortcomings of the former or the energy consumption of the latter. This has been attained by organizing daylit space around glazed courtyards. The generic court can be combined to produce a wide variety of buildings for many uses. Theoretical analysis of designs for office buildings based upon this principle has shown that their annual energy consumption in the British climate would be between one-third and one-half of that of air-conditioned buildings.

In these examples it can be seen how the need to conserve energy has provoked an imaginative response from the designers of buildings and how, no doubt, innovations in services and environmental systems in buildings will be as numerous and significant as they were during the previous two centuries.

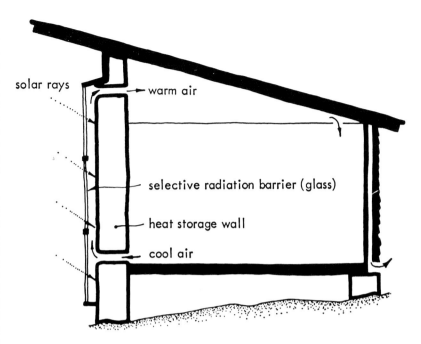

solar rays

warm air

selective radiation barrier (glass)

heat storage wall

cool air

Schematic section of Trombe-Michel House built in the Pyrenees, France. It incorporates a solar wall which stores heat from the sun. This is suitable for area of the house that receive high levels of winter sunshine.

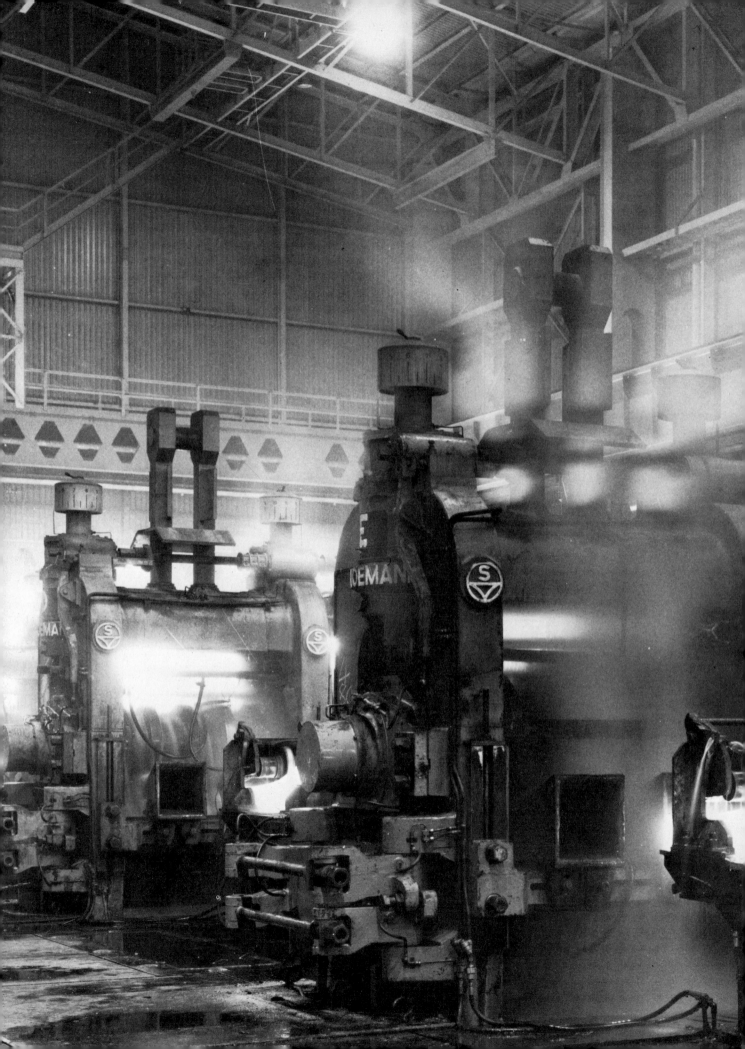

Section 5
Building materials

Building materials

A donkey being used to carry sand to a building site. Changes in methods of transportation have made a great impact on the building industry, but traditional techniques still survive in many parts of the world.

The cast-iron veranda of a school building in Johannesburg (c. 1890). Prefabricated cast-iron components were imported from MacFarlane's iron foundry in Glasgow.

Until the early 19th century, most buildings were constructed and fitted out with easily available, local materials. If the area was well provided with timber, as in Sweden and New England, the buildings were inevitably wooden. If, as in the Cotswold region of Britain, there was an abundance of easily accessible stone, the builder's material would be stone. If, as in the Netherlands, or in much of the Middle East, both wood and stone were in very short supply, it was natural to turn to brick. If wood, stone, and brick earth were all missing, as they were in the prairie region of the U.S., the settlers were forced to make do with walls made of turf.

The exceptions were prestigious buildings such as temples, palaces, monuments, churches, and the homes of wealthy people. For these purposes, fine quality materials were imported from a considerable distance. Luxury, in many countries and for much of history, has consisted in providing an environment which was not built solely of local materials; the exotic was a symbol of power and importance.

Fashion and taste have always played an important part in the choice of building materials. During the 19th century, improvements in canals, roads, and railroads made it economically viable to transport heavy and bulky materials over long distances. The use of materials manufactured far away from the building site became practical for the first time on a large scale. Manufacturers developed elaborate trade catalogs illustrating their products and describing their properties so that designers could choose and specify their goods accurately. Reliable postal services and the rise of telegraphy played a part in easing the flow of information to and from builders and manufacturers.

During this period, a wide range of new building materials was invented or old ones were used in new ways. Inventions and improvements in materials often occurred outside the building industry, but were later assimilated by it. Often the spurts of invention happened during or after major wars or social upheavals, when normal economic patterns were disrupted and shortages and gluts on a large scale influenced the market in building materials. One example of this trend can be seen in the situation after World War II, when the volume of production of aluminum sections and other products had reached large proportions in order to supply the aircraft industry. When this demand lessened after the war, there was a spare capacity in manufacturing potential, which could provide the building industry with aluminum products at a reasonable price.

The tremendous impact of these changes was felt throughout the world, often increasing people's material standards, and the safety and durability of their buildings. In other cases, rapid and ill-considered applications of new materials and techniques led to buildings inferior both environmentally and constructionally, compared to traditional solutions.

Resistance to new methods of building often exists in conservative building legislation, and in the building trades themselves, whose members are often reluctant to adapt their skills to new conditions.

Timber

Until the introduction of structural elements made of iron in the late 18th century, wood was almost the only material available for structural framing, or for those parts of buildings such as beams, trusses, rafters, and joists which had to be capable of withstanding tension and bending.

There are many species of timber used in one way or another in building and they vary greatly in all their properties. They range in density from balsa wood 10 lb./cu. ft. (160 kg/cu. m) to lignum vitae 78 lb./cu. ft. (1,249 kg/cu. m). Timber can normally be classified into two broad groups—the softwoods and the hardwoods. Softwoods are all derived from conifers and are native to the Northern Temperate Zone. Hardwoods are broad-leaved, mainly deciduous trees that are widely distributed throughout the world.

In climates where trees are part of the natural vegetation wood has been used since earliest times as a traditional building material. Generally, the oldest forms of building used timber with a minimum of shaping and forming, often employing complete trunks or saplings in prodigious quantities. This reduced to a minimum the laborious work with crude and simple tools. Where quality timber was naturally scarce, as in Egypt and parts of Arabia, it was a precious commodity which often had to be imported and consequently was only used sparingly for major structures.

The evolution of carpentry through the ages

has normally responded to the introduction of new tools, the growing scarcity of timber, and a better understanding of its structural properties and jointing techniques. New tools have enabled carpenters to undertake more accurate and intensive working of logs to produce specially useful shapes, and have enabled them to consider employing more substantial and complex joints for fastening than lashing and tying, which were almost certainly the earliest forms of timber jointing.

Because of its many varied and competing applications in the construction of houses, ships, bridges, and as a fuel for major manufacturing industries, good building timber has always been in very great demand. Carpenters have therefore been compelled to be more ingenious and economical in their framing techniques. This demand for wood has also stimulated the study of its structural properties in order to achieve economy through efficiency in use.

The Middle Ages

In the English stave church at Grinstead, England, which was built in 1013, the walls were constructed of half-trunks of oak with the split face set inward. Splitting of hardwood with wedges was a process used until relatively late, since sawing with primitive saws was both laborious and time-consuming. Baulks were often finished with the adze.

In Scandinavia and other parts of northern Europe where forests were widespread, a traditional form of construction evolved in which whole timber logs were laid horizontally over one another and notched at the corners. In better buildings straight logs were chosen and shaped slightly to interlock along their length in order to reduce water penetration and drafts. Windows were only incorporated in these buildings in the 17th and 18th centuries since they tend to weaken the structure and require special posts and trimmings. In cruder buildings the gaps between logs would be filled with mud, moss, and earth, a technique of construction that was adopted by American frontier pioneers in their log cabins.

Another form of medieval building is known as post and plank construction, in which the wall consists of a series of heavy boards slotted between even heavier posts or between grooved horizontal members spanning between the posts. The stave churches of Norway dating from the 12th and 13th centuries are of this type. During the medieval period in western Europe, the shape of individual branches was utilized in the construction of roofs. The extreme form of this was the technique known as cruck building, in which roughly squared tree limbs, curving naturally for about a third of their length,

were set upright and socketed at the top into a horizontal beam, producing a framework similar to an inverted boat. The loads were carried directly to the ground. A great many trees were needed for this form of construction in order to find enough limbs with a similar natural curve.

A form of construction often called the "box frame" also has very early origins. Its evolution eventually led to the development of the balloon frame. Here, posts were erected separately and temporarily propped up by permanent diagonal bracing. They were then covered by wall plates that dropped into position over specially prepared joints, and over these spanned a tie beam which locked the posts and wall plate together. Intermediate studs, joists, rafters, and other subframe elements were placed piece by piece into the structure at the same time. This type of building could be made rigid and durable if adequately braced and jointed.

Late and post-medieval timber building

From about the 15th century onward, timber was no longer available in inexhaustible quantities in parts of northern Europe such as Britain, Holland, Denmark, and northern France where it had been a traditional building material. Carpenters responded by using timber more sparingly with smaller structural sections spaced more widely, and with infills of other materials, such as wattle and daub, brick noggings, or planking and boarding. Buildings using secondhand timber, often from old ships, were not uncommon. Oak, which had been the favored timber, was replaced by elm and imported softwoods.

Sawmills were introduced in Germany in the 14th century, and in Sweden in the 16th. These countries began to export timber by water to places where it was becoming scarce. This trade continued to expand through the centuries, and Sweden still continues to be one of the major timber exporters.

Until well into the 19th century the frames of timber buildings used large timber sections by today's standards, but this was often inevitable because of the jointing methods employed. Traditional medieval joints involved cutting out large parts of the cross section of one piece of structural timber in order to joint it to another, thereby weakening it considerably. Variations of this type of framing continued to be utilized until the mid-19th century, even though many attempts were made to reduce the volume of timber used in buildings through developments in the understanding of its structural behavior and refinements in joint design. American settlers in New England and other parts of the U.S. employed large framing members in their buildings, and they adapted this to a wide

Interlocking log construction. Traditional Norwegian building with walls and roof in heavy timber.

Eighteenth-century Norwegian "loft" storehouse at Gransherad. The lower storey under the cantilevered upper storey employs a form of log construction which has changed little from the Middle Ages.

Traditional timber-framed house in Sussex, England. The panels between the framing are filled in with wattle and daub.

range of building applications, including mills, large barns, naval sheds, etc.

The evolution of light timber framing

Mechanical methods of producing cheap screws and nails increased the repertoire of simple wooden joints. Machines for making screws were invented and patented in America in 1760 by Job and William Wyatt. Pointed screws were patented between 1830 and 1840. Nailmaking machines were patented in the late 18th century by both British and American inventors: Jesse Reed's machine of 1807 cut, shaped, and headed nails in a single operation and could make 60,000 per day. As a result, the price of nails fell dramatically.

A further major influential factor in the timber industry was the advent of steam power. Mechanical sawmills, often driven by waterpower, had existed in Roman times and had been used continuously and in increasing numbers after a revival of their use in medieval Europe. Steam power helped to greatly increase the volume of sawn timber. Vast quantities of boards and planks of many sections and profiles were produced and exported throughout the world from countries like the U.S. and Scandinavia, with their extensive softwood forests.

The combination of cheap nails and mass-produced timber sections led to the development of the balloon frame, thought to have been invented by Augustine Deodat Taylor, an architect and builder from Connecticut, who moved to Chicago. The first building to use this form of construction was the small catholic church of St Mary's, built in Chicago in 1833. Thin studs, approximately 2.5 x 3 in. (62 x 75 mm), were closely spaced and nailed to sole and wall plates and externally boarded to form the walls. Joists and battens were nailed to the plates and studs, and diagonal bracing completed the frame. This form of construction could be erected by relatively unskilled workmen using simple tools, nails, and mass-produced standard timber sections, since it eliminated all the skilled work of mortising and tenoning structural pieces of timber.

This technique quickly spread throughout the U.S., especially in rapidly developing areas, and can be considered as the forerunner of many later variations of frame construction, among them the platform frame. Closely spaced studs and other framing members joined together in much the same way as in the balloon frame are extensively used in the production of prefabricated timber panels and other factory-produced timber components. These forms of frame construction have become a standard building method for small buildings in the U.S. and other countries. Balloon-frame buildings were soon mar-keted as packages, with all the lumber cut to length and numbered.

Roofs and large spans in timber

Roof and large-span timber framing of various types evolved from the earliest times. The Romans are known to have made timber frames both for the centering of their large arches and for their bridges. There is a carving of a bridge on Trajan's column (AD 114) in which a statically indeterminate form of truss is shown.

Medieval roof frames were made with many combinations of main frames, purlins, and rafters along with bracing and bridging elements. Sometimes the principal rafters were propped up from the tie beams of box-frame construction, while in other cases collar beams high up in the structure of the roof tied the elements together. Alternatively cross braces, springing halfway up the principals, were used. Many of these roofing techniques relied on the enclosing walls of the building to provide some lateral restraint. Hammer-beam systems were developed in the 14th and 15th centuries, of which the roof of Westminster Hall in London, England, is a good example.

Andrea Palladio (1518–80), in his *Quattro Libri* (1570), shows designs for statically determinate trusses for both roofs and

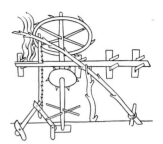

Water-powered saw (c. 1250). From Villard d'Honnecourt's *Sketchbook*.

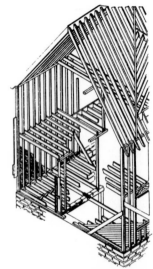

"Balloon frame" house. Only light sawn boards are used, joined together by mass-produced nails. The balloon frame was developed in Chicago in the 1830s.

ABOVE: 19th-century framed building with the spaces between joists, studs, and rafters filled with insulation.

UPPER LEFT: 14th-century roof from Adderbury Church, Oxfordshire, England.

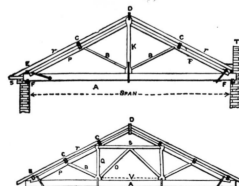

LEFT: Drawings from a 19th-century carpentry manual showing a king post truss above and a queen post truss below.

bridges, but these simple forms were not universally adopted until the work on graphic statics had been done by Ritter, Clerk Maxwell, Culman, Mohr, and Bow in the 19th century. Many of the trusses used in the intervening period had superfluous and redundant members that often prejudiced the efficiency of the truss and certainly made their behavior difficult to predict.

A roofed, crudely triangulated truss bridge dating from 1333 survives near Lucerne in Switzerland (the Kapell Brücke). Another roofed bridge built nearby in 1568 combines a timber arch with the truss in its structure. In 1754 Grubenmann, a Swiss carpenter, built a bridge spanning 394 ft. (120 m) at Schaffhausen. It was constructed with a pillar in the middle since the authorities did not believe such a large span was possible, but the post was removed during the opening ceremony.

After the 16th century, roof trusses in important buildings were commonly of the king-post or queen-post type. The former was used for smallish spans up to 35 ft. (11 m), while the latter could be used in its simple form up to 50 ft. (15 m). Combinations of both types with additional members could span up to 80 ft. (24 m). Larger roofs often had indented and notched arches incorporated into the trussed structure. One of the largest-ever trusses in timber (235 ft./72 m), which was said to have been executed over a riding school in Moscow in 1790, was of this type, although some 19th-century writers have disputed its very existence.

The degree to which savings in timber could be effected by design improvements is illustrated in the design by Inigo Jones (1573–1652) for the roof of St Pauls, near Covent Garden in London, which was built in 1631–38. Its original roof spanned 50 ft. (15m), and had a volume of 273 cu. ft. (8 cu. m) of timber, but the trusses eventually failed through poor jointing. A replacement roof, designed in the 1830s by Philip Hardwick (1792–1870), contained 98 cu. ft. (3 cu. m) of timber. This was possible because many of the joints were no longer simple wooden mortise-and-tenon connections, but included combinations of iron straps and collars that made the joints more secure and rigid and the timber sections lighter.

American bridge building in timber made many advances in the late 18th and early 19th centuries. The earlier trussed bridges were often of the covered type which protected the vulnerable joints from the weather, and in many cases they combined arches with trusses. Timothy Palmer's bridge (1797) over the Schuylkill and Theodore Burr's bridge (1814–15) at McCall's Ferry over the Susquehanna were both of this type. The latter was the longest timber truss ever built in America, spanning 364 ft. (111 m).

Special truss forms invented and adapted by

Late 14th-century hammer-beam roof at Westminster Hall, London; shown in half section and with details.

Medieval timber roof at Rushden, Northamptonshire, England (c. 1500).

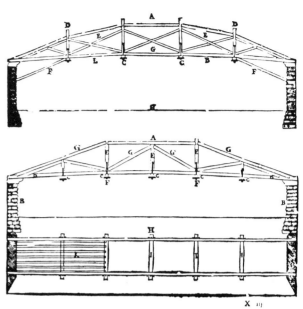
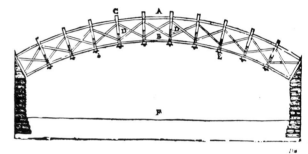
ABOVE: Trusses for roofs and bridges from Andrea Palladio's *Quattro Libri* (1570).

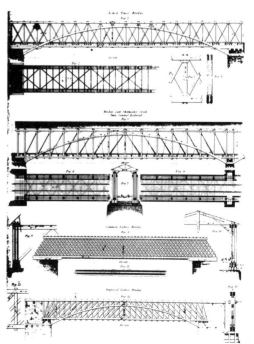
RIGHT: 19th-century American timber truss bridges. Bridges of this type were often protected from the weather by roofs.

Ithiel Town (1820), Elias Howe (1840), T. W. Pratt (1844), Whipple (1840s), and others often combined timber with wrought-iron tension members and cast-iron shoes to simplify jointing. Many of these different truss types were used in railroad bridges, but fire proved a constant hazard and these forms were later translated into iron (see BRIDGES).

Similar developments took place in the evolution of roof trusses. Tension members made out of wrought-iron rods were combined with cast-iron shoes and jointing elements that reduced the amount of labor needed to secure the timber membranes together. Kingbolt, collar beam, trussed rafter (Fink truss), and queen-rod trusses are all trusses of this type.

Attempts to conserve large pieces of timber for the French Navy stimulated the 16th-century architect Philibert Delorme (1512–70) to develop a new system of timber construction for domes using arched ribs of timber in lieu of trusses. These ribs were formed of short lengths of plank placed edgeways and bolted together in thickness, the planks in one thickness breaking joint with those in the adjoining thickness. The Halle au Blé in Paris, France, had an arched timber roof which spanned 120 ft. (36 m), but this burned down in 1802.

Colonel Emy, an early 19th-century French engineer, developed another technique using laminated timber arches, this time with the fibers of the timber coinciding with the curvature. The various planks were bolted or strapped together. The first roof of this type spanned 65 ft. (20 m) and was erected in 1825 at Marac, near Bayonne in France.

Treatment and conversion of timber in the 19th century

During the 19th century many machines for planing, turning, boring, carving, fretworking, dovetailing, mortising, tenoning, and molding timber were invented and improved. These were driven initially by steam, and later by electricity and replaced many of the hand tools commonly used until then by carpenters. Mass production of window frames, doors, sections of baseboard, moldings such as picture rails and cornices, and numerous decorative elements such as turned balusters and valances, were made possible by these developments. Many new types of nail, screw, and bolt were invented, each with its own special properties and a seemingly infinite array of ironmongery items came into the building market, making a very wide range of new joinery items possible.

The range of timber available to carpenters and joiners was vastly increased in the 19th century. The great forests of North America and northern Europe produced softwoods and well-established hardwoods, while the forests of Africa, South and Central America, Asia,

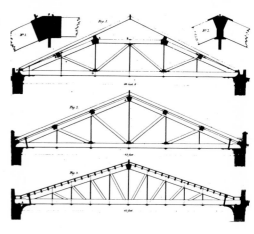

Timber roof trusses with cast-iron shoes and wrought-iron ties.

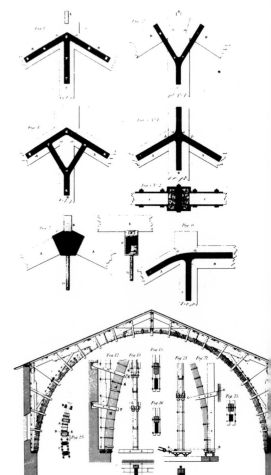

Wrought-iron straps and cast-iron shoes used to simplify and increase the rigidity of joints in trusses.

Arched timber roof at Marac near Bayonne, France (1825), designed by Colonel Emy; shown in section and with details.

Mass-produced turned balustrades for stairways, verandas, and balconies. From 1879 catalog of George O. Stevens and Company, Baltimore.

and Oceania yielded an ever-increasing range of new timbers. Mahogany imported from the Honduras was fashionable during the 18th century for joinery items such as handrails and furniture, but during the 19th century the volume of timber imported and exported grew to the level of major worldwide trade. For example, most of the railroad sleepers used in the Indian railroads of the 1860s were made of Baltic fir which had been creosoted in Britain before being reexported, while teak from India became widely used in the interiors of prestigious buildings both in Europe and the U.S. At the great 19th-century exhibitions and trade fairs, new types of timber were always on show, together with lists of their properties and uses.

Methods of treating timber against insect and fungal attack began to be studied seriously in the mid-19th century. Creosote, which is a by-product of the gas industry, was first used to treat timber in the 1830s in Germany. Other chemicals and techniques of impregnating timber by vacuum processes were developed soon afterward, making it possible to treat timber in depth rather than solely on the surface.

Traditional methods of seasoning timber, which involved slow drying out in protected conditions in the open air, began to cause bottlenecks in the production process. Before timber can be used in carpentry and joinery, it must loose about 50% of its natural moisture content so that it can harden, become stable, and less liable to movement and twisting. Softwood logs may season naturally in three months to one year, while hardwood logs may take considerably longer.

The requisite moisture content in any piece of timber depends on its eventual use. Joinery items normally require very dry, well-seasoned timber. Kiln drying was well established by the 1870s, and this process rapidly spread to many timber-producing areas in an attempt to speed up production. Specialization within the timber industry began to evolve after the mid-19th century when, for example, the manufacture of plywood and other wood-based sheet materials began to develop (see PLYWOOD AND OTHER SHEET MATERIALS).

The 20th century

In the 20th century timber became a scientifically controlled engineering material. Its physical properties are now carefully measured and logged, and it is graded according to its strength and appearance. This was done visually until comparatively recently, but now in many sawmills it is stress-graded by machine and carefully marked from a range of standard symbols, or each batch is given a complete specification, listing all its properties and qualities.

Forest management has become an important part of the industry. Replanting and careful cropping makes timber a renewable resource that can be grown commercially over and over again. This method is now used in most softwood-producing areas, but many tropical hardwoods are still taken from virgin forests, and reafforestation is by no means a common practice.

Tree felling in softwood forests has become increasingly mechanized. Whole areas of trees of roughly the same age are now all cut down at the same time by clear felling. The mechanical chain saw, which replaced axes and manual saws, is being superseded by vehicles known as feller bunchers, which cut trees and lift them to a position where they can be taken up by buckers. These machines, normally crewed by two men, cut off all the branches of a felled tree and stack them ready for collection. They can process two large trees per minute. The trees are then taken by tractor to roadside depots and then transported by trucks to sawmills. In many countries the long-established practice of floating logs downriver to the sawmill is dying out.

At the sawmill, trees are stored in water or under sprays to protect them against fungal growth and insect attack. The logs are taken from storage and studied carefully to determine the best way of cutting them to give the best yield of planks and other sections. They are then cut by frame, circular, or band saws depending on the size of the sawmill and the varieties of log being handled. The cut boards are then passed through edge-trimming saws to render the sides parallel. They are then taken from the mill for seasoning, either by careful stacking in the open air to allow free-air circulation, or in a kiln where they can become seasoned in less than a week. After this process, planks and boards are often cut to fixed lengths, then graded and packed into rectangular bundles held together by plastic or steel straps for easy handling. In many cases

Pre-glazed, factory-produced window units built into a timber frame. An increasing proportion of timber products now arrive on site as finished components.

Loading timber from sawmill depot onto cargo ship.

Various forms of timber connectors used to spread the load of bolts over a wider area within a joint.

Prefabricated, mass-produced timber roof truss with gangnail plate securing joints.

A factory-finished Calder Homes "box-unit" being craned into position in a housing scheme in Britain (1965).

packages are now made up of equal lengths of timber, so that these bundles can be made square at both ends.

Many chemical processes have been developed during this century for treating timber in order to make it more resistant to fungal and insect attack, moisture penetration, and fire. Chemicals may be applied by brush or spray to the surface of the wood, the timber may be dipped or steeped in a liquid bath or the chemicals may be introduced into the wood by pressure impregnation. This century, improvements in timber engineering have simplified jointing techniques still further and have made possible further economies.

New and improved waterproof adhesives, many of them based on synthetic resin products, make it possible to join timber in a way which is analogous to welding in steel, to form permanent bonds and joints. These adhesives replaced earlier ones which were largely made by boiling down and treating organic materials such as hides, bones, starch, and the like, but these were not permanent or waterproof. Joints using resin-based adhesives can be as strong or stronger than the materials they join.

Random lengths of lumber may be united by finger jointing to produce structural sections, which can lead to considerable savings in timber which would otherwise be wasted. Boards can be laminated together to form "glulam" beams or arches. These can be designed in such a way that high grades of timber can be utilized for high-stress areas, whereas other parts of the beam can be made of less expensive material. These types of beams and arches have been widely used since World War II. By using laminated constructions of this type, it is now possible to construct structural elements with rigidity and continuity between beam-and-column elements.

Combinations of boards and plywood are made in order to exploit the properties of stressed-skin construction and these are widely used in the manufacture of girders and beams as well as floor, roof, and wall elements. Timber trusses, using plywood gusset plates, which are nailed and glued to the members, can lead to savings in materials and speed of fabrication. Scientifically controlled nailing, screwing, and bolting, have helped produce efficient joints whose behavior can be carefully designed and predicted. Nailing, bolting, stapling, and screwing are often carried out on the building site where gluing would be impractical. Special elements such as split-ring connectors, and steel shear plates have been developed to spread the load of bolts over a wider area within a joint. The use of nail plates, which are used in a similar way to gusset plates in the manufacture of trusses, has led to the mass production of cheap trussed rafters which are compact, easy to transport, and make use of shorter lengths of timber. Using these modern techniques it is possible to make timber structures of many different forms and large spans.

Many of these developments were the result of research carried out during both World Wars. Timber and labor shortages provided an incentive for new thinking which often led to interesting forms of construction. One such form is the lamella roof, which utilizes short pieces of timber that would otherwise be wasted to make up an arched roof without the use of trusses. Complex shapes based on double curvature geometry, such as hyperbolic paraboloids, have often been constructed permanently in wood, or have been made of wood as shuttering prior to casting in reinforced concrete. Wooden folded-plate structures are becoming widely used. All these techniques use new timber products and fixing methods, combined with new structural ideas.

Timber lends itself to prefabrication, and has been used extensively in many countries for this form of construction. Prefabricated timber buildings have existed since the Middle Ages and became very popular during the 19th century when large and often sudden movements of population took place. The growth of prefabricated timber housing was helped by large government-sponsored projects such as the Tennessee Valley Authority development in the U.S., which started in the late 1920s. Prefabrication in timber has grown dramatically since World War II in many countries.

An increasing proportion of timber products now arrive at the building site as finished components and assemblies. These include wall panels, door sets, preglazed windows, internal fittings, roof trusses, and other items.

Timber is one of the oldest and most adaptable of building materials. Unlike other natural resources, the supply need never be totally exhausted as man has learned techniques of afforestation, thereby ensuring its continued availability. As a building material it is favored for its color, texture, and appearance. For all these reasons, it will continue to play a major role in building for years to come.

Plywood and other sheet materials

A sheet of plywood is essentially a number of thin veneers glued together with the grain of the adjacent veneers running at right angles to one another. Plywood made it possible for the first time to have a thin sheet of wood which was strong, light, and able to be easily curved and fixed without the risk of cracking or loss of strength. It was probably invented in ancient Egypt, where wood was scarce and precious. The sheets contained up to six plies, usually fastened together with wooden pegs.

Until the second half of the 19th century, veneers were nearly always cut with saws, a task demanding great accuracy and skill. In 1844 a factory at Revel, in Estonia, began to make plywood seats for bentwood chairs; and in America, soon after the Civil War, a number of people took out patents for making laminated wood sheets in this way. Early in the 1870s, George Gardner of Brooklyn began making plywood-seated benches for railroad stations and other public places, curving the material by bending it after it had been steamed.

Knife-butting machines became available in the 1890s. By this method, the log was rotated against a knife edge, so that a continuous veneer was produced. These large sheets of veneer, which replaced the narrow strips obtained by the old process, allowed plywood boards to be built up of a size and strength which had been impossible previously. Rapid technical advances were made during World War I to meet the needs of aircraft builders. The waterproof phenolic and other resin adhesives evolved by European and American chemists at this time improved the quality of plywood to such an extent that by the 1920s it had largely lost its reputation of being a cheap substitute for solid wood and had become accepted as a useful material in its own right.

In timber engineering, plywood is particularly suitable for uses where its high-panel shear values, combined with flexural rigidity and light weight, can be fully exploited, as in sheathing to framed buildings, gussets for timber trusses, I-webs and box beams, folded-plate roofs, and stressed-skin panels. It is also widely used as structural flooring, roof decking, wall sheathing, and cladding for timber-framed houses. Shuttering for concrete work is another important market for plywood. The first successful experimental stressed-skin plywood house was erected at the Forest Products Laboratory in Madison, Wisconsin, in the 1930s. A vast number of houses of this type have been erected since that time in many countries.

Other wood-based boards

Since World War II, the forest industries have produced a number of wood-based materials in sheet or slab form which have become commonly used in building and timber engineering. They include, in addition to plywoods and their associated solid-core boards, such as laminated board and blockboard, particle boards, fiber building boards, and wood-wool slabs. Due to their different methods of manufacture, or the admixture of various bonding agents, these panel products possess properties which differ in certain respects from each other as well as from natural wood. There is also a growing production of these materials with overlays of wood veneers, paper, plastics, or metal, or in sandwich form containing insulating cores of

various materials, or combinations of the basic panel products—all designed to meet particular conditions of use.

The development of chipboard (particle board) dates from World War II. Of all the wood-based sheet materials it had the fastest growth in recent decades, rivaling that of plastics. It is made of wood chips, bonded together with synthetic resin and cured under heat and pressure. Wood chipboards may be homogeneous in structure or have surface layers of higher quality or denser texture to improve finishing or strength.

During the past 30 years, the particle-board industry has developed from the original concept of a means of using waste wood to being a major manufacturing industry for which wood is specially grown. Standard grades of chipboard, produced originally for furniture and joinery, were found to be suitable for wall linings, partitions, and the denser grades for structural flooring. Compared with plywoods their shear values are low and their moisture movement high. They are not generally suitable for external work or for conditions of high humidity. Within the last decade or so, structural quality boards have been produced in some countries, as well as boards treated to resist moisture and others with high resistance to the spread of flame.

In fiber building boards the primary bonding is derived from the felting of the wood fibers and their inherent adhesive properties. Additional bonding, impregnating, and other

Interior of a factory manufacturing blockboard, which consists of a layer of softwood battens bonded between two layers of ply.

Chipboard sheets being installed as a floor surface.

Prefabricated plywood wall panel being craned into position on a building.

Framed plywood panels on a house ready to receive external cladding which is in brick veneer on the lower storey.

agents may be added in the course of manufacture. Various processes are used for chipping, pulping, felting, pressing, and subsequent treatment.

Early production of fiberboards was concentrated on highly compressed hardboards and on low-density insulating and acoustic boards. In recent years, however, there has been a greater diversification to meet specific needs in building. New types of board, particularly in the medium-density range, have been produced with good dimensional stability and adequate strength for wall linings and similar uses. Bitumen-impregnated insulating boards are widely used for external sheathing of timber-framed buildings while the denser hardboards, especially of the oil-tempered variety, have been used successfully for structural components.

Mineral-bonded wood-wool slabs were first produced in Austria in 1914. They are radically different in appearance, properties, and end uses from the other purely organic, wood-based boards. They are much thicker and heavier; the portland cement or other hydraulic binder accounting for about half their weight. They are practically noncombustible and have good sound absorption and insulation properties. In addition to their common uses as non-load-bearing partitions, wall linings, and ceilings, the stronger type boards are widely used for roof decking. Steel reinforced slabs are also marketed for the latter use.

The effects of the introduction of wood-based boards of different types into the building industry have been far-reaching. They have revolutionized the production of joinery items, such as doors and kitchen cabinets, made the mobile-home industry possible, and greatly accelerated the development of prefabrication. They simplified timber-framed construction for residential and other buildings and made stressed-skin structures in timber possible. Their use in substitution for solid timber or other building materials has generally reduced the volume and weight of materials, site labor requirements, and construction costs. Their development also enables a much fuller use to be made of forest resources, since in their production, trees and timber species not otherwise suitable for building purposes can be used, as well as tops, branches, and mill waste.

Gypsum board and other sheet materials

Sheet materials based on materials other than wood have also revolutionized certain aspects of building. Gypsum board consists of gypsum plaster sandwiched between two sheets of paper. It originated in the U.S.—Augustus Sackett's first patent was taken out in New York in 1894—and by 1910 this material was being widely used in America.

Until the invention of gypsum board, plastering was a laborious and skilled operation. In frame buildings, plaster had to be applied to laths nailed to studs and joists—bringing wet trades into an otherwise dry form of construction. Gypsum-board ceilings and wall linings can be erected quickly with a minimum of mess. This material also has the advantage of possessing fire-retarding properties.

Improvements in gypsum board have included the use of special paper surfaces capable of holding thin coats of finishing plaster to produce jointless surfaces. Aluminum foil can be bonded to one side of the board to form vapor barriers, as well as increasing the insulation value of a composite wall or ceiling. Specially tapered edges of boards have made it possible to mask and fill joints without the need for a skim coat of plaster.

Straw and other vegetable fibers can be compressed to form useful building boards which are lightweight and also possess insulating properties.

Besides being manufactured to specified performance standards, sheet materials are marketed to agreed sizes. The dimensions and properties of these products dictate the spacing and disposition of framing elements in all types of construction in which they are used.

Stone

Some varieties of stone make excellent building materials, others relatively poor ones. The inferior types have been used locally; the better stones have justified the cost and trouble required to transport them over considerable distances.

Properties of stone vary greatly according to their type, and considerable variations can occur in the properties of different specimens from the same quarry. Durability depends not only on the chemical composition of the stone, but also on the atmosphere in which it is used and its degree of exposure to the elements. Facility of working may be an important criterion in the choice of a particular type of stone. Soft stone with an even grain and no distinct beds may be good for sculpture, whereas a hard stone composed of thin layers that can easily be separated may be appropriate for rubble masonry. Hardness may be a sought-after property in constructions where surfaces are subject to abrasion, or where sharp detailing is required. Hard stones may, however, be brittle and less resistant to chemical decay than other varieties.

The compressive strength of stone may sometimes be important, but most stones used in building are understressed. The weakest sandstones can carry 120 tons/sq. ft. (1,290 tons/sq. m) while at the other end of the scale some granites can support 800 tons/sq. ft. (8,600 tons/sq. m). In the dome of St Peters in Rome for example, the greatest stress in any part of the structure is 15.5 tons/sq. ft. (167 tons/sq. m). Other important factors that guide selection are appearance, which may be an overriding consideration where color, texture patterning, and grain are important, and weight. Weight varies greatly from light rocks such as pumice and vermiculite to dense stones such as granite and basalt.

Types of stone

Geologically, there are three families of stone: igneous, sedimentary, and metamorphic rocks.

Igneous rocks are formed directly from molten materials of the earth's crust. Granites are made up of varying proportions of quartz, mica, and felspar, combined with other mineral materials; they have been popular in building since antiquity. Granites containing a large proportion of quartz are difficult to work, while those with a large amount of mica tend to be weak. Most granites used in building are selected for their hardness and durability. Many can accept a high polish and retain sharp edges for long periods of time. Egyptian granite obelisks such as that from the temple at Thebes, now at Place de la Concorde in Paris, which is 3,500 years old, still show little or no deterioration in their inscriptions. Granite has often been favored for heavy engineering works as well as for surfaces exposed to abuse and wear such as bollards, copings, and paving. Special types of granite have often been selected for their appearance. Syenite from Upper Egypt and porphyritic granite, with its large independent crystals of felspar, are examples of these.

Other igneous rocks used in building include porphyries, which often occur in dykes, and basalts, which are normally dark

Vehicle-mounted mechanical saw being used to cut stone blocks from a rock face.

Eighteenth-century masons at work on a construction site measuring cutting, lifting, carving, and laying stone slabs.

gray or black and are very hard. Mica, which is found in large transparent sheets in Russia, was sometimes used in place of glass, earning the name of Muscovy glass.

Sedimentary stones are composed of either the reconsolidated debris of igneous rocks, or of the fossil remains of living creatures. Sandstone represents the first type, limestone the second.

Sandstones consist of grains of quartz and other materials, cemented by silica, alumina, iron oxide, or other substances. They vary greatly in their color, texture, and other properties. Edinburgh, in Scotland, has a great proportion of its buildings built out of gray Craigleith sandstone. Brownstone row houses, built from a variety of reddish-brown sandstone, became very popular in Boston and New York in the late 19th century. "Laminated" sandstone flags from Yorkshire, in England, have been popular for paving because of their hardwearing surfaces and

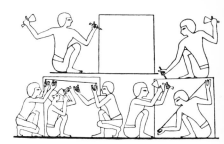

Dressing a stone block. From a tomb at Thebes (c. 1450 BC).

Houses with random rubble granite walls and slate roofs, in Nielson Square, Gatehouse of Fleet, Britain (1812).

St Paul's Cathedral, London, England (1675–1710), built of Portland stone by Sir Christopher Wren.

Moving a stone pillar with levers near a quarry. Transporting and lifting large masonry blocks has always placed limits on builders.

because they are readily split into slabs. Sandstones that laminate easily are also used for tiles or roofing flags. Grits are coarse-grained stones used for engineering work and millstones.

Limestones, which consist of calcium carbonate together with small percentages of clay, iron, magnesia, silica, etc, have been popular over many centuries because of their known hardwearing properties, appearance, and workability. The Great Pyramid in Egypt is built of over 2,000,000 blocks of limestone, some of them weighing more than 1,000 tons. The Romans made extensive use of different types of limestone. Travertine was chosen for its weathering properties and used in exposed parts of structures. Tufa varied in quality, but reliable quarries were in operation by Augustan times; it was used for internal parts of structures. In France and Belgium the famous pierre bleu has been quarried since the 12th century. This limestone is easily worked and stands up well against frost, damp, and smoke. It has an exceptionally high compression strength. Its blue surface becomes white at the points where it is struck by a hammer or chisel, and a wide range of effects can be obtained by different methods of dressing. Caen stone is another famous French limestone, used in many of the Gothic cathedrals and churches. It was imported into England for the construction of Westminster Abbey.

Portland stone, which is a white, hard stone, became very popular for important buildings in Britain from the 17th century onward. It was used by Sir Christopher Wren for St Pauls Cathedral, and has been exported to many different parts of the world, especially British dominions. Bath stone, with its honey-like color, is another well known English limestone. In the U.S., Indiana limestone, which was used in the construction of Washington Cathedral, has been extensively used in building.

Metamorphic rocks originate from igneous and sedimentary rocks that have been transformed under great heat and pressure. Marbles, quartzites, and slate belong to this group. Many varieties of marble have been much sought after for their special color, patterning textures, and ability to accept a high polish. They are normally hard and crystalline, and consist largely of calcium carbonate.

Marble from Mount Pentelikon was used by the Athenians for the buildings on the Acropolis. This marble, which starts off white and acquires an attractive golden patina as it oxidizes, has weathered well over the centuries, but is now faced with rapid deterioration on the Acropolis due to atmospheric pollution.

At Lydia and Caria, in Anatolia, the Greeks quarried a fine white marble for their important buildings in Asia. The translucent onyx marble of Algeria, found in cloudy yellow and brown colors, was used both at Carthage and in Rome. The quarries at Torbole and Carrara in Italy have been worked since classical times. In India, a fine white marble, quarried at Makrana in Rajputana, has been used for civic buildings for a long period; it was the stone used to build the Taj Mahal at Agra. Many countries throughout the world have indigenous marbles with their own coloring and patterning characteristics.

Slate is another metamorphic rock made up of clay sediments that have been hardened by great heat and pressure. Slate splits readily into thin slabs, making it useful for roofing and other purposes.

Quarrying and transporting

Before quarrying was initiated, most stone used in building was obtained from weathered rocks and boulders. The Egyptians pioneered quarrying using wedges, cutting, and drilling to remove the blocks they wanted. These methods have changed little in principle over the ages, but the processes have become increasingly mechanized. In some cases stone is extracted in the open, in others it is mined underground. The use of gunpowder and other explosives has added to the methods of extraction; it is used especially for rocks such as granites which do not have natural bedding planes that can ease removal by wedging and splitting.

In antiquity, large stones were moved by sledges, rafts, rollers, levers, ropes, and ramps. Pulleys were first used in about the 8th century BC by the Assyrians. Wherever possible, stones were moved by water, which made transportation less of a problem. The advent of canals and railroads in the 18th and 19th centuries made the choice of a much wider variety of building stones practicable.

The Romans imported stone from different regions of their Empire from state-owned quarries which supplied large depots with stone. In Rome it was not uncommon for large, standard-sized columns to be produced at a distant quarry for use at a later date in some civic building. Standardized dimensions, accumulation of stocks, and a substantial measure of prefabrication contributed to the possibility of rapid construction of buildings. The vast Baths of Diocletian (AD 298–305) took only eight years to build.

Improvements in hoisting and lifting techniques have contributed to the ease with which stone can be used in buildings. Derricks used by the Egyptians, Greeks, and Romans were capable of lifting weights up to

Side view of the portico of the Parthenon, Athens (5th century BC) built out of pentellic marble.

Cutting thin slabs of marble with a frame saw. From an 11th-century manuscript.

Hand crane for lifting masonry elements.

200 lb. (90 kg). Windlasses, introduced into Europe in about AD 1100, were capable of raising weights of over 1,000 lb. (454 kg). Mechanical power, improved cranes, and techniques of multiplying mechanical advantage have greatly reduced these limitations.

Techniques of cutting, drilling, carving, shaping, and polishing stone have changed little in principle, but there has been an increasing use of machinery, eliminating heavy work. The Romans developed water-powered saws. In the 19th century, steam, and later electrical power, were applied to an ever-increasing variety of machines; these included saw frames, circular saws with carborundum cutting edges, planing and molding machines, and pneumatic hammers.

Masonry techniques

The ways in which stones are combined to form structures vary greatly, from dry stone rubble walling at one end of the scale, where roughly shaped stones are piled on top of each other to form walls without mortar, to carefully executed ashlar work where accurately cut stones are assembled with great precision. The masonry work in the Parthenon in Athens is an example of the latter. Because of the high cost of ashlar work it has often been common practice to build walls and other massive parts of structures as two leaves of ashlar with a filling of rubble, often with header stones to help bind into the core. A large proportion of medieval cathedrals were built in this way. Another technique that contributes to savings is to use a masonry facing with a backing in brickwork.

Where careful attention is paid to the relationship of joints to one another in walls, stronger and slenderer structures can be built. The coincidence of vertical joints should be avoided. (Until about the 2nd century BC the Romans used carefully cut stone blocks without any consistent pattern of staggered joints.) Slender walls can be given greater rigidity by the use of buttresses. Stones may be held together with mortar of various types or by cramps and dowels of metal or wood.

Evidence of different masonry techniques can be as useful as stylistic variations in distinguishing between work carried out at different times in structures that were built over long periods of time. At St Albans Abbey in England, Norman work is distinguished by axe marks, the transitional period by chiseled stones, the early English period by bolster tooling, culminating in the finely scraped work in the Perpendicular period.

Stone arches, vaults, and domes have been used as ways of spanning openings and spaces. The trulli buildings in Apulia in Italy still use an ancient form of corbeled dome for their roofs which are made out of the local

Dry-stone random rubble walling with rubble-on-edge coping.

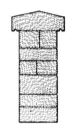

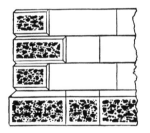

Coarsed rubble wall with coping; front and side views.

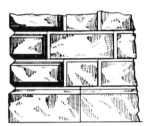

Ashlar walling with chamfered and rusticated quoins and plinths (LEFT) and with rabeted joints and molded quoins (RIGHT).

Royal Crescent, Bath, England (1764–74) by John Wood the younger. Main facade in ashlar; side wall in rubble.

limestone that splits readily into rectangular blocks. The earliest arches and vaults were probably made out of brick in about 3000 BC in Mesopotamia, but these techniques were soon applied to stone. Stone voussoirs in arches, vaults, and domes require a mastery of accurate geometrical techniques. Two-dimensional structures, such as simple arches and barrel vaults, are not so demanding, but the shaping of stones for three-dimensional curved surfaces, such as parts of domes and complex vaults, requires considerable geometrical skill. Gothic cathedrals generally used carefully cut vault ribs with panels made up of smaller, roughly shaped stones between them. In the Renaissance period the art of stereotomy was developed, making it possible to work out and describe the exact shapes of stones for complex vaulting.

Accurate understanding of the forces involved in arched, domed, and vaulted structures made it possible to reduce the mass of material used in masonry structures. Improved understanding of the behavior of these structures can be traced in the evolution of Gothic cathedrals, where masons vied with each other to create lighter, higher, and less massive buildings. A high point in the development of analytical techniques was reached by Antoni Gaudí (1852-1926) in the funicular models he used to examine the likely forces in the church of the Colonia Güell (1898–1914) and later in the studies for the masons of the Sagrada Familia in Barcelona.

Weathering and decay

Stone is subject to erosion by water and wind. Chemicals in the atmosphere, often dissolved in rainwater, contribute to the decay of stone. Some stones are attacked by chemicals given off by lichens and other vegetation, while others are to some extent protected by coverings of these plants. Moluscs attack some forms of limestone that are employed below water. Many varieties of stone harden after they have been quarried and exposed to the air. For important works, where careful selection of stones at the quarry is necessary to ensure their durability, it may be advisable to allow them to weather in the open air for a number of years. This practice was recommended by Vitruvius.

When sedimentary stone is used in building, close attention should be paid to the bedding planes. In walls, these should be laid in the same sense and direction as they are found at the quarry. When this is not observed and stones are used haphazardly, some will weather and decay rapidly through frost action and other weathering due to the easy penetration of water. Badly weathered stone must be replaced, but many chemical preparations have been developed to slow down these processes.

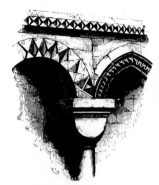

Ornamental arcades, Canterbury Cathedral, England, with stone worked with axe (LEFT–c. 1110) and with chisel (RIGHT–c. 1180).

Cross vault constructed of rough stonework; from medieval castle in Syria.

Injecting epoxy resin into cracks in masonry to counteract weathering.

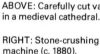

ABOVE: Carefully cut vault ribs in a medieval cathedral.

RIGHT: Stone-crushing machine (c. 1880).

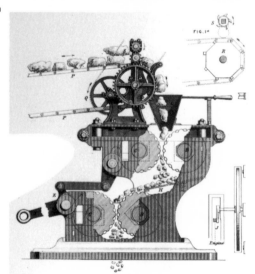

Other types of masonry and artificial stone

Stone is inevitably an expensive material. If there is to be a future for natural stone in new construction, it must lie in more mass production of standard units, and in acceptable techniques which allow ashlar facing to be easily and efficiently attached to steel and concrete frames. Very large stone panels are now available for cladding purposes, and sophisticated systems of attaching thin, non-load-bearing slabs of stone have been developed using noncorrosive clips and dowels made from such metals as stainless steel and bronze.

Crushed stone is extensively used in the

construction industry in connection with road building, and especially as the aggregate component in concrete mixes. Stone used for these purposes is chosen for its particular mechanical properties, and careful grading of sizes often takes place after crushing.

Various artificial or reconstituted stones have been used since the early 19th century. They are made by mixing crushed stones, say granite quarry refuse, with portland cement to create Victoria Stone. Other types of cement and crushed stone may be used in combination with artificial dyes to gain different effects, and many patent systems have been developed. The "stone" is made into objects by casting into molds in much the same way as concrete is formed. It may then be polished or given other surface treatments.

Earth, in various forms

Broadly speaking, builders throughout the world have used earth when they could get nothing else, or where the cost of alternatives has been impossibly high. When the opportunity has arisen to abandon adobe, cob or pisé, or whatever the particular kind of earth-construction may have been, and to go for something more durable and sophisticated, that opportunity has almost invariably been taken. Earth has been very much the poor man's stone, although, like any other material, it benefits from good craftsmanship. It has the great advantage of requiring only very simple tools.

Adobe

Adobe, from the Spanish word *adobar*, to plaster, is the name given to a system of building using sunbaked but unburned earth blocks. Adobe building is widespread in the north of Mexico, in California, and the southwest of the U.S., where the climate is semiarid. A similar system is to be found in New South Wales in Australia, and there are variants in many countries, particularly in the Third World. Almost any soil can be used. It usually contains at least 50% sand and, if the clay content is high, chopped straw or grass fiber are added. The material is first mixed into a sticky mass, often by treading, and then thrown into molds. As soon as it is dry enough not to slump, the blocks, measuring about 16 x 8 x 6 in. (400 x 200 x 150 mm) and weighing about 60 lb. (27 kg), are removed from the molds and laid out on the ground; later they are piled into open stacks for final drying. The quantity of clay in the soil is not critical, because the shrinkage it causes occurs in the drying-out process before the blocks are used. The blocks are laid in mud mortar, sometimes with the addition of a little lime and, as with pisé de terre, they are rendered externally. In eastern England the

Ndebele house built out of sun-dried mud bricks covered with painted mud plaster. Near Pretoria in South Africa.

system was known as clay lump building and was used for domestic and agricultural buildings. An outer skin of brickwork sometimes takes the place of rendering and gives greater protection. Rendered walls are protected from rain by a brick base up to 2 ft. (61 cm) high. In Nigeria, the blocks, formed without molds, are known as *tubali*. They are usually conical and they are laid alternately big and little end outward. The mortar contains horse dung and the blocks are coated with clay as the work proceeds. Two- and three-storey buildings constructed. in this way are quite normal. In adobe building, as with all earth walling, good maintenance is essential for long life.

The adobe used for the Aztec and Toltec pyramids was exceptionally well protected. The body of the pyramid, formed of rubble and adobe blocks, was faced with dressed stone, which in turn was plastered over with brightly colored stucco.

In the southwest of the U.S., artists and do-it-yourself homebuilders have recently revived the Spanish tradition of adobe building with some remarkably beautiful results.

Cob and chalk mud

Cob walls are built from a mixture of earth, straw, and water. The method of building differs from that of pisé de terre. The material is thrown down on a stone or brick base course, without using formwork, and is then trodden and compacted by the workmen, not rammed. Work starts from one end of the wall. The material is thrown to form a course about 1 ft. (30.5 cm) high, care being taken to avoid any vertical joints in the wet material. Once a course is completed it is protected from the weather and allowed to dry out before the next is started. The sides of the wall are leveled and trimmed to the correct width as the work goes on. Like other forms

Houses in Zaria, Nigeria, of plastered mud-brick construction.

of earth wall building, it is a fair weather operation, and for this reason it is usually carried out in the spring and summer.

In the southwest and west of England and Wales, cob was the normal walling material for laborers' cottages in the 17th and 18th centuries. If it was properly mixed and laid and protected by a masonry base and overhanging eaves, it created few problems. Cottages and small houses built long before 1800 are still to be found in good condition. In some central areas of southern England, where only a thin layer of earth covers soft and easily dug chalk, walls have been built of crushed chalk, straw, and water. The process is the same as for cob, but it differs from pisé building in that although lintels are built in as the work proceeds, the openings for doors and windows are not cut out until the walls are complete and any settlement due to shrinkage has taken place. The need to allow each course to dry out before the next is started makes it a slow system. If walls are not completed before the onset of frost or bad weather, they are given a temporary roof of thatch and left until work can be resumed the following spring. Cob walling is no longer carried out commercially, but some enthusiasts have built this way for themselves during recent years.

Pisé de terre

Pisé de terre, or rammed earth, is walling made of compacted earth. It differs from cob and chalk mud in that damp earth is compacted by hand-ramming in climbing formwork which is raised and refilled until the required height of wall is reached. If properly made and laid, it is a hard and durable material. Pisé has been used since ancient times; according to Pliny, Hannibal built pisé watchtowers in Spain. The Romans introduced it into France, but it is found there only in the valley of the Rhône, where two- and three-storey farmhouses several hundred years old are still to be seen, and around La Rochelle in the northwest.

External protection is essential in all but the driest climates. Pisé weathers badly and can be eaten by rats unless precautions are taken. The process is known in most countries of Europe, in Africa, and the Americas, although it mainly occurs where other building materials are scarce. It has had its champions in times of shortages of materials, notably after both World Wars, when officially sponsored experiments were carried out in many countries, including Germany, Belgium, the U.S., and Britain. Although a system of steel formwork was patented after World War II, timber was the traditional material for this, with forms 10–12 ft. (3–3.6 m) long and from 1–3 ft. (30–90 cm) in depth. Walls are usually 1 ft. (30 cm) thick for single-storey buildings and 18 in. (45 cm) for two storeys. Almost any type of earth can be used, provided it does not contain too much clay. An ideal mix consists of 25% clay and 75% small gravel. Lumps have to be broken up and pebbles more than 1 in. (25 mm) in diameter removed. Not more than one day's supply of earth is prepared at any time since the moisture content is critical. Openings for doors and windows are left as building continues and lintels are built in. Mud plaster or a weak mixture of lime and sand is the normal rendering, but sometimes one or more coats of tar, mixed with sand, have been used instead. Colored whitewash has been the favorite form of decoration. To protect walls from rain, a ground course of brick or stone about 1 ft. (30 cm) high is commonly used, and garden walls are capped with thatch or tile. Improved pisé using stabilized earth to protect walls against rising damp and rain penetration, incorporated damp-proof courses strong enough to withstand the impact of ramming.

Wattle and daub

Wattle and daub is a method of wall construction whereby wattles made of vertical timber stakes are interwoven with horizontal branches and daubed with clay. The technique of applying earth materials to a supporting network of light wooden twigs or reeds is common in many forms of primitive traditional building throughout the world. These methods made it possible to make relatively waterproof structures.

Remains of circular Iron Age dwellings using this method of construction have been found in England, where the staves were driven into the ground without framing. In framed buildings the staves, sharpened at both ends, had their upper points inserted in holes bored in the underside of a horizontal timber, while their lower ends were held in a groove gouged from a similar timber. The branches were then woven through the staves to form a stable panel for the clay daub and plaster. The

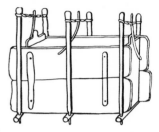

Traditional movable wooden form from Morocco used for the construction of rammed earth or pisé walls.

The blank wall of this building in the Transvaal, South Africa, is built of clay over a close framework of saplings and rendered over by hand with a mud slurry. The wall in the foreground is built of sun-dried clay bricks mixed with straw.

practice died out because of a shortage of dauber craftsmen and the more general use of bricks.

Stabilized earth

Stabilization is a name given to processes which make earth used as a building material stronger and harder, less liable to volume change, or more resistant to water. Earth for pisé de terre, cob, and mud wall building can generally be improved by the addition of stabilizers, of which there are two main types—those which impede moisture penetration, and those which prevent the capillary rise of dampness.

Experimental houses built in Belgium after World War I used hydraulic lime to stabilize crushed brick earth and debris in a pisé de terre system which included reinforced-concrete piers and string courses. The walls were waterproofed and hardened externally with a mixture of benzol, bitumen, resin, and lime. In Britain, the Building Research Station experimented with cement and chalky earth in pisé work, and the U.S. Bureau of Standards and the University of Illinois built in cement-stabilized earth, which was given the name of "Terracrete." The University also produced a bitumen-stabilized adobe block, known as "Butudobe."

In ordinary pisé work, the external wall faces were sometimes "plated" with a thin layer of stabilized earth to give protection against the weather. The stabilized earth was placed down the sides of the form containing the ordinary earth and rammed with it. Bands of stabilized earth, the full thickness of the wall, were placed over the lintels to give added strength.

Bricks

Brickwork uses small building units, often, but not always, made from fired clay and jointed with mortar. Structurally, brickwork is strong in compression but weak in tension. It is excellent for fire resistance. The small size of the unit, together with possible adaptation of joint thickness, not only makes fitting to horizontal and vertical dimensions easy but enables curved work to be constructed relatively simply. Clay suitable for making bricks is widespread in many countries. In the face of modern constructional alternatives, possibly the outstanding characteristic of brickwork is its aesthetic appeal and the fact that its attractive appearance is usually long lasting and combined with low maintenance cost.

History of brick manufacture

The manufacture of clay bricks involves two main processes: shaping the clay and then converting it to durable form. Clay bricks were first made in regions where stone for permanent building was scarce. Brick-building may have commenced in the valleys of the Tigris and Euphrates at least 5,000 years ago, using sun-baked units. For unexposed positions sun-baked bricks were tolerable but for durability and an external finish high-temperature firing is essential. Such fired units were made from very early times, in the Babylonian period. Vast structures such as ziggurats were built solid, with the outer layers of material of fired quality; Birs Nimroud, built by Nebuchadnezzar, was 272 ft. (84 m) square by 160 ft. (48 m) high.

From Mesopotamia brickmaking spread, to Egypt, to Persia, and probably through Persia to India—there was brickmaking near Bombay by 2000 BC. The process probably did not reach or at least become regularly used in China until much later. In Greece the abundance of stone meant brickwork was rare but the Romans used it widely. There were Roman kilns in western Britain in AD 90, and reused Roman bricks are a feature of the tower of St Albans Cathedral near London. With the breakup of the Roman Empire brickmaking stopped in much of Europe. In Britain it recommenced in the 13th century, in East Anglia, possibly after some importation of bricks from Holland. Little Wenham Hall in Suffolk is the earliest known example of that period and was built in the years 1260–80, but certainly by 1300 there was a brickworks at Hull. In America bricks were used by about 1580, with the Dutch having a reputation as good bricklayers.

Until the mid-19th century bricks were fired either in small intermittent kilns with a high fuel consumption, or they were fired in open clamps. In the latter the "green" bricks were stacked on top of a layer of fuel, and after a rough top-covering with old bricks and earth, the fire was started and allowed to burn itself out over a period of several weeks. Large-scale production needed a more economical and better-controlled method, and this came with the introduction in Germany in 1856 of the Hoffmann kiln. This consists of a number of kiln compartments with the fire being transferred from one to another, and heat from the cooling areas being used for warming up later loadings.

Modern developments in brick manufacture

In 1875 at Bridgwater, in England, the first machine that mechanically shaped clay was introduced on an industrial scale. This process, which consisted of extruding a column of plastic clay and then cutting it with wires, has continued as an important method and is still widely used. Further progress in machinery made possible the shaping of much drier clays by forcing them, under high pressure,

Brick cottage in Sussex, England. Bricks of two distinct colors arising out of differing degrees of firing are used to produce a pattern.

Plan and cross section of a Hoffmann kiln. The continuous cycle of stacking, cooling, burning, and drying bricks would be moved consecutively through the chambers of the kiln.

Brickmaking by hand. Removing bricks for stacking and drying.

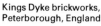

Kings Dyke brickworks, Peterborough, England.

medieval work in Denmark. Long thin bricks are sometimes used today for their aesthetic effect. The Romans mostly used a large, thin unit, more like tiles than bricks, often about 18 × 12 in. (45 × 30 cm) and 1 or 1.5 in. (2.5 or 3.8 cm) thick. The Romans also made triangular-shaped units as a permanent formwork material for masonry walls with concrete cores. Recently, attempts have been made to produce much larger bricks in order to speed up the laying process. Furthermore, few modern bricks are completely solid; with pressed bricks there is a "frog"—an indentation on the top face of the brick—while most extruded bricks have a series of vertical holes through them. These holes help reduce fuel consumption in the firing process by exposing a much larger surface area of brick to heat. Either method reduces weight without seriously affecting strength.

Early bricks tended to be somewhat irregular in shape and size and wide mortar joints were necessary to take up the variations; but nowadays most types of clay and calcium silicate bricks are regular in shape and size and thin jointing is then possible. From about 1600, in England, a particular type of "rubber" brick was made. Being homogeneous in composition and relatively soft enabled it to be rubbed to shape. Arch bricks were rubbed to slightly wedge shape, but rubbers were also used to enable very regular and thin jointed brickwork to be laid. Rubbers and normal bricks were sometimes used as contrasting materials on the same building, as at Hampton Court, England (c. 1690) by Sir Christopher Wren.

Use of bricks in buildings

As a decorative form, glazed bricks have long been utilized. Many of the very early buildings in the Near East were faced with a glazed material, sometimes bricks with a glazed surface, sometimes thinner, glazed tile units. In medieval work some decorative bonding patterns were made more obvious by selecting heavily fired bricks which had fused to a glazed surface. Applied white and colored glazed finishes became popular in the late 19th century for use as light-reflecting surfaces in internal courtyards of large buildings and for wall finishes in such places as dairies and rest rooms where hygienic, easily cleanable surfaces were required.

As with any type of construction, changes in method and results come from two directions: the demand for new building uses and the changes in properties of the materials and in knowledge of how to use them. It must be noted that for brickwork the components are bricks *plus* mortar. In the earliest work, large structures were either solid, or, if built as walls, were very thick. Unnecessarily massive construction was gradually refined by a pro-

into molds. An outstanding example of the advantages this provided was the start of the "Fletton" brick in the Peterborough area of England at the end of the 19th century. By a process of compression, this enabled a clay with a high carbonaceous content to be used, and this provided a large part of the fuel required for firing and therefore a reduction in cost. Fletton bricks now form roughly half of the total output in Britain.

Manufacture of an alternative to clay bricks started in 1881 in Germany. A mixture of lime, silica sand, and water, plus coloring material, was mixed and pressed to shape and then steamed under pressure in an autoclave. This calcium silicate or "sandlime" brick is made in most European countries and in the U.S. and in other countries such as Taiwan. Where good clay is readily available sandlime bricks form a relatively small part of the brick production; in Britain 346 million were produced in 1968.

Brick shapes and sizes

Ease of handling and adaptability have always been important aspects of bricks; and the units have tended to be usually about 4 in. (10 cm) wide, which gives a suitable hand grip, with their length being about two times their width. Much larger sizes have been used from time to time, however: up to 16 in. (40 cm) long in early Persian examples; the "Great Brick" in England following taxation based on numbers of bricks; the "Monk" brick in cess of trial and error plus the occasional introduction of some uncalculated "engineering" improvement. Overall bulk was reduced by forming local buttresses to provide stiffening to long or high walls. An interesting refinement to this is to build a wall which is "wavy" in plan. Walls of this type can be seen at the University of Virginia, Charlottesville in the U.S.

Wall building is a relatively simple process but small building units present a problem when bridging openings. Stepped "corbeled" brickwork was a possibility but the arch was the real answer. The oldest known brick arches date from about 3000 BC. They are the vaulted culverts of the Akkadian Palace at Eshnunna in Mesopotamia. These techniques spread to Egypt and other regions. The Ramseum at Thebes (1292–1125 BC) used brick vaults similar in form to those still used today in the Nile Valley. Brick vaulting has continued to be used and perfected over the ages, especially in areas where timber is scarce. From the earliest times ingenious techniques of laying bricks in vaults and domes were developed which eliminated or reduced the use of elaborate temporary centering. It was used with great confidence by the Romans in combination with concrete, and was developed to a high level by Catalan masons in Spain who carried their techniques to Mexico where these became a vernacular form of construction in some areas.

Throughout the early periods of development in constructional methods the possibilities of decorating buildings simply by the arrangement of bricks was seen. Between the 10th and 12th centuries bricklaying skills and decorative design rose to a peak in such forms as Persian minarets and tomb towers, and in the many and varied treatments of domes, for example the Friday mosque at Isfahan.

After the development of the vault and dome, building changes were gradual and more a matter of design style rather than of fundamental constructional change. Two aspects worth noting for their legacy of memorable architecture are first, the use of brickwork as infill material in timber-frame buildings in European medieval work, and second, the use of rendering as an external finish. The reasons for applying a cementitious material to the external face of brickwork may be either utilitarian or aesthetic. For small buildings, such as most domestic work, quite thin brick walls provide adequate strength but built as solid walling they are not completely proof against rain penetration in wet climates. Their weather resistance can be improved by an applied rendering. In regions such as Scotland, where long periods of wet weather occur, rendered brickwork became, and remains, a characteristic feature.

An alternative solution for full protection against rain penetration is cavity wall con-

Catalan brick vaulting in the crypt of the Güell chapel, near Barcelona, Spain, by Antoni Gaudí.

RIGHT: Ornamental brickwork using bricks of different colors and a variety of bonding patterns.

Vaulted drain beneath palace at Nimrud (c. 8th century BC).

RIGHT: Modern polychrome brickwork used on the Byker Wall, Newcastle-upon-Tyne, England (1977), by Ralph Erskine.

Serpentine or "wavy" brick wall at Lymington, Hampshire, England.

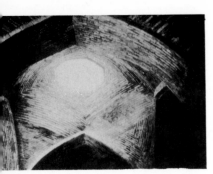

Brick vaulting at Masjid-i-jami in Isfahan (1088).

struction. Scattered examples of this can be found in Europe and North America from about 1880. In Britain cavity construction, at least for small houses, became general by about 1920. In addition to providing better protection against rain penetration, the cavity increased the thermal insulation value of the walling. Thermal insulation can be further improved by filling the cavity with very lightweight insulating materials but care is needed to ensure that this does not spoil the effectiveness of the cavity as a water barrier.

Rendered finishes have also been widely used for aesthetic reasons. Outstanding examples of this can be seen in many regions: in Italian Renaissance palaces and villas; in decorative French work of similar date; or in England in works such as the stucco finishes of the Adam brothers in the 18th century. These coverings were often detailed to imitate expensive stone construction.

The Industrial Revolution brought with it a demand for large factories and warehouses with the need to carry heavy floor loads and to be as resistant as possible to fire. Floors of a composite construction of shallow brick arches, supported on cast-iron beams, were developed in response to this new requirement.

Over the centuries manufacturing methods and constructional knowledge gradually improved but basic scientific research on materials was lacking. Around 1920 research laboratories began to fill this gap, with the Building Research Station at Garston, England, as a notable leader.

Research on the nature and properties of both bricks and mortars had two effects: it pointed the way to making better materials and it gave the rapidly growing work of the structural engineer more reliable data. Quality standards for materials were set up in most industrialized countries and structural codes of practice complemented them. Brickwork structures changed from wasteful rule-of-thumb affairs to properly engineered, calculated work. Although this has affected all types of brick building, its most spectacular effect has been seen in high blocks of appartments and hotels where the small room sizes give a cellular-type building for which brick structure is especially suitable. Some of the earliest work of this kind was in Switzerland, but by 1960 brick structures 18 or 20 storeys high, with thin walls, were widespread. In the U.S. some buildings of this general form used cavity brick walls with the cavity filled with lightly reinforced concrete. This type of wall added to lateral stability, and made feasible the use of wider span floors in conjunction with load-bearing brick walls.

For at least the past 50 years there have been repeated arguments that the bricklaying process is an absurdly antiquated method for producing buildings. Nevertheless, people prefer brickwork, and so far few of the many "prefabricated" home construction alternatives have proved to be very serious competitors. The chief exception is where timber framing is used. Although widespread as a traditional system in the U.S., this has recently achieved some popularity in various European countries. Even so, at least in traditional brick areas, the timber frame is often surrounded by an external brick finish. Nevertheless, the brick industry is aware of competition and of the need for constant development. At the manufacturing end, this has resulted in the closure of many small brickworks and developments such as tunnel kilns, automation within the works, and advances in handling and transportation areas. On the constructional side there has been a good deal of experiment on prefabrication of brick panels—notably in the U.S. and in Holland. This, and possibly more use of reinforced brick construction, seems the likely direction for development in the immediate future.

Terra-cotta

Terra-cotta is most familiar to us in its common use as reddish-brown clay flowerpots. However, in many places in the world it continues to be used for roof tiles on vernacular forms of houses, and, particularly in its glazed form, for decorative purposes. In the past it was extensively used to create splendid and elaborate decorations for buildings of every type.

Terra-cotta is made from yellow to brownish-red clays of a fineness and uniformity somewhere between the clays used in bricks and vitrified wall tiles. The clay is sometimes mixed with fired clay, ground to a powder to reduce the shrinkage of the molded object during firing. After molding the mixture is fired once, if the product is left unglazed, and twice for glazed terra-cotta or faience, the glaze being applied between the two firings. In making faïence, the first firing is carried out at a high temperature and the second at a low temperature.

The Greeks made much use of terra-cotta for roof tiles, cornices, and other ornamentation on buildings, and the Romans also applied the technique to facing slabs and to antefixes at the ends of roof ridges. After the fall of the Roman Empire, the use of terra-cotta declined in Europe, but continued in Turkey, Persia, and the Far East. A great European revival of the craft took place in the 14th and 15th centuries, especially in northern Italy and Germany.

The art of terra-cotta was introduced into England from Italy in the early 16th century. Italian craftsmen made terra-cotta ornaments for Hampton Court Palace and for other great houses of the period. It remained a con-

Terra-cotta facade of the Natural History Museum, London (1887), by Alfred Waterhouse.

The terra-cotta and brick facade of the Russell Hotel, London (1898), by Fitzroy Doll.

Detail of terra-cotta work on the facade of the Russell Hotel, London (1898).

tinental specialty until 1722, when Richard Holt and Thomas Ripley set up a factory at Lambeth, England. In 1769 the premises were taken over by George and Eleanor Coade, who made a very high-quality terra-cotta which became known as "Coade stone". This included a flux and was fired at about 2,012°F (1,100°C). It was very hard, weather-resistant, and greatly superior to ordinary terra-cotta, especially in an urban atmosphere.

In America, Louis Sullivan (1856–1924) used terra-cotta to decorate a number of his buildings including the Wainwright Building (St Louis, 1890–91) and the Prudential Building (Buffalo, 1894–95). One of New York's early skyscrapers, the Woolworth Building (1913), designed by Cass Gilbert (1859–1934), was also clad in terra-cotta. On the West Coast of the U.S., glazed terra-cotta (faïence) medallions and decorative tiles remain popular ornaments for mission-style buildings. This type of decorative glazed terra-cotta is also popular in Italy and Spain, where it is still produced and widely used.

A great deal of old terra-cotta and faïence is still to be seen in large cities, mainly dating from the 19th century. Much of what was produced in Victorian times was fired at too low a temperature. It deteriorated quickly and architects became wary of using it for decorations on a large scale. When well-fired, it has proved durable in urban atmospheres as can be seen in the facades of the Natural History Museum in London (1868), designed by Alfred Waterhouse (1830–1905). The terra-cotta blocks must be relatively small so that distortions which occur in firing do not become too exaggerated. Since the material also shrinks about one-twelfth during this process, all the elements have to be designed with this in mind. In the 19th century it became common to make terra-cotta blocks hollow with walls up to 1 in. (25 mm) thick. The blocks were often designed to fit around structural elements such as iron or steel frames, which added to their fireproofing, or to be closely linked to a brick backing wall. Sometimes the blocks were filled with concrete.

Roofing materials and tiles

Roofs are the parts of buildings most exposed to the elements. The creation of a surface capable of shedding water adequately has always been one of the most taxing tasks for builders in all but the most arid climates. In all other regions, sloping roofs of one kind or another have been used almost universally until relatively recently, when dependable and cheap methods of flat roofing became possible.

The traditional roof forms of any particular region depend on the materials available and

Traditional flat roofs in a village in central Syria.

Conical roof being woven out of straw in southern Mozambique.

the techniques employed in their use, together with their degree of exposure to wind, snow, and rain. Vegetable materials, stone slabs, and animal hides were certainly the first materials used to produce water-shedding surfaces.

Vegetable materials

Thatching is one of the oldest and most widespread forms of roofing. Thatched roofs consist of bundles or layers of reeds, straw, grass, or heather laid onto a sloping framework. The slope and the thickness of the material work together to form a waterproof covering.

Roofs thatched in straw that has been badly threshed often retain seeds which attract birds and rodents, thereby considerably reducing the life of a roof. Mechanical threshing renders straw useless for thatching. A roof expertly thatched with reeds specially grown for the purpose may last up to 100 years.

Thatched roof in southern Mozambique.

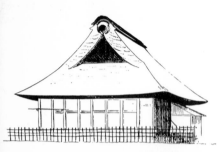

Traditional Japanese thatched roof from a building near Tokyo; from a late 19th-century drawing.

Roofs of plaited palm fronds in northern Mozambique.

Manufacture of roofing slates at a quarry in Wales (late 19th century).

Complex shapes can be created with thatch since it can be laid with ease around valleys, dormers, and ridges in gently flowing lines, without the introduction of other materials. Thatch has a high insulation value but is extremely inflammable.

A combination of layers of overlapping strips of birch bark, covered with grass sod, moss, or other protective layers, was a traditional roofing method in parts of northern Europe, while in areas of the world within the palm tree belt, roofs are often made with "tiles" of woven or matted palm fronds.

Shingles made from split or sawn log drums are an old form of roofing material. Many of the buildings of ancient Rome were shingled. Shingles vary in length and breadth, but are normally wedge-shaped in section. Round or patterned shapes were sometimes cut on the ends of shingles used to roof important buildings in northern Europe, as can be seen on some of the stave churches in Norway. Shingled roofs are light and relatively durable. They may last up to 60 years if they are laid properly to an adequate slope and employ a wood such as western red cedar, which has a suitable grain and oils that inhibit the growth of fungus.

All vegetable-based roofing materials are inflammable to a greater or lesser extent, making their use hazardous in tightly packed urban areas. There are many records of Roman and medieval legislation which restricted or prevented the use of these materials in congested towns.

Stone

Overlapping slabs of stone may form a watertight surface in much the same way as shingles. Slate is a popular roofing stone wherever it is found, since it splits easily along its bedding planes into thin slabs. Slates are generally cut or shaped to regular, often rectangular shapes and fixed to battens by some form of nailing. In some places however, such as in Galicia in Spain, irregular slabs of varying sizes are laid over each other to form very distinctive roofs.

Stone flags or stone tiles made from sandstones or limestone were a common method of roofing in many parts of the world. Roofs covered in these stones are heavier and generally coarser than those covered in slate. The stones were often rounded at the top to reduce weight and were fixed to battens or laths with wooden pegs or animal bones. It is usual to find larger slabs near the eaves with diminishing courses up to the ridge. "Mortar torching" on the inside of the roofing slabs was generally used to prevent wind and snow being blown in through the spaces left between the stones.

Until the advent of cheap and reliable waterborne transportation, slate roofing was restricted to areas close to quarries. In Britain, these 18th-century improvements made slate a strong competitor with most other roofing materials on account of its relative lightness and durability and the fact that it could be laid at a low pitch. Many of the towns that sprang up during the Industrial Revolution are almost entirely roofed in slate.

Clay tiles

Clay tiles are thought to have originated in China. Tiles dating back to 1000 BC are known from remains found at the temple of Hera at Olympia in Greece, and they were introduced into most parts of southern Europe by the Romans. Their use died out in many places with the collapse of the Empire only to be revived in the Middle Ages. These early tiles depend on two elements to achieve a waterproof surface. Wide, slightly curved slabs (*tegulae*) were laid side to side with their ends overlapping the course lower down. The gap between the tiles was covered with overlapping semicylindrical elements (*imbrex*), to prevent water entering the joint. Forms in this family are known as "Normal," "Asiatic," or "Roman" tiles.

Tiling systems based on the imbrex and tegula principle developed characteristic local forms throughout the ancient world south of about 44°N. In some cases, a narrow tegula and a wide imbrex were employed, while in some Roman and Greek forms, the tegula has a flat bottom and vertical sides. The most common and simplest form of this type of tile is one in which the imbrex and tegula are identical, semiconical elements up to 2 ft. (60 cm) long, that fit loosely over each other. This is the traditional tile of the Mediterranean region, and its use extends over large parts of Asia. It is made by various processes—it can be formed over a master mold, it can be made by cutting a cone thrown on a potter's wheel down the middle, as is the practice in parts of India, or it can be made by machine. Decorative molded ridge and eaves tiles were often employed to add interest to roofs. Imbrex and tegula tiles can be laid loosely on a substructure, and in exposed situations stones or other forms of weight may be placed on the roof to prevent tiles being blown off. In China and Japan roof tiles are often bedded in mud.

Clay tiles which are laid in a similar fashion to shingles are the common form in northern Europe. The tiles are laid in regular courses, but in order to protect the joints between tiles, each tile overlaps two others, leaving only about two-fifths of the surface exposed. In order to shed water and snow, roofs of this type have high pitches, generally exceeding 45°. Clay tiles are also used as a cladding for walls. These tiles are thought to have originated as a more durable and less inflammable substitute for shingles. This, the "Germanic"

The slate roofs of a French town in the Loire valley.

Gauge controlling the dimensions of roof tiles, from the market place in Athens (5th century BC).

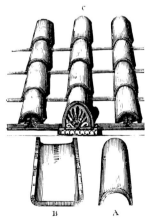

Ancient Greek terra-cotta roofing tiles with decorative 'antefix' closing piece.

or shingle form of tile, is known in England as the plain tile. These tiles often have a marked camber in order to prevent capillary infiltrations, and are held in place by nailing or by ribs molded onto the back of the tile. Because of the double lapping, roofs of this type are fairly heavy. The tiles vary in size from region to region, but their dimensions in England were standardized by Act of Parliament in 1477 as 10.5 x 6.5 in. (262.5 x 162.5 mm).

Steep and complex roof shapes, characteristic of northern Europe, are easily achieved in this tile by using specially formed ridge and valley elements. Varying textures and patterns are possible by using tiles with decoratively shaped, exposed ends and by using multicolored tiles.

Before the Industrial Revolution, a third family of tiles developed. These are known as pantiles or "Belgic" tiles. The tiles are S-shaped and are formed in such a way that each tile covers the side joint with its neighbor. The amount of overlapping is greatly reduced with this type of tile, making roofs considerably lighter than with the plain tile. Roofs can be laid quite safely at lower pitches—down to 30°—thus reducing the amount of roof framing required. This type of tile is best suited to simple roof forms.

The pantile may have developed in Holland or Belgium, where it replaced other types of roof covering. It was introduced into England in the 17th century where it became popular in the eastern counties. Its dimensions were fixed at 13.5 x 9.5 x 0.5 in. (337.5 x 237.5 x 12.5 mm) during the reign of George I. This tile is extensively used in Scandinavia and was introduced into Java by the Dutch. Glazed pantiles have occasionally been popular. A similar form developed, perhaps independently, in Japan. It is interesting to note that the Japanese form laps to the left when viewed from the ground, whereas in other countries lapping is invariably to the right. The Japanese "Yedo" tile, found in Tokyo and Kyoto, is often finished at the eaves in such a way that it looks like an older form of imbrex and tegula tile.

During the 19th century a large number of different, specially shaped, interlocking clay tiles were designed and patented. By the 1830s a variety of nontraditional shapes were available in France. A special form known as the "Marseilles" tile was being widely used in France and exported throughout the world by the 1860s. Most of these new forms were made on accurate molds by hand or produced by machinery. Their profiles are usually complicated, having systems of grooves and ridges intended to reduce the amount of overlap necessary to achieve a weathertight surface. These tiles can generally be laid without risk at considerably lower pitches than the traditional forms, in some cases below 15°. Because they often have a thinner overall

Plain tiles used on a house in Sussex, England. Complex roof shapes are possible in this tile.

Pantile roof.

Shingled roof on a stave church at Borgund, Norway (1150).

Mass production of concrete roof tiles.

The first corrugated-iron roofs were used in these warehouse sheds at the London Dock (1829), by Henry Robinson Palmer.

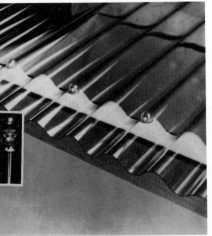

Vinyl corrugated sheeting for roofs. Insert shows drive screw, vinyl washer, and cap.

thickness and require a minimum of overlaps to shed water, these tiles are generally lighter and therefore reduce the roof framing necessary to support them.

Concrete tiles, imitating the forms and colors of traditional and modern interlocking tiles, have been manufactured since the 1920s. These tiles have gained popularity, as they are often considerably cheaper than clay ones and can be made to accurate tolerances. Most tiles used today are laid over a lining of felt or other waterproof material to reduce drafts in the roof space and to reduce the danger of leaks if tiles are dislodged.

Greek tiles made out of marble are known from the 6th century BC, while the use of bronze tiles is recorded by Pliny. The British Houses of Parliament, designed by Charles Barry (1795–1860), are roofed in large cast-iron tiles.

Corrugated sheeting

Corrugated wrought-iron sheeting, invented by H. R. Palmer in 1829 for roofing warehouses in the London Dock, brought a new form of roofing material onto the market. Until his invention, all roofing needed considerable quantities of ever-diminishing pieces of framing to support small scale-like elements that shed the water. Alternatively, roofs had to be close-boarded so that thin metal or other non-self-supporting waterproof sheeting could be laid. Corrugated iron, the first self-supporting light sheet material, required only minimal framing and could be made to span between purlins, thus eliminating rafters, battens, and boarding.

Corrugated iron became even more attractive when it could be given a protective layer of zinc by hotdip galvanizing, which became commercially available in the 1840s. It is an easy material to transport, as it packs together tightly, and can be laid rapidly by relatively unskilled labor—hence its popularity for military hutting, agricultural and industrial building, and any other large enclosures.

Self-supporting, flat, or curved sheeting is made in many other materials with a great variety of surface treatments. Both steel and aluminum corrugated sheets are supplied with specially applied paint surfaces or plastic sheet coatings. Aluminum may have special anodized surface treatments that increase its life and enhance its appearance. Transparent and translucent sheets are made in a variety of plastics, and corrugated asbestos sheeting has been available since about 1910.

Other roofing materials

Sheet lead and copper have been used as roofing materials since antiquity. They can be laid relatively flat, but are expensive and require a flat or boarded surface for support.

Roofs may be waterproofed with hot-laid asphalt or other liquid materials that set to form a watertight surface. Roofing felt, waterproofed with wood tar, was introduced in Sweden in the 18th century. Asphalt roofing became popular in the 19th century when natural deposits of bituminous limestones were commercially developed. Lake Asphalt from Trinidad has a high reputation, together with deposits in France and Switzerland, from which numerous patent compositions emerged.

Bituminous felt is a popular roofing material of relatively recent origin. It is normally laid on flat roofs as a composite of a number of layers, carefully sealed with pitch. Bituminous-felt shingles are a widely used material for sloping roofs.

Plastic-based materials, such as hypalon, have been developed for roofing. In many cases, joints between sheets are welded together on site.

Wall and floor tiles

Decorated wall and floor tiles were made in Egypt as early as the 3rd dynasty. The Assyrians and the Babylonians had glazed wall tiles and bricks in the 9th century BC, with designs painted on the surface with colored glazes. This technique died out in the 4th century BC and neither the Greeks nor the Romans used tiles for decorative purposes.

The Romans made use of both marble and clay tiles for their houses and civic buildings, sometimes with mosaic infilling. The use of glazed tiles was revived in the Moslem countries, as a result of contact with China. The Persians made both luster and mosaic tiles from the 13th century onward, the mosaic consisting of large monochrome tiles cut up into small pieces and reassembled to form complex designs. The method reached its peak in the 14th century and outstanding examples can be seen at Isfahan, in Persia, and at Samarkand, in Uzbekistan.

In Persia, mosaic tiles were gradually abandoned in favor of painted faïence, a fashion which spread throughout the Islamic countries of Asia and North Africa and eventually to Spain, where it can be seen in the Alhambra at Granada and the Alcazar at Seville. Decorated tiles were rarely used in Europe, outside the Iberian Peninsula, until the end of the 12th century, and then only for floors. The Italians particularly favoured marble mosaics, but in northern Europe the cheaper tile mosaics and inlaid tiles were preferred. The inlaid tiles were made by pressing a carved wooden design on the surface of the dried tile before firing and then filling the design with clay of a different color.

Medieval English encaustic floor tiles. A stamp bearing a design in relief was impressed upon them, leaving an ornamental pattern on the tile. Before fixing, the depression would be filled with a clay of another color. Tiles of this type were mass produced in the 19th century.

Design for plaster ceiling from a 19th-century plasterers' handbook.

Decorated glazed wall tiles were made in the Low Countries from the 14th century onward. Majolica tiles were being made there and exported in the early 16th century.

Concern for hygiene in the 19th century led to the development of many forms of mass-produced glazed tiles. These accurately made tiles, with special elements for corners and angles as well as other moldings, made it possible to make easy-to-clean interiors for toilets, dairies, butchers' shops, and the like.

Plaster

For at least 3,000 years builders have experimented with methods of covering masonry and the various types of earth walling with a smooth layer which would protect the material underneath, improve the appearance of the building, and, in some cases, make it easier to apply decoration. Gypsum plaster has often been favored for the purpose.

Calcium sulfate plasters are made either from gypsum or from anhydrate. By calcining gypsum at 266°–338°F (130°–170°C) for about three hours, three-quarters of the water is driven off and plaster of Paris, or hemihydrate plaster, is formed. At 752°F (400°C) all the water is driven off and anhydrous sulfate is produced, much the same result being obtained by grinding the raw mineral anhydrate.

Hemihydrate plaster, made by roasting gypsum in shallow pits, was made by the Egyptians, who preferred it to lime mortar or lime plaster, because wood fuel was scarce and lime burning needs three times as much fuel as gypsum burning. The Egyptians used gypsum plaster to cover brickwork and stone as a means of producing a surface suitable for painting. Both the pyramids at Giza and the tombs at Saqqara contain gypsum plaster used in this way. Gypsum is soluble in water and is consequently a poor material for outside use in countries which have high rainfall. Gypsum plaster was introduced into England from France in the 13th century; hence its popular name, "plaster of Paris."

To make plaster of Paris suitable for building work, small quantities of keratin, made by boiling horn, hoof, or animal hair in caustic soda, have to be added to retard the set. The anhydrous calcium sulfate, on the other hand, needs an accelerator. This resulted, in the early 19th century, in the marketing of a number of brand name plasters. Until the 20th century, however, plaster of Paris was rather expensive, and it was consequently reserved for providing the walls and ceilings of important houses and civic buildings with a smooth, hard finish, sometimes with ornamentation.

These ornamental cornices, ceiling roses, festoons, niches, medallions, and the like were often made in factories and sold to builders via catalogs. For more ordinary purposes, both inside and outside buildings, the plaster was usually made of a mixture of lime and sand, with a little ox or cow hair added as a binding agent, to prevent the plaster from cracking.

During the 15th and 16th centuries, the Italians began to study Roman techniques of plastering and fresco painting. They used a *stucco duro*, a mixture of air-slaked lime and marble dust, with a little gypsum added in order to help it set. Henry VII's Palace of Nonsuch contained stucco decorations, portions of which have been discovered during excavations, still in excellent condition.

Much more elaborate plaster ornamentation was carried out in England from the 17th century onward; the influence of Inigo Jones (1573–1652), Sir Christopher Wren (1632–1723), and Grinling Gibbons (1648–1720) being very marked. Gibbons in particular specialized in plaster sculpture with wires, twigs, and strips of lead used to reinforce the representations of fruit and flowers. English craftsmen included a wide range of substances in the plaster mix, such as fruit juices, beer, cow dung, blood, cheese, milk, and beeswax to improve the quality.

In the mid-18th century, architects began to favor another group of stucco mixtures, known as oil mastics or oleaginous cements. Two of them, patented in 1765 and 1773, were bought by the Adam brothers, who marketed them, first as "Adams Patent Stucco" and later as "Adams Cement." This substance was used as a rendering on the fronts of a number of buildings in Bedford Square and other prestigious London building projects for which the Adam brothers were responsible. Other later compositions were patented by Dihl in 1815 and 1816 and by Hamelin in 1817. Dihl's cement contained linseed oil, lead oxide, china clay, and ground brick, thinned with turpentine. Hamelin used a mixture of powdered limestone, brick dust, sand, lead oxide, and linseed oil. Both of these cements were used by the architect John Nash (1752–1835).

Expanded metal lath has generally replaced the wooden laths that were traditionally used as grounds for plastering. Metal lath is often used in suspended ceilings when complex shapes need to be built up. Plaster can be given various special properties by incorporating special additives; it can, for example, be made with a fibrous content to increase its properties of sound absorption.

Since the end of the 19th century, with the invention of gypsum board and the decline of elaborate decoration, wet-laid plaster has gradually been replaced by this dry material for the lining of interiors in frame buildings. In some countries it is now common to use a portland-cement-based mixture for internal plastering of masonry construction for all but the finest work.

Lime kilns (c. 1870) shown in section; LEFT—empty, RIGHT—ready for firing.

Portable steam-operated mortar mill (c. 1880).

Roman concrete wall using puzzolanic mortar combined with a stone facing and rubble fill, at Hadrian's Villa.

Mortar and cement

The most primitive type of walling consists simply of stones placed one on top of another, with no bonding material to fill the joints and hold them together. This technique, in the hands of a good craftsman, can produce a surprisingly strong wall; but it allows damp and wind to penetrate easily and cannot normally be used for structures much more than 8–10 ft. (2.5–3 m) high. For anything more ambitious or complex, the stones or bricks have to be firmly attached to one another, and mortar is essential. A mortar made from sand and slaked lime was common in classical times, and it was the only kind to be used throughout the Middle Ages and until comparatively recently.

Many experiments were made during the 18th and early 19th centuries in order to discover ways of improving lime mortar. The traditional method was first to burn chalk or limestone, and then to mix it with sand and water. This produced a mortar which hardened as the calcium hydroxide converted into calcium carbonate. The process was slow and affected only the outside of the mortar layer, so that the bond was relatively weak. If some form of siliceous matter, such as volcanic ash or burned clay, was added to the mixture, a much stronger material was obtained, with the whole mass of the mortar becoming more resistant to rain or seawater. The Greeks and Romans made much use of these siliceous additions or *pozzolanas*—the name comes from Pozzuoli in Italy, where a natural source of such material exists in the form of volcanic earth. This is one of the main reasons their structures have lasted so long. The Romans used both natural pozzolanic material and crushed bricks, tiles, and pottery. With the help of these strong mortars, they were able to design buildings with much thinner walls, and to construct arches and vaults with complete confidence.

Water-resistant or hydraulic lime mortar can also be obtained by burning a mixture of limestone and clay, a discovery made in the second half of the 18th century. In 1754, John Smeaton (1724–92) visited the Netherlands and observed the successful use of what was known as tarras mortar—a mixture of pozzolanic earth and slaked lime—in the construction of harbor works and sea defenses. Smeaton experimented with different types of limestone and found that burned Aberthaw blue lias (limestone), which contained clay, produced a cement which hardened effectively. In the construction of the Eddystone lighthouse he used a mixture of Aberthaw lias and pozzolana from Italy.

The demand for stucco as a facing for buildings and the greatly increased scale of civil engineering works—canals, harbors, bridges, and railroads made it essential to have reliable hydraulic cements. Inventors realized their opportunity, and by 1850 three types of cement were available for making a mortar which would set in a damp atmosphere or in the absence of air. These were the so-called natural hydraulic cements, the artificial or proprietary cements, and the true portland cements.

The natural cements were made by burning stone which contained a suitable mixture of lime, alumina, and silica. Several patents for these cements were taken out between 1790 and 1830. One of the best known was James Parker's (1796). It was made from the septaria or nodules of argillaceous limestone which were found in large quantities along the northern shore of the Thames estuary, particularly around the coast of the Isle of Sheppey. So much of this material was taken from the foreshore that in 1825 the British government prohibited any digging closer than 50 ft. (15 m) from the cliffs. Parker misleadingly called his product Roman cement. After his patent elapsed in 1810, Roman or natural cement used material from several sources along the eastern and southern coasts of Britain. Parker's cement was used a good deal by both Thomas Telford (1757–1834) and Marc Isambard Brunel (1769–1849).

In 1818 the American canal engineer, Canvass White, discovered a source of natural hydraulic cement near Sullivan, New York. These local deposits of clayish magnesian limestone became known as cement rock. White used it originally for facework on the walls and aqueducts of the Erie Canal, but by 1850 it had been adopted throughout the U.S. for many different kinds of structure. Between 1819, the date of White's patent, and the middle of the century, natural cement was discovered at a number of sites along the eastern seaboard, in Kentucky, and Illinois. The best product came from Rosendale, New York. Rosendale cement proved superior to White's cement, and it was the most favored by builders and engineers until the establishment of an artificial cement industry in America during the 1870s.

The artificial or proprietary cements were made by mixing limestone or chalk with clay or shale, by rule-of-thumb methods, and then burning the mixture at a temperature of between 2,012° and 2,372°F (1,100° and 1,300°C), which achieved partial vitrification of the material. Joseph Aspdin's so-called portland cement, for which he was granted a patent in 1824, probably belongs to this group although the patent says nothing about kiln temperature. All that is known of his method is that he burned a mixture of limestone and clay at a temperature high enough to produce a glassy clinker. When this was ground to a powder, it yielded a stronger and more reliable cement than had hitherto been achieved. It is impossible to say whether Aspdin was the first to produce portland cement, but he certainly originated the name, choosing it because of its resemblance to Portland stone which enjoyed a high reputation for quality.

By the 1850s the true portland cements constituted a third, distinct group. They consisted mainly of calcium silicates, produced by raising the material to a temperature of around 2,500°F (1,370°C) in order to have as complete a reaction as possible between the lime and the silica, although this was discovered by trial and error and the chemical processes were not understood at the time. The first reliable portland cement was produced in 1845 by I.C. Johnson in Kent, England. Portland cement works, to be economically viable, had to be close to readily available sources of chalk and clay.

Cement kilns were originally of the intermittent type. The first continuous kiln, the Dietsch kiln, was introduced from Germany in 1880. The first successful rotary kiln, which allowed the raw materials to be fed into the kiln as a slurry without the need for previous drying, was not in operation until 1900.

Scientific cement production was pioneered in Germany. By 1875, the best German cements were reaching compression strengths 80% higher than in 1860 and tests made in 1885 showed a further increase of 60% compared with 1875. Without these greatly improved cements, reinforced concrete construction would have been difficult.

Since the 1880s, a number of specialized portland cements and cement additives have been developed. These include cements that harden rapidly, evolve less heat of hydration than normal, resist various forms of chemical attack, have water-repellent or waterproof characteristics, and many other properties. White cement is manufactured in special kilns from white china clay and white limestone. Colored cements can be made by adding pigments to white or gray portland cement. Cements may be made resistant to rapid deterioration in moist atmospheres by coating the particles with water-repellent film—the film is rubbed off when the cement is mixed with the aggregate and hydration takes place normally. Additives to cement include products that make mixes more workable, requiring less water to be used. Accelerators increase the rate of setting and strength development. Retarders, on the other hand, reduce the rate of setting and have various useful applications.

High-alumina cements were evolved to develop early strength, resist certain chemicals, and resist high temperatures. These cements have been misused in some situations which have led to structural failures. Supersulfated cements also have special chemical-resistant properties and evolve comparatively little heat in setting which makes them useful for mass concrete in hot climates.

Concrete

Roman concrete

The invention of concrete is normally attributed to the Romans who revolutionized building by its use. It is conjectured that its special properties were discovered near Putoli (modern Puzzoli near Naples) where the inhabitants used local volcanic dust or sand as a filler combined with lime mortar. This "filler" material improved the quality of the mortar by making it set harder and faster as well as underwater. Similar types of sand are found over a large area of Italy, south of Lake Bolsena, and their varying properties became known by slow experiment over the years. A concrete made with similar mortars and stone aggregate was used from ancient times on the Greek island of Thera, but its use did not spread throughout the Greek-speaking world.

In 199 BC, the harbor of Putoli was built using pozzolanic material which allowed the cement to set underwater. Concrete was used in many buildings and civil engineering projects, and by the Augustan era its use was generalized. Roman bridges often consisted of an arch of single stone thickness with spandril walls built of masonry coursed as headers and stretchers, with the headers projecting into the core of the structure which, when completed, was filled with concrete to form a well-bonded monolithic mass. This method greatly reduced the labor in making the falsework, by using elements that only had to be strong enough to support one ring of stones at a time and could be used again, an advantage since shuttering would invariably be made out of timber, which was often scarce. The technique of building concrete arches also vastly reduced the work in producing precise masonry.

By the end of the Imperial regime, the Romans were using concrete in bridges and other structures without relying upon any other masonry structural members. Large Roman walls were commonly built with two facing skins of stone or brick with a core of concrete. A special form of flat triangular brick was often used with its point projecting into the concrete core in order to bond it in. These became especially popular after Tiberius, when large centralized brickyards could provide regular supplies. This facing was known as "opus testacium." Other types of facing commonly used form the bases upon which Roman structures can be classified and dated—these include "opus mixtum" and "opus reticulum." Roman concrete was normally made by pouring pozzolanic cement on layers of small broken stones (*caementa*) and repeating the process until the structure was filled.

Baths and utilitarian buildings such as warehouses and markets had been vaulted in

Ruins of the Baths of Caracalla in Rome (AD 211–217). The piers have lost their brick facing skins, exposing the concrete core.

concrete since the 1st century BC. The fire which destroyed large parts of Rome in AD 64 in Nero's time helped establish regulations which attempted to impose strict controls on construction. Buildings were restricted in height to 70 ft. (21 m), they had to be structurally independent of one another, and the use of inflammable materials was severely restricted. Concrete was the material suited to answer these needs. New concepts of space made possible by the use of concrete were evolved and absorbed over a very short period, completely revolutionizing the forms of enclosed structures except in the most conservative areas of religious building where traditional forms persisted, often as a thin veneer over a concrete core. Domes, vaults, and extensive walls became the elements of a new architectural vocabulary which was used with great skill and confidence on a large scale that would have been difficult in any other material.

In the period after the fire, the use of concrete spread to an ever-increasing range of buildings including the famous Golden House of Nero, built in AD 64, where the architects Severus and Celer used concrete walls, domes, and vaults to great effect.

The Pantheon, which was built in AD 118–128, illustrates the degree to which Roman concrete construction could be taken. The hemispherical dome has an internal diameter of 143 ft. (43 m), which continued to be the largest span for this type of structure until well into the 20th century. At the base, the walls are 20 ft. (6 m) thick, narrowing to 4 ft. (1.2 m) at the crown where a 30 ft. (9 m) oculus was open to the sky. The aggregate material (*caementa*) was varied at different levels in the construction. At the lower levels, large tufa stones were incorporated to help bear the heavy load, while further up lighter stones were used culminating in pumice and hollow clay jars, which considerably reduced the burden of the dome. At the crown, the material has two-thirds of the weight per unit volume of the material in the lower parts of the structure. There are complex systems of brick arches that go through the body of the dome to strengthen it, a slightly unusual feature in Roman concrete construction of this date.

Large public buildings such as the many baths, amphitheaters, forums, and basilicas were built out of concrete. The Baths of Diocletian erected in AD 302 are a good example. For smaller buildings, including domestic apartment complexes, the surviving remains of Ostia, largely built after the end of the 1st century AD, provide an idea of the extent to which concrete was employed.

Concrete continued to be widely used until the end of the Roman Empire. The vaults and arches at the lower levels of St Sophia in Constantinople (AD 540), are made in concrete.

The Eddystone Lighthouse (1756) built by John Smeaton using carefully chosen hydraulic mortars. It was reerected at South Point of Hoe in 1882.

The revival of concrete

During the Dark Ages, the use of concrete died out except in isolated areas where its working was passed from father to son. Interest in concrete was revived after 1414 when a manuscript of Pollio Vitruvius was discovered in a Swiss monastery. Vitruvius, who completed his famous books on architecture in about AD 27, discussed the properties of concrete and the various forms of pozzolanic earths at considerable length. The first edited version of his work was printed in 1486. Fra Giocondo (c. 1433–1515), who edited a text of Vitruvius which was published in Venice in 1511, used pozzolanic mortar in the pier of the Pont de Notre Dame in Paris in 1499. He claimed to have made the first recorded use of concrete since Roman times. Old workings of pozzolanic material in the area between Koblenz and Cologne may cast some doubt on this statement. This material, known as Rhenish trass, was used for many of the protective works of the Low Countries.

John Smeaton (1724–92), who began to erect the Eddystone Lighthouse in 1756, needed a strong hydraulic mortar to help bind and point his dovetailed blocks of Portland stone. He conducted extensive research into the properties and methods of making mortars and cements and experimented with Italian pozzolana and Rhenish trass which he had seen used in Holland. Both types served him well and his meticulous reports on the work were useful to later engineers. Mass concrete was first used on a large scale in Britain at the West India Dock built in 1800 by William Jessop (1745–1815), who was the son of Smeaton's principal assistant on the Eddystone Lighthouse. It is probable that the cement employed in this case was Parker's "Roman cement" which was first marketed in 1796. It contained natural argillaceous limestone which was found in nodules in tertiary strata near Northfleet where Parker set up a factory to burn and grind the material.

Natural hydraulic cement was discovered in the U.S. by Canvas White in 1818 and was extensively used in the docks, abutment works, culverts, and other constructions associated with the Erie Canal. His ''water lime'' cement continued to be used until the 1890s. Similar deposits were found in many other countries making the use of these cements widespread.

Portland cement was invented by the Englishman Joseph Aspdin (1779–1855) in about 1811. He made his artificial cement by burning a controlled mixture of clay and limestone. He patented his process in 1824 and moved his factory from Wakefield near Leeds, to Gateshead on the Thames. Obadiah Parker from New York developed a similar cement to Aspdin's in the 1830s and built a number of houses in which monolithic walls of that material were used. Isaac Charles Johnson improved (1844 patent) on Aspdin's cement by heating materials of closely controlled chemical composition to a sintering temperature of 2,550°F (1,398°C) evolving a more reliable product, very similar to the type generally used today.

Concrete became a popular material for the foundations of buildings and for other civil engineering works. Early in the 19th century, French engineers used massive concrete blocks (344 cu. ft./9.7 cu. m) for a harbor works at Algiers and Louis-Joseph Vicat (1786–1861), a French polytechnician, developed methods of testing the properties of hydraulic limes; these he published in 1818 and 1828. These scientific studies form the basis of many tests still used today, and made it possible for manufacturers of cement to control the quality of their products carefully and accurately. Other French scientists made further contributions to the theoretical knowledge of cement and its properties throughout the century. Of these, A. L. Lavoisier (1743–94), Le Chatlier, and Feret deserve special mention.

In the U.S., natural cements, or portland cement imported from Britain, continued to be used until David O. Saylor began to manufacture artificial portland cement at Copley, Pennsylvania, in 1871. Many manufacturers of ''artificial'' building stone had sprung up in the U.S. from the 1850s onward, and this was one of the most popularly used forms of the material. G. A. Frear, who set up a company in Chicago in 1868 to manufacture his patent blocks, made many forms including decorative elements and trim that would have been very expensive if hand-carved in stone. His products were used in many Chicago buildings and his influence spread as far as San Francisco. His ''stone'' behaved well in the Chicago fire of 1871, so

In-situ concrete house under construction; from an American drawing of 1886.

Michele's machine for testing cement briquettes (c. 1870).

Precast concrete blocks imitating masonry building elements (late 19th century).

Concrete block-making machine shown with three blocks ready for removal (early 20th century).

The suspension bridge over the Menai Straits, Wales (1815–26), by Thomas Telford. Iron bars are used in the concrete of the abutments.

his business continued to prosper. Burnham and Root's Phoenix Building in Chicago (1885–86) was built with his materials. Other important manufacturers of artificial stones included E. L. Ransome, who set up a company in San Francisco in 1868, and the New York and Long Island Coignet Stone Company which was established in the early 1870s and held the American rights to Coignet's patents.

Reinforced concrete—the early experiments

The combination of concrete, which normally consists of well-mixed cement, sand, and stones, with some form of reinforcing of other materials, has an obscure early history. Concrete by itself is weak in tension, and many attempts were made to remedy this by embedding other materials in the body of structural elements. Ralph Dodd, a British engineer, took out a patent in 1818 for including wrought-iron bars in concrete. Thomas Telford (1757–1834) used iron bars in the concrete abutments of the Menai Bridge (1825) to tie them together more firmly and in 1829 Dr Fox, also in England, developed a method of filling the space between iron girders with this material, and he patented this in 1844.

In the 1850s, a large number of inventions were patented for combining iron with concrete. W. B. Wilkinson, an engineer and inventor from Newcastle, invented a system of using old wire colliery ropes and iron bars in concrete beams in 1854. He built a cottage in about 1865 which was made entirely of concrete reinforced in this way. When it was demolished in 1954, the reinforcement was found to be in the correct lower portion within the beams and slabs, so as to resist efficiently the tensile forces which occur in bending. In 1849, Lambot made a boat of concrete reinforced with iron rods, which was exhibited at the Paris Exhibition of 1854. This boat lasted

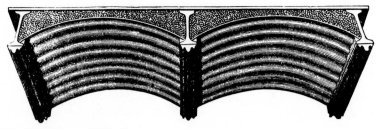

Flooring system combining iron joists and corrugated-iron arches with concrete fill. From the catalog of the Philadelphia Architectural Iron Company (1872).

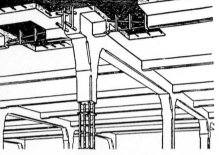

Reinforced-concrete system by François Hennebique (1892), showing positions of reinforcing bars in cutaway sections.

Weaver and Company granary and flour mill in Swansea, Wales (1895–97), by François Hennebique.

St Jean de Montmartre, Paris (1894–1903), by Anatole de Baudot.

well into the 20th century, but the techniques used in making it were not widely exploited for some years to come. François Coignet in his first of many patents (1855) invented a system of combining concrete with iron joists.

Joseph Monier (1823–1906), a Parisian gardener, began to use iron mesh as reinforcement in concrete flower tubs in 1861. He extended his very successful method, which he patented in 1867, to other articles such as containers, pipes, and railroad sleepers. These products were exhibited at the Paris Exhibition of 1867, together with reinforced-concrete articles by Coignet, and were noted by many visitors. Monier built an arched 52 ft. (16 m) bridge in 1875, but his floor slabs, in which he did not position the reinforcement efficiently, were not a great success.

Work on grading of cements and aggregates undertaken in France made possible the production of reliable mixes, without which all these products would have had unreliable properties.

In the early 1850s, Thaddeus Hyatt (1816–1901), a British lawyer who had established himself in the U.S., experimented independently with the use of flat iron bars which were perforated at intervals to receive transverse round bars. He concluded that these would perform best in the tension zone of the concrete beam or slab. Hyatt made many remarkable deductions about the behavior of reinforced concrete, advocating the use of T-beams and remarking on the similar coefficients of thermal expansion of iron and concrete. He continued his work in England where he conducted tests in Kirkaldy's laboratory between 1876-77. He published the results of his work in a book *Experiments with Portland cement concrete* which was one of the first scientific investigations on the subject that was widely circulated.

Reinforced concrete appeared in the U.S. after the Paris 1867 Exhibition and was influenced by French research, even though there had been some independent work in that country including S. T. Fowler's system for walls with reinforcing made out of bolted timber grillages patented in 1860, and Charles Williams' walls (1868) reinforced with iron straps, inspired by reinforced brickwork techniques for grain silos. Concrete and metal floor slabs made by filling in between joists over corrugated-iron plates were patented by J. Gilbert in 1867—a remarkably similar

method to one invented by the British corrugated iron contractor, J. H. Porter, in the late 1840s, which in turn was a development of a system that had been used by William Fairbairn with arches of flat plate between beams in his refinery building of 1845.

Between 1871–76, William E. Ward built his own palatial residence at Chester, New York, in which all the structural elements were in reinforced concrete. The building was designed by Robert Hook, and the system of construction was inspired by the patents of Monier and Coignet. This large house, which cost over $100,000 at the time, was thoroughly tested structurally by having large loads imposed on the floors before it was completed. Only small deflections were noted in the slabs, but this building remained an isolated phenomenon in the U.S. for about 15 years.

Reinforced concrete—early structures

François Hennebique (1843–1921), one of the great French pioneers of reinforced concrete, began work in 1879. In his early work he established the best position for reinforcing within a concrete section. In beams this was found to be in the lower portion where tensile stresses are at their strongest. He may have known of Hyatt's work, and certainly took advantage of the theoretical research done in France in the previous decades. Hennebique patented his ideas in 1892 and these were all used over Europe and parts of the Americas. By 1900 he himself had been responsible for over 3,000 structures in reinforced concrete, of which at least 100 were bridges. The others were mainly industrial buildings. Hennebique maintained a reputation of being an excellent contractor and able businessman. In the 1890s he built a villa for himself in Bourg-la-Reine in which he demonstrated some of the unique structural possibilities of concrete even though the house was decorated in the idiom of the time. The villa employed large cantilevers over the streets and had various roof gardens at different levels. His Charles VI mill, built in 1895, was one of the first buildings to use a repetitive, unadorned, reinforced-concrete grid of columns, beams, joists, and slabs, that was to become an essential part of the idiom of modern architecture.

Anatole de Baudot (1834–1915) designed and built the church of St Jean de Montmartre in 1894. It had slender concrete columns and vaults and was enclosed by thin walls. It was an early example of a reinforced concrete, non-utilitarian public building. Baudot, a pupil of Henri Labrouste (1801–75) and Eugène Emmanuel Viollet-le-Duc (1814–79), was an enthusiast for the invention of a new architecture and believed in the necessity to create new architectural forms appropriate to modern materials.

Auguste Perret (1874–1954), who trained at the Beaux Arts, dedicated himself to the use of concrete from the early years of his career as an architect. His first building in reinforced concrete was built in 1890. The famous apartment block, which he built for himself at the Rue Franklin in Paris, in 1903, was an inspiration to young architects in that its plan was not encumbered with load-bearing walls since the structure consisted of columns, beams, and slabs. His Garage Ponthieu in Paris, 1906, is also often quoted as a seminal work. Perret had a long career and designed many, often very big buildings of many types. In his later years he was sometimes condemned for his return to a stiff classical vocabulary, but his church of Notre Dame du Raincy (1922–23) developed an inspired use of columns and vaulted slabs with large expanses of glazed non-load-bearing external walls that have been widely admired.

In the 1880s, theoretical knowledge about the behavior of concrete structural elements developed very rapidly. G. A. Wayss, having seen Monier's exhibit at the Antwerp Exhibition of 1879, bought the rights for the process for use in Germany where M. Koenen (1849–1924) participated in initial familiarization tests on the material. Koenen published the first analysis of the behavior of reinforced-concrete beams in 1886. Between 1888 and 1894, Edmond Coignet (1850–1915), son of François Coignet, collaborated with N. de Tedesco on research which led to the derivation of expression for the strength of beams based upon elastic behavior. This work was the foundation upon which many of the computation methods used today are based.

In the U.S. in the 1880s, concrete construction began to be considered seriously. Numerous patents were taken out on many forms of beam, arch, and slab, and a large range of reinforcing elements from expanded metal to combinations of bars and meshes. Concrete foundations, poured around grillages of metal beams, became common for large buildings, and experiments with precast beams were initiated. P. H. Jackson from San Francisco patented a method of making prestressed concrete beams in 1886, but these beams were not entirely successful.

Ernest L. Ransome (1844–1917) was born in England and later moved to the U.S. in the 1860s, after having gained some experience in the use of concrete from his father, who was an iron founder and manufacturer of special cements. Ransome went directly to California where he eventually worked as superintendent of the Pacific Stone Company. In the 1880s he began to patent various improvements to reinforced-concrete construction, including systems of expansion joints and the employment of twisted iron bars to increase the bond between them and the concrete.

In 1884 Ransome began to build structures or parts of structures largely made of reinforced concrete, starting with the Arctic Oil Works at San Francisco and a mill for Starr and Company at Wheatport in California. This mill, built in 1885, had an entire structure of reinforced concrete. In 1888 he built a large floor made up of a series of arches joining a series of beams together. He built the first concrete bridge in the U.S. in 1889 and in the same year designed the structure of the four-storey Academy of Sciences building in San Francisco, in addition to work carried out at the Borax Works at Alameda, California, where he cast beams, slabs, and joists as homogeneous structural elements. One of Ransome's largest 19th-century works was a factory for the Pacific Coast Borax Company at Bayonne, New Jersey (1897–98), where he broke away from the tradition of building with small windows, common in other forms of masonry construction.

In the 1890s, the Ransome Engineering Company received commissions throughout the U.S. and Ransome's reputation as a designer and engineer grew. He invented and improved many processes and techniques commonly used today in reinforced-concrete construction. Among these were patents for glass lenses or prisms cast in concrete as a method of lighting basements (1891 and 1894), techniques of extending floor slabs, uniting reinforcing bars and cantilevers, as well as a system of standardized and reusable formwork units (1902 and 1909).

In Europe, Tony Garnier (1869–1948), while a scholar in Rome, designed a whole city in which a large proportion of the buildings were concrete structures (1901–04). This visionary scheme for *Une Cité Industrielle* anticipated remarkably accurately many of the forms that would become part of the general vocabulary for designers in concrete. In his career after the completion of this project, Garnier had the opportunity of building many buildings along lines similar to those he had established in this work. His *Grands Travaux de la Ville de Lyon* of 1919, the Grange Blanche Hospital (1915–30), and the Etats Unis residential district (1928–35) are a few among these.

Large American factory buildings in reinforced concrete, built by Ransome and other pioneering engineers, with their simple lines, large windows, and repetitive structures, became known and admired by European avant-garde architects including Walter Gropius (1883–1969), Le Corbusier (1887–1966), Erich Mendelsohn (1887-1953), and others who also drew inspiration from American and Canadian grain or cement silos from the period around the turn of the century. In 1913, Walter Gropius published photographs of these factories and silos in the annual of the Deutscher Werkbund as examples of

Apartment house at 25, Rue Franklin in Paris, France (1922–23), by Auguste Perret.

Notre Dame du Raincy, Paris, France (1922–23), by Auguste Perret.

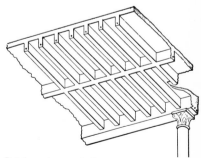

Reinforced-concrete floor construction. Diagram of system used by E.L. Ransome in the Borax Works, Alameda, California (1899).

Standard column-and-slab house frame: Dom-Ino project (1914) by Le Corbusier.

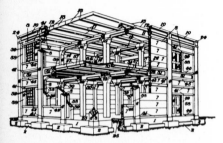

Conzelman system of precast concrete construction.

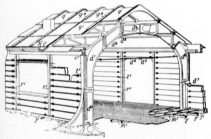

Patent drawing for single-storey buildings entirely built of precast elements, by Bessonneau and Besnard (1917).

Van Nelle Factory in Rotterdam, Holland (1927–28), by J.A. Brinkman and L.C. Van der Vlugt.

modern architecture to be emulated by progressive designers. Le Corbusier eulogized them in his book *Vers une Architecture* of 1923: ". . . they show us the way and create plastic facts, clear and limpid, giving rest to our eyes and to the mind the pleasure of geometric forms. Such are the factories, the reasssuring first fruits of the new age." In the Dom-Ino housing project of 1914, Le Corbusier postulated a regular grid of columns supporting flat slabs between which "free plan" forms could be made with non-load-bearing walls. He had become familiar with concrete through his many contacts including Auguste Perret, for whom he had worked in 1908, and the circles of the Deutscher Werkbund, whom he had been in contact with in 1911–12.

There were many important buildings built in Europe before World War I in which reinforced concrete was used with confidence and innovatory skill. Among these, mention must be made of Max Berg's Centenary Hall at Breslau (1912–13) which employed a huge ribbed dome spanning 213 ft. (65 m), and Matté Trucco's vast five-storey Fiat-Lingotti works in Turin of 1915, which incorporated an automobile testing track on the roof.

The high standard set by Hennebique throughout his career both in his buildings and his bridges, were applied by other engineers in Europe from before the turn of the century onward. Among his great bridges, mention must be made of Pont Neuf Châtellerault (1898) which spanned 164 ft. (50 m), and the Risorgimento Bridge in Rome (1911), spanning 328 ft. (100 m). Great care was taken to use the right proportion of stone, cement, sand, water, and in the placing of reinforcement and the pouring of concrete, so that the behavior of the structures could be more accurately predicted. This was helped by the researches of R. Foret whose work of 1892 established a quantitative basis for determining the final strengths of concrete using different amounts of water in the mix. Engineers had always had to compromise with this variable, as wet concrete is easier to handle and coax into molds and into spaces between reinforcing elements. In most cases, drier concrete will develop a higher strength, but it requires careful placing to avoid leaving voids in the structure, and this could only be achieved by fairly labor-intensive ramming.

The work of Mörsch, Wayss, and Freitag in Germany, both in theoretical fields and in actual structures, helped widen the use of concrete. Hennebique was followed by engineers such as Cottancin and Chandy in France, but the work of the Swiss engineer Robert Maillart (1872–1940) stands out among the second generation of reinforced-concrete innovators. His brilliant bridges began with a fairly modest span of 98 ft (30 m) at the Inn Bridge at Zuoz in the Engadine in 1901. In this bridge he used a cellular structure, but his intuitive understanding of the material led him to simplify its form. Hitherto, engineers had normally used concrete in systems of beams, joists, and columns, similar in concept to timber and steel structures where linear forms are the elements of the vocabulary. Maillart made slabs perform as structural elements in their own right either as flat or curved planes. He continued to exploit and perfect the use of slabs in many of his major works. In the Schwarzenburg Bridge of 1933, Maillart developed the ideas he had been working with over the years to a high level of perfection. The bridge consists of an arched concrete slab in the form of an inverted catenary, which is just thick enough to resist compressive stresses without buckling. From this arch springs a series of vertical slabs to carry the curved roadway which acts as a stiffening girder for this fully integrated structure.

Maillart's work included many buildings, a large proportion of them industrial, in which he explored many new possibilities of concrete construction. Beamless floor slabs for industrial and other buildings were first developed by C. A. P. Turner in the U.S. who published an article in 1908 entitled *The mushroom system of construction*. The first building to use this form of slab was the Bovey Building in Minneapolis. In the slab, the reinforcing was arranged radially near the columns which had large capitals to help reduce the stresses around them. By this technique it was possible to avoid the use of beams which could interfere with the daylighting of interiors, the disposition of services, and the total useful height of each floor. Maillart's experiments with flat concrete slabs for floors also date from 1908, but he had used them in bridges earlier. Maillart's columns employed capitals to spread the load but their design, and the way he disposed of the reinforcement, were slightly different. They were employed in a warehouse he designed in Zurich in 1910. Turner's system was used in the Van Nelle factory in Rotterdam (1927–28) by J. A. Brinkman (1902–49) and L. C. Van der Vlugt (1894–1936).

Pier Luigi Nervi (b. 1891) studied engineering at Bologna and graduated in 1913. His first important work was the Communal Stadium at Florence (1930–32) in which he created an architecture out of the naked structural elements of raking concrete beams and cantilevered curved staircases. This work was much admired by avant-garde modern architects. Nervi executed a number of military aircraft hangars between 1935 and 1941, built up of curved lamella networks of load-bearing joists, with which he was able to lighten the structure considerably. In the 1940s he undertook extensive research into prefabrication of precast concrete elements, prestressing techniques as well as thin concrete components that derive their strength from their forms. He designed a series of immense roofs for a large range of different buildings, among which the roof of the Exhibition Hall at Turin (1948–49) stands out. It consisted of a large barrel vault made up of precast undulating components. He used similar forms for the ribs of the large shallow dome at the Palazzo dello Sport in Rome (1958–59), capable of accommodating 15,700 spectators. Many of Nervi's forms are derived from careful observation of structures in nature. He has exploited forms that give strength through shape, and in the Pirelli building in Milan (1958) he worked with the architect Gio Ponti (b. 1891) to produce a structure that reduces dramatically in cross section throughout the height of this tall building.

Prestressed concrete

In the early years of this century, further research was carried out by Stussi and Whitney, among others, on the properties of reinforced concrete under load. This led to the evolution of ultimate load theory, by which it was possible to design beams and other structural elements in which the steel and the concrete would begin to fail together at the same load. High-tensile steels became available, capable of taking up to many times the tensile stresses permitted for mild steel, but even with special precautions it was difficult to control the cracks that develop with concrete under heavy loading.

Eugène Freyssinet (1879–1962) was among those who solved this fundamental problem by eliminating the tensile stresses in the concrete by stretching the reinforcement so that when released it would impose a compressive stress over the whole concrete section throughout the life of a structure over a wide spectrum of loadings. Earlier attempts at prestressing concrete beams had failed, through lack of thorough investigation, since among other things concrete shrinks while curing, and steel tends to stretch. Pioneering work in this field had been undertaken in many countries. In the U.S., P. H. Jackson had patented the idea in 1886. Doehring, a Berlin builder, invented a similar technique in 1888 for floor panels. Other attempts were made by Emperger (Hennebique's North American agent), Ritter, and Rabut who had been one of Freyssinet's teachers.

Freyssinet began his practical work on prestressing in parts of structures about 1908 when he used a prestressed concrete tie to unite the abutments of a bridge over the Allier. His investigation into prestressing became known in 1927 at the same time as the Belgian G. Mengel published his parallel work. From 1933 onward, prestressing was used in many structures, some of them large

civil engineering works by Freyssinet. Hewet in the U.S. introduced a system in 1923 of using lubricated steel bars in concrete, that were subsequently prestressed. This technique was mainly used for large cylindrical tanks. Dirchinger and Finsterwalder in Germany developed the use of post-tensioned ties in bowstring bridges and lattice girders.

Freyssinet's Elsby Bridge (1949) was made up of elements precast in a nearby factory. The bridge, with its shallow arch span of 248 ft. (76 m), used all the most advanced techniques; many of them developed by his firm, including special jacks and anchorages for pretensioning cables. The steel used was up to five times as efficient as mild steel at half the cost. Freyssinet developed and perfected many other branches of concrete construction. As director of Enterprises Limousin (1913–28) he designed the famous airship hangars at Orly (1916–24), 205 ft. (63 m) high and 984 ft. (300 m) long, built out of precast curved and specially shaped elements. Vibrated concrete was first used in these structures making possible a much higher quality of concrete through thorough compaction. Freyssinet was to continue using carefully conceived, repetitive precast elements in many of his structures, often steam-cured in their molds to speed up production. His company was responsible for a wide variety of works ranging from bridges and airports to dams and many buildings of all types throughout the world. In Germany, Hoyer, who adopted the Freyssinet system, introduced piano wires as reinforcement, capable of very high tensile strength. Because of the very good bond that these wires develop with the concrete, on account of their favorable area to cross section ratio, no anchorages are needed.

For many years Freyssinet's systems were a virtual monopoly in France. In Germany, where prestressed bridges became extremely popular to replace the vast numbers destroyed in World War II, many engineering and contracting firms developed their own systems of anchorages for cables, jacking systems, and other special refinements.

Prestressing can be achieved in two ways. Bars or cables can be stretched with jacks and then released once concrete has been poured around them (pretensioning), or voids can be left in precast concrete elements to allow cables to be threaded through and stretched to the right level (post-tensioning).

Pretensioning is extensively used in the production of factory-made building components, such as beams, up to very large spans, and other prefabricated elements, among them, street light posts, railroad sleepers, etc, made under carefully controlled conditions. It is often combined with steam curing which speeds up the hardening of the concrete, making it possible to reuse the mold rapidly. (In countries with a warm dry climate

Pirelli Building, Milan, Italy (1958), by Gia Ponti and Pier Luigi Nervi.

Prestressed concrete joists used in conjunction with hollow tile blocks ready to receive concrete topping.

pretensioning is especially popular as it can take place outdoors.)

These techniques make it possible to use much lighter building components as the constituent materials are used much more efficiently. Larger spans or heavier loads can be accommodated with far less material and structural depth than conventional reinforced concrete.

Shell structures

Shell structures, in which the thickness of the material is slight in relation to surface area, exploit the shape of constructional elements to gain rigidity. In 1925, Walter Bauerfeld, an engineer in the firm Dykerhoff and Widmann, designed a hemispherical dome for a planetarium for the Zeiss company at Jena. It had a diameter of 52 ft. 6 in. (16 m) and the concrete shell was only 1.2 in. (30 mm) thick. By the mid-1930s a number of industrial commercial exhibition and sports buildings had been built using shell concrete construction. These included the barrel-vaulted market hall at Frankfurt am Main by Martin Elsässer of Dykerhoff and Widmann (1927)— the first large building using shell concrete; repair shops at Bagneux by Freyssinet (1928–29); a foundry at Milan by Giorgio Baroni; the racecourse roof at La Zarzuela (1935) by C. Amiches, L. Dominguez, and Eduardo Torroja (1899–1961); the pelota court at Madrid (1935) by Torroja, and aircraft hangars by Dykerhoff and Widmann at Munich (1938–40).

Maillart's cement hall at Zurich (1938–39) aroused interest in the slender forms that could be achieved by this form of construction, and after World War II non-rectangular curved forms became acceptable as elements in the vocabulary of modern architecture.

Concrete shells have been exploited brilliantly by a large number of architects and engineers. They have been used in many forms: as barrel vaults, domes, warped surfaces such as hyperbolic paraboloids, and countless other shapes. Water towers, cooling towers for power stations, and roofs for buildings are among the many uses to which this form of construction has been applied. The expensive formwork needed for molding complex shapes has often led to the use of repeated precast components, in many cases subsequently prestressed, or to the use of geometries such as those found in hyperbolic paraboloids where surfaces are generated by straight linear elements. Inflatable formers have been used as temporary structures to support the concrete of some shell roofs. Sprayed concrete, a process developed in the 1920s, lends itself to some types of shell construction since large amounts of rapidly hardening material can be delivered accu-

Restaurant Xochimilco in Mexico: Shell roofs under construction; by Félix Candela and Ordónez (1958).

rately over a large area.

Outstanding postwar examples of shell concrete construction include the Cosmic Rays Research Laboratory at Mexico City (1951) by Jorge Gonzalez Reyna and Félix Candela (b. 1910); the Kresge Auditorium M.I.T. in Cambridge, U.S. (1954–55), by Eero Saarinen (1910–61); the Palazzo dello Sport in Rome (1957) by Pier Luigi Nervi (b. 1891); and the church of San José Obrero (1959-60) in Monterrey, Mexico, by Félix Candela. The competition design for the Sydney Opera House (1956) by Jørn Utzon (b. 1918) was originally intended to be built as a shell structure, but after extensive analysis, carried out by the engineers of Ove Arup in London, this was found to be impractical and the vaults were made of carefully shaped blocks of precast reinforced concrete that were post-tensioned on site.

Other significant developments

An important contribution to making the use of concrete feasible for large works was the evolution of methods of mixing and delivering it mechanically. The hand-operated concrete mixer was invented by Louis Cézanne in 1854. During the remaining years of the 19th century, improvements were made on mixers and larger and larger models became available, eventually using mechanical power to drive the mixing drum. For larger building programs, when concrete is mixed on site, it is common to find large batching and mixing plants where the quality of concrete is controlled accurately by regular tests. The manufacture of ready-mix concrete off the building site was made possible by the development in the U.S. of lorry-mounted transit mixers which appeared in about 1926. This technique has become popular all over the

Carey and Latham's concrete mixing machine (c. 1890). Sand and balast are delivered into the mixing cylinder by the bucket and chain where they are joined by a controlled volume of cement and water.

Messet's patent concrete mixer (c. 1890). This machine was hand operated.

world, and is especially useful for buildings of a medium size or where sites are restricted and the local mixing of the concrete is impractical.

Concrete is normally delivered to the necessary part of the structure by means of skips carried by cranes or by the use of mechanical hoists. Relatively inaccessible parts of structures can be concreted by pumping the materials into the right position through flexible hoses. Truck-mounted pumps can easily be connected to ready-mix delivery vehicles. In countries where labor is cheap, delivery is generally by means of wheelbarrows, buckets, and ramps—even for large structures.

Sprayed concrete, in which the correct amount of water is mixed with the other constituents in the nozzle of a cement gun, was developed after World War I. Concrete delivered by this method can be very dense and with a low water content. This technique is used for resurfacing worn and weathered concrete, repairing damaged structures, forming embankments, and so on. It also lends itself to shell structures where a thin compact layer of relatively dry, rapid-setting concrete is required.

An almost infinite range of different formwork systems exists. For in-situ work steel, wooden, or plywood shuttering is used in combination with systems of adjustable props and struts. Formwork represents a major cost item in concrete construction, which has stimulated the development of modular systems of reusable shuttering elements that can be adapted to a wide variety of jobs including curved work. The surface of the shuttering can be of great importance if the concrete is to be visible since it imparts its texture to it. Le Corbusier and many modern architects favored the appearance of concrete with the imprint of shuttering boards with a rough grain (*béton brut*), while in other situations a smooth finish has been desirable, and carefully faced metal, ply, or sometimes plastic shutters have been used. Shutters with absorbent linings of material such as wallboard can help by slightly reducing the water content of the concrete, which may have been high as an expedient for efficient placing. This technique also eliminates air bubbles forming on the surface. Other methods of extracting excess water include a vacuum process whereby suction mats are applied to the surface of the concrete as soon as it has been poured. Where it is desirable to use less water, vibration is used in order to compact and consolidate concrete. Vibrators can be mounted on the shuttering or they can be in the form of vibrating pokers that are introduced into the material in its wet state.

Tall structures such as concrete chimneys, oil rigs, elevator cores, and silos often employ systems of moving formwork that gradually

Modern mobile transit mixer being loaded with raw materials.

Concrete floor being cast with a pumped delivery of concrete from a mobile pump unit fed from a transit mixer.

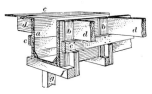

Timber shuttering for a reinforced-concrete beam.

Simple board shuttering for in-situ concrete beams and slabs in Istanbul, Turkey.

Modular plywood shuttering with steel stiffeners being prepared for the casting of walls in an apartment building.

climb the structure so that continuous pouring can take place without construction joints. These forms are often combined with steam curing which accelerates the development of strength in the concrete. Multistorey structures with repetitive plans may economize on formwork by having all the floors cast on the ground and fitted into position up the columns by means of hydraulic jacks. This technique became popular for multistorey apartments and other buildings after World War II.

Formwork for factory production of components can make elements to close tolerances. Steam curing, closely controlled mixing, prestressing, and carefully designed forms have contributed to making concrete prefabrication of components fast and reliable.

Precast concrete performs many duties in construction. In the USSR after World War II the prefabrication of large units on a vast scale, including whole boxlike room sections of apartments, was initiated as a solution to chronic housing shortages. Whole wall, floor, and roof units have been produced in many countries often incorporating ducts and conduits for all the utilities. Frame elements such as columns, beams, trusses, flooring systems, and portal frame units in precast concrete are also widely used for all types of buildings.

Joints in precast concrete have to be carefully designed to ensure the stability of structures. It is often difficult to achieve in precast concrete the degree of continuity and monolithic construction possible in concrete cast in situ. Dimensions of precast components are controlled largely by the size of lifting machinery and restrictions on sizes of components that can be transported.

Concrete cast in situ lends itself to the creation of forms with continuity of structure and complex special shapes. It has been used in an inventive way by many 20th-century architects to achieve forms impossible in any other material. Frank Lloyd Wright's Johnson Wax Building at Racine, Wisconsin (1936–39), where he used cantilevered mushroom roof forms; the Falling Water House at Bear Run, Pennsylvania (1936), with its large cantilevered terraces; and the Guggenheim Museum designed in 1943-46 and built in 1956–59, are a small sample of significant buildings designed by one architect using this method. Countless other architects have contributed to innovations in forms and techniques by using this relatively new material.

From the time reinforced concrete began to be used, vast amounts of scientific research have been undertaken to improve the quality and predictable behavior of all the ingredients and methods by which they are selected, graded, and tested.

Since World War II there have been great improvements in concrete-mix design, making it possible to achieve much higher compressive strengths. Special designs of reinforcing bars vastly increase the bond between them and the concrete, reducing cracking and permitting a more efficient use of the steel.

Cements with special properties have been developed. Among these, high-alumina cement develops its strength very rapidly and heats up considerably during the process, enabling work to be carried out in frosty conditions. Unfortunately, concrete made with this cement is liable to crack more readily than other concretes. Concretes that heat up very little during setting are useful in work where large volumes of material are needed, so that the heat evolved during hydration can be controlled. Concretes that do not shrink but swell on setting were developed in France during World War II for repairs to foundations of damaged buildings. They are also very useful in conjunction with pretensioning techniques.

Aggregates can be used to give concrete special properties. Lightweight aggregates made from foamed blast-furnace slag, expanded clay, sintered pulverized fuel ash, and pumice, can reduce the weight of concrete components and increase their thermal insulation. No-fines concrete, which uses specially graded aggregates without employing sand in the mix, makes it possible to use less cement and very much lighter formwork comprised of frames supporting wire mesh panels. The concrete, which does not contain much water, can be held in these without leaking out. This technique was developed in Europe in the 1920s and has been extensively used in housing schemes when two-storey wall units can be poured in one operation. The cavities between the stones help to increase the insulating qualities and form a barrier to capillary water infiltration.

Aerated concrete, made by including additives which evolve gases during setting, are useful for blocks and other building elements where good insulating properties are required. Various chemicals can be added to concrete mixes to accelerate or retard setting, while other chemical additives can increase the workability of the mix.

Good concrete design involves careful examination of many factors outside purely structural considerations. Since concrete is placed as a liquid material, there will be joints at places where different pouring programs meet. Expansion joints have to be used to take up movements in all but the smallest structures. The detailing of joints and surface finishes are of paramount importance in concrete structures where these are visible.

Concrete with a surface imprint of boarded

Precast wall panels in a factory yard in southern France.

Different concrete finishes. A panel with exposed aggregate against a wall with a surface imprint of wooden shuttering boards.

Reinforced concrete and shuttering of various types in the construction of a theater at the Barbican in London, England (1976).

formwork has already been mentioned. Other textures and patterns can be included in the design of molds. The surface layer of concrete may be removed to expose the aggregate by bush hammering or by using retarders in the concrete close to the shuttering. Expensive white concrete can be placed as a thin skin on an ordinary concrete body by pouring both mixes into a mold with a separator between them which is withdrawn as the formwork is filled. Masonry skins and other finishes such as tiles and mosaics can be cast onto concrete by including them in molds. Many of these finishing techniques lend themselves to pre-casting in factory conditions where the manufacturing process can be carefully controlled.

Asbestos

Asbestos is a fibrous mineral. Fire-resistant fabrics made from it were available in the 1870s, and it was used in combination with cement as a fireproofing material to case vulnerable structural elements and as insulation to boilers. Asbestos sheets, which were made by combining asbestos with canvas cemented to a surface layer of felt, produced a compact, flexible roofing material resembling leather and this came into use in the 1880s. Lagging for heating and other pipes in asbestos dates from this period.

In 1893 Ludwig Hatschek, an Austrian textile manufacturer, carried out experiments in mixing short fibers with various bonding agents in an attempt to produce a new type of building material. In 1900 he succeeded in making the first asbestos-cement sheets, using a modified Fourdrinier papermaking machine, fed with a slurry composed of 15% asbestos and 85% cement. By 1910, asbestos cement was being manufactured on a large scale in no less than ten countries.

Corrugated roofing sheets were in use during World War I, and by the 1920s factories were producing sheets of the same size and profile as those familiar to us today. The first asbestos-cement pipes were made in England at Widnes in 1927, and by 1930 public health authorities in many countries had approved asbestos-cement water pipes as an alternative to the traditional cast iron. They were shown to have a resistance to corrosion superior to iron, and their joining could be flexible enough to absorb settlement and vibration. Asbestos-cement sewer pipes, gutters, roof decking, and drainpipes, introduced after 1940 have met with similar success.

Asbestos-cement products are bulky and heavy in proportion to their value, consequently transportation costs are always a matter of concern. For this reason, manufacturing has been decentralized as much as possible, a system which also helps to reduce the danger of breakage in transit.

Asbestos products are still used extensively in the building industry for their fireproofing properties. Asbestos can be used in many forms, ranging from boards that can be sawn in much the same way as wood, to a form that can be sprayed directly onto structures.

In recent years there have been two developments which have considerably reduced the popularity of asbestos cement. One is the now proved health hazard to workers both in the mine and in the factory, and the other the decolonization of most of Africa where many of the mines are located.

Glass fibers are becoming a strong competitor for some of the duties normally performed by asbestos, especially when combined with cement to produce glass-reinforced concrete (G.R.C.).

Iron

Iron artifacts are known to have existed before 2500 BC, but its smelting and use on a significant scale date from the end of the Hittite Empire in the Middle East (c. 1200 BC).

Early iron smelting was a laborious process in which small spongy masses of iron mixed with slag "blooms" would be removed from a furnace after many hours of hard work at the bellows. This directly reduced wrought iron was then worked at the forge to produce useful objects. It could be converted into steel by combining it with carbon by further labor-intensive processes.

Iron made by these methods was a precious metal and its use in building remained slight until other processes of production were developed. Improved and longer-lasting tools with harder and sharper cutting edges were manufactured. These made the working of wood, stone, and other metals a great deal easier and more accurate than with tools of bronze or stone.

Iron was used for fastening timber and stone building elements together and for strengthening parts of structures, as well as for ironmongery. Wrought-iron gates and screens date back to antiquity, and medieval cathedral builders used iron frameworks in their traceried windows to support the leading which held the stained glass in place. They also used iron for the elaborate hinges they made for church doors. Iron-reinforced masonry was introduced in some 13th-century cathedrals in the Ile de France.

The development of smelting

Blast furnaces capable of reaching the melting temperature of iron (2,786°F/1,530°C) were only developed in Europe, near Liège in Belgium, in the 14th century although they had existed in China as early as the 4th century BC. These furnaces were charged with

Corrugated asbestos warehouse roofs.

Medieval wrought-iron gate screen—Sé Velha, Lisbon, Portugal.

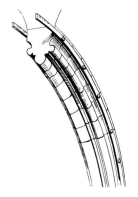

Wrought-iron bands reinforcing a masonry vault rib, Ste Chapelle, Paris, France (c.1240).

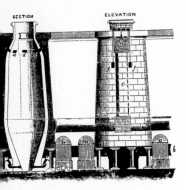

Blast furnace (c. 1870); shown in section and in elevation.

Nineteenth-century puddling furnace shown in section. The coal is burned in a chamber separated from the iron.

Puddling furnace—stirring the metal with an iron paddle to expose it to the gases within.

The Minerva Iron and Steel Works, Staffordshire, England, in the 1870s. A typical enterprise of the period.

ore, charcoal, and fluxes and their combustion was aided by water-driven air blowers.

When one of these furnaces was tapped it could produce up to 1 ton (907 kg) of pig iron: a brittle form of iron with a high carbon content and a crystalline structure containing a number of impurities. Pig iron was converted into wrought iron by remelting it in a charcoal-fired hearth "finery," in which oxygen combines gradually with the carbon found in pig iron, reducing it to a more fibrous and workable form. Iron made in this way was then shaped into bars and strips by the use of tilt hammers and slitting mills, and this malleable material could then be made into objects by the process of forging. Alternatively, the pigs could be remelted and poured into sand molds to form useful objects of cast iron: a crystalline and more brittle form of iron.

Blast furnaces used up prodigious quantities of charcoal which contributed to the rapid depletion of forest in areas where the iron industry developed. In Britain, Abraham Darby (1677–1717) succeeded in smelting iron with coke in 1709 at his iron foundry in Coalbrookdale, Shropshire. In 1784, Henry Cort (1740–1800) developed a "reverbatory or puddling" furnace in which pig iron could be converted into wrought iron without picking up impurities from the coal which was burned in a separate compartment.

James Watt's rotary power steam engine, developed in the 1780s, made it possible to drive all the machinery necessary for large-scale iron production, thus allowing all the parts of the manufacturing process to be brought close together. Previously these had been dispersed to take advantage of water-power.

All these inventions and innovations contributed to making iron a much cheaper and more readily available material, and were followed by many further developments in the 19th century that made it possible for blast furnaces to produce well over 1,000 tons (1,016 metric tons) of iron per day.

Wrought iron up to the end of the 18th century

Wrought iron continued to be used for railings, gates, and other fittings in increasing quantities after the Middle Ages but its method of manufacture relied upon the craft of the blacksmith and changed little until the end of the 18th century, when small rolled sections became available.

Architectural wrought iron reached its height in the late 17th and 18th centuries, with the work of French master craftsmen such as Daniel Marot (1660–1752), Jean Tijou, and Jean Lamour. In 1693, Jean Tijou published a pattern book entitled *A New Book of Drawings* which had a great influence on English wrought-iron designs in the 18th century.

The structural use of wrought iron in combination with masonry was revived in France when Claude Perrault (1613–88) built the eastern facade of the Louvre in 1667–70. This technique was refined by Jacques Soufflot (1713–80) in the giant portico at the Panthéon in Paris (1770–72). In 1779, Soufflot also developed a self-supporting iron roof, spanning 51 ft. 8 in. (15.8 m) for a gallery of the Louvre. This roof was made up of forged bars assembled with many collar joints similar to those used in the portico. Experiments with combinations of forged bars forming beams and joists for floors were made from 1782 onward by M. Ango. Victor Louis combined these uses of iron with a revival of Roman hollow pot vaulting, introduced at the time by St Fart, in his designs for the Théâtre du Palais Royal in Paris (1785–90). This remarkable building was built entirely of incombustible materials to reduce the risk of fire, which had caused many deaths in similar Parisian buildings.

The development of cast iron as a structural building material 1700–1850

Cast iron came into use in buildings gradually. Until the 18th century it was mainly used for decorated firebacks and other domestic fittings. In 1714, stout cast-iron railings were erected around St Paul's Cathedral in London. From that time cast iron became an increasingly economic substitute for wrought-iron railings and balconies. These elements could be mass produced in sand molds from their wooden patterns. Trade catalogs from the late 18th century onward helped market these products far and wide.

The structural use of cast iron was pioneered in Britain. In 1706, Christopher Wren (1632–1723) used slender cast-iron columns to support a gallery in the House of Commons, so as not to obstruct the view on the floor of the House. Similar columns were occasionally used in churches later in the century. In the 1750s, John Smeaton (1724–92) initiated the use of cast-iron parts for mill machinery and claimed to have made cast-iron beams.

The first successful iron bridge was erected over the River Severn at Coalbrookdale in Shropshire, England, in 1779 by Thomas F. Pritchard, John Wilkinson, and Abraham Darby III. The various components are joined together with dovetails and pegs, reminiscent of carpenters' joints.

Thomas Paine (1737–1809) arrived in England from America in 1787 with the intention of having a cast-iron bridge made for crossing the Schulkill at Philadelphia. The bridge was to be assembled from interlocking cast-iron voussoirs. He abandoned the project after initial trials, but a bridge to a similar design

Wrought-iron gates at Hampton Court, England. From Jean Tijou's *A New Book of Drawings* (1693).

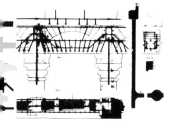

Masonry reinforced with wrought iron—portico of the Panthéon, Paris, France (1770–72), by Jacques Soufflot.

St Michael's in the Hamlet, Liverpool, England (1814), by John Cragg and Thomas Rickman. A brick building with Gothic decoration in cast iron.

was erected over the River Wear at Sunderland in 1796 by Rowland Burdon.

By 1800 iron bridge building was firmly established in Britain. The Pont des Arts in Paris (1801–03), designed by L. A. de Cessart (1719–1806) and J. Lacroix Dillon (1760–1806), was the first iron bridge erected in France. (See also BRIDGES.)

Toward the end of the 18th century it became imperative for British mill owners to find ways of making multistoreyed buildings incombustible, since many of the expensive timber-framed textile mills had burned down with great loss of life and property. There was an early attempt in 1792–93 at a mill in Milford, built by William Strutt (1756–1830), where iron columns were used to support heavy timber beams with brick arches spanning between them. The timber was cased in incombustible materials. Soon after this in 1796, a flax mill at Shrewsbury, built by Charles Bage, had a complete iron frame within the sturdy brick walls. The beams were 11 in. (275 mm) deep cast-iron flats 1.125 in. (28 mm) wide, integrating an 8 in. (200 mm) skewback on the bottom chord. They spanned 9 ft. 6 in. (3 m) between cast-iron columns of X-section and carried segmental brick vaults of 10 ft. 6 in. (3. 2 m) span which sprung from the skewbacks. The floor was leveled over these arches to provide a working surface. Bage evolved formulas for calculating the strength of columns and beams based on full-scale tests.

In the years 1799–1801, James Watt (1734–1819) built a mill at Salford in which spans of 14 ft. (4 m) and arches of 9 ft. (3 m) were used. The columns in this case were hollow Doric cast-iron tubes. By 1803 cast-iron roof trusses were being used in some of these buildings. This form of structure became generalized for incombustible buildings, not only for factories and warehouses but in many cases for civic buildings. Cast-iron beams were used in the floors of the British Museum, designed in 1824 by Sir Robert Smirke (1780–1867). Experiments by Thomas Tredgold and William Fairbairn (1789–1874) in the early 19th century contributed to the evolution of cast-iron beam shapes. The mathematician Eaton Hodgkinson (1789–1861) established the ''ideal'' form for cast-iron beams in 1827–30, in which the material is distributed in the most economic way. These beams employed far less material for a given load than the original Bage design.

In 1813–16, John Cragg (1767–1854) and Thomas Rickman (1776–1841) built churches in Liverpool, England, with iron roofs and extensive Gothic cast-iron decoration, eliminating the work of the stone carver. Cast-iron railings, balconies, verandas, porches, brackets, window frames, and countless other useful objects were produced and sold in ever greater quantities to builders

The Iron Bridge at Coalbrookdale, England (1779).

Cast-iron bridge at Sunderland, England (1796), erected by Rowland Burdon using interlocking cast-iron voussoirs.

throughout the world. After the end of the Napoleonic wars in 1815, the price of iron dropped dramatically and many new constructional applications for it were found.

At the Quadrangular Storehouse at Sheerness (1824–29) Edward Holl used cast-iron beams and joists supporting flagstone floors, spanning between them to make a lighter building than would have been possible with brick arches, to suit the bad soil conditions. This technique was adopted in cast-iron prefabricated barrack frames which were ordered for the West Indies in 1826 by the Duke of Wellington (1769–1852) who was at the time Master General of the Board of Ordnance.

The rapid expansion of the railroads from the 1830s onward demanded the erection of many bridges, and in Britain a large proportion of these were in cast iron, either in the form of arches or beams. Beams were favored in many cases because they exert no lateral thrust and do not rise in the middle. Single castings were limited by handling and casting

West Indian barrack system with cast-iron verandas and floor structure (1826).

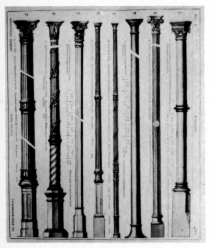

Cast-iron columns from a catalog of the Coalbrookdale Company (c.1880).

Interior of Oxford Museum, England (1855–60), by Dean and Woodward.

techniques to about 50 ft. (15 m). As the demand for wider spans grew, engineers invented many new techniques, including beams made up of a number of sections bolted together, and cast-iron beams trussed with wrought-iron elements in order to exploit the tensile strength of that material. By the late 1840s many forms of combined beams had been tried, some ending up in the collapse of buildings and bridges, as in the Dee Bridge disaster of 1847. This led to strict regulations and considerable caution in the use of cast iron in situations where it was subject to bending. Metal bridges on the truss principle based on the designs of Howe, Pratt, Whipple, Bollman, Fink, Warren, and others, were developed in America in the 1840s, the first being the Highway Bridge over the Erie Canal in 1840. These structures often combined cast and wrought iron and wood, and there were also a number of failures.

After 1850, cast iron was used less and less in structural elements subject to bending, except over relatively short spans. The Crystal Palace of 1851 had cast-iron columns, but all the girders except the shortest 24 ft. (7 m) spans were in riveted wrought iron, which by this time had become relatively cheap.

Cast iron continued to be employed for all the decorative purposes mentioned above and its use extended to the production of whole facades of buildings and components for internal framing which became popular along the eastern seaboard of the U.S. as well as in the cities of the Midwest. James Bogardus (1800–74) and Daniel Badger were two New York iron founders who specialized in this field from the 1850s onward. From this time the range of cast-iron goods used in buildings expanded still further to include parts for all the utilities, from lampstands to drainpipes.

In Britain during the 1850s, there was a public reaction against the use of visible iron elements in important buildings. This had been partly stimulated by the appearance of a much disliked exhibition building that was erected in 1856 in South Kensington, London. This building had been bought from a contractor, C. D. Young, who specialized in prefabricated structures, because public funds were strained by the Crimean War. Another blow to confidence in the use of iron in buildings was the spectacular burning of the New York Crystal Palace in 1858.

John Ruskin (1819–1900) objected to mass-produced molded elements on the grounds that they degraded the nobility of craftsmanship. He persuaded the architects Thomas Dean (1792–1871) and Benjamin Woodward (1815–1861) to use iron columns with wrought-iron decorative capitals, each made individually by a blacksmith, for the Oxford Museum building of 1855–60. The opposite view was held by some architects and designers such as Owen Jones (1809–74)

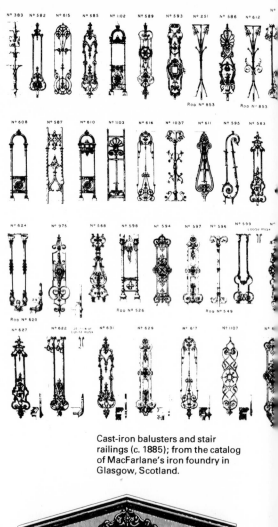

Cast-iron balusters and stair railings (c. 1885); from the catalog of MacFarlane's iron foundry in Glasgow, Scotland.

Cast-iron decorative window hood, front and side views. From the Buffalo Eagle Iron Works, New York—catalog of 1859.

who, in collaboration with the Derbyshire iron founding firm of G. Handisyde, constructed a number of buildings in cast iron including an ornamented kiosk that was exported to India in 1870. The use of mass-produced components and the creation of a new architecture based on the potential applications of iron construction were enthusiastically advocated by William Vose Pickett in a number of pamphlets published in the 1840s and 1850s.

From the mid-1840s onward, iron founders and other contractors in Britain and Europe expanded their markets to cover the supply of railroads, manufacturing plant, public utilities, complete buildings, and prefabricated components, to countries throughout the world, especially in the rapidly developing areas of South America, Asia, Australia, and Africa. In these parts there was little or no resistance from established building crafts; trained labor was in short supply but was often skilled in working with iron in ship and railroad workshops, mining, and other engineering trades.

The development of rolling mills and wrought iron after 1800

Rolled sections found applications in many branches of industry from boiler plates to tie rods. Between 1800 and 1820 wrought-iron angles were first rolled—perhaps in France for shipbuilding—while T-sections appeared between 1828–30. Attempts to roll rail sections in wrought iron to replace more brittle and heavier cast-iron ones succeeded in the 1820s with the appearance of the small 2.25 in. (56.25 mm) deep Birkenshaw Rail. The familiar form of rails used today was invented in 1831 by R. L. Stevens of the Camden and Amboy Railroad of New Jersey. After many trials he persuaded a Welsh rolling mill to produce the first batch of 3.5 in. (87.5 mm) rails. By 1846 four American mills were rolling rails of this profile.

Wrought-iron rods and straps were used as tie members in timber structures and later in combination with cast iron. One of the earliest examples of the use of rolled wrought iron in compression members of a truss was in the roof frames of London's Euston Station (1835–39), where Robert Stephenson (1803–59) used T-sections for the principal members. Experience gained through experiments in developing structural framing in iron for his shipbuilding ventures during the 1830s led William Fairbairn (1789–1874) to deduce in 1839 that the "strongest and most suitable for the support of decks" was the wrought-iron I-section. Similar deductions had been reached through theoretical testing by A. Duleau, a French engineer working in academic seclusion at the Ecole Polytechnique in the 1820s. The conclusions of these experiments were not widely circulated until 1854 when William Fairbairn published a book entitled *The Application of Wrought and Cast Iron to Building*. Because of the limitations of the rolling mills, beams and girders which had come into use in the 1840s generally consisted of combinations of angle and T-sections and iron plates riveted together, often to form large box sections. These played a major part in the research undertaken by Stephenson and Fairbairn in 1845, culminating in the erection of the Britannia Tubular Bridge of the Menai Straits (1849), with clear spans of 460 ft. (140 m).

The first I-section made from a single piece

Rolling stands for wrought-iron bars.

of iron appears to have been rolled in France in 1848, to the orders of the French consulting engineer C. F. Zores who had secured a sufficiently large order to satisfy the mill owners that the large cost of new machinery was justified. Zores intended his I-sections for use mainly in buildings. A patent for a similar form was obtained in 1844 by Messrs Vernon and Kennedy of Liverpool for shipbuilding, but it is not clear whether or not they ever manufactured.

Channel or half-beams were rolled and joined back to back for use as beams for the structure of the U.S. Assay Office Building in New York (1853).

By 1862 techniques of rolling iron had advanced to the point of being able to make beams with a depth of 3 ft. (90 cm) with top and bottom flanges 12 in. (30 cm) wide and up to 40 ft. (12 m) in length while plate could be made up to 8 ft. (2 m) wide, 4.5 in. (112.5 mm) thick and 16 ft. (5 m) long. However, the manufacture of elements of this size in bulk with reliable properties was only possible with the introduction of steel-making processes that rendered the volumetric limits of the puddling process obsolete.

The invention of corrugated wrought iron in 1829 produced an entirely new building material whose potential was rapidly exploited for roofs and walls in lightly framed industrial, agricultural, and temporary buildings.

In England in 1741 a bridge 70 ft. (21 m) long, with wrought-iron chains, was built over the River Tees. In 1801, James Finley (c. 1762–1828) erected a bridge, with 70 ft. (21 m) spans, and towers to support the wrought-iron chains, over Jacobs Creek, Pennsylvania. From that time many suspension bridges were built using chain cables such as Telford's

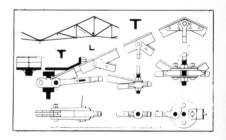

Rolled wrought-iron roof trusses of the original Euston Station, London, England (1835), by Robert Stephenson.

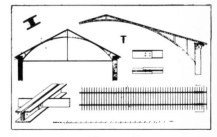

Riveted angles and flats used in roof framing in France in the late 1830s.

Menai Bridge (1826), which had a 570 ft. (173 m) span. In 1816 a pedestrian bridge spanning 408 ft. (124 m) and using six wire cables was built by White and Hazard in America to connect their ironworks to the area across a river where their labor force lived.

In France this technique was exploited for many of the 114 suspension bridges built from the 1820s to the early 1840s. The first French bridge to have iron cables was erected in 1824 by the Seguin brothers. L. J. Vicat (1786–1861), who had been an assistant to C. L. M. H. Navier (1785–1836), suggested systems of sheathing cables in metal envelopes to protect them from corrosion, and in 1834 he invented a process of spinning cables in the air to make wire rope, rather than relying on wires being draped side by side. This technique was discovered independently in 1842 by John Roebling (1806–69), when he used it to carry a canal over the Allegheny River. This was the first of many bridges erected by this method by Roebling and his son, which include the Cincinnati Bridge (1867) and the Brooklyn Bridge (1883) which spanned 1,600 ft. (488 m) and used steel in the spun cables. Steel cables have since been used extensively in the construction industry for suspension bridges, suspended and membrane roofs, diagonal bracing to light structures, in pretensioning concrete, and as cables for elevators.

Cast and wrought iron in buildings in the 19th century

Apart from the buildings and bridges already mentioned, countless other structures were built during the 19th century in which cast and wrought iron were used in new ways.

In industrial buildings, numerous innovations in fireproofing and beam, column, and floor design were made. A wide range of different truss-roof shapes were evolved combining cast iron with timber, timber with wrought iron, cast and wrought iron, and finally rolled wrought iron on its own. The cast-iron roof over Maudslay's Machine Workshop (1832) was a much publicized example. When first erected it collapsed through fracture in the ribs, but was still considered by contemporary commentators to be a good design. Efficient truss types were evolved by American and European railroad engineers. Camille Polonceau (1813–59) used trusses of the same form as the American Fink truss in his sheds for the Paris-Versailles railroad of 1837. These had timber principals, cast-iron struts, and wrought-iron ties. This form became very popular in Europe. The trusses at London's Euston Station (1835–39), designed by Stephenson, used rolled wrought iron for the first time. North-light trusses in wrought iron were used in Britain by William Fairbairn for his mill at Saltaire (1853). In 1840, Fairbairn built a small two-storey flour

Corrugated iron wall and roof, Prince Albert's Ballroom, Balmoral, Scotland (1851).

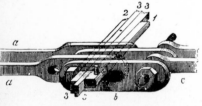

Detail of a link in the wrought-iron chains of Telford's Menai Suspension Bridge, Wales (inaugurated 1826).

Anchorages for steel cables being constructed for suspension bridge at Staten Island, New York (1963).

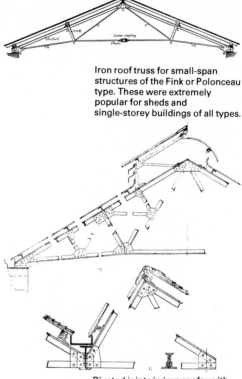

Iron roof truss for small-span structures of the Fink or Polonceau type. These were extremely popular for sheds and single-storey buildings of all types.

Riveted joints in iron roofs; with skylight and ventilators (TOP) and supporting slates and glazing (BOTTOM) in a north-light roof truss.

James Bogardus' factory in New York (1848–49), constructed with facades entirely formed in cast iron.

mill which was ordered from Constantinople. This building had a complete iron frame, and all the enclosing walls were made in the same material. It was followed by an order for a large woollen mill also for Turkey.

James Bogardus (1800–74) developed his famous cast-iron framing system in his own factory in New York in 1848–49 at the corner of Center and Duane Street. In this, and many buildings built on a similar principle, he used a range of standardized columns and spandril girders which fitted accurately together by bolting, after the meeting surfaces had been machined. In 1855–56, he built two shot towers, the height of the first 175 ft. (53 m), the second 217 ft. (66 m), in which he used iron frames on an octagonal plan with infill panels of brick; pioneering freestanding tall-framed structures in iron. In about 1865 Daniel Badger, who had been in the cast-iron business in New York before Bogardus, built cast-iron grain elevators for the Pennsylvania Railroad and the U.S. Warehousing Company to the designs of his engineer G. H. Johnson. The St Ouen Dock Warehouses near Paris, designed in 1864 by Hippolyte Fontaine, consisted of a six-storey block measuring 630 x 82 ft. (192 x 25 m), constructed with hollow brick arches spanning between wrought-iron beams carried on columns at 13 ft. (4 m) centers. This building pioneered the use of the multistorey iron frame, no longer reliant on masonry walls for its rigidity. On the external enclosure, the iron framing is visible, and supports infill panels of brickwork. Rigidity is achieved through the column-and-beam connections. The Boat Store at Sheerness in Britain (1858–60) by Col. Greene is a four-storey, iron-framed building also relying on rigid column-and-beam connections for its stability. The external walls have long horizontal windows between the columns with corrugated iron spandril panels under them. In this building much of the floor framing is made in timber.

The Menier Chocolate Factory (1871-72) at Noisiel-sur-Marne, designed by Jules Saulnier (1817–81), is another early example of the complete iron frame in a multistorey building. In this building the new structural ideas are an integral part of the architectural design. The facades which consist of non-load-bearing decorative brick walls are overlaid with a system of diagonal braces that confer rigidity to the building frame. These structures, with non-load-bearing external walls, were the forerunners of the Chicago frame buildings of the 1880s, and other later skyscrapers.

The first wrought-iron columns in the U.S. were developed by David Reeves of the Phoenix Iron Company at Phoenixville, Pennsylvania, in 1864. These and other types made up of rolled section became popular when confidence in cast-iron columns reached a low ebb after the collapse in 1860 of the Pemberton Mill at Lawrence, Mass., which killed 200 people.

Railroad stations, especially in large cities, were often covered with iron roofs of innovatory design. Euston Station, London, had wrought-iron roofs covering individual tracks. In 1849–51, Richard Turner used a single arched span of 153 ft. 6 in. (47 m) to cover six tracks, three platforms, and a roadway at the Lime Street Station in Liverpool, England. This roof was of light wrought-iron construction with principal rafters approximating standard rail sections in shape, strutted with wrought-iron members

and tie rods. The covering was corrugated iron. An iron-framed barrel-vaulted shed was used in the Gare de l'Est in Paris (1847–52). Paddington Station in London (1852–54) by I. K. Brunel (1806–59) is an example of an early large terminus shed with a roof made up of a series of barrel vaults. It incorporated a transept. The decoration of this utilitarian structure was entrusted to the architect M. D. Wyatt (1820–77). In 1863 Jacob Hittorf (1792–1867) replaced a smaller station building for the Gare du Nord in Paris, built in 1847 by Leyonce Reynaud (1803-80), with a large train shed using Polonceau trusses spanning 115 ft. (35 m) with aisles of 57 ft. (17.5 m) supported on delicately decorated columns.

In 1868, W. H. Barlow and R. M. Ordish designed the large roof for St Pancras Station in London which comprised a single wrought-iron lattice arch springing from track level and spanning 240 ft. (73 m), with a height at the center of 82 ft. (25 m). This was to remain the largest span in an iron roof until the Galérie des Machines of 1889, by Victor Contamin (1840–98) and Ferdinand Dutert (1845–1906), with its three-pin portal frames spanning 375 ft. (114 m). In the U.S. the first all-iron arched train shed covering a number of tracks was built at Cleveland, Ohio, in 1865–66 by B. F. Morse with a span of 180 ft. (55 m). The largest single-span train shed in the world was Broad Street Station, Philadelphia (1892–93), designed by the engineers Joseph M. Wilson and Brothers. The clear span of 300 ft. 8 in. (92 m) is made up of a three-pinned, multi-centered arch.

Market halls became common in many cities throughout the world in the 19th century. An early iron example was the market hall at La Madeleine in Paris, built in 1824. The Hungerford Fish Market in London (1835), designed by Charles Fowler (1791–1867), was an elegant, freestanding, single-storey, cast-iron pavilion with a central raised section to admit light and air. The roofs were pitched at shallow angles over the main span with cantilever sections beyond the two rows of columns which carried the roof gutters. In 1854 Victor Baltard (1805–74) and Felix Callet (1791–1855) began to build the great central market of Paris with its many three-tiered iron and glass pavilions and covered streets. This complex, which enclosed a vast area, was demolished in 1973. Baltard and Callet's pavilions were less innovatory from a structural point of view than rival designs submitted by Hector Horreau (1801–72) and M. Flachat (1802–73), which had wider spans and more forward-looking structures.

Covered shopping arcades were built in many European cities. In Paris the earliest ones date from the 1770s, and many of the 19th-century ones have iron and glass roof framing. Generally these have fairly modest spans, rarely exceeding 20 ft. (6 m). Examples

Boat Store, Sheerness, England (1858-60), by Col. Geoffrey Green.

Paddington Station interior, London, England (1852-54), by Isambard K. Brunel.

St Pancras Station, London, England (1868), by W. H. Barlow and R. M. Ordish.

Hungerford Market, London, England (1868), by Charles Fowler.

include the Passage du Grand Cerf (1824–26) and Passage Jouffrey (1845–47). The Galleria Vittorio Emanuele in Milan, built by Giuseppe Mengoni (1829–77) in 1867, is an example of this form of building on a grand scale. The central cupola is 160 ft. (49 m) above ground, at the intersection of two covered streets, one 643 ft. (197 m) and the other 344 ft. (105 m) in length with roofs at 88 ft. (27 m).

Commercial buildings using iron became widespread in the 19th century. By the end of the 1830s many shop fronts and beams above large shop fronts were being erected in cast iron in European and American cities. In 1835, J. L. Mott built a foundry in New York specializing in the manufacture of cast-iron store fronts. A four-storey commercial building erected at No. 50 Watling Street, London (c. 1843), incorporated cast-iron beams and columns in its upper two storeys, making large windows possible. In the U.S. the Lorillard Building in New York (1837) has cast-iron columns extending through the first two storeys, combined with cast-iron beams at both these levels. In 1846, Daniel Badger moved to New York from Boston, where he had been making iron store fronts since 1842. Between the 1850s and 1870s he manufactured facades and other components for a prodigious number of buildings, mostly between two and six storeys, to be erected in the U.S. and for export. These buildings were normally composed in a repetitive Venetian Renaissance style with storey heights varying between 9 and 14 ft. (3 and 4 m) and bay widths around 6 ft. (2 m). The five-storey Haughwout Building at the corner of Broadway and Broome Street, built in 1857, is a good example.

James Bogardus built many cast-iron facades and other building components, including buildings for the California gold rush and a scheme for the New York Crystal Palace with a suspended roof in iron and glass (1852). He supplied cast-iron columns and beams for the Harper and Brothers Printing Company Building (1854), designed by J. B. Coliers. This building was the first in the U.S. to use wrought-iron joists to support the brick arches that spanned between them. These were rolled by the Trenton Iron Works.

In Britain cast-iron fronts similar to those manufactured in the U.S. were made for export to the expanding colonies and other markets, as well as for use at home. The Jamaica Street Warehouse in Glasgow (1855) and Oriel Chambers in Liverpool, designed by Peter Ellis (1804–84) in 1864, with its repetitive bay windows foreshadowing some of the Chicago buildings of the 1880s and 1890s, are both good examples. In Paris, the Bon Marché department store (1873–76) designed by Louis-Auguste Boileau (1812–1896) with Armand Moisant and possibly Gustave Eiffel

(1832–1923) as engineer is a tour de force in its mastery of cast- and wrought-iron construction. This large building is in some places up to six storeys high; the basement covered an area of 30,000 sq. ft. (2,790 sq. m). Its deep plan is perforated by light wells enclosed in iron and glass roofs and bridged by "passarelles." Monumental staircases and balconies with decorative iron work enrich these light-flooded spaces. This building formed a model for many subsequent large department stores in cities throughout the world. The office building at 24 Rue Réaumur in Paris (1904–05), which is attributed to Georges Chedanne, uses riveted wrought-iron columns and sheet iron spandrils in a totally fresh way, appropriate to the material and without any clear allusion to previous architectural styles. Around this time, Art Nouveau architects, among them Victor Horta (1861–1947), Hector Guimard (1867–1942), and Frantz Jourdain (1847–1935), used wrought iron forged into sinuous shapes as decoration and structure, often with the acceptance of rivets as part of the decorative scheme. The Maison du Peuple in Brussels (1896–99) by Horta, and the Samaritaine department store by Jourdain illustrate this trend. Eugène Emmanuel Viollet-le-Duc (1814–79), the French architectural theorist, had advocated the use of exposed and riveted iron in his lectures.

Public buildings and places of assembly and entertainment were among the first building types to employ iron structural elements. The House of Commons gallery in London (1706) and the Théâtre du Palais Royal (1758–90) have already been mentioned. Foulston's Theater Royal in Plymouth (1811–13) was probably the first public building in Britain to employ cast- and wrought-iron framing on a large scale. John Nash (1752–1835) used iron columns and other elements in the Brighton Pavilion (1818–21). The Chesnut Street Theater in Philadelphia (1818–24), designed by William Strickland (1788–1854), was the first U.S. building to use iron columns. In this building the slenderness of cast-iron columns and decoration made possible by molding techniques were exploited architecturally. The John Travers Library in Paterson, New Jersey (1846), initiated the use of cast-iron floor beams in the U.S. These beams spanned 16 ft. (5 m).

In the 1840s, many public buildings in Britain employed iron in roofs, columns, and beams, but these were often covered in other finishes or otherwise hidden. Bridgewater House (1847–57), designed by Sir Charles Barry (1795–1860), and the Museum of Economic Geology (1847–48), designed by Sir James Penethorne (1801–71), both had cast-iron roofs over their galleries. In the sorting room of the General Post Office in London (1845), designed by Sydney Smirke (1798–1877), the cast-iron arched ribs with their

Dittenhofer Building at 427–429 Broadway, New York. A typical multistorey cast-iron front (1870).

ABOVE: Le Parisien Building at 24, Rue Réaumur in Paris, France (1904–05), by Georges Chedanne.

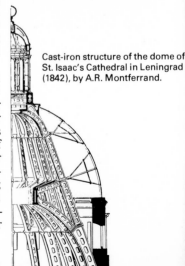

Cast-iron structure of the dome of St. Isaac's Cathedral in Leningrad (1842), by A.R. Montferrand.

bolted connections were clearly visible internally. Sir Charles Barry's Houses of Parliament (1840–56) employed iron members discreetly in many parts of its structure, but had an iron roof covering of large cast-iron plates. The Bibliothèque Ste Geneviève in Paris (1843–50), designed by Henri Labrouste (1801–75), was probably the first large public interior to use cast and wrought iron from floor to ceiling. The external walls are of masonry construction. The reading room measures 278 x 69 ft. (85 x 21 m) and is divided down the middle by a row of 16 columns which support two rows of perforated arched ribs which carry the roof.

In Labrouste's Bibliothèque Nationale (1858–68) the reading room is enclosed by nine domes supported on light, semicircular iron ribs, which are carried on slender cast-iron columns. The book stacks, which were designed to house 900,000 volumes, are arranged on five floors including the basement, and the space is flooded with light from a glass roof by the use of gridiron open-work floors. Thomas U. Walter (1804–87) designed additions to the U.S. Capitol in Washington using iron extensively as a fireproofing measure in the interiors of the Library of Congress room (1851–52) and the new wings (1852–54). Roof trusses, ceilings, wall paneling, window frames, and the like are all made of iron.

Many public buildings in the 19th century employed large domes over major spaces, and there was a tendency to build the framing of these in iron, which was often far cheaper, lighter, and quicker to erect than masonry and much safer than wood, both in durability and incombustibility.

The Bourse de Commerce in Paris, which had been the Halle au Blé before its wooden dome burned down, was covered with what was probably the first large iron-framed dome (1806–1811), designed by François-Joseph Bélanger (1744–1818) and Brunet. The 51 cast-iron ribs were cast at Creusot, and when in position were held together by wrought-iron rods and straps. In 1842, August Ricard Montferrand (1786–1858) designed an elaborate cast-iron framing system for the dome of St Isaac's Cathedral, St Petersburg, Russia. The components were cast under the direction of William Handisyde, a British engineer who worked in Russia for many years. The Coal Exchange in London (1846–49), designed by James Bunstone Bunning (1802–63), had over its main court a large hemispherical dome composed of richly decorated, cast-iron ribs framing elaborately painted panels. This was followed in 1854 by Sydney Smirke's large dome over the reading room of the British Museum, London. The timber dome of the U.S. Capitol was replaced in 1856–64 with a larger more impressive cupola designed by Thomas U. Walter (1804–87), with engineer-

ing by M. C. Meigs and Schoenborn. All the structural elements in this dome were made of cast iron. Later in the century wrought-iron framing was used in the construction of large domes such as the one over the Albert Hall in London (1867–71) by Captain Fowke (1823–65), which had a span of 185 ft. (56. 5 m).

Large glasshouses completely framed in iron were built from the early years of the century. The Musée d'Histoire Naturelle in Paris (1833–34), by C.R. de Fleury, is an early example of an extensive building of this type. The iron and glass structures are built against a masonry wall. Another important glasshouse completly framed in iron is the Palm House at Kew, built in 1845–47 by Decimus Burton (1800–81) and Richard Turner. It is a freestanding building with tightly spaced iron glazing bars enclosing barrel-vaulted spaces with apsidal ends. Many of the great exhibitions of the 19th century took their inspiration from these glasshouses. The London Crystal Palace of 1851, by Sir Joseph Paxton (1801–65), was the first and although it used iron for its major structural components, its floor, gutters, glazing bars, and other parts were largely made in timber. Hector Horreau (1801–72) and Richard Turner had both proposed all-iron and glass designs in the competition for this building. Many important exhibition buildings and structures followed, including the New York and Dublin Exhibition of 1853, the Paris Exhibitions of 1855, 1867, and 1878, culminating in the construction, for the 1889 Paris Exhibition, of the Galérie des Machines and the Eiffel Tower (1887–89) which reached a height of 1,000 ft. (300 m), and was completed in just 26 months.

Churches and chapels using iron components and structural elements could in many cases be built considerably cheaper than those using conventional construction. The three churches erected in Liverpool between 1813–16 by Thomas Rickman (1776–1841) and John Cragg (1767–1854) have already been mentioned. Among these St George's in Everton (1813–14) was the one in which iron was used most extensively. In 1842–43, Edward Blore (1787–1879) erected a chapel at Buckingham Palace, London, with cast-iron columns and decorative trusses. In 1855, Louis-Auguste Boileau (1812–96) and the engineer A. L. Lusson completed the church of St Eugène in Paris, in which cast and wrought iron was used extensively for the structure and interior of the building. It was built for a quarter of the cost of competitive designs, and its interior, with its slender structural members and lofty proportions, albeit in a Gothic style, was much admired for the new spatial possibilities it offered. Notre Dame du Travail, also in Paris (1901–1903), by Jules Astruc (1862–1935), created an interior in which the structural wrought-iron

Wrought-iron gates (1905), by Hector Guimard.

Detail of ironwork from the Bibliothèque Ste Geneviève, Paris, France (1843–50), by Henri Labrouste.

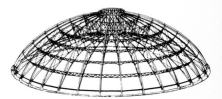

ABOVE: Dome structure of the Albert Hall, London, England (1867–71), by Captain Fowke with Groover and Ordish as engineers.

elements with all their rivets and diagonal bracing were visible without extra adornment.

In 1876, an iron spire was built on the tower of Rouen Cathedral to replace a wooden one which had burned down. The cathedral thus became the second tallest building in Europe, second only to Cologne Cathedral. Cast-iron and wrought components for churches were exported by missionary societies and congregations to British and later to French colonies. The church at Macquire Street in Sydney (1853) had a cast-iron front probably made by the contractors Charles D. Young and Company of Edinburgh and London. Samuel Hemming, a corrugated iron building manufacturer, exported many churches to the colonies and erected hired temporary churches for congregations in Britain. The temporary church of St Paul's, Kensington (1855), with seating for 800 people, is an example of one of his larger buildings. Because these structures were popular in Britain, the Ecclesiological Society, which had set itself up to control the forms and use of objects used in the ritual of the Church of England, commissioned the design of an iron church which was published as a model in 1856. Many of the internal details of this building were borrowed from Dean and Woodward's Oxford Museum of 1855. The exterior was of simple, corrugated iron.

Other types of iron buildings made for export became extremely popular after the 1830s. King Eyambo of Calabar ordered a two-storey iron palace from William Laycock of Liverpool in 1843. Industrial sheds and other buildings such as the market for San Fernando, Trinidad (1848), and a sugar factory for Barbados (1846), were exported to the West Indies and other places by J. H. Porter (1824–95) during the 1840s. E. T. Bellhouse (1816–81) of Manchester exported a number of buildings and sheds to California during the gold rush in 1849 and when his exhibit was admired at the 1851 Exhibition by Prince Albert, he was commissioned to build a corrugated iron ballroom at Balmoral in Scotland, which still stands. Bellhouse's contracting, like many of his competitors, was not limited to buildings. In 1858 he built the gas works and supply system for port Buenos Aires in Argentina. Daniel Badger in New York exported buildings abroad, including a large sugar storage shed for Havana in Cuba in the 1860s. These contractors built railroads, water supply systems, port facilities, factories, and many other civil engineering works throughout the world. The British West Indian barrack system of 1826 has already been mentioned. The French also experimented with prefabricated, military buildings for their colonies. In 1845 the French Navy erected a small experimental hospital building in wrought iron in Guadeloupe to designs by A. Romand. Perhaps the most sophisticated 19th-century system for prefabricated buildings in iron was developed in the 1880s in Belgium by Joseph Danly. He used a wrought-iron structure, cast-iron connecting pieces, and modular pressed-iron cladding panels forming walls with internal ventilation. This system was intended for buildings in the Congo.

Cast-iron lighthouses were erected in many isolated places throughout the world. The lighthouse at Morant Point in Jamaica, erected by Alexander Gordon in 1841, was constructed of 135 flanged cast-iron plates bolted together on the inside of the structure which was 108 ft. (33 m) high. A skeletal wrought-iron lighthouse structure was erected by Richard Walker, a British contractor, for the American Government on the Florida Sands in 1851.

In Britain in the early 19th century the development of seaside resorts resulted in the building of many promenade piers. One of the first to use an iron structure was the pier at Brighton, built by Captain Samuel Brown in 1822 as a suspension structure using wrought-iron chains with spans of 255 ft. (78 m). The total length was 1,136 ft. (346 m) but parts of it were destroyed in gales in 1833 and 1836. Many other piers followed, but they were generally built as a deck supported on cast-iron columns often founded on iron screw piles: Gravesend Pier, built in 1844, is a good early example.

Steel

Steel is a form of iron containing a measured proportion of carbon (0.15–1.5%). The properties of steel can be carefully controlled: it can be made highly elastic and ductile, hence its general replacement of wrought iron, or it can be made of great hardness and durability. Many other properties can be given to steel by alloying it with other metals.

In 1854, the Englishman Henry Bessemer (1813–98) accidentally developed a process of

Eiffel Tower, Paris, France (1887–89), by Gustave Eiffel.

King Eyambo's Palace exported to Africa in 1843.

Prefabricated corrugated-iron buildings, mainly for export to the colonies, at Samuel Hemming's works near Bristol, England (1854).

converting pig iron into steel by blowing air through molten iron. The air "burned" out a large proportion of the carbon normally found in the pig iron, leaving behind steel. His results became widely known and considerable interest was shown by many people in the iron trade. The invention of this technique is a controversial subject since the American William Kelly (1811–1888) contested Bessemer's claim, proving in a patents case that he had made "refined iron" by a similar process in 1851. After initial setbacks, Bessemer was able to make steel from phosphorus-free ores in his "converters" which, soon after their invention, could make 5 tons (4,536 kg) of steel in less than half the time needed to make 560 lb. (254 kg) of wrought iron.

During the late 1850s Charles Werner von Siemens (1823–83) developed ways of reducing the fuel consumption of furnaces by recovering heat from combustion gases. His furnace, which was gas-fired, found applications in other industries requiring high temperatures for their manufacturing processes. In 1863 the Frenchman Pierre Emile Martin (1824–1915) successfully produced steel in a Siemens furnace. This gave rise to what is known as the "Siemens–Martin" or "open hearth" steel-making process, in which the iron is raised to a very high temperature in contact with a suitable slag, which decarbonizes the iron, converting it into steel. By 1870 the open-hearth process was being employed in a number of countries. This method had the advantage of allowing the use of a greater variety of iron ores, as well as scrap, and since the operation of the process was much slower than the Bessemer converter (6–15 hours compared to half an hour), the quality of the product could be carefully controlled and monitored.

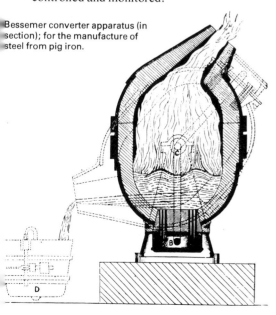

Bessemer converter apparatus (in section); for the manufacture of steel from pig iron.

Steel rails were first rolled in 1860, and rolling mills making other sections soon took advantage of the fact that they were no longer restricted by the puddling process, and could roll sections containing a large volume of material. Wider and taller rolled products could now be made on a large scale.

Rolling mills developed enormously in the 1850s and 1860s by becoming increasingly mechanized. In 1862 the continuous mill was introduced, in which the piece to be rolled would pass from the first roll stand near the furnace through a succession of roll stands, until the desired form emerged. Previously, pieces had been passed back and forth across a stand of rolls, with much loss of heat and wastage of labor.

The increase in production of blast furnaces and steel mills led to further mechanization, and hence the initiative passed from Europe to the U.S. where mechanized pig iron casting and furnace changing featured among the many new important developments. By 1890 the U.S. was the world's largest iron producer. The electric motor began to be used in rolling mills from 1880 onward. At first its use was confined to light duties, but gradually it superseded the steam engine as a more easily controllable, less bulky, and less awkward source of power.

The universal mill was invented in 1897 by the Englishman Henry Gray, but it was first used in Germany in 1902. In this type of mill, rollers act on all sides of a section simultaneously, resulting in a product with truer dimensions and with surfaces more accurately faced. At first, it was used to roll plate, but was later employed for the universal beams, columns, and channels so well known to the construction industry.

The first steel bridges were a group of three erected in Holland in 1862 with spans of up to 121 ft. (37 m). Further bridges constructed in Holland, which employed Bessemer steel, showed signs of unreliability. Steel was regarded with suspicion in the construction industry, even though it had been used in the construction of a steel-hulled warship for the French Navy in 1874. Improvements in steel production, aided by metallurgical research, gradually raised confidence in the material, but wrought iron remained the main structural form of iron until about 1890. In the U.S. the first bridge made entirely of steel was the Glasgow Bridge (1872) over the Alton River in Missouri. This made use of steel whipple trusses, with spans of 314 ft. (96 m), produced in an electric arc furnace. Extensive tests were carried out on this material before it was accepted, and even though one of the spans collapsed during erection it was not as a result of the quality of the steel.

In 1877 the British Board of Trade authorized the use of steel in construction engineering which led to its use in the famous Forth

View from control pulpit of slabbing mill at a modern steel works.

Finishing mill at a steel works.

Forth Bridge, Scotland (1882–90), by J. Fowler and B. Baker.

Bridge in Scotland, built between 1882 and 1890 by Sir John Fowler (1817–98) and Sir Benjamin Baker (1840–1907). The main structural members were hollow steel tubes constructed out of riveted plates. The two center spans were 1,710 ft. (521 m) long.

Steel-framed buildings

In the late 1880s the multistoreyed steel frame using riveted and bolted steel columns, beams, and joists was developed in Chicago. The Home Insurance Building (1884–85), designed by William Le Baron Jenney (1832–1907), was the first Chicago building to use steel in part of its structure. In the upper six of the ten storeys, Jenney was permitted to use steel beams in place of wrought iron as originally specified, and these were supplied by the Carnegie Phipps Steel Company of Pittsburg. This building had originally been designed as a complete freestanding frame, but the authorities insisted that some of the loads be carried by the party walls. In the second Leiter Building (1889), Jenney was able to use an independent steel cage to support the whole structure, and the eight-storey facade was no longer carried on heavy masonry walls.

The Auditorium Building in Chicago (1887–89), designed by Dankmar Adler (1844–1900) and Louis H. Sullivan (1856–1924), although partly of masonry construction, used iron and steel framing with impressive confidence in the creation of many of its large and elaborate volumes. The complex, containing 63,350 sq. ft. (5,891 sq. m), included a large theater, offices, hotel rooms, and other accommodation. The Tacoma Building (1888–89), designed by Holabird and Roche, had street elevations completely framed in iron, and all the junctions between the structural elements were riveted, conferring extra rigidity to the structure. Baumann and Huell's Chamber of Commerce Building (1888–89) was the first building of this type employing no structural masonry at all.

The Reliance Building (1890–94), designed by Daniel Hudson Burnham (1846–1912) and John Wellborn Root (1850–91), is an outstanding example of a freestanding Chicago steel tower with a marked vertical emphasis and a facade of steel and glass which anticipated many 20th-century buildings. In the structure, two-storey steel columns were used, which made the erection of the frame very rapid. The structural cage of the final ten storeys was erected in 15 days.

By the time of the 1895 depression, engineers and architects in Chicago had created many structural innovations to make the framing of these steel-framed buildings more reliable, efficient, and economical. These innovations include the development of bed-rock caissons, fully riveted structures, and the use of wind-bracing elements integrated into the structure. Adler and Sullivan's Carson, Pirie, Scott Store (1899–1906), is an example of the mature form of "Chicago construction."

In New York the erection by the Keystone Bridge Company of Gustave Eiffel's iron internal framework for the Statue of Liberty (1883–86) fired the imagination of many designers, with the potential that this type of structure offered.

Iron framing for tall buildings began to be used tentatively in New York before the end of the 1880s, but New York lagged behind Chicago in structural innovation. The first application of iron framing in New York was the 11-storey Tower Building at 50, Broadway (1888–89), designed by Bradford Gilbert. The Manhattan Life Insurance Company Building (1893–94) by Kinball and Thompson, used a completely framed structure for its 17 storeys. From then onward, buildings of ever-increasing height were erected. By the end of the century, a 30-storey building had been built. Improvements in developing more rigid framing connections and other technical developments enabled the construction of such buildings as the 58-storey Woolworth Building (1912), by Cass Gilbert (1859–1934), and the 102-storey Empire State Building (1931), designed by Shreve, Lamb, and Harrison.

From the early years of this century, many advances were made in the construction of tall, steel-framed buildings. After World War II, the use of masonry, which had been the normal external cladding, began to be superseded by curtain walling, which had been proposed by Mies Van der Rohe (1886–1969) as the appropriate cladding for tall buildings as early as 1920–21 in his project for glass-clad skyscrapers. Typical examples of buildings as early as 1919-20 in his project for construction include the Alcoa Building in Pittsburg (1953), by Harrison and Abramovitz, with its stamped aluminum cladding panels, and Lever House in New York (1954) by Skidmore, Owings, and Merrill (S.O.M.). Precast concrete as a cladding material became popular in the 1960s, and the Pan Am Building in New York can be taken as an example of this type of construction.

Masonry cladding had contributed to the stiffness of structures, but with lighter walling systems other methods had to be found for bracing tall structures against wind and seismic loading. Elevator and service cores, which had been used for a long time as stiffening elements, had greater demands placed on them. More reliable rigid connections between columns and flooring elements were also developed. Plant floors, carrying air conditioning and other services, were distributed at a number of levels through

Steel frame of the Unity Building, Chicago (1892), before concealment by external cladding. Architect C.J. Warren.

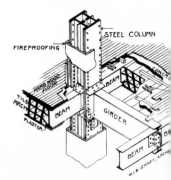

Detail of steel column and girder structure; designed by William Le Baron Jenney for the Fair Store Building in Chicago (1892).

Diagram of the steel structure of the Statue of Liberty in New York (1883–86), designed by Gustave Eiffel.

Woolworth Building, New York (1912), by Cass Gilbert.

the height of the building, and these were made into rigid structural elements by the introduction of extensive diagonal bracing that would have been obstructive to windows on other floors. The use of these braced plant floors contributed considerably to the stiffness of the entire structure. Another form of stiffening which has become popular is the use of diagonal bracing throughout the height of the building, making the whole structure into a rigid tube. The John Hancock Building in Chicago (1968), designed by Fazlur Kahn of S.O.M., is a well-known example of this type of construction.

The Sears Tower in Chicago (1974) which reached a height of 1,460 ft. (445 m), was designed by S.O.M. as a further development of the same principle, being built up of nine minitubes each 75 ft. (23 m) square which reduce in number through the height of the building. These externally braced towers produce great savings in steel when compared to other types of framing for buildings of a similar height.

Normal techniques of fireproofing steel structures involve the use of some form of casing to the structural members, either in the form of concrete cast around the members or of some other kind of protective material. These systems evolved after a number of iron-framed buildings burned down in the 1830s and 1840s: until then, they had been presumed fireproof. The Chicago fire of 1871 was a further stimulus to the development of fireproofing techniques. A complete system of fireproofing using terra-cotta elements was patented in Britain by Whitchord in 1873. The disadvantages of fireproofing techniques are that they may add considerable weight to the structure, increase the time taken for installation, and tend to make the structural steel elements invisible by burying them in some type of casing. An alternative fireproofing method has been developed which allows the steel frame to be left unclad, by constructing it of tubular members containing water fed from tanks, which cool the structure in the event of a fire. This alternative method is employed in the 64-storey United States Steel Corporation Building in Pittsburgh and the Centre Pompidou in Paris (1976), designed by Piano and Rogers.

Other developments in the use of iron and steel

After the mid-19th century, metallurgical science advanced at a rapid rate and many new alloys of iron were studied and developed. An example of this progress can be seen in the invention of tungsten-alloy steel in 1868, which revolutionized steel-cutting tools, by providing harder, more durable cutting edges for drills, lathe bits, etc.

Welding and associated processes, which

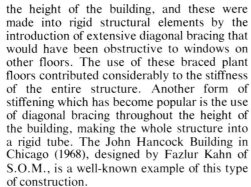

Lever House in Park Avenue, New York (1950), by Skidmore, Owings, and Merrill. The building is clad in a glass curtain wall.

John Hancock Center, Chicago (1968) by Skidmore, Owings, and Merrill. Constructed with external bracing throughout the height of the tower.

Alcoa Building, Pittsburgh (1953), by Harrison and Abramovitz; clad in stamped aluminum panels.

Large-span tubular steel beam at the Centre Pompidou, Paris, France (1976), by Piano and Rogers.

Castellated steel beam supporting a steel roof deck and a ceiling.

developed toward the turn of the century, increased the range and potential of iron and steel design and fabrication. The melting of metals by electric arc was invented by the Russian Barnados in 1887, with further developments following shortly afterward. Acetylene was discovered in the U.S. in 1892, and by 1900 the first serviceable oxyacetylene torch was already in use. It is possible to reach very high temperatures (5,548°–6,268°F/3,100°–3,500°C) by means of these techniques, thereby allowing the iron or steel (melting point 2,668°F / 1,500°C) to join together by fusion at the welded joint. There is no reason for these joints to be weaker than the parent material, although welded joints were initially regarded with suspicion. Welded structures began to be used toward the end of World War I. The first all-welded bridge in the U.S. dates back to 1927.

Flame cutting of iron and steel evolved at about the same time as the welding techniques described above. The heated steel does not melt under the jet of oxygen, but oxidizes rapidly to form a clean cut. These cuts can be made quickly even through thick pieces of material. Flame cutting is cheap, fast, and efficient for producing any desired shape in steel. Cutting torches can be guided by hand but the process also lends itself to automation.

Flame cutting combined with welding makes it possible to fabricate structural members that have forms tailored to the loads they have to carry. Cleaner, lighter structures are therefore possible. They are less subject to corrosion in hidden joint areas and the simpler forms reduce the amount of maintenance work required.

It became feasible to cut standard rolled sections and produce efficient structural shapes which were economic in their use of material, by giving them extra depth where required and by removing steel where it was not needed. Castellated steel beams are an example of such a form.

Between the wars, structural sections made out of cold rolled steel strip or sheet began to be used. The sheet is deformed into desired shapes by rolling or bending. Useful sections can also be made by building up combinations of these cold rolled sections by spot welding, a process that can be highly mechanized.

In 1935 the French engineer E. Mopin built a complex of 1,200 apartments using structural members of this type in the building frames. These sections are employed today in many parts of the construction industry. Good examples are the Z-purlins, used as spanning elements over roof trusses, and the members of grids that support suspended ceilings. Steel sheet has many further applications in building: it is used for the fabrication of ventilation and air-conditioning ducts; it is made into expanded metal used in concrete and in plastering, and it may be

stamped into objects such as electrical junction boxes, or formed into protective conduit for electrical wires. The spanning characteristics of corrugated sheet steel are utilized in the construction of some reinforced-concrete floors, in which the steel contributes to the load-bearing properties of the deck and acts as permanent shuttering, resulting in speedy erection and savings in formwork. Mild steel bars have been used in ever-increasing quantities since before the turn of the century for reinforcement in concrete construction. These bars, especially in larger sizes, often have deformed surfaces to increase their bonding with the concrete. Bars may be welded to form premade reinforcement cages, meshes, or mats; alternatively they can be made into the desired shape on the building site, by bending and site assembly.

Wide-flange beams and columns were developed soon after World War I in Luxembourg. These have wider and thicker flanges than normal sections and the flanges are of uniform thickness, making the connections, splices, fixings, and joints much simpler and cleaner. In the U.S., which pioneered their use on a large scale, these larger and heavier sections were rolled with varying widths of flange for the same depth of beams, simplifying the range of connection details needed in complex structures.

Numerous steel products began to be made and used in the period between the wars, among them steel window frames and steel tubes. Seamless steel tubes were first manufactured by the Mannesman brothers in 1885. Steel tubes, normally 2 in. (50 mm) in diameter, became very popular for scaffolding, especially when combined with patent connections which made rapid erection and removal possible. Galvanized steel water pipes and electrical conduit were marketed extensively, combined with a large array of accessories to form connections and junction parts. Steel tubes are strong in compression and were therefore used in the construction of columns and trusses, and later as members for space frames. At first, tubes with a circular cross section were the only ones available, which made jointing by welding fairly complicated since the tubes required cutting to complex shapes, but square and rectangular tubes became economically viable for use in buildings in the late 1950s and early 1960s. These can be used to make very clean and light-looking beams and trusses.

Developments in metallurgy and structural theory have imposed stricter demands on the performance of iron and steel alloys, and have led to advances in structural design; special high-tensile steels are available which have a permissible tensile stress much higher than normal mild steel. These steels were used in the construction of Bailey bridge panels during

World War II, and many improvements have since been made in the properties of these steels.

Stainless steel was developed accidentally in 1912. This alloy, which does not corrode, normally contains about 10–14% chromium. It is used in building products such as kitchen sinks, facing panels, bolts, tension wires, etc.

Other steel alloys such as "Weathering," also known as "Cor-ten," have been adopted by architects, engineers, and builders for their special properties. "Cor-ten", which is a copper-steel alloy, develops a tenacious oxide coating which obviates the necessity of painting. Ordinary steel emerges from the mill in a form that will corrode rapidly if not treated. Many paint systems as well as plastic and metal coating systems, such as galvanizing and sherardizing, have been developed to improve its resistance to oxydization and to enhance its appearance. Steel may also be stove-enameled, which gives a resistant and smooth vitreous coating which can be easily cleaned. Intumescent paints swell up under intense heat, producing a layer of insulation around structural elements. This is a relatively recent form of fireproofing.

Steel framing has been used by many architects in remarkably personal and individual ways, displaying different attitudes to the material and contrasting aesthetic ideas. This often becomes most apparent in designs for houses, where budgets and other limitations can be more relaxed. This divergence of stylistic approach is exemplified in the works of four major 20th-century architects.

Jean Prouvé (b. 1901) was trained as an art-metalworker. He has dedicated himself to the study of construction using light metal prefabricated components and has designed a wide range of houses and other buildings in which he abandons formalistic attitudes and replaces them with an empirical analysis of practical problems of assembly and use. Production and manufacturing techniques, as well as forms derived from the automobile and aircraft industries, are often a major source of inspiration to him. This is clearly seen in such buildings as the Abbé Pierre House and the experimental houses at Mendon (1954).

Mies Van der Rohe (1886–1969), in his Farnsworth House in Fox River, Illinois (1946–50), uses standard rolled steel H-sections as the framing elements of a welded frame, supporting horizontal roof and floor slabs which contain a rectilinear glass enclosure. Obsessive attention was paid to visual detail in this carefully perfected structure, which demanded an extremely high level of craftsmanship.

Charles Eames (1907–78) built his own house at Santa Monica, California (1949), using prefabricated steel components such as light lattice joists, beams, and window units ordered direct from manufacturers' catalogs.

He used these in a highly original way, with close attention to detail, color, and proportion to create lively and airy spaces—some of them reminiscent of old Japanese houses.

Bruce Goff (b. 1904) has always made use of materials in an inspired way, often unencumbered by conventional preconceptions. In 1949 he utilized war surplus components, taking the curved steel framing members of Quonset huts as structural elements for the Umbrella (Ford) House. Instead of using them to create barrel-vaulted enclosures, for which they had been originally designed, he combined them radially with other materials to form a remarkably original internal space, which was covered on the outside with wood shingles.

Copper

Copper began to be worked c. 3500 BC in China, and before 3000 BC in Mesopotamia. It was in common use in Egypt by 2600 BC where copper cutting tools have been found dating from the Old Kingdom. It is possible that they were hardened by the addition of arsenic, or that traces of arsenic and other hardening metals occurred naturally in the ores. Until the 15th century AD, the demand for copper was small, the requirement being mainly for making bells and utensils.

It was occasionally employed in building and in sculptural additions from weather vanes to large figures on account of its durability and the ease with which it is worked. Hinges cast in copper were sometimes used in the Middle Ages. Basel Cathedral, rebuilt in 1356, has a copper roof with plates of two different colors forming a patterned surface. Statues made of copper repoussé work held over a supporting core were not uncommon in the Middle Ages. The great dragon of the Ghent belfry, measuring 12 ft. (4 m) from tip to tail, was brought from Constantinople, to be installed at Ghent in 1204. Copper lends itself to gilding and enameling, but if left unprotected it will develop a patina ranging from light green to almost black.

Between 1450 and 1550, however, the development of bronze cannons and mortars completely changed the situation; improved production methods had to be discovered and the scale of operations changed. By 1650 a large proportion of the copper used in Europe was coming from the Falun mine in Sweden.

The use of copper for roofs, gutters, rain leaders, sculptures, etc grew rapidly. Palladio's basilica in Vicenza has a copper roof (1549), and the quadriga sculpture on the Brandenburg gate in Berlin was made in this material.

In the 19th century, mechanically embossed copper was extensively used for ornamental work. Copper, because of its ductility, is easier to work and form than most other

Tubular steel members in a space frame; Vienna Airport, Austria (1960), by Klaudy, Hoch, and Schimke.

School building at St Michel-sur-Orge, France (1967–68), by Jean Prouvé, using light metal structure.

Farnsworth House, Fox River, Plano, Illinois (1945–51), by Mies van der Rohe.

Umbrella (Ford) House at Aurora, Illinois (1949), by Bruce Goff.

Basilica at Vicenza, Italy (1549), by Andrea Palladio. The roof is covered in copper sheeting.

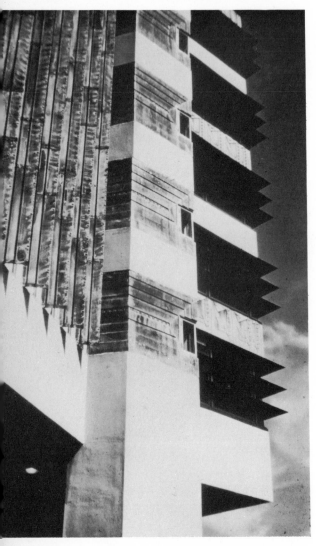

Price Tower, Bartlesville, Oklahoma (1953), by Frank Lloyd Wright. Panels and fins are clad in embossed copper.

metals. Sheet copper would be stamped a number of times between zinc dies until the desired shape of pilaster capital, decorative coffer or spandril panel emerged. Frank Lloyd Wright's Price Tower in Bartlesville, Oklahoma (1953), has stamped copper cladding panels which were used as permanent formwork for reinforced-concrete backing. Where copper is to be used for an individual work, the sheet can be beaten to the required form. The Statue of Liberty in New York (1883–86), designed by Frédéric Auguste Bartholdi (1834–1904), is made up of copper sheeting supported on an immense and elaborate steel framework. Copper is easily soldered and can be lapped and screwed. Another method of working copper is to shape the sheet material in a lathe by forming it over a circular block of the desired shape. Many conical, spherical, and other shaped lamp fittings have been made in this way.

Facsimiles of three-dimensional work can be made by electro-deposition (electro-types) in which a layer of copper is deposited over a core in another material.

The unprecedented market for brass, an alloy of copper and zinc, brought about by the 18th-century growth of the Birmingham metal trade's use of engine and machine building, led to an enormous expansion in copper mining, quarrying, and smelting. Copper is used with other metals to form a wide range of alloys. Copper oxide is used in the coloring of glazes, and copper salts are an important ingredient in many timber preservatives.

In the building industry, copper has continued to be used, in the 20th century especially, for hot-water storage tanks, water pipes, electric wiring and cables, and considerable quantities have continued to go into sheets for roof covering and flashings.

Brass

Brass, an alloy of copper and zinc, was probably a Greek discovery, made early in the 1st millennium BC. At that time, and for many centuries later, it was produced by heating copper bars embedded in charcoal and powdered zinc ore. It has been widely used in the East since antiquity. In the West it was of minor importance, except as a soldering agent, until the late 18th century, when the new engineering and machine building industries began to require it in large quantities for fittings of all kinds.

From that time onward, four main markets can be distinguished in addition to engineering. The growth of population and prosperity during the Victorian period greatly increased the demand for hinges, castors, knobs, handles, hooks, and window fasteners, and a high proportion of these were made of brass. The improvements in sanitation which took place in the middle of the 19th century made it

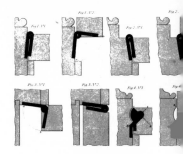

Hinges, stays, and other door and window furniture are often made in brass. The illustration shows sections of various hinge installations from a late 19th-century carpentry textbook.

Yale lock (1889): precision-made lock mechanism in brass.

Ornamental zinc roofing and trim; from the catalog of Frederick Braby and Company (c. 1890).

necessary to extend the manufacture of brass faucets, plugs, washers, pipes, and valves. The mounting investment in household and industrial property made the more prosperous sections of the community more security minded and more anxious to buy locks with which to safeguard their possessions and, although many of these locks were of iron or steel, brass was often preferred, especially for domestic premises. Electrical fittings, joinery screws, and other precision-made elements are often made in this material.

Brass is easily cast, lends itself to being worked on a lathe, can be extruded without difficulty, and in its sheet form can be formed by stamping and embossing. Because of its durability and resistance to corrosion, it is often used in naval applications and in buildings with severe exposure to sea air. Window gauzes can be made out of brass to give them a longer life.

Zinc

Zinc in a relatively pure form is thought to have been discovered in India in the 14th century, and in the West early in the 16th, and it may have been known in China even earlier in the Christian era. Zinc-bearing ores have been used since antiquity in the making of brass. Unrefined zinc was being imported by the Dutch and Portuguese from the Far East in the 17th century. European production of zinc in quantity did not begin until the 18th century and did not become important until the 19th, after the production of galvanized iron had become well established.

In western Europe, and especially Britain, the primary use of zinc continued for a long time to be as an alloy with copper to make brass, but from the late 1830s the process of coating iron or steel with the metal, known as galvanizing, became increasingly important. At first iron was coated by dipping it in vats of molten zinc after it had been treated to remove grease and dirt. Toward the end of the 19th century galvanization by electrolysis became commercially viable. Zinc compounds are also used in protective paints and other corrosion-resistant finishes.

Zinc roof covering was pioneered in Europe and by the mid-19th century the roofs of many buildings in cities such as Paris comprised the material. The Vieille Montagne Company, which was established near Liège in Belgium before the 1850s, was a very large producer of zinc for roofing and embossed ornamental work. Zinc roofing is normally laid over a boarded base, but corrugated sheets were made self-supporting and were used in the 19th century. It has also been extensively used for embossed ornaments, for flashings and other waterproofing elements, as well as for fixings for slates and other roofing elements continually exposed to the weather.

Bronze

Bronze, which is an alloy of copper combined with tin or antimony, is thought to have originated in the Caucasus in about 3000 BC. It was used for ornaments and tools in the latter half of the 2nd millennium BC in Mesopotamia, and its use is recorded in Egypt from about 2200 BC. In China, bronze working techniques were developed to a very high level of perfection in the period between the 14th–3rd century BC, and it was employed in the manufacture of a wide range of objects from cooking utensils to statues and bells.

In Greece, large lifesize statues were cast in bronze, culminating in the superb works of the 5th century BC. Many of these statues were made by the "cire perdue" or "lost wax" technique, in which a clay core is covered in wax which is molded to the appropriate shape and then covered in a further outer layer of clay. The wax is melted out, leaving a space between the two layers of clay which is then filled with molten metal. This technique required great skill and control both in the making of molds with their appropriate vents and supports, as well as in the production of sufficient quantities of molten metal at the correct temperature. Bronze tiles, cladding, and fixing elements and other objects connected with building were also made by the Greeks.

The Romans extended the use of bronze to include bronze pillars and monumental doors for civic buildings, large statues like the equestrian statue of Marcus Aurelius, as well as utilitarian objects such as water pipes, locks, keys, and oil lamps. Bronze window frames were used in Pompeii, and the Pantheon in Rome (AD 118–128) had bronze roof trusses in its portico which, at the time of the Renaissance, were dismantled in order to utilize the metal for the manufacture of cannons. The Pantheon also had bronze doors, which still remain in position.

After the collapse of the Roman Empire, bronze working on a large scale continued in Byzantium but the technique was largely lost in the rest of Europe. Large bronze doors were made for Mainz Cathedral in Germany in AD 998, and in Italy Byzantine craftsmen made bronze church doors in Verona, Rome, Amalfi, and other places between the 10th and the 12th centuries. These doors, which represented considerable investment both in skill and money, were often made of cast, repoussé, and engraved bronze elements fixed to a wooden core. This tradition of monumental bronze doors continued, culminating in the great bronze doors cast by Lorenzo Ghiberti (1378–1455) for the Baptistery of Florence Cathedral (1425–52), which marks a high point in design and metalcraft.

Renaissance sculptors revived the Roman art of making large equestrian statues by

Bronze doors in the Baptistery of Florence Cathedral, Italy. These were designed by Andrea Pisano in 1336.

Seagram Building, New York (1958), by Mies van der Rohe. Curtain wall in glass with bronze mullions and facade elements.

perfecting sophisticated techniques of bronze casting. The Gattamelata statue at Padua (1453) by Donatello (1386–1466) and the Colleoni statue in Venice by Andrea del Verrocchio (1435–88) are good examples. Other semi-architectural elements made in bronze since the Renaissance include railings, church fonts, and other pieces of church furniture such as lecterns. The famous Baldacchino over the high altar at St Peter's Rome, created by Giovanni Lorenzo Bernini (1598–1680), was made of bronze.

Bronze does not corrode easily, and develops an attractive, maintenance-free patina and is therefore favored for external ornament, fittings, and other building elements where cost is not a major consideration. Because it can be melted down easily, much early architectural bronze in Europe was seized by invading armies in various wars, in order to make coinage, cannons, and other objects.

Bronze lends itself to rolling and extrusion. Elements made in this way are often used as framing members for shop fitting, and for ornamental work such as cornices and beading which may have sections appropriate to these processes. In some cases the bronze used for extrusion contains a proportion of zinc to increase its ductility, making it more like brass. Bronze window frames are made from extruded or rolled sections and are popular when they can be afforded. The Seagram Building in New York (1958), designed by Mies van der Rohe (1886–1969), has a complete curtain-walling system made out of superbly detailed bronze components.

A wide range of special bronze-based alloys is produced for their special distinct properties. These include phosphor bronze, which is popular for load-bearing fixings, and gunmetal, containing zinc and sometimes lead, which is popular for casting. Bronzes with 6% tin are suitable for cold working and drawing, while those with 10–18% tin are suitable for casting and are a popular material for bells.

Lead

Assyrian and Babylonian masons used molten lead for anchoring clamps and fixings in masonry. Lead was used by both the Greeks and the Romans as roofing sheets, for pipes, and for lining water tanks. For these the metal had to be cast in flat trays or beaten out to the required thickness, so that only relatively small sheets were available. Pipes were made by wrapping lead sheets around a wooden spindel and soldering the seam. In Rome the lead strip used for pipemaking was required by law to be of a standard width, about 4 in. (100 mm). This is the first known example of engineering standardization. The Romans, incidentally, were well aware of the dangers of lead poisoning.

During the medieval and Renaissance period, the Roman practice of soldering pipes with molten lead was continued, but from the 17th century onward the joint was closed with tin and a soldering iron, the lips of the seam being beveled before soldering was undertaken. Joining, in Roman times, was effected by melting a lead casing around the junction and then hammering to shape.

Mills for rolling sheet lead were introduced in England in the late 17th century and in France in about 1730. Until the late 18th century, lead was the only metal to be rolled industrially. Lead tubes were first extruded in 1820, by Thomas Burr, at Shrewsbury, England, using a hydraulic ram to force the metal through the die. A patent for a similar type of process had been taken out in 1790 by John Wilkinson (1728–1808).

In many instances, lead-bearing ores are found combined with silver, and a workable method of separating the two metals was first discovered by the Romans. Improved techniques for doing this were developed in Bohemia in the early 15th century, so increasing the profitability of combined lead/silver mining operations. In general, lead is one of the few metals to have been in adequate supply from Roman times onward. The introduction of blast-furnace smelting of rich ores in the 19th century allowed the great increase in demand to be met.

Until the 19th century, lead was almost the only reliable material that could be used for relatively flat roof areas in the damp and rainy climate of northern Europe. Many Gothic cathedrals and churches have lead-covered roofs. Gutters, flashings, and rainwater goods such as hoppers and drainpipes were often made of sheet lead since it is a material that can easily be formed into fairly complex shapes.

Ornamental lead drain leaders became popular in England from the 15th century onward. In France ornamental roof ridges and finials were used on important buildings at the same time.

Because of its low melting point, lead can be cast with greater ease than other metals and was often used for sculpture. André Le Nôtre (1613–1700) employed many lead sculptures in his design for the gardens of Versailles (1662–90). These developed a soft white patina which blends in with the vegetation.

Lead is comparatively expensive, but because of its ability to resist corrosion it is often used today in combination with other materials to provide reliable waterproofing at critical joints between building materials. Lead is sometimes alloyed with antimony to make it harder.

The great stained-glass windows of the Middle Ages employed lead cames to hold the different pieces of colored glass together. The H-shaped strips of lead were fitted snugly

Lead used as a waterproof covering over a masonry wall and domes in Jerusalem.

Lead cames holding stained glass in position in a Gothic window.

around the glass panes to make a waterproof seal. These strips were then soldered at the joints.

Heavy-duty underground electrical cables are in many cases sheathed in lead to protect the copper conductor from corrosion.

Aluminum

The commercial production of aluminum dates from the 1890s. The cupola of the S. Gioacchino in Rome was covered with aluminum sheet in 1897, but this was an architectural freak. Until the outbreak of World War I aluminum was used mainly for pots and pans. Output increased considerably during the war as a result of military demand, but for the first postwar decade the market continued to have a kitchenware emphasis. The building industry began to think seriously about the possibilities of aluminum in the early 1930s, when it became possible to produce aluminum glazing bars reasonably cheaply by the extrusion process. This method allowed more complex and more efficient shapes to be designed, incorporating condensation channels and glazing systems into one extrusion. There was the additional great advantage that the tedious task of painting could now be entirely avoided.

Apart from glazing bars, most of the aluminum that went into buildings before 1939 was used for decoration. The Canada Life Building (1895) in Montreal has an aluminum cornice; the German Evangelical Church (1927) in Pittsburgh was given an aluminum spire; and the Rockefeller Center, New York, has cast aluminum spandrels. Aluminum windows were used occasionally when finances permitted.

After World War II both the aircraft and aluminum industries found themselves with surplus production capacity. Prefabricated aluminum housing was one answer to the problem. By 1948, 78,000 aluminum bungalows had been built in Britain. Most of them used aircraft scrap, which was in plentiful supply at the time.

During the same period in the U.S., aluminum houses designed by Charles Goodman for private purchasers were built in large numbers by National Homes at the company's three regional factories. Fifty-three different models were available and the price was kept low by using modular components. Ribbed aluminum sheets were also used during the 1940s for the external cladding of prefabricated schools.

Even scrap aluminum is an expensive material, and the designer has to use the minimum quantity that will give the strength he requires. Aircraft designers are used to approaching problems in this way, because the need to save every ounce of weight is of great importance in aeronautical engineering.

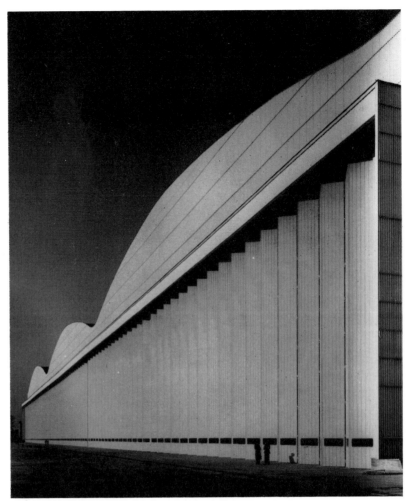

ABOVE: Aluminum doors of Brabazon Assembly Hall, Bristol, England (1947). Because of its light weight, aluminum was ideally suited for the construction of these huge doors.

LEFT: Aluminum door furniture.

Houses constructed in aluminum consequently had the benefit of very advanced production techniques.

During the 1940s a great deal was learned about ways of using aluminum in structural engineering. During the 1950s and 1960s, the range of applications of aluminum in the building industry was greatly widened, from structural components to foil insulation. Aluminum roof-trusses are also increasingly used, especially in corrosive atmospheres and horticultural buildings, where low maintenance costs and long life more than offset the higher original costs.

Aluminum has been successfully used for bridge building. The world's first all-aluminum bridge, the Arvida Bridge over the Saguenay River, Quebec, was completed in 1950 and has a main span of 250 ft. (77 m). The Grasse River Bridge at Massena, New York, has a 100 ft. (30 m) center span of aluminum, which weighs only 53,000 lb. (24,062 kg), compared with 128,000 lb. (58,112 kg) for the adjoining steel spans of the same dimensions.

The use of aluminum instead of steel is often the only way of adding another storey to an existing structure. This is well illustrated at the Ford Motor Company's Rotunda Building in Dearborn, Michigan. The old Rotunda Building would not carry the 160 ton framework of a steel dome, but easily supported an 8.5 ton aluminum geodesic dome. Aluminum tubes are very popular for geodesic domes and for space-frame structures.

Aluminum alloys of many kinds are produced each with its own special properties. Some of these, for example, are highly ductile, increasing the range and size of products that can be made by cold forming.

Aluminum sheeting has been increasingly used since World War II for cladding, roofing, suspended ceilings, trimmings, and many other parts of buildings. In order to increase its rigidity, sheet aluminum is often deformed by rolling or stamping. (Prefabricated storey-height aluminum cladding panels were used in the 30-storey Alcoa building designed by Harrison and Abramovitz in 1953). Aluminum sheets are sometimes bonded to other materials such as plywood polyurethane, to produce rigid, light, sandwich panels.

Many different surface finishes are possible for aluminum products, ranging from untreated mill finishes to surface bonded layers of plastic films. Anodized aluminum is produced by artificially increasing the depth of oxide film on the surface of a piece of aluminum by electrolysis. This process creates a harder, more stable surface. This oxide film can be dyed with a wide range of colors.

Glass

Early glass

The manufacture of glass is of considerable antiquity. It was invented in about 2000 BC, probably in Syria, but for many centuries its use was confined to vessels and decoration. The site of a glass factory was discovered at Tel-el-Amarna by Sir Flinders Petrie (1853-1942), who dated it at 1400 BC.

By the end of the 3rd century BC, techniques of forming vessels by glass blowing and of making crown glass were common in the eastern Mediterranean. Glassmaking

flourished in Roman times, and part of a large window frame with cast glazing has been found in Pompeii. Roman window glass was cast, and was full of distortions and imperfections; the largest piece known to us was 3 ft. 8 in. x 2 ft. 8 in. (1,100 x 800 mm). Roman window glass was not transparent. The windows were composed of small, thick, translucent panes of a greenish or bluish color. Even in Roman times, glazed windows were very exceptional and alabaster, mica, and shell continued in use until the glassmakers were able to produce an economic, color-free glass, in about AD 200. The Roman Empire spread glassmaking skills over a large area, but when it collapsed only France and parts of Germany continued to use the techniques which had been developed. In northwest Europe, where protection against cold and rain was welcome, the houses of the medieval period had wicker lattices, sometimes covered with oiled linen and parchment.

The first recorded glazing of an English church is at York c. AD 670. But nothing is known about the glass employed. By this time lead strips or "cames" had already been used to join glass in windows. The use of metallic oxides to color the glass greatly increased the scope of the church glaziers, and fragments of 10th-century pictorial windows still survive.

From the Middle Ages to the Industrial Revolution

The period from the early 12th century to the end of the 15th century was the great era of stained glass. As building technology developed throughout the Gothic period, windows increased in size and the art of the glazier became more important. The discovery, in the 14th century, that the addition of silver nitrate made glass yellow, came at a time when church builders wanted their buildings to be flooded with light, and this led to glazing techniques using white and yellow glass to allow the daylight to come through the great perpendicular windows. The Reformation largely put an end to church patronage of stained glass, but the newly powerful courts had commissioned lavish windows during the 15th century. Then the art of stained glass declined with the coming of the Renaissance, and the glassmakers turned to the more mundane problem of making glass for ordinary domestic windows.

The first sheet glass was made in Germany in the 11th century and the process remained a major glassmaking technique until the 20th century. The method was for the glassmaker to blow a glass sphere on the end of a hollow blow iron and to swing this to and fro so that the sphere elongated to a sausage shape. The ends were opened out to form a cylinder, and this cylinder was, in turn, cut with shears down its length, reheated, and flattened in a

Stained glass windows in Notre Dame d'Auxonne, France (16th century).

Blowing and swinging cylinder glass (c. 1850).

Spinning a blob of glass in a "flashing furnace" in the making of crown glass.

furnace to form a sheet of glass. Glass made in this way was known in England as broad glass. It was considered to be inferior to crown glass and was used in the houses of poorer people.

In the 14th century, the glass industry of Normandy achieved preeminence with crown glass. This technique involved the spinning of a blob of glass on the end of a rod until it had formed a flat disc or "table," which could be anything from 3 ft. 4 in. to 5 ft. (1-1.5 m) in diameter. The larger sizes demanded great strength and skill on the part of the glassmaker. The disadvantages of this process were firstly that the thickness was far from uniform and this caused distortion, secondly that large pieces could not be obtained without an eye in the middle where the glass had been sheared off the punty, and lastly that there was a lot of wastage, as the glass was made circular and most windows needed rectangular panes. However, the brilliance of the surface of crown glass is unmatched by glass made by any other technique before or since, as it is the only method of glass manufacture where the glass does not come into contact with any other material when it is cooling. Crown glass was introduced into England from Normandy in the late 17th century; and its name originated from the habit of a London glassmaker, John Bowles, of embossing a crown in the center of each pane.

The old Roman technique of casting glass was reinvented in France at the end of the 17th century by Bernard Perrot, at the Royal Glass Works. Perrot made glass of unprecedented size by pouring it into frames where it was spread evenly with rollers. The glass was then stuck on a stone surface with lime, and ground and polished. The process was laborious, but for the first time builders had at their disposal plate glass in large sizes without blemishes. This also made possible the manufacture of large mirrors, free of distortion, which became very popular in interior decoration, especially in palaces and public buildings such as theaters.

The Industrial Revolution

As glass with relatively few distortions became available at a reasonable price, so the nature of the window changed from the medieval idea of something to look at, to the modern one of something to look through. At the beginning of the 17th century, windows were fixed in small panes in lead cames between stone mullions, a system in which the quality of the glass was not critical. By the end of the century, the double-hung sash window was normal, with much larger panes of glass in big windows subdivided by thin glazing bars.

Plate glass was first made in Britain at St

Helens in 1773, but it was too expensive for general use. However, in 1832 Georges Bontemps (1799-1884) and Robert Lucas Chance improved the production of sheet glass in England. They utilized the continental cylinder process, but improved on it by using a larger cylinder, by cutting the cylinder with a diamond cutter, and by flattening it onto glass instead of sand-covered iron plates. There was a great demand for the new sheet glass, and by 1851 Chance was able to supply 1 million sq. ft. (92,900 sq. m) of it for the Crystal Palace at Hyde Park.

During the 19th century, plate glass gradually came down in price—by 1849 it was about one-quarter of the cost it had been five years earlier—and many owners of Georgian town houses ripped out their glazing bars and put one piece of plate glass in each sash. When Victorian architects made bold use of the newly available plate glass in commercial buildings the results could indeed be impressive, as, for example, in the buildings of the "Chicago School" of the last two decades of the 19th century. The Chicago window, usually consisting of a very large square of plate glass flanked by double-hung sashes, gave a distinctive style to the city.

Cellar lights, thick slabs of cast glass fixed into stone or into an iron frame, were being installed from about 1850 onward. The more familiar metal-framed sidewalk light, with cut squares of glass fitted into a metal grid, dates from the 1880s. More recently, the practice has been to fit smaller lenses into reinforced concrete. This combination withstands heavy traffic without damage. During the 20th century, new developments in the glass industry have provided the architect and the builder with a series of important technical innovations. The invention of wired or "Georgian" glass came from Pilkingtons in 1898. In the event of breakage, the mesh, embedded in the center of the glass, supports the fragments and greatly reduces the risk of injury from falling splinters.

The 20th century

Glass bricks and blocks made their appearance in Germany at the turn of the century, and were quickly adopted by modern architects as part of their new vocabulary. Bruno Taut (1880-1938), for example, used this new material extensively in his Cologne glass pavilion of 1914, creating hitherto impossible spatial and visual effects.

In 1900 the Germans began making what was then called "glass silk," that is, glass fiber, a material not developed commercially until the 1930s. It is now extensively used for thermal, acoustic, and electrical insulation. It is the reinforcement element in glass reinforced polyester (fiberglass), and in glass reinforced concrete. It can be produced as a

Nineteenth-century casting table for the manufacture of plate glass.

Victorian window with large sheets of glass.

Large steel windows with louvered opening sections; Casa del Popolo, Como, Italy (1931), by Terragni. The concrete roof over the window is cast with glass lenses.

Maison de Verre in Paris, France (1928–31), by Pierre Charreau. The walls are constructed of glass blocks.

Float glass manufacturing plant.

Daily Express Building in London, England (1932) clad in transparent and opaque (vitrolite) glass. Designed by Owen Williams.

"wool," or it may be woven into a fabric.

Foam glass was developed by the Pittsburg Plate Glass Company. It is used in the form of cellular lightweight blocks for insulation and is produced from molten glass by "foaming" or aerating, by the evolution of internal gas under pressure at high temperature.

In 1904, a Belgian invention by Emile Fourcault enabled flat sheet glass to be drawn direct, avoiding the double process of blowing a cylinder and then flattening it. Fourcault's process involved using a tank of molten glass. The sheet was drawn through a fireclay slot below the level of the glass in the tank, and the glass was then drawn up over rollers into an annealing chamber. The following year the American company, Libby-Owens, improved on the Fourcault process by running the glass onto rollers revolving at different speeds. The Pittsburg Plate Glass Company further improved the process by containing the edges of the glass.

The making of plate glass was similarly mechanized, first by the Bicheroux process, involving rollers, and then by Ford at Detroit, who discharged molten glass straight onto the rollers to obviate the casting process. Pilkingtons subsequently adapted the Ford process to the larger sizes of glass required for building, and evolved a continuous process to include the polishing. Finally, in 1959, Pilkingtons rendered the whole process of plate glassmaking obsolete by their float glass process, where the glass is floated onto molten tin, a surface so flat that polishing is no longer necessary.

Toughened or annealed glass is a 20th-century product which greatly extends the range of uses to which glass can be put. It is made by heating the glass until it becomes almost plastic, and then cooling it rapidly by means of cold-air jets. The resulting prestressed glass is resistant to shock and heat, being able to withstand loads up to five times that of normal glass. It is widely used in industry, for vehicle windows, for frameless doors, shop display windows, partitions, and in all other situations where conventional glass might be considered hazardous. Annealed glass must be made to an exact size, with all holes and shaping, since it cannot be cut or worked after heat treatment.

Laminated glass is another form that is resistant to breakage. Two sheets are cemented with an interlayer of polyvinyl butyral. If breakage occurs, the fragments are held in place by the interlayer.

Glass is now produced with many special properties. Opaque glass has been popular for cladding, fittings, and shop signs since the 1930s. There are two types of opaque glass; one having opacity and color throughout the thickness of the glass, and the other having a colored ceramic fused onto one surface. Surfaces of glass sheets may be treated during manufacture in various ways. Patterned glass normally has one face imprinted in a rolling process with a texture or pattern giving various degrees of obscuration and diffusion. Diffuse reflection glass is specially prepared to reduce reflection by having a slight texture on both its surfaces.

Glazing systems

The great conservatories and exhibition buildings of the mid-19th century were totally glazed structures. Glazing-bar systems, employing putty to hold glass in place, were replaced by glazing systems where glass was held in place with clips and other fixings, cutting down manufacturing and installation costs. The London Crystal Palace of 1851 had wooden glazing bars, made on the building site by sophisticated milling machines. Iron sections were being used in many conservatories soon after the end of the Napoleonic Wars in 1815, but they required regular painting to keep them from rusting.

Metal glazing systems became commercially practical after the introduction of the hot rolling and extrusion processes for the manufacture of clip-together, "patent" glazing bars, combined with methods of eliminating painting, such as galvanizing, lead sheathing, or by using a metal such as aluminum which is not subject to rapid corrosion. These systems were originally developed for industrial and horticultural buildings.

Iron, steel, and concrete structural frames made it possible to use non-load-bearing walls in multistorey structures. Architects, wishing to break with the tradition of heavy masonry buildings, used glass walls to clad their framed buildings. The Fagus Factory (1911), designed by Walter Gropius and Adolf Meyer has a corner made up of windows, which would have been impossible in a building without a framed structure. Mies van der Rohe (1886-1969), in his Berlin skyscraper project of 1919, designed a 20-storey building entirely clad in glass, with the independent floor planes clearly visible behind the transparent skin.

After World War II, glass curtain walling became popular for many buildings, especially skyscrapers. Industrial glazing systems were improved and adapted, and new forms emerged. Special injection-molded, synthetic rubber gaskets were developed to provide weathertight seals as well as protection against excessive vibration, expansion and contraction, and heat loss or gain through the glazing bars.

Suspended assemblies of toughened glass have made it possible to glaze very large apertures in buildings without the use of frames or glazing bars of any kind. The glass skin is stiffened by glass fins that are fixed to the frame of the building, and the gaps between the sheets are filled with a transparent polysulfide or silicone sealing compound.

Special forms of glass

Special types of glass were developed for glass-clad buildings in order to make their internal environments bearable, and to reduce the costs of heating and air conditioning. Solar control glass is specially manufactured to reduce solar heat gain either by absorption, reradiation, or by reflection, while still allowing a high proportion of visible light to be transmitted. Carefully controlled tints of color may be added to the surface or to the body of the glass. Reflective solar control glass is made by the vacuum deposition of a thin metallic film on the inner surface of one of two panes of glass which are then laminated together to give complete protection from abrasion and atmospheric attack. The properties of this kind of glass can be varied by controlling the type of metal deposited, typically gold or bronze, and by controlling the thickness of the film. Photosensitive glass may soon become practical for buildings. This type of glass, commonly used for eyeglasses, changes its transparency with the amount of light falling on it.

Double glazing, with clear or tinted glass, can reduce energy costs considerably by increasing insulation. Factory-sealed, double-glazed windows are manufactured in many forms, either with the edges closed, by the panes being fused together, or by the use of another material as a separator.

Glass may be curved and shaped on reheating to form domes and curved windows and may be worked in various ways for special effects. These include silvering to produce mirrors, engraving and brilliant cutting, sandblasting, and etching.

Paint

Using the knowledge and skills already available in Mesopotamia, the Egyptians made considerable use of paints for decorative and protective purposes. By about 1300 BC colors were obtained from finely ground mineral pigments and the base was usually tempera, prepared with a glue, gum gelatin, or egg albumen. The Egyptians, like the Romans, also used encaustic painting, either by mixing beeswax into the pigment, which they applied with a brush, or by spreading it as a protective coating over the surface to be covered. For a short period the Egyptians also used transparent varnishes made with resin from pine trees, sandarac, or the mastic tree. The fresco technique, using pigments dissolved in limewater and spread over a freshly plastered wall, was first used toward the end of the Hellenistic period and was common practice in Roman times.

In the Chou dynasty (1169–255 BC) the Chinese developed lacquer work using the sap of the lacquer tree. This technique spread to Japan and reached a high level of perfection during the Ming dynasty (AD 1368–1644). In some cases it was used to protect vulnerable structural timber units from the elements, but its use was restricted by laborious methods of preparation and application.

In Europe, oils were used for varnish in special situations such as high-quality external work as early as the 6th century AD. The oils form a thin protective film on drying. In the 11th century, a monk, Theophilus, developed a recipe for varnish prepared by dissolving resins in hot oil. The resins commonly used were amber and copal. Pigments, usually of mineral origin, were added to the varnish. Linseed oil was the oldest and most commonly used oil, but castor and fish oils were also employed. These types of paint application became more widespread during the Renaissance, and by the 1640s driers were being used in paint preparations. In 1750, Alberti of Magdeburg mentioned the use of turpentine as a thinner.

The paint industry, in the sense of the supply of ready-made products, dates back to the end of the 18th century. Until that time, the situation was not greatly different from what it had been 2,000 years earlier, with each craftsman preparing his own material. Varnish factories were established in England in the 1790s and in France and Germany 30 years later. Once these developments had taken place, very few changes occurred until the 20th century. The marketing of tested, reliable paints in quantity is a recent development.

In water-based paints, the water acts as a vehicle for carrying the film-forming and coloring ingredients. The procedure established by the Egyptians and Hebrews—water and whitewash mixed with milk curds (calcium caseinate) as a binder—continued to be used until the 1800s. In colonial America, the casein came from skim milk. At the end of the 19th century, the powder paints for mixing with water consisted of glue-bound clays or whitings, sometimes with inorganic pigments added.

The film-forming constituent in modern paints is the binder or resin. Solvents or thinners may be added to the binder to make it flow easily. Other constituents include driers which assist in the hardness of the film; pigments which produce color; stabilizers which neutralize the destructive effects of ultra-violet rays in sunlight; fillers which improve the mechanical properties; and plasticizers which assist in the hardening of the film or give the paint flexibility.

The mechanics of film forming were not understood until World War I. It then became possible to synthesize film-forming products. These synthetic resins, which included the

The Palm House at Kew, London (1845–47), by Decimus Burton and Richard Turner.

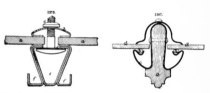

Sections of 19th-century patent glazing bars.

Suspended assembly of toughened glass. Willis, Faber, and Dumas Building, Ipswich, England (1975), by Foster Associates.

alkyd resins, became commercially available in the 1920s. Research into the chemistry of synthetic rubbers during World War II stimulated the development of latex resins for use in water-based paints.

The pigment in a paint film has mechanical as well as decorative properties. It controls transmission of moisture through the film and screens out harmful light rays. Some pigments, such as red lead, are corrosion inhibitors. Metallic pigments, for example stainless steel, zinc, and aluminum, are widely used in corrosion-resisting paints.

Extenders and fillers improve the physical and mechanical qualities of the dry film by preventing cracking and increasing the resistance to abrasion. They also control gloss or flatness. Common fillers include chalks, clays, talcs, and silicas. Nowadays fungicides and insecticides are often added to house paints.

An important 20th-century development has been paints for concrete. These have to be free flowing to cover the rough surface, yet they must not penetrate too deeply. They must be water- and alkali-resistant. Cement water-based paints consist of portland cement, lime, and often a siliceous aggregate, such as sand or gravel. A small quantity of calcium stearate is added to provide a water-repellent film.

Water-based paints containing acrylic and polyvinyl acetate resins are widely used as masonry coatings. They have good resistance to alkalinity and water. Paints based on chlorinated rubber and on polymethane and epoxy resins give tough, durable coatings. They are used, for example, on swimming pools and floors.

Intumescent paints swell or bubble on heating to form an insulating layer. Coatings containing brominated compounds release gases that exclude air from the surface and so extinguish flames.

Bitumen and asphalt

In Mesopotamia, bitumen occurs naturally as seepages in valleys and as rock asphalt in the mountains. In ancient times it was used as a mortar for bricklaying and for waterproofing walls, tanks, and floors, and it continued in use until Greek and Roman times, when building methods changed and lime-based mortars supplanted bitumen. After that, the smaller demand for waterproofing materials could easily be met by wood tar or pitch, so the traditional sources of bitumen and asphalt were forgotten.

Bitumen was valued as a waterproofing material in ancient Egypt, although the quantity used annually was probably never very large. It began to be extensively marketed in Europe in the 19th century, when improved transportation made it possible to organize regular and large-scale imports from the principal world source: the huge Pitch Lake in Trinidad. Shipped in the form of hard blocks, it was widely used during mid-Victorian times for paving, basement flooring, flat roofing, and damp courses.

Felts made from a mixture of animal and vegetable fibers are impregnated with bitumen and given various surface treatments according to their intended use. They were developed in the late 1870s and have been widely used for flat and sloping roofing as a waterproofing material either in the form of sheet, generally supplied in rolls, or as shingles. There are special bituminized felts which are used as underlining to other roofing materials such as tiles. Special felts were developed toward the end of the 19th century for lining reservoirs and dams to reduce seepage.

Kraft paper, saturated or coated with bitumen, is used for lining light-frame buildings on the inside of their weathering surface. This lining reduces drafts and inhibits the penetration of moisture. Saturated stout paper may be used as a damp course.

Rock asphalt began to be used as a waterproof road-surfacing material during the second half of the 19th century. This useful material consists of limestone impregnated in its natural state with bitumen. It is mined at several places in Europe, the most important being Ragusa in Sicily, Limner in Germany, and the Val de Travers in Switzerland.

Rock asphalt is sold either as a powder or as mastic. The powder is used mainly for road surfacing. It is laid hot and constant traffic is needed to keep it compacted. The mastic is prepared at the works by heating the powder with a small quantity of refined bitumen and then casting the mixture into blocks. At the site, the blocks are broken up and heated for several hours, sand and a little bitumen being added to the mix. Properly laid, this produces an excellent waterproof coating. Successful asphalting is dependent on skilled, conscientious workmanship. Poor results are usually caused either by not carrying out the work quickly and continuously, or by laying the mastic on a damp surface. It is extensively used for roofing, flooring, and tanking below ground.

Plastics

In the 19th century scientists made the basic discoveries about plastics, but it was not until the late 1930s that commercial production began. The range of available plastics is now very wide. The basic differences between the various types are increased during manufacture by the inclusion of plasticizers, to make the material softer and more flexible; of

Main Stadium at Munich Olympics (1968–72). Roof detail by G. Behnisch and Partners, with Frei Otto. Transparent skin of plexiglass.

Retractable PVC roof of a skating rink at Conflans-Sainte Honorine, France (1972), by Frei Otto.

Factory-made cladding panels in plastic for the GLC Building in Paddington, London (1966–67).

antioxidants, to prevent weakening by light or heat; of fillers, to add extra toughness; and of coloring agents. In addition to this, most plastics can be foamed, to produce lightweight materials.

The main types of plastics and their uses are:-

Decorative laminates. These are available as durable veneers or with the veneer bonded to a core such as masonite. The veneer is a hard, dense material, made by impregnating layers of paper with synthetic resin and then subjecting them to heat and pressure.

Epoxide resins. Before use, these are mixed with hardeners. They have excellent mechanical properties and chemical resistance. They adhere to concrete, glass, metal, and timber and are used to reinforce concrete.

Melamine formaldehyde. This comes in the form of moldings and of a surfacing liquid for decorative laminates.

Nylon. This is the general name for thermoplastic materials known as polyamides. Nylon is tough and resists abrasion. It is much used for window and door fittings, and for castors.

Phenol formaldehyde. This group of materials is used for making, for example, electrical fittings and toilet seats. Fillers have to be incorporated in the mix and these are chosen to suit the kind and quality of product required. Fiberglass, paper, nylon, asbestos, cotton, and sawdust, are all used. Phenolics fade in sunlight and are produced only in the darker colors.

Polystyrene. This is a cheap material, easily molded and with a good surface finish. It is often blended with synthetic rubber to improve its properties. It is available in sheet form, which is excellent for concrete shuttering, for moldings, which range from wall tiles to toilet cisterns, and as a cellular material. Expanded polystyrene, styrofoam, is a good cheap insulating material; it is supplied as sheets, boards, or moldings.

Polyethylene. This can be bought as film, sheet, moldings, or extrusions. The film is widely used for damp courses, for on site protection of materials and machinery and for sheltering building workers against the weather. Molded and extruded, polyethylene is used for drainage systems.

Polyvinyl chloride (Vinyl). This can be flexible or tough and rigid, according to the fillers and plasticizers that are employed. Vinyl is used for rooflight sheeting—these panels are virtually unbreakable, they are light and easy to fix and they can be made to exactly the same profile as other corrugated sheeting. Floor coverings, tiles, rainwater pipes, guttering, and handrail coverings are other popular uses of this material. Vinyl-coated metals are used for cladding for industrial buildings.

Polyvinyl fluoride. Supplied as a film, this can be bonded to a variety of materials to provide a stain-resistant surface.

Reinforced plastics. Any type of plastic material containing a reinforcing agent is a "reinforced plastic," but the term has come to be reserved for moldings produced for polyester and expoxide resins reinforced with fiberglass. These very strong moldings are particularly useful for casting complicated concrete shapes.

Urea formaldehyde. This is used for bonding plywood, for door fittings, and for a cellular insulating material that can be put into wall cavities as a foam.

In general, plastics are lightweight, combustible (their main disadvantage), resistant to corrosion by chemicals, and adversely affected by sunlight. They have a high strength to weight ratio, but they are much more flexible than other structural materials, and because of this they cannot be used as load-bearing structures in the usual way.

One of the earliest large, all-plastic enclosures was the U.S. Exhibition Building in Moscow, erected in 1959. It consisted of interconnected fiberglass umbrellas. In some situations, the lightness of the material is important. In France, the Société des Chantiers Réunis Loire-Normandie erected a market hall at Fresnes in 1964 which consisted of 18 large umbrella elements containing translucent parts. The structure occupied a site with unsatisfactory soil conditions that would have implied prohibitive foundation costs for most other structural solutions. Space structures using pyramidal plastic elements, combined with aluminum or steel struts, have been erected with success either as simple space-frame decks or as barrel vaults or domes. Plastic foams bonded to plastic, plywood, or metal sheet make sandwich panels that are extremely rigid and combine spanning properties with insulation. Corrugated metal roofing with a foam backing is popular for industrial and agricultural buildings.

In the 1960s Dow Chemicals developed the "spiral generation" technique of forming enclosures, in which boards of rigid framed plastic are bent, placed, and bonded into position by an erecting device. Disaster relief shelters have been built by "Bayer" in which urethane foam is sprayed onto an inflated balloon which is removed after the plastic has set, leaving a rigid structure.

Plastic products will continue to replace components in traditional materials in the building industry. Extruded sidings for frame buildings are relatively maintenance-free when compared to painted wood. Plastic plumbing is considerably easier to work with than conventional types. In many cases these new products are handled by the traditional trades. When whole structures are involved however, these often have to be produced and erected by specialist crews, often from outside the building trades.

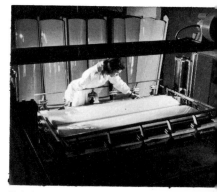

Polystyrene sheet being removed from a thermoforming machine after being molded into a bath panel.

Rigid plastic panel being removed from a compression molding press.

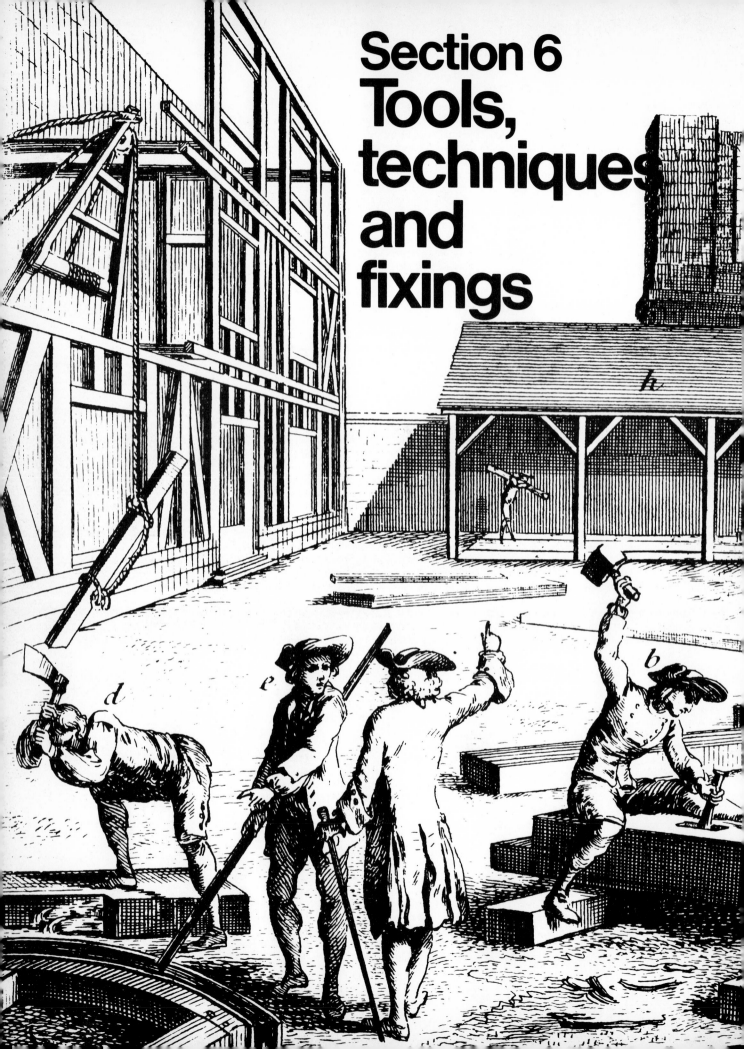

Section 6
Tools, techniques and fixings

Tools

The hand tool existed in a rudimentary form as far back as the Stone Age, and it is still widely used today, often with surprisingly few fundamental differences from its counterpart in antiquity. But power-driven machinery was also used in the ancient world: water wheels, windmills, and treadmills were pressed into service for a variety of building operations, particularly for the moving, lifting, and processing of heavy items. Heron of Alexandria, in the 1st century AD, described a variety of basic mechanical devices used in building: the screw, the wedge, the windlass, the lever, and the pulley. Writers such as Vanoccio Biringuccio (1480–c.1539) and Georgius Agricola (1494–1555) vividly portray how quite massive machinery had come into use in medieval times, mainly in mining but to a not inconsiderable extent in building. Denis Diderot (1713–84), in his *Encyclopédie*, shows how mechanization had further developed by the mid-18th century. After that the pace of events, which had not changed much for nearly 2,000 years, was quickened by the introduction, in comparatively rapid succession, of the steam engine in the 18th century and the internal combustion engine and electric motor, both in the second half of the 19th century.

In some cases the introduction of mechanical power in place of human strength involves rather little change in the basic design of the tool concerned: the modern electric drill is little different in principle from the bow drill of the early Egyptian or Roman carpenters, except that it involves continuous rather than intermittent rotary motion. Much the same may be said of the saw. Early mechanical saws, such as that depicted by Villard de Honnecourt (c.1250), copied the to-and-fro motion of the traditional handsaw. Continuous motion came much later, with the circular saw and the band saw.

On the other hand, the introduction of power led to some new kinds of machine, such as the milling machine for producing flat surfaces, invented by Eli Whitney in 1818. The advent of steam, giving a compact source of power far greater than any available in the past, had another important consequence—it made feasible the undertaking of much heavier jobs than had been previously possible. James Nasmyth's steam hammer, invented in 1839, was, for example, far more powerful than the water-powered tilt hammers that preceded it: in building it found an important application in driving piles for foundations. Another important consequence of the availability of very powerful machinery was that it made possible new methods of forming materials into standard shapes suitable for use in building. In this way, the first massive iron building units were cast, for use in the famous iron bridge at Coalbrookdale, England (1777–79). Later powerful rolling mills were used to form iron into strip, sheet, and girders.

The consequences of the availability of steam

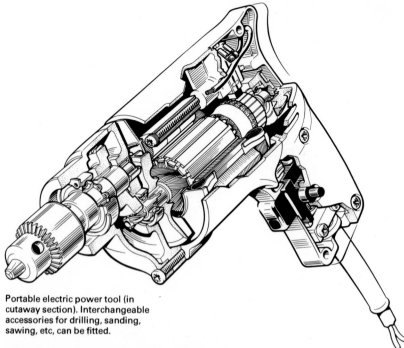

Portable electric power tool (in cutaway section). Interchangeable accessories for drilling, sanding, sawing, etc, can be fitted.

and other engines need some qualification. In ancient times extraordinary feats—which would tax the resources of a modern building contractor—were accomplished because of an abundance of cheap human labor. Egyptian temples incorporated stone blocks weighing hundreds of tons and a block in the quarry at Baalbec, squared up but never used, weighed over 1,000 tons (907 metric tons). The difficulty with manpower is concentrating it: there is a limit to the number of men who could be deployed, for example, to gather around and lift the drums used to construct the columns of Greek temples. Lifting devices of one sort or another became essential. The steam hammer made possible a rapid succession of powerful blows that no amount of organized manpower could have contrived.

In short then, the developments in buildings went hand in hand with developments in technology as a whole, although these have not occurred universally at an equal rate. Even today, builders in India and other countries where money is scarce and labor plentiful, use methods similar to those depicted in medieval pictures. Similarly on many modern building sites, huge containers of ready-mixed concrete stand next to laborers using the traditional method of mixing sand and cement with the aid of a shovel. The building industry is something of a paradox in being at one and the same time both conservative and progressive.

Carpentry

Until the comparatively recent advent of powerful machinery for the mass production, usually off site, of wooden components for the building industry, such as circular saws, band

Reciprocating mechanical saw c. 1850 used for cutting thin planks of timber.

Raising the obelisk in St Peter's Square, Rome (1586), using ropes, pulleys, and capstans. Directed by the papal architect Domenico Fontana.

saws, and shapers, the basic woodworking tools remained basically the same for 2,000 years. The replacement of bronze by iron, and iron by steel, gave keener and more durable cutting edges and enabled work to be done more quickly, but the tools themselves changed remarkably little.

The hammer

Basically the hammer is a striking tool like the axe or adze, and the power of its blow is a function of the weight of the head and the length of the handle. It may be applied directly, as in driving a nail or gouging a piece of iron, or indirectly, as in tapping a chisel or striking an iron wedge to split wood. Hammers vary greatly in weight, from the half-ounce (14 g) head used for veneer pins and small tacks, to the 10 lb. (4.5 kg) and upward of the ordinary sledgehammer.

The shape of the head is often varied to provide a function different from that of merely striking a blow. One of the commonest variants is the claw hammer, which enables nails to be withdrawn as well as hammered in. Metal-working hammers commonly have a round boss as well as a flat head.

Closely allied to the hammer is the mallet, used when a more cushioned blow is required or to avoid striking soft surfaces, such as wood, with hard metal. It too varies greatly in size from the small bench tool used, for example, for knocking out the blades of planes, to the heavy beetle or maul, with a head the size of a small cheese, which is used for driving wedges.

Axes and adzes

These two fundamental tools have much in common. Both have sharp cutting edges on a relatively large head, allowing a heavy blow to be struck. In the axe, used mainly for chopping across the grain and for splitting, the cutting edge is parallel to the shaft. In the adze, used mainly for rough trimming along the grain, the blade is at right angles to the shaft. The first axes were of stone, the axe head being lashed to a knee-shaped piece of wood or antler: this was a curious design for the crook was an obvious point of weakness. In Europe this cumbersome design continued into the Bronze Age. Bronze axe heads had a square or round socket for the shaft that continued upward from the blade (palstaves), still necessitating the angled handle. It has been suggested that this is why European bronze axe heads are surprisingly small—the weakness of the handle would not allow a very heavy head. Expense may have been another factor; stone axes continued in use throughout the Bronze Age. By contrast, the now universal shaft-hole axe was introduced at a very early date in Mesopotamia.

Homer's Odyssey (8th century AD) describes the use of both bronze and iron axes

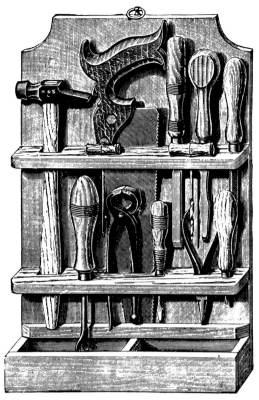

Tool kit and tool rack c. 1900.

ABOVE: Tack and claw hammers c. 1900.

BELOW: Hammer heads c. 1880 as used by engineers, mechanics, and boiler makers.

Masons and carpenters using mallets, bolsters, pit saws, hand saws, axes, pincers, and braces in construction of a stone bridge and wooden fortifications during the siege of a city c. 1460.

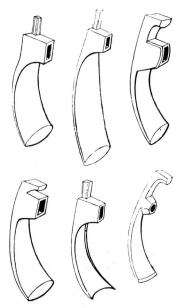

Adze and hammer combination heads c. 1880.

Quirk O.G. and plough planes c. 1900 with wooden blocks and cast-steel irons.

(and adzes) and the way in which iron was tempered by heating it strongly and then plunging it into cold water. Bronze had the advantage over flint in having a longer life before becoming blunt, and because it was three times denser, in giving a much heavier blow for a given size. But ordinary iron has no great advantage over bronze; the secret of the popularity of the iron axe was that early blacksmiths discovered, quite empirically, that if a little carbon was worked into the iron it turned to steel, which can take a much harder and keener edge. Analyses of Roman axes show that many contain steel. Very often, a steel cutting edge was welded over an iron base, though a few were made entirely of steel. By medieval times the axe with a steel cutting edge had reached essentially its present form, though with a variety of regional shapes. It was the most widely used of all woodworking tools and virtually no representation of carpenters at work failed to depict it.

The history of the adze was similar. As in the axe, the cutting blade was at first lashed to the end of a knee-bend shaft, but even by the 1st millennium BC this had been considerably improved upon. Instead of being lashed, the blade was secured by an iron collar, made tight by driving a wedge of wood between collar and blade. As we shall see later, this was the way in which blades were secured in early planes, and it is possible that the wedged adze was the forerunner of the plane.

A working carpenter would be changing frequently from axe to adze and it is not surprising therefore that hybrid tools appeared at an early date. One half of the head would comprise an axe and the other an adze. Occasionally, again, a hammer head might be partnered with an axe or an adze.

The plane

With the axe and adze alone the ancient craftsmen could accomplish a surprising amount. For example, the trunk of a tree could be squared up to form a main timber without recourse to any other tools; the timbers of many old houses show the adze marks very clearly. If a really smooth finish was required, however, the surface had to be planed rather than chipped; that is to say, thin layers of wood had to be shaved off with a sharp blade held at an angle to it. For small surfaces this could be done with a chisel, carefully pushed along either with the hand alone or tapped with a mallet, though the chisel was used also for chopping wood out. But for heavier work a plane was used: in effect, this is a chisel blade permanently locked in a rectangular block of wood with a hole roughly in the middle of its base to allow the blade to extend very slightly beyond it. The hole in the base is carried through to the top of the plane as the mouth, and is sufficiently large to admit the shavings, which are thus thrown up to the top of

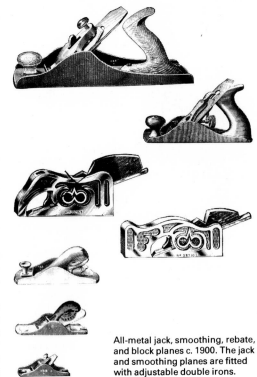

All-metal jack, smoothing, rebate, and block planes c. 1900. The jack and smoothing planes are fitted with adjustable double irons.

the block, or stock as it is called.

The next major development came at the end of the 18th century, with the introduction of the double iron, probably invented in England. In this an upper iron (the break iron) is attached to the cutting iron to stop the wood tearing. Finally, in the mid-19th century, better methods of adjusting the blade, usually incorporated in a cast-iron stock, were devised. A screw clamp replaced the wedge, and the lateral set of the blade was controlled by a lever which could be clamped when the setting was made.

Generally speaking, there are three principal types of plane. The relatively short jack plane is used for coarser work; the longer trying plane gives the work a better figure, testing it for straightness and accuracy; and the short smoothing plane gives a final finish. Space does not permit description of the wide variety of planes devised for special purposes, such as rebating, grooving, and molding, but all of these work on essentially the same principle of fixing a cutting iron firmly in a stock which is pushed away from the operator—though a few early planes were drawn toward the worker, like a spokeshave—while held against the work. Rough edges were trimmed up with a file and for a final finish, abrasive sheet, apparently introduced about the beginning of the 19th century, would be used. Abrasive sheet consists of powdered emery, glass, or sand glued to paper or cloth. This was a logical development of the use of abrasive powder, worked with a stone block, which was commonly used in Egyptian times.

The saw

Like the axe and the plane, examples of the saw are found among the earliest tools known, though flint blades with serrated edges were probably used as knives or scythes rather than saws. However, as plenty of true saws survive from the Copper and Bronze Ages, it is demonstrably a tool that has been in use for some 5,000 years. In appearance, the oldest saws closely resemble their modern counterparts but the tool has in fact gone through a steady process of evolution in both design and construction.

Copper and bronze are relatively soft metals and the teeth of saws made in these materials were, therefore, set so that the cutting stroke was that in which the blade was pulled toward the operator. With very few exceptions, modern saws cut on the thrust action and this practice was followed when iron, or iron/steel saws were introduced in Roman times. An alternative method of preventing buckling of the blade was to keep it under tension in a frame, as in the modern hacksaw or bush saw. Large frame (or bow) saws were in common use in Roman times, mainly for such jobs as cutting planks. The tension of the blade was maintained by twisting a double strand of string and using a wooden toggle to stop it unwinding: such saws are still sold today. The biggest saws of this kind were worked by two or three men, as were long wide-bladed saws with a handle at each end.

These two-handed saws were commonly used vertically, one man standing above the work and the other below it, although sometimes there were two men below and one above. In England and Denmark this was commonly done by using a saw pit, which avoided the necessity of lifting very heavy timbers. Elsewhere, however, it was usual to get the necessary clearance by lifting the wood onto a trestle.

Another method of conferring rigidity in the blade was to provide it with a stiffener along the back, as in the modern tenon saw. The disadvantage of this was that the stiffener was too wide to go into the kerf (the narrow slot cut into the wood by the blade) and the depth of cut was therefore limited. This defect was put to practical use in the so-called grooving saws, virtually unknown in Britain, but used on the continent of Europe from the 18th century. In these the depth of the blade—set into a wooden stock—is very small indeed, say one-quarter or one-half inch (6 or 12 mm). Such saws are very convenient for repetitive cutting of grooves of a fixed depth.

The operative parts of the saw are, of course, the teeth. In the earliest saws these were irregular and set in the plane of the blade. Pliny records that even in his day it was usual to make the teeth project slightly left and right in turn, though he did not fully understand the reason for this. The main purpose is to widen the kerf,

so that the blade is less likely to jam in it as the cut gets deeper. Saws of this type have to be reset from time to time; that is to say, the teeth have to be rebent and sharpened with a file on their cutting edge. A skilled craftsman can set a saw with a hammer, laying the edge of the blade along a piece of rounded iron. In inexperienced hands, however, this is liable to break the teeth off, it is better to use a special saw-setting tool.

In the saws described, emphasis has been laid on rigidity, but for cutting curves narrow flexible blades are required. Some of these are of conventional design, as in compass saws; in others (pad saws) the interchangeable blade is fixed in the handle with a locking screw.

In the earliest saws the teeth were very simple, and known as peg teeth: they were simply a succession of V-shaped teeth, sometimes with a flat bottom to the gullet between them. From a very early date they were "raked" in the direction of the cut. Teeth shaped like an inverted M appeared toward the end of the 15th century. Later, the Great American tooth was introduced, with three points in each group instead of two as in the simple M-type saw. The "count" of teeth—i.e. the number of points per inch—and the amount of "set" depends on the nature of the work to be done. Generally speaking, coarser teeth and a wider set are used for ripping wood along the grain rather than for cross cutting.

The drill

Holes need to be made in wood for many purposes, but particularly to receive the thongs, dowels, or nails used to hold the finished job together. The simplest of all tools for this purpose is the awl, simply a metal spike set in a wooden handle and thrust into it. It was soon realized, however, that a quicker and neater job could be effected if the tool was rotated while it was applied to the work.

In northern Europe, at least from Viking times, the auger was widely used. This was an iron rod spoon-shaped or twisted at the end to enable it to cut into the wood and throw out the waste. It was either square ended, to fit into a wooden handle at right angles to the shaft, or—especially for the larger sizes—had a ring-shaped end through which a toggle bar could be put to turn it. The familiar modern gimlet is the direct successor of the ancient auger.

Surprisingly, the auger seems to have been unknown in ancient Egypt, although a number of surviving bits indicate that it was widely used by the Romans. The Egyptians relied on the bow drill, in which rotary motion was imparted by looping the string of a bow around a cylindrical piece of wood holding an iron bit. As the bow was pushed to-and-fro the tool turns, though only intermittently, and with an idle return stroke.

Another very old boring tool was the breast

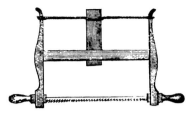

Bow or frame wood saw. A finely-toothed blade fixed in a wooden frame is tensioned by twisting a piece of catgut or thong.

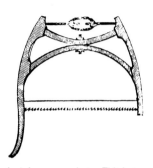

Arch frame wood saw. This is a modification of the bow saw which became popular in the U.S. in the 19th century.

Tenon saw with fine toothed blade, used for cutting across the grain of wood.

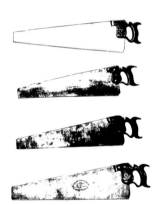

American hand or panel saws with coarse teeth, used for ripping wood along the grain.

Roman bow drill with iron bit.

An auger and gimlet maker c. 1526. The tips of the tools were wrought not filed.

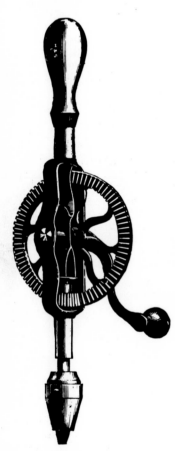

Hand drill. Mass-produced drills of this type, using interchangeable bits, became popular in the 19th century.

auger. In this the shaft of the auger had a freely rotating boss at its upper end and short projections at the side. The workman pressed the tool into the work with his chest, at the same time turning it with his hand by means of the side pieces. About the beginning of the 15th century the now familiar carpenter's brace appeared. In some respects it resembled the earlier breast auger, but the pressure was applied with one hand and the rotary action with the other, this being made possible by the crank-shaped stock.

In about the beginning of the 19th century hand drills appeared in which continuous rotation was effected by a cranked handle, turning the bit through beveled gears. In the larger sizes, there was something of a reversion to the breast auger. The operator pressed the tool into the work with his chest and steadied it with a short handle at the side.

What the teeth are to the saw, the bit is to the drill. The advent of improved steels in the 19th century made possible a keener cutting edge and a longer life. Much ingenuity was shown in the design of bits to ensure that they engaged easily with the wood, usually by means of a short screw thread at the end, and that the helical groove in the shaft discharged the waste smoothly.

Masonry

Just as the Roman carpenter would see nothing to surprise him in the toolchest of his modern counterpart, so the ancient mason would see surprisingly little change over the last two millennia. Perhaps the main difference he would note, with satisfaction, was that shaping was no longer done by simply pounding a large stone with a small one. The basic tools remain the scabbling axe, the wooden mallet, the metal chisel, assisted—as with the carpenter—by a try square and plumb line. Stones that had cleavage planes were split with wedges. Sometimes wooden pegs were used; when soaked with water these expanded and exerted great pressure. When such simple means failed, saws were used which were very similar to those already described for cutting wood. To assist the cutting of hard stones, sand or mixtures of sand and steel filings were sprinkled under the teeth of the saw.

There is evidence that hand saws up to 15 ft. (4.5 m) in length were used by Roman masons, but as early as the 4th century, the poet Ausonius refers to sawmills located in the valley of the Ruiver, used for cutting stone required for buildings in the Imperial City of Treves. Rough surfaces were smoothed by patient rubbing with blocks of hard stone, working in a mixture of sand and water.

Abrasives were also used in drilling stone, when the bit was commonly turned with a bow drill, as for wood. Roughly speaking, the amount of work that has to be done in drilling a

hole is proportional to the amount of stone to be removed. This led, again at a very early date, to the use of tubular bits, used with an abrasive. These drilled out a neat core of stone; if the thickness of the stone was greater than the length of the bit the stone could be reversed and bored from the other side. This use of abrasive offset the softness of the drill.

An alternative method of drilling holes is by simple percussion: a pointed tool is repeatedly hammered into the stone, rotating it a little between each blow, until a hole of the necessary depth is made. This method is still widely used today: for example, to bore shot holes in quarrying and to make small holes in walls to receive plugs.

With such simple tools masons cut blocks of stone, from quite small ones up to giants weighing scores of tons, with such precision that far from requiring mortar the joints between them were barely visible. The bricklayer's problems were rather different. Although local standardization of size was general, early bricks tended still to be rather irregular, so rather thicker layers of mortar were used than are usual today. This was applied, at least from Roman times, with a diamond-shaped trowel identical with those used by modern bricklayers.

Masons using wheelbarrow, plummet level, hoist, and plumb line in the construction of a wall. From a 13th-century manuscript.

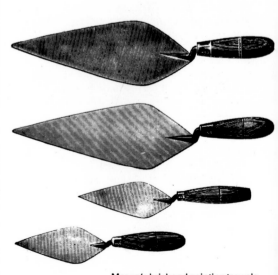

Mason's brick and pointing trowels c. 1900. From a trade catalog.

Glazing and sealing

The glazier's tools, like those of the mason, have remained very simple. Technically, glass is a supercooled liquid: that is to say, its properties are uniform in all directions unlike wood or many stones, which have a pronounced grain. In cooling, tensions are set up and if these are relieved, fracture easily results. Such relief may be imparted by scratching the surface and glass may readily be cut to shape by scratching the desired lines of fractures and then applying slight pressure. For a clean break, however, a single uniform scratch must be made. The ideal way of achieving this is with a diamond, but the normal glass cutter consists of a stout blade with a small sharp-edged wheel at the end made of hard steel. The blade has a series of rectangular notches along it, corresponding to different thicknesses of glass. These are used for breaking off pieces of glass that have not parted freely in the first instance, but the skilled workman will rarely have occasion to use them.

Glass is notoriously brittle and cutting must be done on a flat table, usually covered with a blanket to even out minor irregularities. To ensure a straight and accurate cut, the tool is guided by a simple straight edge or T-square. Toughened glass, however, must be cut, ground, and drilled in the factory before it is treated to give it its special properties. The other main tool of the glazier is the simple putty knife, used to apply the seal that holds the glass firmly in position in its frame.

Many new types of seal, other than putty, are now used in building. These include mastics, which in some situations may be applied by knife. In many cases the compound is injected into the joint between glass and frame or any other gap between building components with a pressure gun similar to the grease gun commonly used in lubricating machinery. Seals using gaskets and employing zippers require special tools and lubricants to force the zipper into position. When taped joints are used, special devices have been developed to facilitate this process.

Finishing and ancillary trades

Plastering

Plastering, in a crude form, was one of the earliest building crafts, for wattle daubed with mud or clay was one of the earliest constructional materials. Plastering of a much more sophisticated kind is portrayed at a very early date, however, and surviving murals from Pompeii show the use of a rectangular float with a handle on the back, such as those supplied by builders' merchants today. It was, indeed, to form a suitable surface for such murals that much plastering work was done, although the smooth rendering of rough surfaces was also

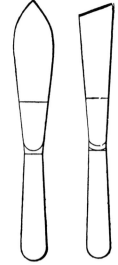

Glazier's knives used for application of fresh putty (LEFT) and for removal of old, hard putty (RIGHT).

Using a tool with a small cutting wheel at the tip to cut through glass.

Brushes for the application of paint and varnish.

important. The plastering of walls with lime, made by roasting limestone in kilns to expel carbon dioxide, dates from at least 2500 BC. For walls that were to be decorated, calcined gypsum, sometimes mixed with a little lime, was preferred. Externally, stucco—a mixture of lime and sand—was used to give inferior brickwork or masonry a superficial resemblance to a massive stone wall.

Apart from the float, mentioned above, the plasterer requires few basic tools, and these are very simple. Plaster is applied from a hawk, a board no more than about 1 ft. sq. (30 cm sq.) with a stout handle underneath. It is applied with a trowel, which has a flat steel blade about 12 in. (300 mm) long and 4 in. (100 mm) wide, with a handle turned back parallel to it. Before the plaster sets, the surface is finally leveled with a darby, a two-handled board about 4 ft. (1.2 m) long and 4 in. (100 mm) wide. Usually, a smooth surface is desired for the application of paint or paper but sometimes other finishes such as tiles are to be applied. In that case the plasterer gives a rough finish to the work by means of a scratcher, a simple wooden tool carrying about six pointed projections.

In framed buildings plaster was traditionally applied to closely spaced wooden laths. These have generally been superseded by substitutes such as expanded metal lath. This material can be shaped to the many complex forms which may be required in decorative ceilings.

Plasterboard and other sheet materials are now extensively used in the lining of interiors of buildings. Plasterboard may be given a complete coat of fine plaster to cover nail heads or joints. In many cases, however, the edge of the board is slightly feathered, to allow for taping and a small application of plaster or other material to bring the surface up to a smooth plane. Joints in other sheet materials may be treated in a similar way, allowed to remain visible, or given a cover strip of some kind.

Painting

The traditional tool for the application of paint is the brush, made in a variety of sizes and shapes according to the nature of the work to be done. The whitewashing of walls, for example, calls for a large, relatively coarse brush. Most domestic woodwork, coated with oil paint to preserve and decorate it, called for a much smaller brush with finer bristles; the endless task of painting the glazing bars of windows called for a finer brush still, with the end of the bristles often cut on the slant, to make working at an angle easier.

The ancients used reeds as brushes. The Egyptians soaked the ends in water to separate the fibers. Bundles of hog's hair were tied to sticks and, later, tufts of bristle were set in holes in a stick and secured with pitch.

The brushmaking industry was well developed in Europe by the 15th century. In the

Decorator's chisel knives or strippers.

Paraffin and petrol blowlamps c. 1900 used for metalwork and paint stripping.

U.S. brushmaking was a household industry until the early 20th century, but important technical developments had taken place long before this. An American patent of 1830 for a paintbrush in which the bristles were held in the handle with pegs was ahead of its time, but pointed the way to commercial developments which became significant later in the century. All the split ends of the bristles were exposed, which increased the paint-carrying capacity of the brush and made it soft. Animal bristles were always used until after World War II. A high proportion of brushes are now made from nylon and other synthetic fibers. With the advent of emulsions, latex and other nondrip interior finishes, rollers, instead of brushes, have become popular.

Spray painting was introduced into the automobile industry in the 1920s, but has not become widely popular on the building site. One reason is that the drift of the spray is not easy to control, and is unpleasant to the worker unless properly ventilated spraying booths can be used.

The blacksmith

Until recently, the blacksmith was a familiar craftsman on all the bigger building sites, his main task being to repair and sharpen the various tools that have been described. Additionally, he would make the various iron fitments that came increasingly into use as iron became more plentiful. Iron dowels and clamps, set in lead, were used to secure masonry in the 1st millennium BC, and bronze and iron fittings were used for doors and windows. His requirements were simple: a forge with bellows to improve the draft, anvil, tongs, and hammer. As with many other craftsmen, however, the work of the blacksmith increasingly moved away from the site to workshops at a distance. Although some architectural ironwork for special purposes is still handforged, most of it is now mass produced in factories.

Finally, mention must be made of the pick and shovel, basic tools for site preparation until the advent of modern earth-shifting machines and still widely used for minor works and for major ones where labor is cheap and plentiful. The wheelbarrow is traditionally associated with the pick and shovel: it is clearly depicted in some of the frescoes of ancient Egypt.

Powered tools

Until the beginning of the 20th century hand tools prevailed on most building sites. Medieval pictures, it is true, show cranes, and other devices for lifting and hauling operated by treadmills or windlasses but these would be encountered only in the building of major works. In the main, the building worker relied heavily—as he had done from time

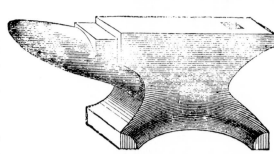

Blacksmith's anvil. Metal objects, held with tongs, are worked and shaped on the anvil with a hammer.

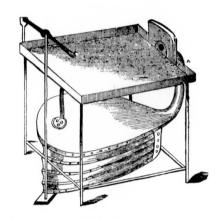

Portable forge c. 1830.

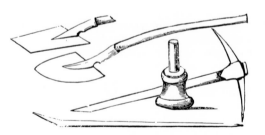

Shovels, pick, crowbar, and rammer c. 1870.

immemorial—on picks, shovels, and wheelbarrows to prepare the site and traditional hand tools for the construction work. However, the advent of the internal combustion engine and of small but powerful electric motors at about the turn of the 20th century began substantially to change this. The internal combustion engine could be used in two ways. Firstly, it could be used to provide compressed air, as for the pneumatic drill and certain other hand tools. Secondly, it could be a direct source of power, as in cement mixers, hoists, and pumps. It can be used in remote situations where no mains electric supply is available. Where there is an electric supply the electric motor is especially suitable for small hand tools, because it is clean and light, and requires only a thin electric cable in place of the heavy hose of the pneumatic devices. Its adoption involved some changes in the nature of certain tools. For example, although the electric jigsaw with a reciprocating blade is widely used—for cutting laminates, for example—the natural role of the electric motor is to provide continuous rotary action. Thus the electrical successor of the handsaw is more often than not the circular saw or, for rougher work, the chain saw. Sanding machines may be based on a continuously rotating abrasive disk or on vibrating orbital motion. The principle of the drill, however, was unchanged, for as we have seen it had long been based on continuous rotations; electric hand drills began to become available shortly before the beginning of the 20th century.

The electric drill lent itself to a variety of interchangeable accessories that could be fitted into the chuck at will. One such accessory was the sanding disk already mentioned. Others were the wire brush, useful for cleaning up iron and other rough surfaces; rasps; and hole cutters. For many purposes the high speed of the electric drill is an advantage resulting in fast work and a clean finish. In some circumstances, however, it is a disadvantage, for example when drilling very hard materials such as stone or brickwork with bits tipped with specially tough steel. At high speeds excessive heat is generated, resulting in the tool losing its sharpness. To obviate this, drills with reduction gearing were introduced. As a further refinement, so-called hammer drills appeared. In these, the rotation of the drill bit is combined with a hammering action that helps to break down the hard material at the bottom of the holes; this is, of course, the action of the pneumatic drill, widely used in site preparation.

The hammer was adapted for power in two ways. Firstly, the necessary repetitive blows were struck by a reciprocating mechanism actuated pneumatically or by an electric motor. Secondly, the so-called hammer gun was introduced. For massive bolts the power is derived from a small explosive cartridge. More commonly, however, compressed air or the release of a powerful spring is used. As we shall see later, mechanical riveting hammers were introduced soon after the middle of the 19th century, though they did not come into general use until the turn of the century.

Surveying instruments

Although small buildings were doubtless built on an ad hoc basis, the more ambitious projects had to be constructed to a carefully preconceived design. The execution of this demanded surveying techniques of a fairly high order. As early as the 3rd century BC land surveying in Rome was in the hands of the

gromatici, so called from their use of the *groma* as a sighting instrument to fix lines of orientation. This consisted of two wooden arms placed at right angles in a horizontal plane. Plummets were suspended from the ends of the arms.

A more elaborate instrument was the *chorobates*, used for leveling. Essentially, this was a four-legged table, with two plummets hanging from each side. The table was judged exactly level when all eight lines coincided with marks on the side of the table. According to Vitruvius (active 46–30 BC) the leveling process was often assisted by a water trough in the center of the table, with a horizontal groove on its inner surface. When the water surface coincided with this groove the table was level. Vitruvius also describes the use of a glass tube containing a bubble of air, clearly closely akin to the modern spirit level.

Another important surveyor's instrument, described (and perhaps invented) by Heron of Alexandria in the 1st century AD, was the *diopter*. This was essentially a theodolite combined with a water level. The rule of the *diopter*, some 6 ft. (1.8 m) long, was provided with an eyepiece at one end and an objective at the other. It was mounted on a table which could be turned both horizontally and vertically. However, it was far too sophisticated and elaborate an instrument to have been used in everyday building.

The modern theodolite, a precision optical instrument that can be regarded as a successor to the *diopter*, appeared about the middle of the 18th century. Accurate angular measurements can be made in the horizontal and vertical planes with the aid of vernier scales, and the incorporation of a compass allows precise bearings to be taken. In its various forms, the theodolite is an essential part of the modern surveyor's equipment for the layout of constructional sites. Except for the very largest of these, such as long ridges, line-of-sight observations suffice, the curvature of the earth being negligible. For distances above about 300 yd. (270 m) a small correction may have to be made to allow for this.

Key points in the construction, such as anchor bolts, are located by first plotting their position on a coordinate grid and then marking them in the field by proceeding from a convenient reference point by means of measurement by tape and transit sighting.

In marking out the site a system of reference points must be established that will not be disturbed during the progress of the work. Orientation in the horizontal plane is achieved by means of the compass or by relating to fixed points. Differences in level are determined by the use of a leveling staff. The latter—varying in length between 10 and 16 ft. (3–5 m)—is graduated in feet or meters with their subdivisions. To determine the difference in height between two points on the site, the staff is held

Surveying in the Middle Ages with sighting instruments.

Surveying in the 18th century using a modern type of theodolite.

Measuring chain, land chain, and measuring tape.

Techniques

With few exceptions, materials used in building have to be subjected to various preliminary treatments to convert them to a manageable form, and during the course of history, several principal techniques have evolved to effect these changes. A few techniques, such as oxyacetylene cutting of steel, are comparatively recent innovations, but most are direct developments of techniques practiced in ancient times. The modern lathe, for example, does not differ in principle from the pole lathe depicted in Egyptian frescoes, though the improvements are enormous.

For many purposes materials may have to be subjected to several finishing techniques before being ready for use: wood may first have to be cut to manageable size and then turned on the lathe; small metal components may be made by first rolling the metal into sheet and then stamping out the desired shapes.

The revolution in the mechanization of woodworking is epitomized in the block-making machinery designed at the beginning of the 19th century for the Portsmouth Dockyard in England by Marc Isambard Brunel (1769–1849). At that time the Royal Navy required more than 100,000 pulley blocks a year. Brunel's other machines automatically cut the blocks of wood roughly to shape, bored holes for the sheave pin and the start of the mortise for the sheave, cut the slot in the sheave, formed the outer surfaces with a shaping engine, and cut the groove for the rope with a scoring machine. Only the final assembly and polishing remained to be carried out by hand. The plant enabled the work of 110 skilled men to be done by 10 unskilled men, and there was an annual saving to the Admiralty of £17,000 per annum for a capital outlay of £54,000.

Similar machinery was devised for making virtually all the basic wooden components for the building industry. The introduction of the steam engine as a source of power, and of improved cutting steels for the tools made it possible to work more quickly and to handle heavier work. In Britain a considerable incentive was given by the urgent needs of the Great Exhibition building of 1851. Traditional woodworking methods could not have done the necessary work in the time: for example, some 200 mi. (320 km) of sash bar were required for the roof alone. Charles Tomlinson, in the 1866 Appendix to his *Cyclopaedia of useful Arts and Manufactures* states that the machines devised primarily for the urgent needs of the Crystal Palace had by then found their way into general use. They included machines for precision cutting of handrails and rafters, as well as sash bars. There was even a machine for painting the sash bars so that they were delivered to the site ready for use. A machine for cutting mortises and tenons was used for preparing sash frames and sash bars for assembly, and mechanical planes were commonplace. In effect, wood could be produced shaped to virtually any

upright at the two points and readings are taken on each while the telescope of the theodolite (or dumpy level) is turned in a horizontal plane. The difference between the two readings—the back sight and the foresight—gives the difference in elevation. Accurate measurements may now be made using electronic instruments, and laser beams are becoming increasingly popular, both for measuring and leveling. Various techniques of accurate photographic survey have played a part in reducing and supplementing the work of the surveyor in the field.

The *groma*, the *chorobates*, and the *diopter* enabled the work to be set out with precise vertical and horizontal alignments. As it developed, however, the working craftsman would rely very much on the plumb line to keep it vertical and the try square to ensure that the angles were true.

Finally, mention must be made of the rule and dividers. Being generally made of wood, few rules have survived to the present day, but they are widely illustrated among craftsmen's tools. A few bronze folding rules survive from Roman time. Units of measurement were, of course, standardized only recently and there were considerable local variations. However, this did not matter greatly provided that all craftsmen on a particular site were using the same units. Allied to the rule was the ungraduated straightedge, which could be used to check that a line was true, or to mark out straight lines on wood or stone. For marking out measurements dividers were used, and these have remained virtually unchanged to the present day.

ABOVE: A transit, an extremely accurate theodolite for surveying over great distances, c. 1900.

BELOW: Block-shaping machine. Portsmouth Dockyard, England (1801-08), used to produce blocks for ships' rigging. Invented by M.I. Brunel, these steam-powered machines could do the work of 50 men with only 4.

section, precision cut to length, and with precut parts of joints ready for immediate engagement with other members.

Casting and molding

Since they depend on the same principle, these two techniques may conveniently be considered together. In either case the desired shape is imposed on the material while it is in a liquid or plastic form, which it retains on solidification. In the casting of metals, solidification occurs simply on cooling, this is true also of certain plastics. In the case of some other materials, such as plaster or concrete, solidification depends on internal chemical changes.

The simplest form of metal casting is sand casting. In this, sand is packed in a suitable container around a pattern of the component desired. The compacted sand is then divided into two halves so that the pattern can be withdrawn, and the cavity is filled with molten metal. When this has cooled and solidified, the mold is reopened and the casting removed. If a hollow casting is desired, a core of appropriate shape must be used.

There are a number of important factors which must be taken into account when considering the technique of casting metals. First and foremost, the pattern must be so designed that the finished product parts easily from the mold—this precludes, for example, undercut parts and makes difficult the inclusion of thin parts, such as fins. Since virtually all metals shrink on solidification, the casting will be slightly smaller than the pattern, and allowance must be made for this. An allowance must be made too for a further reduction in size if the casting is to be subjected to machine finishing. As the solubility of gases is greater in liquid metals than in solid ones, gas tends to be produced as the metal cools, and this may cause undesired porosity. Again, the final metallurgical structure of the casting is determined by various factors, including rate of cooling; consequently the outsides of castings, which cool fastest, may differ significantly in properties from the interiors.

Special green molding sands are used for this process. They are usually mixed with a little clay and various additives to ensure their stability when the pattern is withdrawn. Commonly, the sand is preheated to drive off excess

ABOVE: Machine manufacture of sash bars for the roof of Crystal Palace, London, for the Great Exhibition of 1851.

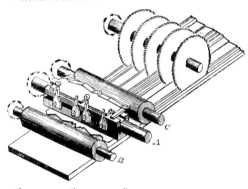

RIGHT: Sash bar machine detail. Wooden plank passes first over rotary cutters then through a set of circular saws.

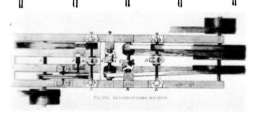

RIGHT: Plank (shown in section) cut into four sash bars.

RIGHT: Gutter cutting machine c. 1851 used at Crystal Palace, showing series of cutters (A–D) set at various angles.

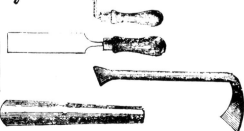

RIGHT: Step-by-step formation of gutter as timber passes across cutting machine.

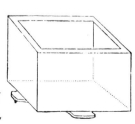
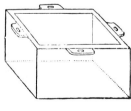

Mold-making kit consisting of pair of molding flasks, molding trowel (side and top views), molding wire, and runner stick.

ABOVE: Inspecting a rigid plastic component removed from a compression molding press.

BELOW: Injection molding machine producing plastic containers.

moisture which, giving rise to steam, might cause surface defects.

A major advance, made during World War II, was shell molding. In this process the sand is mixed with a thermosetting resin. The pattern is heated, and the mold mixture immediately in contact with it sets solid, thus producing a thin shell corresponding exactly to the shape desired.

Another important form of casting is die casting. In this, the liquid metal is forced under high pressure into a metal mold by means of a piston. By varying the details of the technique, the method can be used both for metals of low melting point, and for those nonferrous metals which have a higher melting point, such as aluminum. Die casting has the advantages of rapid production, close tolerance, good surface finish, and high strength due to rapid cooling. In centrifugal or spin casting the molten metal is distributed and forced into the shape of the mold by centrifugal force.

While the casting of metals is a technique which goes back to the Bronze Age, the molding of plastics, which is a basically similar technique, is a recent innovation. The first commercially successful plastic, Baekeland's Bakelite, was introduced in 1906, and since that time the ever-increasing range of plastic items and components indicates its importance in industry generally. Plastics fall into two main categories. First there are the thermosetting plastics; these are akin to ceramics in that once heated they retain their shape permanently. The second category, thermoplastics, resemble asphalt or lead; they can be repeatedly melted and solidified by heating and cooling. Both types are commonly made into the required shapes by molding. In the early days, only quite small items could be made in this way but now comparatively large ones, such as water cisterns and garbage cans, can be molded. Although the material is relatively cheap, molds are expensive and the technique lends itself primarily to the manufacture of large numbers of identical items, such as electrical switches, sockets, door furniture, buckets, and so on.

In compression molding the plastic, in powder form, is forced into the mold under pressure and this is then heated. Thermosetting plastics can be removed almost at once but thermoplastic materials must be cooled beforehand to avoid distortion. In injection molding, the method most widely used for thermoplastics, the liquid material is forced under pressure into the cold mold. As it cools rapidly it can be removed quickly, thereby favoring high production rates.

For most plastics, the difference in temperature between softening and decomposition is so small that some form of pressure must be applied in the molding process. A few, however, such as highly plasticized cellulose acetate-nitrates, are sufficiently stable to be melted and cast in molds like molten metals. As plastics became available in sheet form, vacuum-molding techniques were developed. In these, the softened sheet is sucked by vacuum against a surface having the desired profile. The method is cheap and quick and is used, for example, in making decorations and display signs.

Plaster was widely used as an interior finish for walls at a very early date (see PLASTER). Basically, it is made by calcining gypsum, a naturally occurring form of calcium sulfate. When this is powdered and mixed with about one-fifth of its weight in water, it quickly sets hard. If this process is allowed to take place in a mold, a variety of decorative fixtures, such as rosettes, cornices, and medallions can be made and used to give an artistic finish to interiors.

Much more important than plaster in the present context, however, is concrete. Like plaster, the setting of this material depends upon an irreversible hydration process. Today, it is used in enormous quantities for a wide variety of purposes. In most applications, whether on site or off site, it is shaped to its final form by some kind of molding.

Extrusion

Extrusion is allied to casting and molding in that the material is shaped by means of a fixed-shape former. The technique is exemplified by the squeezing of toothpaste from its tube, when it emerges as a long cylinder. Most metals can be similarly extruded through a die, either hot or cold, but the technique may not be economic for some high-strength alloys because of the great pressure required. In hot extrusion, the heated slug of metal is placed in a die having an orifice of a shape corresponding to the cross section of the desired product. Hydraulic pressure is then applied and the hot metal is forced steadily out through the die. Some lubricant is necessary to facilitate the process. The speed of extrusion depends on the metal—some aluminum alloys must be extruded slowly while some copper alloys and lead may emerge at speeds up to 1,000 ft./min. (304 m/min.). Cold extrusion, which was first applied to steel in about 1930, has been called cold forging, because the great forces exerted on the metal are similar to those exerted in the forging process.

Plastics, because they are so easily extruded when softened by moderate heat, are widely shaped by this technique. By using suitably shaped dies a variety of tubes, angles, rods, and the like can be quickly and cheaply manufactured. Extrusion is also used in brickmaking, the soft unfired clay being extruded as a rectangular length and chopped into appropriate lengths by wires.

Carving

Casting, molding, and extrusion are all processes appropriate for repetitive work. Very often, however—though less now than in the past—decorative work in wood or stone is required for particular buildings or for restoration work. The basic tools are gouges, chisels, and a mallet, although the skilled craftsmen undertaking elaborate work will require a great variety of these for roughing out, finishing, incising letters, undercutting, and so on. The design to be carved is first transferred to the wood and then painstakingly excised, great care being necessary to observe the grain. Whether working in wood or stone, the texture of the material must be noted; the finer the work the closer the texture must be.

Cutting

In much cutting work the basic technique is sawing, as it was in very early times. Over the centuries, however, saws have been greatly improved in both the scale of work that can be undertaken and in their constituent materials. Although it is established that water-powered saws were used by the Romans in the 3rd century AD, and one is depicted in a French manuscript of the 13th century, they made little overall contribution until somewhat later. Widespread interest was first shown, not surprisingly, in countries where wooden buildings were commonly used, such as Germany

Large diameter PVC plastic pipe emerging from an extrusion die.

and Scandinavia. The biggest user, however, was to be America, where a serious shortage of labor, combined with a need for ambitious projects, favoured mechanization. The great timber bridges that carried the railroads westward in the wake of the log cabins are symbolic of the rapid growth of a new nation that set great store by labor-saving techniques.

The technique of power sawing developed along two main lines. In one, the reciprocating action of the hand saw was imitated, with a number of blades being mounted parallel to each other in a frame. They could be individually renewed or sharpened as the need arose. The circular saw, introduced around the end of the 18th century, was an improvement in two respects: firstly, it avoided the necessity — inherent in all reciprocating devices—of having repeatedly to reverse the motion of moving parts, and secondly its cutting action was continuous because there was no idle return stroke. Against this, the need for a central axle limited the depth of cut that could be made, just like the stiffener along the back of the tenon saw. The depth of cut could be roughly doubled, however, by arranging for the saw to be taken around the wood to be cut, as in the pendulum saw made for the Portsmouth Dockyard, England, in about 1803. The band saw overcame this effect and, because of the narrowness of the blade, could be used to cut curved shapes such as those required for ships' timbers. For a long time, however, no satisfactory method of making the blades was found. Riveting was unsuitable, because of the extra thickness of the join, and brazing was unreliable. Eventually, in the mid-19th century, continuous blades were made. This was made possible by the adaptation of the technique developed for making steel tires for railroad rolling stock.

Such techniques suffice for cutting wood and stone, although for the latter the cutting action of the teeth needed to be reinforced by

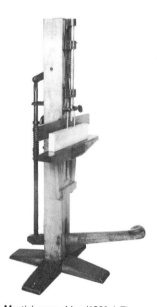
Mortising machine (1860s). The mortise is carved out by a chisel operated by a foot press.

Pendulum saw (1803) being used to cut a tree log.

application of an abrasive powder. Soft metals too, such as lead and copper, could be cut by similar methods. Steel and similar hard metals presented a problem as they increasingly came into use, for the ordinary saw blade could make little or no impact on them. Special cutting tools were therefore introduced, to be held against the moving metal. The development of steels with a high carbon content, and later the introduction of tungsten steels—often with a finely powdered tungsten carbide disseminated through the matrix—greatly increased the cutting power and enabled cutting speeds to be increased fivefold in the early years of the 20th century. Before the introduction of thermal cutting at the end of the 19th century, iron and steel had to be cut by laborious techniques using saws and red-hot metal or large mechanical shears. These techniques often bent the metal being cut, which then required extra work to straighten it out.

Meanwhile, a new cutting technique was developing, namely thermal cutting. In this technique, cutting of metal is affected either by melting or by chemical reaction with oxygen at a very high temperature. The necessary heat may derive from an oxyacetylene flame or from an electric arc. The point of heat application can be so strictly localized that a clean cut can be effected. This method of cutting is widely employed in the erection of structural steelwork where the same technique, using a high-intensity flame or arc, is used for welding.

Drilling

Much drilling is still done by methods based in principle on traditional techniques. The main developments have been the introduction firstly of special alloy steels to drill tough metals, and secondly of drilling machines with powerful motors, making it possible to undertake far heavier work than could be considered in the past. Such drills were of particular importance in building in the heyday of riveted steel construction, but this was to a great extent superseded by welding.

Medieval builders were remarkable for the lack of attention they paid to foundations, but in more recent times the literally fundamental importance of this has been fully realized. In the context of building in its broadest sense, boring now extends to the drilling of holes in the ground to receive concrete piles, to carry the weight of the structure. For this purpose large augers are used, which are virtually no more than scaled-up versions of the traditional carpenter's tool. For constructional work they may be 1 ft. (30 cm) or more in diameter. Smaller sizes may be driven from the power take-off of a tractor, but for major works a specially built mobile drill is employed.

As previously discussed, drilling is no more than a way of making a hole, and in the earliest days this was done simply with a pointed awl.

More recently, the electric arc has been used to replace the boring machine: in this process the work forms one electrode and the boring bar the other. The boring bar is fed into the work as the hole deepens and debris is flushed away with a nonconductive liquid. An advantage of this method is that its application is not limited to circular holes, as the hole will follow the cross section of the boring bar, though it will always be a little larger.

Grinding

Grinding is a technique for removing material by the action of an abrasive and was used by early Egyptian masons for smoothing the surface of building stones. Today, grinding is effected by the action of a power-driven grinding wheel, with the abrasive grains of the wheel making a succession of minute cuts on the surface. The grinding agents are usually silicon carbide or alumina, which are so hard that they can be used to machine the hardest steel. A great deal of heat and dangerous dust is generated in the process, and some form of fluid is necessary to dissipate these. Oil, or oil emulsions, are commonly used for this purpose. Although a wheel is the commonest grinding tool, modern grinding machines have many variants to allow different types of work to be done. In some machines the work is moved across the abrasive, in others the reverse is done, while as a further variant, a planetary motion can be used. In addition to surface grinders, there are internal grinders designed to grind the surfaces of holes to a precise tolerance.

Hand-operated saw c. 1885 with a continuous steel blade.

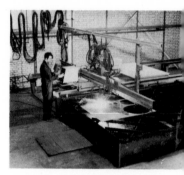

Automated thermal cutting of steel sheet.

Chipping rock debris from a drill auger used to excavate holes in frozen terrain for the erection of steel supports for an oil pipeline.

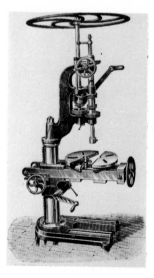

Heavy-duty, hand-operated pillar drilling machine c. 1900. By virtue of a system of gears and flywheel, relatively heavy drilling work could be undertaken.

Mortising, boring, and tenoning machine c. 1900.

Filemaker using sharp-edged hammer to cut each line of the file (c. 1417).

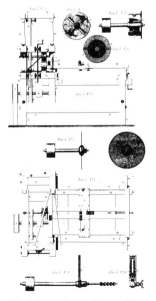

Mortising and tenoning machine c. 1870 (elevation and plan) with details of cutting, shaping, and drilling elements.

Milling

Milling is another technique for removal of metal, in which a rotating multitoothed cutter, made of special alloy steel and of a profile corresponding to the shaping to be done, shapes the surface of the work as it is passed across it. Most milling machines are horizontal or vertical, according to the axis of rotation of the power spindle, but universal machines can be fixed at any desired intermediate angle.

The origins of milling techniques have already been discussed at the beginning of this section.

Turning

Like milling and grinding, turning in a lathe is a technique for removing material from a workpiece by rotating it under power against a suitable cutting tool. This process can be used for wood, stone, or metal. In simple lathes the craftsman holds the tool and urges it against the rotating surface, varying its proximity to the axis of rotation according to the shape desired. Modern lathes are much more sophisticated; for example, the cutting tool is moved radially or longitudinally by a mechanical device. In the turret lathe, first introduced in the 19th century, a succession of cutting tools, which are mounted on a rotation turret, can be brought into operation without needing to reset the machine. Apart from general turning, lathes can be used to cut screw threads of any pitch by varying the longitudinal speed of the tool against the rotation of the workpiece.

Grinding, milling, and turning are essentially finishing techniques. They are used to give a final form to products already roughly shaped by casting, cutting, or other methods. This is done by removing excess metal by moving the workpiece repeatedly over some form of cutter. In their simplest form these processes are done manually, but modern machine tools, capable of handling very massive items, are fully automated. They have the common disadvantage of removing metal, however, which is wasteful. These techniques fall within the field of precision engineering.

There is another set of techniques which shapes metal essentially by brute force. These were the techniques most profoundly affected by the advent of the steam engine in the 19th century. This technological innovation not only enabled far greater force to be applied, but provided the means for handling the heavy products involved. In forging, for example, the traditional equipment was still the blacksmith's hammer and anvil. Later, as heavier work was demanded, this was reinforced by the water-powered tilt hammer, although its use was limited by the fact that it could be raised only a short distance. In 1839, when the steamship *Great Britain* was being designed, it was proposed that it should have a paddle shaft 30 in. (750 mm) in diameter (screw propulsion was

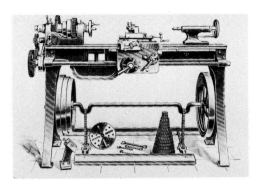

Industrial lathes c. 1900. Prior to electrification, lathes were invariably worked by a treadle or were belt driven.

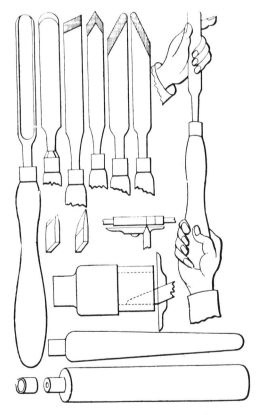

Representative forms of gouges and chisels used for wood turning c. 1900.

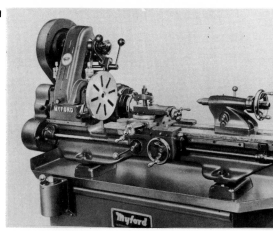

LEFT: Stone baluster being worked on a lathe.

RIGHT: Electrically operated modern bench lathe.

ultimately adopted). There was at that time nowhere in the world where such a shaft could be forged and this led the Scottish engineer James Nasmyth (1808–90) to devise his famous steam hammer.

Rolling

This is a metal-shaping technique in which deformation is effected by heavy rollers. It can convert ingots into strip, rod, sheet, or special shapes such as corrugated sheet for galvanized iron roofing. This method can be used with hot or cold metal, but the latter gives better dimensional tolerance and a smoother surface. Rolling is normally a repetitive process, the ingot being repeatedly passed back and forth to reduce its thickness progressively. One of the first to use a rolling mill appears to have been the Swedish ironmaster Christopher Polhem (1661–1751), who employed this technique early in the 18th century. Thereafter, progress was rapid. In 1861, 20 ton (18,144 kg) plates of 1 ft. (30 cm) in thickness—destined for use as naval armor—were rolled in Sheffield, England, and by the end of the century, Krupps could reduce a 130 ton (117,936 kg) ingot to sheet of the same thickness, measuring roughly 40 x 10 ft. (12 x 3 m).

One limitation of the use of iron is the readiness with which it rusts unless the surface is protected. The introduction of rolling led to an important development in this respect. In the early 19th century, thin iron sheets were coated with a thin layer of tin to form tinplate, the manufacture of which remained virtually a British monopoly until 1890, when America introduced the McKinley tariff. In 1829, R. Walker of Rotherhithe, England, under licence from the inventor, H. R. Palmer, introduced rolled corrugated iron sheeting for roofs, the corrugation giving additional strength. Eight years later the French chemist, M. Sorel, improved this product by coating it with a protective film of zinc to form so-called galvanized iron. Galvanized roofing sheets

began to be manufactured in Wolverhampton, England, in 1838 and were soon used in enormous quantities throughout the world. Galvanized wire, including barbed wire, was also extensively used for fencing, especially in the huge new cattle-raising areas.

The availability of heavy rolling equipment had other consequences for the building industry. Toward the end of the 19th century, rolled mild-steel joists became available in various sections. One of the commonest was the I-section joist with thick flanges. This made an excellent beam but its stiffness was much reduced when it was bent in the plane of the flanges; it was, therefore, unsuitable for use as a column. For this purpose composite sections were made by riveting several of the available standard ones together (Gray system). Later, improved rolling techniques made it possible to make wide-flanged beams in which the flanges, instead of tapering, were of uniform sections throughout. These were first rolled in Luxembourg at the very beginning of the 20th century, and were soon afterward manufactured on a large scale in the U.S.

For many constructional purposes and particularly with compression members, tubes are favored. These are used especially for exposed structures, such as bridges and pylons, because their wind resistance is much lower than that of

Sheet mills for tinplate rolling, early 20th century.

corresponding flat-surfaced units. Originally, tube was made by bending strip metal over a mandrel bar and then welding or brazing the joint, or alternatively by extrusion. The latter technique was used from about 1820 for soft metals such as lead, which until recently was widely used for plumbing. However, copper and its alloys proved more difficult and their extrusion was not achieved until the latter part of the 19th century. Steel is an even harder and more intractable metal, and seamless steel tubing was not available until the Mannesmann brothers perfected an entirely new rolling technique in 1885.

Forging

Traditionally, forging is associated with the shaping of red-hot iron by hammering on the blacksmith's anvil. To a limited extent this still prevails, as for example in the making of custom-made grates and grills, and similar small architectural items. Some large forgings, too, are still made by beating hot metal with mechanical hammers: the importance of the introduction of James Nasmyth's steam hammer in 1839 has already been mentioned. This method has the advantage of being appropriate for a large range of sizes and having low tooling costs, but it is too slow for long production runs.

Today, the technique of forging is much more widely interpreted. In particular, it generally implies the shaping of metal against some kind of former or die. The necessary force may be applied by hammer blows—for example, using a drop hammer—but this can equally well be achieved by using the silent pressure of a hydraulic press. For very large forgings, forces up to 50,000 tons (45,350 metric tons) may be necessary. Presses usually have vertical rams but may also have horizontal ones, permitting simultaneous forging in several directions. The forces now available make it feasible to forge cold metal in this way, but in practice this can be done only for small parts because of the strain hardening that results in all cold-working operations.

Forging is a less versatile way of shaping metals than casting, simply because solid metal cannot be made to flow under pressure with anything like the readiness that liquid metal will flow in a mold. Articles for forging must be free of the recesses, sharp corners, small holes, and the like, which present no problem in casting. As in casting, however, any sort of under-cutting must be avoided, as otherwise the finished article cannot be removed from the die.

Bending

Like most metal-working techniques, the bending of metals has developed from a simple to an elaborate one. Malleable metals can be bent simply by laying them across the edge of some rigid base, such as an anvil, and then hammering along the line of the edge until the required degree of bending has been achieved. For simple articles with a high margin of strength this is satisfactory, but on long production runs, where metal must be conserved, problems arise.

A bend is a potential source of weakness; metal on the outside is stretched and on the inside compressed. As a result, wrinkling or cracking may result. Further, except with the very softest metals such as lead, some degree of springback occurs after bending. To allow for this, the metal must be slightly overbent, so that when pressure is released the required degree of bending is permanently assumed.

Short straight bends are generally carried out in mechanical presses, but for long sections strip is passed through a series of rollers, each one bending the metal a little further than its predecessor. In wiper bending, the stock is progressively forced by a roller against the curvature of a former. It is particularly useful for bending tube, where there is a risk of flattening on the bend, so obstructing the bore.

Most of the bending techniques described above are carried out without heat. When it was necessary to bend a large rolled iron section to a special shape, as was often the case in the framing of iron ship hulls, these elements were heated to a high temperature and bent against pegs on a special floor. In buildings, elements of this type may be used for the framing of domes and other curvilinear work.

Stamping

Sheet metal can be formed by stamping, in exactly the same manner as addresses can be impressed on writing paper by an embossing machine. It is a process much used in the mass production of such objects as small cups or box-shaped items. Basically, the sheet metal blank is forced into the die by means of a piston which squashes the metal against the walls of the die; it is not necessary for the metal to be heated. The process is, therefore, not limited to articles of circular cross section. Except for very simple items, stamping is not usually completed in one operation. The usual practice is for the desired shape to be impressed by successive deformation involving different tools.

As in bending, stamping presents difficulties in attaining the desired distribution of metal. Flanges, for example, are compressed on the inside and stretched on the outside, the sheet metal blank is stretched in the stamping process and the walls of the finished product may be too thin or of insufficiently uniform thickness. Stamping, therefore, has the advantage of being able to impart quite elaborate shapes to sheet metal but, like most other metal-working techniques, it at the same time alters its mechanical and physical properties.

Joints and fixings

Joints and fixings are of great importance in architecture for two reasons. Firstly, virtually every building is an assembly of units which must be securely joined together. Secondly, every joint is a potential source of weakness and it is important to choose the one most appropriate to a particular situation. Joints can be a source of weakness for several reasons. First and foremost, they may be intrinsically weak from the outset because too much material may have been sacrificed in making them. On the other hand, they may be adequate at the outset but become weaker with time. It is commonplace to see iron screws and nails in old woodwork, for example, that have corroded to such an extent that there is virtually no metal left; glues may disintegrate through the action of moisture or molds; metal fatigue may be a source of breakdown in rivets or welds. Again, joints may be adequate for their original purpose but not for new ones as, for example, when bridges are rebuilt to carry heavier loads. Even unfixed joints may be a source of weakness; for example, rafters may rot through at the ends where they enter damp walls, and yet be completely sound for the rest of their length. Generally speaking, joints and fixings are potentially the weakest points of most buildings and it is not surprising, therefore, that much ingenuity has been shown in devising them.

Lashings

One of the simplest ways of joining things is to lash them together, and we have already noted that the heads of tools such as axes and hammers were originally lashed to the shaft. It was natural that this device should be carried over into construction, especially for temporary structures. In the 4th millennium BC, Egyptian workers used to build temporary wooden huts in the fields. The walls were of roughly cut planks, overlapping and with holes bored along the edge so that they could be laced together. Later, quite elaborate systems of knots were evolved for joining timbers together, especially for use on board ship or for construction of military devices in the field. Of more direct relevance to buildings, medieval pictures of builders at work clearly show scaffolding lashed together at the joints, the lashing often tightened with a tourniquet. Although, as we have noted, this has now been almost entirely superseded in the Western world by steel-tube scaffolding, it is still common elsewhere to use lashed wooden poles.

Homogeneous joints for wood

At a very early date woodworkers devised a range of joints dependent on one member being so cut and shaped that it interlocks with complementary shaping of another. No alien material was included in them. These will be termed homogeneous joints.

When such joints first came into use is doubtful, but many of those used in modern woodwork are to be seen in Egyptian coffins of the Old Kingdom, that is about 2500 BC. The construction of Stonehenge suggests that carpenters of northern Europe were equally familiar with some of the basic joints for woodwork at much the same time. Each upright stone of the main circle is topped with two tenons, which engage with corresponding mortises cut in the lintels. The latter are further secured by having fishtail joints at their ends: the whole structure therefore interlocks firmly both horizontally and vertically. This mode of construction is generally regarded as a translation into stone of earlier methods used in wooden architecture.

In devising joints for woodwork two primary considerations had to be kept in mind if maximum strength was to be achieved. Firstly, supposing that members of equal size had to be joined, roughly the same amount of wood should be sacrificed from each part. Secondly, the joint should be as unobtrusive as possible; ideally, it should not be apparent at all. The latter consideration would, of course, be of greater consequence in doors, windows, etc, than in roof timbers which, even if exposed, are not normally open to close inspection.

Homogeneous joints may be subdivided into two main classes. In one, the joint is so constructed that only the two members concerned have to be shaped to fit. In the other, a third piece of wood is involved: this might be in the form of a simple dowel or a wedge-shaped key interlocking with the two principal parts. Three main types of joint are involved in the building:

(1) Joints for lengthening timbers, by connecting them end to end.
(2) Framing and bearing joints used in trusses, floorings, etc.
(3) Joints for ties and braces.

Timbers may be joined end to end simply by putting them in position and securing them with short pieces of wood overlapping the joint and secured with nails and bolts. Such a joint is, however, weak as well as inelegant and consequently scarf joints were introduced at an early date. In effect, the two pieces to be joined are tapered off equally but in opposite senses and the two sloping ends put face to face. The joint is then secured with screws, bolts, or straps. This is the very simplest type of scarf, however; by 2000 BC Egyptian carpenters were using more sophisticated variations. In one, wedge-shaped mortises were cut in the two halves of the scarf and butterfly-shaped cramps of hard wood were carefully cut to fit exactly into these. Another early alternative was to cut corresponding wedges and grooves to receive them on the opposed faces of the scarf. The French scarfing joint known as the *traits de Jupiter* was so called because the jagged cut of the scarf, seen from the side, resembled a flash

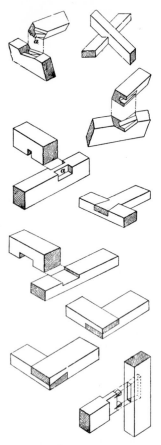

Common simple homogeneous joints used in carpentry. Combined with pegs and dowels these joints could be made rigid.

of lightning. The size of the scarf—i.e. the overlap—depended on the wood used. Tredgold's *Carpentry* recommended that in oak, ash, or elm the length of the scarf should be six times the depth of the beam and in deal it should be twice as long again. These figures were for unreinforced joints; if bolts and indents were used a reduction of two-thirds could be made. Mechanically produced finger jointing, combined with adhesives and techniques of lamination have largely replaced these forms of joint.

The scarf joint was, of course, used for extending horizontal timbers subject to bending forces. For vertical timbers, subject only to longitudinal compression, some form of mortise-and-tenon joint would suffice, though sometimes these were quite elaborate. Thus the top of one unit might be castellated, engaging with a corresponding cross-shaped tenon on the bottom of the next higher one. Where boards had to be joined edge-to-edge wooden dowels, inserted in holes bored in the edge, were used. Even at an early date these were sometimes glued into place and this was particularly useful if the fit was bad; a good craftsman, however, would work with such precision that the two pieces could be simply tapped together.

For framing and bearing joints some form of mortise-and-tenon joint is used, the tenon often passing right through the mortised piece. Tenons are usually made one-third of the thickness of the timber they are cut from.

Mortise-and-tenon joints may be held rigid simply by cutting them so tightly that they are quite firm when driven home. If the structure is likely to develop strains that might draw the tenon from the mortise, some means of anchoring the joint is necessary, and ancient craftsmen developed three methods that are still used today. Firstly, the tenon may be made so long as to extend beyond the far end of the mortise, when a cross pin or wedge can be inserted in a hole or slot in the extended part. Secondly, if the mortise extends right through the wood, a cut is made across the end of the tenon before it is inserted: when in place, a narrow wedge is driven into the crosscut slightly to expand the end of the tenon. This device is still commonly used today for fixing the heads of axes, hammers, etc, on their hafts: most ironmongers sell suitable iron wedges in various sizes. If the mortise did not go right through the wood an ingenious technique known as foxtailing was used. As before, a cut was made across the end of the tenon and a narrow wedge was inserted loosely in this. The tenon was then hammered home with a mallet and as the thick end of the wedge met the bottom of the mortise the wedge was driven into the cut in the end of the tenon, making a very secure joint.

For making cross joints for ties and braces the most satisfactory is the simple halving joint. As its name implies, each piece is cut out to half its depth, to the width of the crosspiece.

Dovetail joints are often seen used for this purpose, but are not really satisfactory because as timber ages it shrinks more across the grain that it does longitudinally. Dovetail joints—again of great antiquity—are, therefore, best reserved for joints in which the grain in both pieces runs in the same direction. The advantage of the dovetail is that the joint pulls tighter under strain, but formerly its construction demanded considerable skill on the part of the craftsman. Tody, dovetails and other joints can be cut mechanically with precision. For cheaper work corner locking was used: this resembles the dovetail but the interlocking parts were cut square instead of wedged.

Masonry joints

Masonry joints developed from existing practice with wood. Where very massive units were used, carefully shaped beforehand to rectilinear proportions, their own weight would keep them in position, but for smaller ones some means was necessary to prevent them shifting once they were in position. Most of the wood joints just described have their parallel in masonry.

We have noted that in the construction of Stonehenge massive mortise-and-tenon joints were used. These were very laborious to construct, especially the tenons where almost the whole end of the block had to be pounded away to leave the tenon standing proud. By classical times resort was already had to some form of simple dowel. A common device was to drill matching holes in the two opposing surfaces ready to receive a dowel made of bronze, iron, or oak. When metal dowels were used they were commonly set in lead, poured into place in the molten state. None of these materials were, however, entirely satisfactory. Wood became brittle and lost its strength; iron was liable to corrosion which caused unsightly staining and, worse still, caused expansion which might in fact end up by disturbing the structure rather than holding it firmly in place. These difficulties were avoided by using stone cramps—often with a butterfly section, as in early woodwork—fixed in position with cement.

Again on the evidence of Stonehenge, tongued and grooved joints were used at a very early date. Early masons also used what were called joggled joints, in which a wedge-shaped tenon at the end of one block fitted into a correspondingly shaped mortise on another. However, this too was very laborious and wasteful of stone, and an easier way was simply to cut a mortise, open at the top, in each block and drop a dowel joggle into this when the work was assembled.

For special purposes very elaborate dovetailing and doweling systems were used, as in John Smeaton's Eddystone lighthouse, built in 1759 to replace two earlier structures, one of which had been swept away in a storm (1703) and the

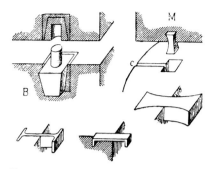

Masonry joints (5th century BC). Metal dowels were used for vertically stacked elements, cramps for those arranged horizontally side by side.

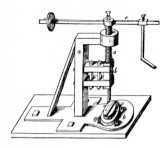

Nail cutting machine, early 19th century.

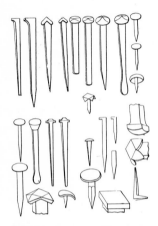

Nails for carpentry and masonry c. 1860s — floor brads, clasp nails, chair nails, tacks, and sprigs.

other, a wooden structure, by fire (1755). Smeaton realized that he could not transport to the Eddystone Rock blocks of sufficient size to withstand Atlantic gales by their weight alone. He therefore had every course prefabricated on land from stone blocks, each cut to dovetail with its neighbor until the circle was complete. The joints were then grouted with strong hydraulic cement, thus forming virtually a solid disk of stone. The stones of each course were drilled on their upper and lower surfaces so that they could be secured with heavy oak dowels. The whole structure resembled a Chinese puzzle and, indeed, when a new lighthouse was built in 1882 it was possible to dismantle Smeaton's structure and reassemble it on Plymouth Hoe.

Nails

The variety of nails is such that they could form the subject of a book in itself. All probably derived from the wooden pins used from the most ancient times to secure mortise joints, join planks, and so on. From this it would be an easy transition, as copper, bronze, and iron successively became available, to nails of modern type, and these in fact survive from Egyptian times. When soft metals were used it was first necessary to bore a hole, somewhat narrower than the nail, to receive it; this also helped to avoid splitting the wood. In Egyptian work, the heads of exposed nails were often decorated. Nails were used not only for joining pieces of wood together, but until the 16th century were invariably used to fix such attachments as hinges, locks, and bolts.

The earliest nails must have been hand forged, but by the end of the 15th century nailmakers were improving their output by driving short lengths of iron rod through slightly smaller holes in an iron plate. In effect, this is the reverse of making them by drawing wire through a die. Mechanical sawing of wood was first extensively used in America because of the exceptional use there of wood in buildings, bridges, etc. Vast numbers of nails were required for this type of constructional work, and it is not surprising therefore that it was in America that nailmaking machinery first appeared. The first patent was taken out by Ezekiel Reed in 1786 and some ten years later machines capable of producing 500 ready-pointed nails a minute were in use. By 1830 manufacture of nails in eastern U.S. alone far exceeded that in Britain, which for many years had exported nails to all parts of the world. In the mid-18th century some 60,000 people were engaged in the nail trade in Birmingham alone, on a cottage industry basis, making some 200 tons (181 metric tons) of nails per week.

In appropriate circumstances, nails make excellent fastenings, but they have two major defects. Firstly, like all ironwork they are liable to rust, and some woods contain corrosive substances which accelerate this process: nails taken from old woodwork are often no more than a fraction of their former size. Corrosion of the fixing nail is one of the commonest causes of failure in slate roofs. Secondly, as the nail is basically a smooth cylinder it puts up little resistance to tension along its length. Various devices were, and are, used to overcome this defect. If the nail was long enough to go right through the wood, until stopped by its head, it could be clenched by bending the protruding part flat with a hammer, preferably across the grain of the wood. Another method was to make circular corrugations around the shank of the nail to make it less easy to withdraw. More recently, the screw nail has been introduced. This has a fluted shaft rather like a stick of barley sugar: as the nail is driven home the fluting makes it rotate and engage with the wood, like a screw.

Most nails have a flat or slightly domed head, to stop them going any further once the whole shank has been driven home. This is not always desirable, however. For example, it may not be desirable for the head of the nail to show. Again, iron nails rust so easily that they are liable to stain through when wood is painted. In such cases, headless nails are used which can finally be driven a little below the surface of the wood with a punch. The resulting hole can then be filled with some sort of stopping, to give a completely smooth surface.

As every handyman knows, nails are liable to split wood, especially when driven in near the end of a piece. This can be obviated by drilling a guiding hole first or, more simply, by using nails of oval or rectanguar cross section, driven in with the longer axis parallel to the grain. Finally, mention must be made of the staple, which is essentially a double-headed nail. This is used in enormous quantities of fixing wire and netting fences to wooden posts.

Smaller wire staples similar to those used to secure papers together are used in fixing sheeting materials to framing. Small staple guns are spring activated, but in larger models compressed air is used. Special steel masonry nails have been developed for fixing to brick, stone, and concrete. In many cases these are driven home by explosive cartridges used in special nailing guns.

Nails are available in an immense variety of shapes and sizes, from tiny sprigs used to hold glass in place, to 6 in. (150 mm) nails for roof timbers. Variations in shape reflect particular uses to which they are put. Thus nails with domed heads and slightly twisted shanks are used for securing corrugated-iron roofs; flat-headed nails are used for roofing felt, to make it less likely to tear free, and for slates.

Nail plates, made from sheet steel that has been stamped to produce a sheet with many spikes, are used on surfaces of timber members to join them rigidly together. They are used extensively in mass-produced trussed rafters.

Screws and bolts

The principle of the screw was well known in the ancient world, but only as a mechanical device to obtain a gearing effect, as in presses. Not until the 16th century were they used in carpentry, and not at all widely so until the 17th. At first they were of uniform bore throughout their length, making it necessary first to drill a hole sufficiently deep to receive them. The modern type of screw, tapering to a point, with a relatively coarse spiral thread along about two-thirds of its length, was not in general use until the beginning of the 19th century. The slotted head brought into use (c. 1812) a new hand tool, the screwdriver. More recently this has found a rival in the cross-head (Phillips) screw, requiring a driver with a correspondingly shaped head; this is particularly suitable for use with power-driven screwdrivers. Apart from size, screws differ mainly in the shape of their heads. If, as is commonly the case, the screw must be flush with the surface of the work, a countersunk head is necessary: this requires a rose countersinking drill bit to make the depression necessary to receive it.

The great advantage of screws over nails is that they are extremely resistant to longitudinal tension. For most purposes they are made of iron, but like nails they are then liable to corrode. This is doubly disadvantageous: apart from being weakened, rusting makes it difficult to remove a screw should this be necessary. For high-quality work brass screws may be used.

The larger sizes of screws, requiring considerable force to insert, often have square heads so that they can be tightened with a wrench rather than a screwdriver. In certain circumstances it is desirable to use a screw in the same way as a dowel; this gives a stronger joint and one less liable to lengthwise movement. For such situations double-ended screws are available. Special screwing and bolting techniques have been developed for use with new sheet materials such as chipboard.

For many purposes—such as pinning scarf or mortise joints—a steel rather than a wooden pin or iron nail is required. For this purpose threaded bolts have been used since the 18th century. The coach bolt is a typical example. This has a broad head, immediately beneath which the shaft is of square cross section slightly larger than the shank: this ensures that when driven firmly home into the hole bored to receive it, the bolt will not turn in its socket. This makes it easy to screw a nut, usually with a washer beneath it, onto the other end of the shank, of which only the protruding length is normally threaded.

Specially designed torque bolts and wrenches have been developed which make it possible to tighten the bolts to an exact and predetermined degree. Other types of nuts and bolts with locking devices that prevent them becoming loose through vibrations are also widely used.

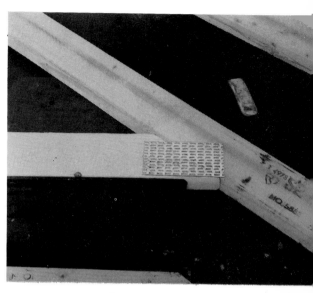

"Gangnail" connector plate. The metal nail plate is laid across the joint of this prefabricated roof truss and then driven home under mechanical pressure.

Bolts are used extensively for fixing structural elements together and for securing components such as cladding panels to the frame. Because of problems of tolerance between components and movement of various kinds, a large variety of special fixing systems have been developed that allow for fixing bolts to be used with a minimum of on-site adaptation to the components. In these fixing systems the final position of the bolt may be adjusted to varying degrees in all three dimensions.

Riveting and welding

As for so many of our modern building techniques, the joining of metals by rivets can be traced back to classical antiquity: a surviving example of riveted work from Crete dates back to about 750 BC. Often the heads of the rivets were decorated to improve the appearance of the finished article. Until comparatively recently, however, riveting was not used in major constructional work. Whatever the use, the principle is the same. The shank of the rivet passes through holes drilled in the parts to be joined, fitting snugly, with a short length protruding. According to circumstances, the head of the rivet is then held immovable against either an anvil or a heavy hammer, while the free end is hammered to mushroom it out and stop it retracting. A final hammering with a punch having a recessed head gives a neatly domed finish. To give a smooth finish, the end of the rivet was driven into a countersunk hole in the work, any excess metal being ground down or struck off with a chisel.

Where extra strength is required, the two parts to be united may be joined not directly to each other but to a third unit, such as a piece of

Nails and tacks — cut taper blanks for glazing and linoleum fixing; cut tacks for carpets and upholstery; grimp pins and grooved tacks for heavy-duty upholstery work.

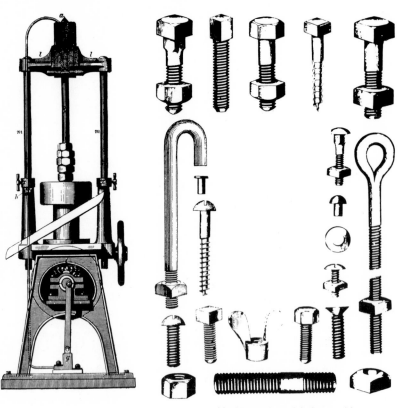

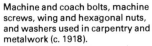

Machine and coach bolts, machine screws, wing and hexagonal nuts, and washers used in carpentry and metalwork (c. 1918).

Proprietory fixing blocks specially designed for butt jointing composition boards. Phillips screws are being driven with a mechanical screwdriver.

Nut cutting and tapping machine c. 1870 (front and side view).

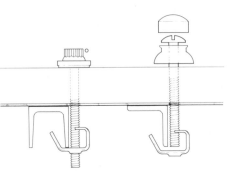

Clip and bolt cladding fasteners allow for considerable tolerance in the fixing of corrugated sheeting to rails or purlins.

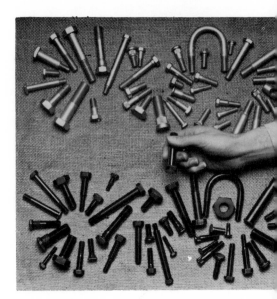

A range of bolts of different shapes and sizes suitable for a wide range of applications.

heavier plate or angle iron. Rivets are commonly of mild steel, but brass and copper have been extensively used and—since the introduction of aluminum sheet for constructional purposes—also aluminum. To avoid galvanic corrosion, it is desirable to make rivets of the same metal as those being joined.

The smaller sizes of iron rivet can be worked cold, but for larger ones it is easier if they are inserted red-hot—necessitating a portable hearth—and the work completed before they cool. This method of working has an additional advantage: as the metal cools it contracts, thus forming a particularly tight union.

For work in light gauge sheet metal, pop rivets, which were developed in the aircraft industry, are often used. These rivets, which are applied to the work from one side only are often made out of aluminum, and are placed and secured in position by a special tool which pulls a steel pin that is incorporated in the rivet, pulling and distorting its foot so that it tightens together the pieces being fixed.

Riveting became of major importance with the advent of wrought iron, and later steel, as major constructional materials. A notable early example was the Britannia tubular railroad bridge over the Menai Straits, designed by Robert Stephenson (1803–59), which was completed in 1850. In this construction over two million rivets were used, securing some 10,000 tons (9,072 metric tons) of wrought iron and 2,000 tons (1,814 metric tons) of cast iron. This was a reasonable advance: the longest spans over the water were 459 ft. (140 m) compared with a maximum of 31 ft. (10 m) previously achieved in wrought iron. It was the forerunner of tens of thousands of plate-girder bridges throughout the world, many still in use.

With the construction of massive riveted structures such as this, some mechanization of the process became desirable, but this was confined largely to making the rivet holes. For the Britannia Bridge, for example, a riveting machine was devised that could punch 8,500 1 in. (25 mm) holes a day in .75 in. (19 mm) plate. This work had to be done before erection and difficulties were encountered in getting the holes aligned, especially where several thicknesses of metal had to be joined; the channel had to be reamed out before the rivet—up to 5 in. (125 mm) long—could be driven through. This was unsatisfactory, in that some of the metal plates were then inevitably a loose fit and undesirable strains were set up. The closing of rivets, however, was still largely done manually. A riveting hammer worked by compressed air was introduced about 1865, but it was half a century before such machines came into general use. Hydraulic riveters came in about 1871, and portable riveters of this type were specially designed for building the 54,000 ton (48,988 metric ton) Forth Bridge (1882–90). For this bridge all rivet holes were drilled and not punched as for Britannia.

Apart from the problem of alignment, the mere drilling of the rivet holes was a source of weakness: in the mid-19th century it was supposed that a riveted structure was only about half as strong as it would have been had it been possible to make it in one piece. That so few major failures resulted was largely a consequence of the high margin of safety allowed for in the design, often as much as fourfold; trouble came when old structures were exposed to much bigger loads than they had been designed for.

The inherent weakness of riveted structures is avoided, in theory at least, in welded ones, in which there is no discontinuity in the metal. Forge welding, in which hot metals are joined by hammering or pressing, was familiar to Egyptian artisans: it was used, as previously noted, for putting a steel cutting edge on an iron tool. In its usual modern form, however, the edges of the parts to be joined are actually melted so that they fuse together on cooling. In practice this is usually done by shaping the edges to be joined to a V: molten weld metal from a welding rod is then allowed to flow into the resulting valley. This requires an intense and compact source of heat, such as is provided by an oxyacetylene or oxyhydrogen flame. In arc-welding the metal to be melted forms one electrode of an electric arc: the other electrode is made of nonconsumable material, such as tungsten or carbon.

Another modern form of welding, particularly suitable for sheet metal work, is electric-resistance welding and spot welding. In these the parts to be joined are pressed in contact and then heated to near melting point, so that they coalesce, by passage of a powerful electric current. The process is, therefore, analogous to the old forge-welding process. In flash welding, by contrast, the parts to be joined are held lightly in contact while an intense electric current is passed through spaced-out contact points. Local melting occurs almost instantaneously at these bridges, and flashing is continued until a layer of molten metal has been formed, when the two surfaces are firmly pressed together to complete the weld.

Electric welding processes lend themselves to joining at intervals—like riveting—as in spot welding, or to forming a continuous seam that is completely gastight and watertight. In the latter case rotating electrode wheels are used. If very easily oxidizable metals are to be welded, such as aluminum, oxygen must be excluded and the welding carried out in an inert atmosphere such as argon or helium.

As we have noted, welding is in theory an ideal way of forming a joint because it creates what is in effect a single homogeneous structure. In practice, however, this is not achieved: the intense heat of the process causes local changes such as brittleness, and stresses in the metal which are a potential source of weakness. Such defects became very apparent in some of

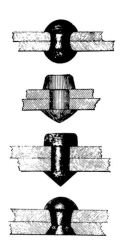

Rivets—capended; pan- and cheese-headed hammered; and countersunk.

Semi-tubular rivet machine in operation.

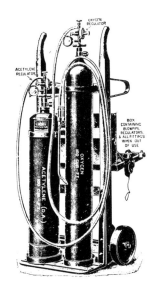

Oxyacetylene welding trolley c. 1900.

Sophisticated, carefully controlled double jointing welding work on oil pipeline in progress.

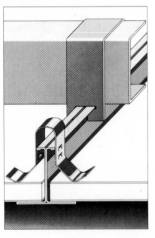

Metal clip and rail fittings for fixing plasterboard or other sheeting materials in the construction of suspended ceilings.

the Liberty ships built during World War II.

Both riveted and welded structures, therefore, have their defects. Although riveted structures are still being erected, there has been a decisive swing to welding since the 1920s, the main factor being economic. The amount of labor that could be saved by welding was enormous, and it was no longer necessary to carry large stocks of differently shaped steel rollings. There were no rivet holes to drill out and align by reamering, and no great stocks of variously sized rivets to hold. Above all—although designers were slow to recognize this—scientific principles of design could be applied without the constraints necessarily imposed by the use of riveted girders. In particular, it made possible the development of the box girder with its great torsional strength. This was used, for example, in the Severn Road Bridge in England (1966). Although this was of almost exactly the same length as the Forth Road Bridge, completed only two years earlier, it required only half as much steel. But the most historically significant event had occurred some two years earlier still. In 1962, the first welded box girders were used in a reconstruction of Isambard K. Brunel's famous suspension bridge at Chepstow, built in 1852.

Pieces of metal may be joined together by soldering and brazing. In these techniques the pieces of metal are united by an alloy which is more readily fusible, so the composition of the solder varies with the metals to be jointed. These processes are of considerable antiquity but the availability of portable blow torches and electric soldering irons have made these techniques considerably easier.

Adhesives and sealing compounds

Adhesives of animal or vegetable origin were used by ancient craftsmen, but generally only for making relatively small articles for indoor use, such as furniture. This was doubtless because such glues tend to become brittle with age and are not resistant to moisture. Adhesives became important in building only in comparatively recent times, when a variety of tough and durable synthetic products became available. Their main role has been in laminated structures, of two main kinds; these are respectively sheets and beams.

Plywood is familiar to everybody as ordinary three-ply sheet, though more complex sheet laminates are also made. In these, the differing strengths of wood along and across the grain can be evened out by bonding them alternatively at right angles. If desired, the outer plies can be of better quality wood than the others.

Similarly, beams of great size can be built up from comparatively thin strips. As the highest stresses occur in the top and bottom laminations, the central ones can be built of random lengths of relatively poor quality

material; weight for weight, large beams can be built up as cheaply as small ones. By the use of cramps and formers before the glue sets large beams of curved section can be built up; spans of up to 200 ft. (60 m) can be achieved. Such beams have better fire resistance than trusses of similar strength made of smaller pieces. Laminates also permit maximum utilization of the full strength of the wood. In ordinary woodwork, unless very carefully selected, the occurrence of knots and other defects may mean that little more than half of the theoretical strength can be achieved. In laminates, even allowing for the chance of imperfections in adjoining plies coinciding, the theoretical strength can be much more nearly attained.

Synthetic glues are also increasingly used to secure joints of various kinds including those in quite heavy members such as roof trusses—thereby replacing nails or screws. The glue may be applied directly to the joints in the ordinary way, or a plywood gusset may be cut to reinforce the joint.

Even with the best craftsmanship, not all joints are self-sealing and wind and rain are liable to find their way in. Adhesives play a double role in both securing and sealing joints, but sealing compounds based on bitumen have been used from the earliest times, especially in Mesopotamia where it occurred naturally and formed a useful sealing compound for the very porous bricks used there. Genesis tells us that Noah's Ark was sealed within and without with pitch. The Romans used wood tar and in northern Europe this was in constant use until the 19th century. One of the biggest outlets was in roofing felt, apparently first introduced in Sweden in the 18th century. With the rise of the gas industry in the 19th century, coal tar became available as a cheap and efficient substitute, and by the end of the century the roofing-felt industry, especially in Germany and the U.S., had an annual output measured in hundreds of thousands of tons. In America, the growth of the petroleum industry and the increasing exploitation of natural gas, led to coal tar being largely replaced by similar residues from petroleum distillation.

Another important and widely used sealing agent is ordinary glaziers' putty. Soft putty is made basically from raw linseed oil and whiting; for a harder and quicker drying putty boiled linseed oil is used. Today, with availability of a large range of prefabricated metal-framed windows, often of the double-glazing type, increased reliance is placed on various forms of weatherstrip.

In glazing with lead cames little or no putty is used, and a longer lasting seal is obtained. This is how the stained-glass windows of the Gothic cathedrals were assembled from small panes of glass. In the 19th century a large variety of "patent" glazing bars were developed mainly for horticultural and commercial buildings, in which the glass is held in position by clips and other devices and sealed against drafts with strips of material such as greased cord or rubber. These glazing techniques have been much improved in the 20th century.

A wide variety of mastics is now used to seal glass and other materials into frames. With large sheets of glass, the material is normally held in position by blocks of some relatively soft material such as lead to prevent fracture due to the buildup of local stresses. The glass may also be suspended. In these types of glazing, the mastic acts purely as a sealing compound.

Plastic mastics are nonsetting compounds and are mostly based on oil, bitumen, polyisobutylene, butyl rubber, and other products. They may be applied by putty knife or by injection gun. These are used in relatively sheltered positions where large amounts of movement are not expected. They are employed in some forms of glazing and for sealing gaps between door and window frames and the structure or cladding of a building. Elastic mastics are those which set after application and include polysulfide rubber, rubber, silicone, polyurethane, and some butyl rubber mastics. Many of these are supplied in two parts which react chemically and are thoroughly mixed prior to application, which is often done by means of a pressure gun similar to those used for applying grease to machinery. Many of these mastics are used in glazing, normally with some backing material against which they are forced. Some, such as silicone mastics, which may be made transparent, may seal the joint between two sheets of glass without any other material intervening.

Mastics are widely used in other situations where seals are necessary. The trend toward dry construction has greatly extended their use. The choice of mastics depends on a large number of factors including the materials to be joined, exposure, and expected movement. Some mastics may need to be protected against sunlight as ultraviolet rays accelerate chemical breakdown, and this may have implications on the shape of junction details between components.

Gaskets, made of natural or synthetic rubber or plastic, extruded to a particular form, depend upon being held in compression rather than upon adhesion to the components to be sealed. Many different forms are available for glazing and other sealing duties, ranging from very simple profiles to complex shapes. Gaskets can grip sheet materials such as a sheet of glass, or fasten around a linear element such as the flange of a steel angle. These are often made in two parts; the main body of the gasket gripping the material while the ''zipper'' element, which acts like a linear wedge, exerts pressure so as to maintain a tight joint. Before the zipper is forced into position by a special tool, the gasket is fairly flexible, making it easy to force around the edges to be joined. Intersections of gaskets can be weak spots, but it is possible to weld most gasket materials, and this is normally done off site.

Many other products exist for sealing joints between building components against water, moisture, and draft penetration. They include a range of adhesive tapes with various treatments and sponge material impregnated with water-repellent material, such as bitumen.

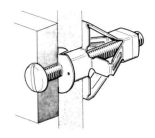

Hollow wall fixing – the legs of the anchor fold out as the bolt is tightened.

Miscellaneous fixings and fastenings

When the main structure of a building is complete, a great range of additional fittings are still required before it is completely ready for occupation.

The evolution of methods for dry assembly of components, which began in the 19th century and has accelerated ever since, has introduced an almost infinite range of special fixing and joining devices into the building industry. In many cases the fixings form a basic and integral part of the range of components used in a particular system. In suspended ceiling assemblies for example, special fixings may be required by particular fixing problems posed by the structural components used. In these assemblies, the components must be joined together and allowance made for leveling off.

In finishing trades it is common only to be able to work from one side, especially where objects are being fixed to walls. Where walls consist of hollow partitions made of modern sheet materials, a wide choice of ingenious toggle-screw systems exist that simplify the operation. Some of them have captive nuts that can pass through the holes drilled in the sheet, but then fold out. Others have threaded elements that crush on the inside of the partition when the screw is tightened. Such fixings can support considerable weights; they are often used in shelving systems.

Many techniques of concealed fixing exist, but with dry construction it is often difficult to conceal joints. These are often carefully considered and incorporated into designs by carefully coordinating their appearance on various surfaces.

Light fixings for use in masonry can be made by the use of screws and screw anchors, which expand in position when the screw is driven in. These were formerly made out of whittled wood, but today mass-produced screw anchors of plastic or other materials are in general use. Where heavy fixings are required, expanding bolts are generally used; these automatically expand after insertion as a nut on the protruding shank is tightened.

Hollow wall fixing. The assembly can pass through a narrow hole drilled in the wall lining. As the screw is tightened, the anchor is forced open on the inside face of the lining, securing the fixing.

Index

Aachen, 32
Aalto, Alvar, 98, 115, 213, 215
Aarhus University, 63
Abbas I, Shah, 42-3, 60
Abbasids, 42
Abu, Mount, 46
Abydos, 155
acoustics, 204, 213-14
Adam brothers, 40, 82, 246, 251
adhesives, 234, 312-13
Adler, Dankmar, 107, 274
Adler and Sullivan, 110, 203
adobe, 241
adzes, 291-2
Aegean, 26, 79
Affleck, Desbarats, Dimakopoulos, Lebensold, Sise, 114
Afghanistan, 21
Africa, 14, 20, 29, 35, 46, 85
Agra, 46
Agrigento, 27
Ahlschlager, Walter, 139
Aihole, 46
air conditioning, 108, 203, 216-18, 223
airports, 70-5, 93
Alacaluf, 15
Albert, Prince Consort, 116, 163, 272
Albert, C., 98
Alberti, Leone Battista, 36, 37, 89, 158, 196
Albini, Franco, 130
Alexandria, 183
Algeria, 93, 238
Alphand, Jean, 60-1
alpine climate, 14
Alps, 17
Al'Ubaid, 21
aluminum, 281-2
Amaravati, 46
Amarna, 78, 88
Ambasz, Emilio, 225
Amiches, C., 260
Amiens, 34
Amsterdam, 38, 61-2, 140, 142-3
Anatolia, 18, 154-5, 238
Andes, 25-6
Anthemius of Tralles, 31, 32, 169
Antwerp, 148-9
An-yang, 44
apartments, 86-8 see also houses
Arabia, 12, 14, 21, 42, 229
Arabs, 89
Aranjuez, 90
Archaemenids, 22
Archer, Thomas, 37
arches, 171-2, 196; brick, 245; bridges, 64, 65; Gothic, 34; Islamic, 42; Roman, 184; Romanesque, 33
Archimedes, 94
arctic climate, 14
Arizona, 20
Arkwright, Richard, 94-5, 97
Arminghall, 19
Armstrong, Sir William, 207, 210
Arp, Jean, 47
Art Nouveau, 47, 48-9, 270
Arts and Crafts movement, 83, 84
Arup, Ove, 170, 260
Aryans, 45
asbestos, 263
Ashur, 22, 23
Asia Minor, 27, 42, 94
Aspdin, Joseph, 252-3, 255
asphalt, 250, 286
Asplund, Erik, 52
Asplund, Gunnar, 63, 139
Assyria, 21, 22, 56, 77, 88, 155, 238, 250, 280
Astruc, Jules, 271
Athens, 27-8, 32, 94, 238, 239
Atkinson, Frank, 113
Atkinson, William, 215
Atlanta, Georgia, 93

Atlas Mountains, 84
Atwood, Charles, 110
Augsburg, 201
Australia, 143, 241
Austria, 38, 39, 41, 82, 92, 159
Avebury, 19
Awazo, Kiyoshi, 53
axes, 291-2
Ayrton, Maxwell, 117
Aztecs, 25, 241

Babylon, 22, 56, 88, 155, 243, 250, 280
Bad Gastein, 92
Badami, 45
Baden Baden, 92
Badger, Daniel, 99, 102-3, 266, 269, 272
Baekelmans and Bilmeyer, 148-9
Bage, Charles, 95, 102, 265
Bagenal, Hope, 213
Baghdad, 42
Bagnaia, 57
Bahamas, 62
Bahrain, 13
Baker, Sir Benjamin, 66, 274
Baker, Sir Herbert, 46, 122
Baljeu, 50
balloon frames, 230
Baltard, Victor, 269
Baltimore, 68
Banham, Reyner, 52
banks, 114-16
Barcelona, 114, 240
Barlow, W.H., 69, 269
Barnard, Henry, 125
Baroni, Giorgio, 260
Baroque, 37-8, 40, 57, 82, 90-1, 136-7
barrows, 19
Barry, Sir Charles, 41, 203, 250, 270-1
Bartholdi, Frédéric Auguste, 278
basilicas, 30, 31, 185
Bassae, 28
Bath, 60, 91
Battista, Giovanni, 39
Baudot, Anatole de, 256-7
Bauerfeld, Walter, 260
Bauhaus, 49, 50-1, 99
Baumann and Huell, 274
Beaman, S., 69
beams, 172-3, 177
Beauvais, 34
Beccaria, Cesare, 150
Beckford, William, 41
Behmisch and Partners, 143
Behrens, Peter, 47, 98
Bélanger, Francois-Joseph, 271
Belgium, 16, 34, 49, 53, 92, 118, 160, 238, 242-3, 249, 263
Bell, Alexander Graham, 106
Bell, Andrew, 124
Bellhouse, E.T., 103, 272
bending techniques, 305
Benedictines, 32-3
Benin, 12
Bentham, Jeremy, 151
Beranek, Leo, 214
Berg, Max, 258
Berkeley, 62, 128
Berlin, 51, 58, 129, 139, 215
Bernini, Giovanni Lorenzo, 37-8
Bessemer, Henry, 272-3
Bethlehem, 30
Bibienas, 136
Billings, John S., 205
Birmingham, 128, 205-6
Birs Nimroud, 243
bitumen, 286
bituminous felt, 250
Blackburn, William, 150-1
blacksmiths, 296
Blenheim, 38, 58, 82, 90
Blore, Edward, 271
Blouet, Abel, 152
Bockum, 16
Bogardus, James, 99, 102-3, 266, 268-9, 270
Bogazkoy, 154
Bohm, Gottfried, 52

Boileau, Louis-Auguste, 113, 270
Bolivia, 24, 26
Bologna, 114, 126, 127
bolts, 309
Bonatz, Paul, 48
Bonomi, Ignatius, 96
Bontemps, Georges, 283
Borland, McIntyre, and Murphy, 143
Borromini, Francesco, 37, 38
Boston, 92, 93, 111, 123, 204, 207, 213, 237
Boucher, François, 39
Boucicaut, Aristide, 113-14
Boullée, Etienne Louis, 40
Boulton, Matthew, 95, 96, 102
Bourges, 34
Bournville, 61
Brabant, 101
Bragdon, Claude, 52
Bramah, Joseph, 115
Brasilia, 61, 122
brass, 278-9
Brazil, 59, 67, 122
Breuer, Marcel, 49, 122
Brialmont, 160
bricks, 174, 243-6
Bridgeman, Charles, 58
bridges, 64-7, 170, 180, 196, 231, 258-9, 264-8, 273-4, 282
Brinkman, Johannes Andreas, 98, 258
Bristol, 69, 115
British Columbia, 17, 19
Brittany, 63
bronze, 279-80
Bronze Age, 18-19
Brown, Lancelot "Capability," 58
Brown, Captain Samuel, 272
Brüchsal, 39, 82, 90
Brugelmann, J.G., 97
Bruges, 114
Brunel, Isambard K., 65, 66, 68, 69, 148, 162, 205, 269
Brunel, Marc Isambard, 252
Brunelleschi, Filippo, 35-6, 147, 169-70, 175
Brunet, 271
brushes, 295-6
Brussels, 113, 117, 118, 270
Buckhout, Isaac C., 69
Buddhism, 43, 44, 45-6, 57, 85
Buffalo, 62, 93, 106, 117
Bugniet, 150
Buhen, 155
building methods, 193-7
Bukhara, 42
Bunning, James Bunstone, 271
Burdon, Rowland, 265
Burgee, John, 218
Burle Marx, Roberto, 59
Burlington, Lord, 37, 82
Burnet (Sir John) and Partners, 217
Burnham, Daniel Hudson, 110, 113, 121, 274
Burnham and Root, 110, 208, 255
Burr, Theodore, 231
Burr, Thomas, 280
Bursa, 91
Burton, Decimus, 271
Bush, Lincoln, 70
Butterfield, William, 41
Byzantium, 30-2, 43, 181, 185, 193, 279

cable nets, 180, 188
Cadbury Brothers, 61
Caen, 34, 127
Cairo, 13, 34, 85
caissons, 177
Calais, 92
California, 59, 62, 63, 113, 131, 241
California, University of, 128, 134
Callet, Felix, 269
Cambridge, 34, 126-7, 128
Cameroons, 14
Campbell, Coler, 37
Canada, 20, 114, 138, 143
Canberra, 122
Candela, Felix, 260
Candilis, Josic and Woods, 129
Canterbury, 33, 34, 200

cantilevers, 187
Caracalla, 144
Carcasonne, 158
Carlo, Giancarlo de, 63
carpentry, 229-33, 290-4, 306-7
Carrier, Willis, 217, 218
Carthage, 79, 238
Cartwright, Edmund, 94
carving, 301
Caserta, 90
Cassan, Urbain, 70
Casson, Hugh, 118
casting, 299-300
castles, 33, 154-61
Catal Hüyük, 18, 154
cathedrals, *see* churches
cement, 252-3
cemeteries, 63
centering, 194-5, 196
Central America, 24-5
Cessart, L.A. de, 265
Ceylon, 13, 85
Chaldeans, 22
chalk mud, 241-2
Chambers, Sir William, 58
Chance, Robert Lucas, 283
Chandigarh, 46, 61, 122
Chartres, 34
Chatsworth, 58
Chaux, 97
Chavin, 25
Chedanne, Georges, 270
Chemnitz, 113
Chermayeff, Serge, 132
Chicago, 177, 188, 230, 255; airports, 74, 93;
    1893 Exhibition, 117; movie theaters, 139;
    multistoreyed buildings, 110, 111, 190, 191,
    274, 275; offices, 106-8, 218; railroad stations
    69, 70; services, 203; shops, 113; theaters,
    138; university, 128; warehouses, 103, 104;
    windows, 283
Chichén Itzá, 25
China, 43, 44-5, 263; building materials, 243-85
    passim; chinoiserie, 58; gardens and parks,
    56, 57, 60, 86; Great Wall, 155; houses, 20,
    85-6; palaces, 89; primitive architecture, 18,
    20
chipboard, 235
Christianity, 30
Church, Thomas, 59
churches, 185-6; Byzantine, 31-2; early
    Christian, 30; Gothic, 34-5; Gothic Revival,
    41, 265; iron used in, 271-2; mosques, 42, 43;
    Rococo, 39; Romanesque, 32, 33; stained
    glass, 280-1, 282
CIAM, 51-2, 53
cities, Greek, 28; parks, 60-2; Roman, 29
civic buildings, 120-3
cladding, 274
Clarke, Gilmore D., 62
CLASP system, 193, 197
Classicism, 39-40
Claude, George, 215
Claude Lorrain, 40, 58, 82
clay tiles, 248-50
climate, 12-14
Coade, George and Eleanor, 247
cob, 241-2, 243
Cobb, Henry Ives, 128
Cockerell, Charles Robert, 115
Cody, J.C., 129
Coehoorn, 159
Coignet, Edmond, 257
Coignet, François, 256
Colbert, J.B., 94
cold climates, 14
Cole, Henry, 116
Coliers, J.B., 270
Cologne, 34
Colombia, 24
columns, 173-4, 177; early Christian, 30;
    Egyptian, 23; Greek, 26-8, 183; Romanesque,
    32
concrete, 253-63; beams and slabs, 173; bridges,
    64, 66-7; frames, 191; prestressing, 197,
    259-60; reinforced, 255-9; shells, 187; tiles,

250; walls, 174
Constantine, Emperor, 30, 33
Constantinople, 30-2, 43, 156, 169, 180, 185, 193,
    254
Constructivism, 49-50, 51, 87
Contamin, Victor, 97, 269
Cooke and Wheatstone, 106
Copeau, Jacques, 138
copper, 250, 277-8
Le Corbusier, 46, 48, 51, 52, 61, 63, 84, 88, 122,
    129, 161, 213, 222, 257-8, 261
Cordoba, 42
Corinth, 27
Cort, Henry, 264
Cortona, Pietro da, 38
Costa, Lucio, 122
Courbertin, Pierre de, 142
courtyards, 12, 26, 77-80, 81, 84-5
Coventry, 91, 114
Cragg, John, 265, 271
Cramford, 95
cranes, 194
Crete, 26, 27, 73, 78, 89, 200
Crompton, R.E.B., 99
Cross, Hardy, 170
Crusades, 157
Crystal Palace, 47, 60, 90, 117, 161, 163, 177,
    187, 190, 197, 266, 271, 283, 284
Ctesiphon, 23, 42, 181
Cubitt, Joseph, 68
cutting, 301-2
Cuzco, 26
Cyprus, 18, 34, 77

Daestrum, 27
Dallas, 73
Damascus, 42
Dan, Dam, 53
Danly, Joseph, 272
Danube region, 17
Daoust, A., 143
Daphni, 32
Darby, Abraham, 65, 264
Darby, Abraham III, 264
Davy, Sir Humphry, 202
Dean, Thomas, 266
deformation, 167-8
Deir-el-Bahari, 24
Delhi, 46
Delorme, Philibert, 37, 232
Delos, 28, 79
Denmark, 17, 62, 63, 156, 229, 244
Desguliers, Dr, 202
design, structural, 168-71
Dessau, 50
De Stijl, 50
Didyma, 27, 28
Dihl, 251
Dirchinger, 259
Dodd, Ralph, 255
Doehring, 259
Doesburg, Theo van, 50
domes, 174-5, 178; Byzantine, 31; concrete, 254;
    Florence Cathedral, 169-70, 175; frames, 179;
    geodesic, 163, 179; iron-framed, 271; Roman,
    28, 30, 175, 184; Turkish, 43
domestic architecture, 76-88
Dominguez, L., 260
Dordogne, 15
Dorians, 26
Dornach, 50
Downing, Andrew Jackson, 210
Dowson, Philip, 133
drainage, 200, 205, 211
Dresden, 144
drills, 293-4, 297, 302
Duboy, Emile, 87
Duiker, Johannes, 140
Dulles, airport, 75
Dupin, 95
Duquesney, François, 69
Durham, 34
Dutert, Ferdinand, 97, 269
Dyckerhoff and Widmann, 66, 100, 260

dynamics, 166
Dystus, 78

Eads, James, 65, 66
Eames, Charles, 130, 277
earth, building with, 241-3
earthquakes, 192-3
Eastman, George, 138
Eckbo, Garrett, 59
Eclecticism, 47, 49
Ecuador, 24
Edinburgh, 60, 115, 237
Edirne, 43
Edison, Thomas, 107, 138, 210
educational buildings, 124-35
Eesteren, Cor van, 50
Eftu, 24
Eggert and Faust, 70
Egypt, 43
Egypt, ancient, 182-3; building materials, 229-86
    passim; building methods, 193, 194;
    fortifications, 155; gardens, 57; hospitals, 146;
    houses, 77-8, 84; palaces, 88; primitive
    architecture, 12, 17, 21, 23-4; structural
    elements, 171, 181; tombs, 63
Ehrenkrantz, Ezra, 126
Eiffel, Gustave, 110, 113, 117, 270, 274
Eindhoven, 130
Eirmann, Egon, 118
Eisenmann and Smith, 112
electricity, 99, 204, 210-11, 217, 222
elevators, 207
Ellis, Peter, 270
Elsässer, Martin, 260
Emberson, John, 139
Emdrup, 62
Emy, Colonel, 232
Ephesus, 27, 28
Epidaurus, 28, 136, 141, 146
Erickson, Arthur, 63
Erskine, Ralph, 84
Eshnunna, 245
Eskimos, 15-16, 20
Essex, 18
Etruscans, 28, 29, 79
Euboea, 78
Evans, Oliver, 98
exhibition buildings, 116-18, 163
Expressionism, 49-50
extrusion, 300-1
Eyck, Aldo van, 52

factories, 94-100
Fairbairn, Sir William, 96, 102, 170, 256, 265,
    267, 268
Faraday, Michael, 99
farms, 80
Fatehpur Sikri, 46
Feininger, Lyonel, 51
Ferdinand and Contamin, 117
Feret, 255
Ferris, Hugh, 111
Fertile Crescent, 16, 77
fiberboards, 235-6
Filarete, Antonio, 147, 148, 158
Fin de Sièclism, 48-9
Fink, A., 97
Finland, 14, 48, 70, 213, 215
Finley, James, 67, 267
Finsterlin, 50
Finsterwalder, 259
fireproofing, 192, 208, 275
Firuzabad, 23
Fisker, Kay, 63
fixings, 306, 308-13
Flachat, M., 269
Flanders, 89, 101
flats, 86-8
Fleury, C.R. de, 271
floor tiles, 250-1
floors, 175-6
Florence, 35-6, 37, 57, 81, 89-90, 96, 120, 147,
    169-70, 175, 196, 258, 279
Florida, 62
foam glass, 284
Font-de-Gaume cave, 15

Fontainebleau, 37
Fontaine, Hippolyte, 103, 269
Fontana, Carlo, 150
Forest, R., 258
forging, 305
formwork, 194-5, 261-2, 263
Fort Worth, 73
fortifications, 154-61
Foster Associates, 131, 219
foundations, 64-5, 176-7
Fourcault, Emile, 284
Fourneyron, Benoir, 97
Fournier, Charles, 87, 95
Fowke, Captain, 271
Fowler, Charles, 269
Fowler, Sir John, 66, 274
Fowler, S.T., 256
Fox, Dr, 255
frames, 189-91; portal, 187; rigid, 177; roof, 230;
  space frames, 178-9; steel, 274, 277; timber,
  229-30
France, airports, 73; apartments, 87, 88;
  Baroque, 38; bridges, 66-7; building materials,
  229-87 passim; churches, 32, 185; civic
  buildings, 120-1; exhibitions, 116, 117; farms,
  80; fortifications, 158, 159, 160; gardens and
  parks, 57-8, 60-1; Gothic architecture, 34,
  178; hospitals, 146-8, 149, 204; hotels, 92, 93;
  houses, 81-4; industry, 94, 97-8, 104-5;
  modern architecture, 48-9; movie theaters,
  139; multistorey buildings, 189; museums,
  131; palaces, 90; primitive architecture, 15;
  prisons, 150, 152; railroad stations, 69, 70;
  Renaissance architecture, 37; Rococo, 38-9;
  services, 201, 215; shops, 112-13, 269-70;
  tents, 163; theaters, 137; universities, 126,
  127; warehouses, 190
Frankfurt am Main, 70, 74, 92, 260
Franklin, Benjamin, 201
Franklin, Kump and Falk, 220
Franzen, Ulrich, 133
Frear, G.A., 255
Freeman Fox, 67
Freysinnet, Eugène, 66, 170, 187, 197, 259, 260
Friesen, Gordon, 149
Fry, Maxwell, 125
Fujii, Hiromi, 53
Fuller, R. Buckminster, 118, 163, 179
Fulton, James B., 142
Furness, Frank, 41
furniture, office, 108
Futurism, 50

Gabo, Naum, 49, 62
Galileo, 167, 168
Gallen-Kellela, 48
Gandhara, 46
gardens and parks, 56-62, 86
Gardner, George, 235
Garnier, Charles, 137, 204
Garnier, Tony, 48, 257
gas, 210-11
Gaudí, Antoni, 48, 240
Gaynor, J.P., 113
GEC, 107
Geneva, 122
Gentofte, 63
geodesic domes, 163, 179
Georgian houses, 82
Gerloo, Vanton, 50
Germany, airports, 73; Baroque, 38; bridges, 66,
  67; building materials, 229-87 passim;
  cathedrals, 32; churches, 185; farms, 80;
  fortifications, 157, 160-1; Gothic architecture,
  34, 35; Greek Revival, 41; houses, 82-4;
  industry, 94, 97, 98, 101, 104; inns and hotels,
  91-2; modern architecture, 47-9, 50-2; movie
  theaters, 139; museums, 129; offices, 108, 109,
  219; palaces, 90; parks, 60; prefabrication,
  162; primitive architecture, 29; railroad
  stations, 68, 70; Rococo, 39; schools, 124-5,
  208-9; services, 200-1, 210-11, 215; shops, 113;
  sports buildings, 143, 144; theaters, 137, 138,
  214; universities, 127
Gesellius, 48

Gestetner, 106
Ghiberti, Lorenzo, 279
Gibberd, Frederick, 61
Gibbons, Grinling, 251
Gibbs, James, 37
Gibson, John, 115
Giedion, Sigfried, 51, 110
Gilbert, Bradford, 70, 274
Gilbert, Cass, 41, 110, 247, 274
Gilbert J., 256
Gillinson and Barnett, 145
Gilly, Friedrich, 40
Giocondo, Fra, 254
Giorgio, Francesco di, 158, 159
Giulio Romano, 36
Giza, 23-4, 63, 251
Glasgow, 115, 190
glass, 280-1, 282-5, 295
Glass Chain group, 50
glass fiber, 283-4
glasshouses, 271
Godin, André, 95
Goff, Bruce, 52, 277
Goodman, Charles, 281
Gordon, Alexander, 272
Goslar, 89
Gothic architecture, 34-5, 109, 110, 169, 171-2,
  178, 181-2, 185-6, 240
Gothic Revival, 41, 82-3
Gramme, 99
Granada, 56, 89, 250
Gray, Henry, 273
Great Britain, 178-9; apartments, 87, 88; banks,
  114-16; Baroque architecture, 38; bridges, 65,
  67; building materials, 229-87 passim;
  cemeteries, 63; churches, 32, 33; civic
  buildings, 121; exhibitions, 117-19;
  fortifications, 154-61; gardens and parks,
  58-62; Gothic architecture, 34; Hadrian's
  Wall, 155; hospitals, 146-50, 205-6, 216-17;
  houses, 80-4, 210-12, 223; industry, 63, 94-7,
  101-2, 103, 105; inns and hotels, 91-2, 93;
  laboratories, 132-4; modern architecture, 47,
  48, 52; movie theaters, 139, 140; museums,
  129-31; Neo-Classicism, 40; offices, 106-8,
  217, 219; primitive architecture, 18-19, 29;
  prisons, 150-1; railroad stations, 69-70, 269;
  Romanticism, 41; schools, 124, 125-6, 208-9,
  221-2; services, 201-25 passim; shops, 112,
  114, 270; sports buildings, 142, 144-5; tents,
  163; theaters, 137, 138; universities, 63, 126-7,
  128-9; warehouses, 190
Greece, 142, 192
Greece, ancient, 91, 183; building materials,
  238-80 passim; building methods, 193; civic
  buildings, 120; fortifications, 155-6; gardens,
  57; houses, 78-9; markets, 111-12; primitive
  architecture, 26-8, 29, 77; services, 201; sports
  buildings, 141, 142, 144-5; structural
  elements, 171, 178; theaters, 136;
  warehouses, 101
Greek Revival, 41
Greene, Colonel, 102, 269
Greene, Herb, 52
Greenough, Horatio, 52
Gremer, Lothar, 214
grinding, 302
Gropius, Walter, 50, 51, 98, 99, 125, 138, 257-8,
  284
Group Zo, 53
Guarini, Guarino, 38, 90, 182, 186
Guatemala, 24, 25, 163
Guimard, Hector, 270
Gulbarga, 46
Guptas, 46
Gutbrod, Rolf, 118
Guthrie, Sir Tyrone, 138
gypsum board, 236

Hacilar, 154
Hakra, 23
Halaf culture, 18
Hall, Edwin T., 206
Halprin, Lawrence, 59, 62
Hamburg, 70, 73, 162, 200-1

Hamelin, 251
hammers, 291, 297
Handisyde, William, 271
Hanseatic League, 101
Hansen, Sven, 63
Hara, Hiroshi, 53
Harappa, 45
Hardwick, Philip, 68, 102, 231
Hargreaves, James, 94
Haring, Hugo, 52
Harlow, 61
Harriman, Alonzo J., 220
Harrington, Sir John, 201
Harrison, Wallace K., 122, 218
Harrison and Abramowitz, 274
Hartley, Jesse, 102
Harvard University, 127
Hatschek, Ludwig, 263
Hatsheput, Queen, 24
Haussmann, Baron Georges, 60, 120-1
Hawksmoor, Nicholas, 37
heating, 201-25 passim
Helsinki, 70, 73, 115
Hemming, Samuel, 272
Henman, William, 205-6
Hennébique, François, 98, 256, 258
Herculaneum, 39, 82
Hertzberger, Herman, 109
Hewet, 259
Hierakonpolis, 155
Hildebrandt, Lucas von, 90
Hinduism, 46, 85
Hitchcock, Henry-Russell, 52
Hittites, 154-5, 263
Hittorf, Jacob Ignaz, 69, 269
Hodgkinson, Eaton, 170, 265
Hoffmann, Josef, 49
Holabird and Roche, 110, 274
Holden, Charles, 70
Holl, Edward, 102, 265
Holland, 34, 50, 61-2, 70, 82, 109, 112, 130,
  148-9, 229, 243, 249
Holt, Richard, 247
Honduras, 24, 25
Hood, Raymond, 110-11
Hook, Robert, 256
Hoole, Charles, 124
Horreau, Hector, 269, 271
Horta, Victor, 49, 113, 270
Horyuji, 43
hospitals, 146-50, 204-6, 215-17
hot climates, 12-14
hotels, 69, 91-3
houses, 76-86, 161-2, 163, 209-12, 222-5
Houston, 114,144
Howard, Sir Ebenezer, 61
Howard, John, 150-1
Howe, Elias, 232
Howe, William M., 69
Hoyer, 259
Huaca del Sol, 26
Hungary, 34, 157-8
Hyatt, Thaddeus, 256

Ictinus, 28
Imhotep, 23
Incas, 25, 26, 159
Inchtuthil, 146
India, 12-14, 19, 20, 44, 45-6, 56, 146, 233, 238,
  243
Indian Ocean, 13
Indians, North American, 20
Indonesia, 20, 45
Indus Valley, 45
industry, 63, 94-105
Ingelheim, 89
inns, 91-2
Institut de Recherche et Coordination
  Acoustique/Musique, 214
International Modernism, 47, 51-2
Iraklion, 73
Iran, 20, 84, 90-1, 192; see also Persia
Iraq, 13, 16, 20, 34
iron, 263-72, 275-6; beams, 173, bridges, 64, 65,
  66; cast, 264-7, 268-72; columns, 174;

corrugated, 161-2, 250; frames, 177; roofs, 186; wrought, 264, 267-72
Iron Age, 19, 154, 242
Ise, 43
Isfahan, 42-3, 60, 181, 245, 250
Isidorus of Miletus, 31, 169
Islamabad, 46
Islamic architecture, 41-3, 56, 84-5, 171-2, 175, 181
Isozaki, 53
Israel, 161
Italy, 109; banks, 114; building materials, 238, 239-40, 247, 250, 251; churches, 32-3, 185, 186; factories, 100; fortifications, 157-9; gardens, 57; Gothic architecture, 34-5; hospitals, 147; houses, 79, 82; museums, 130; palaces, 89-90; primitive architecture, 18, 27; prisons, 150; railroad stations, 70; Renaissance architecture, 35-7; Roman architecture, 28-30; Romanticism, 41; sports buildings, 143; theaters, 136-7; universities, 63, 126-8

Jackson, P.H., 257, 259
Japan, 14, 43-5, 52-3, 56-7, 60, 86, 91, 118, 163, 248-9
Jarmo, 16
Java, 249
Jebb, Joshua, 152
Jefferson, Thomas, 62, 128
Jekyll, Gertrude, 59
Jellicoe, G.A., 62, 63
Jenney, William Le Baron, 107, 110, 274
Jericho, 16, 18, 154
Jerusalem, 30, 33, 42
Jessop, William, 254
Johnson, G.H., 99, 269
Johnson, Isaac Charles, 253, 255
Johnson, Philip, 111, 138, 218
joints, 306-8
Jones, Inigo, 37, 82, 137, 163, 231, 251
Jones, Owen, 266
Jonval, 97
Jourdain, Frantz, 270
Juvarra, Filippo, 90

Kahn, Albert, 99-100
Khan, Fazlur, 111, 275
Kahn, Louis, 46, 132, 133
Kahn and Jacobs, 218
Kallmann, McKinnell, and Knowles, 123
Kandinsky, Wassily, 51
Kannel, Theophilus van, 208
Kansas City, 74
Karlsruhe, 60
Karnak, 24, 172
Kassel, 60
Katsura, 57
Kaufman, Oscar, 139
Kelly, William, 273
Kent, William, 58
Khafagae, 77
Khajuraho, 46
Khirokitia, 18, 77
Khorsabad, 22, 77, 88, 200
Kikutake, Kiyonori, 52
Kinball and Thompson, 274
Klee, Paul, 51
Knap, Georgia, 222
Knight, Richard Payne, 58
Knights Hospitalers, 91
Knights Templars, 33
Knossos, 26, 89
Koenen, M., 257
Korea, 43
Krak des Chevaliers, 157
Krakow, 127
Krayl, 50
Kroll, Lucien, 53
Kurokawa, Kisho, 52
Kyoto, 43, 57, 249

laboratories, 132-5
Labrouste, Henri, 256, 271
Lacroix Dillon, J., 265
Laloux, Victor, 70

Lamb, Thomas, 139
Lambot, 255-6
Lamour, Jean, 264
Lancaster, Joseph, 124
Lanchester, F.W., 163
Lapidus, Morris, 93
Laplanche, M.A., 112-13
Lariboisière, 148
Lasdun, Denys, 52, 138
lathes, 303
Latrobe, Benjamin Henry, 40, 115
Laugier, Abbé, 39
Laurens, Thomas, 148
Lavoisier, A.L., 255
Laycock, William, 272
Leathart and Granger, 215
Lebon, Philippe, 96
Le Chatlier, 255
Leclaire, Joseph, 98
Ledoux, Claude Nicolas, 40, 97
Leeds, 96, 106
Lees, Frederic, 222
Le Havre, 70
Leipzig, 70
Le Marec and Limousin, 70
L'Enfant, Pierre, 60
Leningrad, 63
Le Nôtre, André, 57-8, 280
Leonardo da Vinci, 158, 167, 170
Le Roy, Jean Baptiste, 148
Letchworth, 52, 61, 83
Le Vau, Louis, 90
lighting, 107-8, 130-1, 200-21 passim
Lindgren, 48
Lipp, Friedrich, 215
Lisbon, 38
Lissitzky, Eliezer, 47, 49
Liverpool, 68, 106, 115, 190, 265, 269, 270, 271
liwans, 12, 22-3, 84-5
Llewelyn-Davies Weeks, 149
Lloyd and Morgan, 144
Lockwood and Mason, 96
Lodoli, Carlo, 39
London, apartments, 87; banks, 114, 115; bridges, 65-6; exhibitions, 47, 116-18, 163; gardens and parks, 60, 62; houses, 52, 82; inns and hotels, 91, 92; MARS plan, 52; movie theaters, 140; museums, 131; offices, 106, 107, 217; Olympic Games, 142; prisons, 150-2; railroad stations, 68-9, 187, 269; schools, 124; sewers, 201; shops, 112, 113; theaters, 137; university, 128; warehouses, 102, 190; water supply, 201, 211
long houses, 17, 20, 76, 77, 80
Loos, Adolf, 47
Los Angeles, 113, 121, 138-9
Lossow and Kuhne, 70
London, John Claudius, 59
Louis, Victor, 264
Lowell, Francis Cabot, 98
Lucerne, 130
Luckhart Brothers, 50
Lumbe, John, 94
Lusson, A.L., 271
Lutyens, Sir Edwin, 46, 122
Luxor, 183

McArthur, John Jr, 148
MacCormac and Jameson, 225
McGrath, Raymond, 222
McHarg, Ian, 62
McKenzie, Voorhees, and Gmelin, 217
McKim, Meade, and White, 70, 203, 204
Mackintosh, Charles Rennie, 49
McLeod and Ferrara, 220
Madrid, 144
Magdalenian culture, 15
Mahabalipuram, 46
El-Mahun, 77-8
Maillart, Robert, 66, 98, 103, 170, 173, 258, 260
Maki, Fumihiko, 52
Malevitch, Kasimir, 49
Malta, 34
Manchester, 108, 112, 115, 215

Mannerism, 57
Mannesman brothers, 276
Mansart, Jules Hardouin, 65, 90
Maori, 19
Marani, Rounthwaite, and Erickson, 218
Mari, 22
Marinetti, Filippo, 50
Marot, Daniel, 264
Marseilles, 61, 84, 88
Marshall Lefferts and Brother, 103
Martin, Leslie, 118
Martin, Pierre Emile, 273
Martini (Eugène) and Associates, 62
masonry, 239-40, 294, 307-8
materials, building, 229-87
Mathura, 46
Matthew, Robert, 118
Matthew (Robert), Johnson-Marshall and Partners, 63, 218
May, Ernst, 52
Maya culture, 25
measuring equipment, 194
Medinet Habu, 88
Mediterranean, 13, 18, 20, 93
Melbourne, 143
Melos, 78
Mendel, John, 58
Mendelsohn, Erich, 50, 98, 113, 139, 215, 257
Mengoni, Giuseppe, 112, 270
Merv, 23
Mesopotamia, 29, 84; building materials, 240-86 passim; fortifications, 155; primitive architecture, 12, 17, 18, 21-3, 77; structural elements, 171
Messel, Alfred, 113
Metabolism, 52-3
Mexico, 24-5, 241, 245
Mexico City, 143
Meyer, Adolf, 98, 99, 284
Meyer, Hannes, 51
Meyerhold, Vsevolod, 138
Miami, 93, 139
Michelangelo, 36, 37, 90
Michelozzo, 36
Mies van der Rohe, Ludwig, 47-51, 52, 110, 111, 128, 274, 277, 280, 284
Mijares, R., 143
Milan, 32, 70, 107, 259
Miletus, 28, 156
milling, 303
mills, 94-7, 98, 101-2, 265
Mills, Edward D., 118
Milton Keynes, 61
Minoa, 101
Miró, Joan, 122
Mistra, 32
Modern Movement, 50, 53, 88, 98, 113, 121, 125, 139, 140, 220, 222-3
Moguls, 46, 56
Mohenjo-Daro, 45
Moholy-Nagy, Laszlo, 51
Moisant, Armand, 270
molding, 299-300
Moller, C.F., 63
monasteries, 32-3, 46, 91, 94, 146
Mondrian, Piet, 50
Monier, Joseph, 256, 257
monsoon climate, 13
Monta, Mozuna, 53
Montalembert, 159
Montaner, Domenech i, 48
Montferrand, August Ricard, 271
Montreal, 114, 118, 143, 188
Montuori, E., 70
Monumentalism, 47-8
Moore, Henry, 122
Mopin, E., 276
Morandi, 67
Morewood and Company, 103
Morgan, Enslie, 221-2
Morocco, 13, 14, 93
Morris, William, 83
Morse, B.F., 269
mortar, 252
Moscow, 112, 213, 231

mosques, 42, 43, 46
motels, 93
Mott, J.L., 270
movie theaters, 138-41, 214, 215
multistorey buildings, 107, 109-11, 188-92, 274-5
Munich, 115, 137, 143, 180, 188
Murdock, William, 96
Muscat, 13
museums, 129-32
Muthesius, Hermann, 49
Mycenae, 26, 27, 101, 155, 171, 175, 182

Naiku, 43
nails, 308
Nalanda, 46
Nara, 45
Nash, John, 60, 83, 112, 251, 270
Nasmyth, James, 290, 304, 305
National Romanticism, 48
Navier, C.L.M.H., 167, 170, 268
Nelson, George, 108
Nelson, Paul, 149
Neo-Classicism, 39-40, 41, 82 .
Neoplasticism, 50
Nervi, Pier Luigi, 100, 122, 143, 187, 188, 258-9, 260
Neufchâteau, Baron de, 116
Neuhaus and Taylor, 114
Neumann, Johann Balthasar, 39, 90, 182
Neutra, Richard, 59, 220
New Brutalism, 52
New Delhi, 46, 122
New Lanark, 87, 95, 96
New Mexico, 20, 224, 225
New Orleans, 144, 161
New York, airports, 74, 178; apartments, 87;
    banks, 115, 116; civic buildings, 121, 122;
    exhibitions, 117, 118; factories, 99; hospitals,
    149, 216; hotels, 92, 93; houses, 237; movie
    theaters, 139; museums, 106-8,
    206, 207, 217, 218; parks, 61; railroad stations,
    69, 70; services, 203; shops, 113, 270;
    skyscrapers, 41, 110-11, 192, 247, 274;
    warehouses, 102-3, 190
New York Five, 53
Newcastle, 68, 84
Newcomen, Thomas, 95
Niemeyer, Óscar, 122
Nigeria, 12, 20, 67, 241
Nimrud, 22, 77
Nineveh, 22, 77
Ninos, 101
Nissen, Captain, 162
Nocera, 30
Nogachi, Isami, 122
Normandy, 283
Normans, 156-7
North America, 18, 20, 35, 58; see also United
    States of America
Norway, 229, 248
Norwich, 131
Nubia, 155
Nysa, 23

Oakland, 62, 131-2
offices, 106-9, 206-8, 217-20
Olbrich, Joseph Maria, 49
Olifant, 67
Olivetti, 108
Olmecs, 25
Olmsted, Frederick Law, 60, 61, 62, 121
Olympia, 27
Olympic Games, 141, 142-3
Olynthus, 78
Oman, 13, 21
Ommayads, 42
Ontario, 149
Ordish, R.M., 69, 269
Oregon, 62
Organicism, 52
Orkney Islands, 18
Orthodox Church, 32
Osaka, 43, 118
Osmunden and Staley, 62
Ostia, 79, 86, 101, 189, 254

Otaka, Masato, 52
Otis, Elisha, 103, 104, 107, 110, 113, 207
Ottawa, 122
Otterlo, 51
Otto, Frei, 118, 143
Owen, Robert, 87, 95, 96
Oxford, 126, 127, 128

Pacific, 20
Padua, 127
Paestum, 39
pagodas, 43-4
Paine, Thomas, 65, 264
paint, 285-6, 295-6
Paiute Indians, 15
Pakistan, 13, 14, 46
palaces, 21-3, 26-30, 39, 44-5, 77, 82, 88-91, 129
Palestine, 16, 34, 157
Palladian School, 37, 40, 82
Palladio, Andrea, 36, 37, 66, 81-2, 90, 136, 230-1,
    277
Palmer, Henry Robinson, 103, 161, 250
Palmer, Timothy, 231
Panovsky, Erwin, 35
Panthéon, 175, 184, 254, 279
Paris, airports, 73; apartments, 87; banks, 115;
    civic buildings, 120-1; exhibitions, 116, 117,
    163; factories, 98; Gothic architecture, 34;
    hospitals, 147, 148, 204; houses, 82; industry,
    103; museums, 131; Olympic Games, 142;
    palaces, 90; parks, 58, 60-1; railroad stations,
    69, 70, 269; Rococo, 38-9; shops, 112-13,
    269-70; theaters, 137; universities, 126-8;
    warehouses, 104-5, 190; water supply, 201
Parker, Barry, 83, 212
Parker, James, 252, 254-5
Parker, Obadiah, 255
Parkinson, John and Donald, 113
parks and gardens, 56-62
Parma, 136
Parthians, 22-3, 84
Pasargadae, 22
Pataliputra, 45, 46
pavilions, 163
Paxton, Sir Joseph, 47, 58, 60, 61, 112, 116-17,
    163, 271
Pei (I.M.) and Associates, 114
Peking, 20, 45, 60
Pelae, 28
Penethorne, Sir James, 270
Penizzi, Baldassare, 115
Penn, William, 92
Pennsylvania, 149
Percy, Dr John, 203
Pergamun, 28
Perpendicular style, 34
Perrault, Claude, 90, 264
Perret, Auguste, 48, 98, 257, 258
Perronet, J.R., 196
Perrot, Bernard, 283
Persepolis, 22, 88, 172
Persia, 12, 21-3, 42, 46, 56, 60, 84, 88, 91, 116,
    243-6, 250; see also Iran
Persian Gulf, 13, 21
Peru, 24, 26, 159
Peruzzi, Baldassare, 36
Pevsner, Antoine, 49
Phaestos, 26
Phantasts, 50
Philadelphia, 69, 92, 108, 113, 115, 148, 151-2,
    264, 269
Philae, 24
Piano and Rogers, 131, 275
Picasso, Pablo, 117, 122
Pick, Frank, 70
Pickett, William Vose, 266
piers, 173-4
Piraeus, 101
Piscator, Erwin, 138
pisé de terre, 241, 242, 243
Pittsburgh, 69, 191
planes, 292
plaster, 251, 295
plastics, 286-7
Plaw, 83

Plymouth, 92, 114, 148
plywood, 234-5
Poelzig, Hans, 47, 49, 50, 98, 138, 215
Poland, 32, 34, 127
Polonceau, Camille, 97, 268
Pomeranzev, 112
Pommersfelden, 82
Pompeii, 39, 57, 82, 112, 279, 282
Poncelet, 97
Ponti, Gio, 259
portable buildings, 161-3
portal frames, 177
Porter, J.H., 103, 256, 272
Portman, John, 93
Portugal, 35, 38, 46
Post-Metabolism, 53
Post-Modernism, 53
Potsdam, 58
Poussin, F.H., 152
Poussin, Nicolas, 40, 58, 82
Powell and Moya, 138, 216-17
powered tools, 296-7
Poyet, Bernard, 148
Prairie School, 52
Pratt, T.W., 232
pre-Columbian architecture, 24-6
prefabrication, 161-2, 197, 234, 262
prestressing, 197
Pretoria, 122
Price, Sir Uvedale, 58
Priene, 28, 78-9, 156
primitive architecture, 14-20, 76-81
prisons, 150-3
Pritchard, Thomas F., 264
Probst, Robert, 108
Procopius, 31-2
Prouvé, Jean, 277
Providence, Rhode Island, 69
Ptolemies, 24
Pueblo culture, 18, 20
Pugin, Augustus Welby, 41, 47
Putoli, 253
Pylos, 26
pyramids, 23-4, 25, 26, 183, 238

Quickborner Team, 108

Raben, John G., 113
Radicalism, 49
railroad stations, 68-70, 92, 186-7, 269
Ramee, John Jacques, 127-8
Ransome, Ernest L., 99, 103, 255, 257
Raphael, 36, 90
Ratingen, 97
Ravenna, 30, 31, 158
Red Sea, 13
Redden, John S., 113
Reed, Jesse, 230
Reed and Stem, 70
Reeves, David, 269
Reid, John Lyon, 126, 203, 221
Reims, 34
Reinhardt and Sossenguth, 70
Renaissance, 35-7, 57, 81-2, 89-90, 110, 129,
    136-7, 175, 240, 246, 279-80
Renard, Bruno, 97
Renkioi, 148, 162, 205
Rennie, John, 65, 102
Repton, Humphrey, 58, 60, 112
Revett, Nicholas, 40
Reyna, Jorge Gonzalez, 260
Reynaud, Leonce, 69, 269
Rheims, 70
Rhineland, 17, 33
Richardson, Henry Hobson, 41, 47, 69-70
Richelieu, Cardinal, 57
Rickman, Thomas, 265, 271
Rietveld, Gerrit, 50
Ripley, Thomas, 247
riveting, 309-12
Riyadh, 143
Robbia, Luca della, 147
Roberts, Henry, 87
Roberts, John, 143
Robertson, Leslie, 111

Robinson, William, 59
Robson, E.R., 125, 208-9
Roche, Kevin, 131-2
Rococo, 38-9, 40, 82, 137
Roebling, John A., 65, 67, 268
Rogers, Isaiah, 92
rolling, 304-5
Roman Catholic Church, 32
Romand, A., 272
Romanesque, 32-5, 41, 181
Romanticism, 40-1, 82-3
Rome, 37, 70, 90, 143, 147, 150
Rome, ancient, 23-32, 184-6, 188-9; apartments, 86; bridges, 64, 65; building materials, 230-85 *passim;* building methods, 193, 195; cemeteries, 63; civic buildings, 120; fortifications, 155; gardens, 57; hospitals, 146; houses, 79-80; industry, 94; influence of, 35, 36, 40, 41, 82; inns, 91; markets, 112; palaces, 89; services, 200, 201; sports buildings, 141-2, 144-5; structural elements, 171, 174-6, 178, 181; theaters, 136; warehouses, 101
roof gardens, 56, 62
roofs, 186-8, 230-2, 247-50, 268, 269-70, 279
Root, John Wellborn, 274
Rosen, Recamier, Gutierrez, and Valderde, 143
Ross, A.W., 113
Rotterdam, 49
Rovehead, 148
Rumford, Count, 201
Ruskin, John, 41, 47, 266
Russia, 15, 17, 41, 49, 63, 87, 88, 112, 161, 262

Saarinen, Eero, 63, 132, 133, 260
Saarinen, Eliel, 48, 70
Saarinen/Ammann and Whitney, 178
Sabine, Wallace Clement, 204
Sahara, 18
St Leonards, 93
St Louis, 70, 93, 103, 117, 142, 196
Salford, 95
Salonica, 32
Salt, Titus, 95, 96
Saltair, 96, 268
Samarkand, 250
Samarra, 34
Samos, 27
San Antonio, 108
San Diego, 149
San Francisco, 92, 103, 117, 255, 257
San Jacinto, 59
Sanchi, 45
Sangallo, Antonio de, the Elder, 158
Sangallo, Giuliano da, 90
Sanmicheli, Michele, 158
Sansovino, Jacopo, 127
Sant'Elia, Antonio, 50
Saqqara, 23, 172, 183, 251
Sasaki, Dawson, Demay Associates, 62
Sassanians, 21, 23, 42, 84, 171-2, 181
Saulnier, Jules, 97, 269
savanna climate, 13-14
Savot, Louis, 201
saws, 293, 301-2
Saylor, David O., 255
scaffolding, 194-5
Scamozzi, Vincenzo, 36
Scandinavia, 17, 32, 63, 129, 229, 230, 249
Scarpa, Carlo, 130
Scharoun, Hans, 50, 52, 214
Scheerbart, Paul, 50
Schinkel, Karl Friedrich, 47, 95, 102, 125, 129
Schlemmer, Oskar, 51
Schmidt (R.E.), Garden, and Martin, 103
Schoffler, Schlontach, and Jacobi, 215
Scholasticism, 35
Schönbrunn, 90
schools, 124-6, 208-9, 220-2
Scotland, 34, 48, 63, 87, 115, 127, 158, 237, 245
Scott, Sir George Gilbert, 47, 69
screws, 309
sealing, 295, 312-13
Sears, 104
Seattle, 73, 74
Seguin brothers, 268

Sehring and Lachmann, 113
Selinus, 27
Semites, 21, 22
Semper, Gottfried, 47, 203
Serlio, Sebastiano, 36
services, 200-25
Settignano, 57
Seville, 94, 250
sewers, 200-1, 211
Shahjahanabad, 46
Shaw, Richard Norman, 47, 48, 83, 207, 210
Sheffield, 88
shell structures, 177-8, 187, 260
shingles, 248
Shook, John B., 69
shops, 111-14, 269-70
Shreve, Lamb, and Harmon, 111, 274
Shrewsbury, 95, 102
Shushan, 116
Shute, John, 37
Siberia, 15
Sicily, 27, 29, 79, 89
Siemens, Werner von, 99, 273
Silchester, 91
Simpson, James, 211
Sinan, Koca, 43
Skansen, 130
Skidmore, Owings, and Merrill, 108, 111, 116, 216, 218, 274
skyscrapers, 107, 109-11, 191-2
slabs, 172-3
slate, 248
Slovenia, 154
Smeaton, John, 252, 254, 264
Smirke, Sir Robert, 265
Smirke, Sydney, 270, 271
Smith, T.R., 209
Smithson, Peter and Alison, 52, 88
Snow, John, 211
Soane, Sir John, 40, 115, 210
solar heating, 224-5
Soleri, Paolo, 52, 53
Sompting, 33
Sonck, Lars, 48
Sørensen, C. Th., 62, 63
Soufflot, Jacques Germain, 186, 264
South Africa, 67
South America, 13, 24, 25-6
space frames, 178-9
Spain, 29, 32, 35; Baroque, 38; building materials, 245, 247, 248, 250; churches, 185; fortifications, 158-9; gardens, 56; Gothic architecture, 34; hospitals, 147; houses, 85; inns, 91-2; modern architecture, 48; Moorish architecture, 42; palaces, 90; sports buildings, 142, 144; universities, 127
Speer, Albert, 48
Split, 30, 89
sports buildings, 141-5
Sri Lanka, 13, 85
stabilized earth, 243
Stacchin, Ulisse, 70
stained glass, 280-1, 282
Stam, Mart, 49
stamping techniques, 305
Starrett and Van Vleck, 111
statics, 166
Statler, Ellsworth M., 93
Steegman, P., 63
steel, 64, 66, 67, 174, 272-7
Steelcase, 108
Stein, Clarence, 61
Steiner, Rudolf, 50
Stephenson, Robert, 66, 68, 170, 267, 268
Stevens, R.L., 267
Stevens and Hunt, 113
Stevenson, J.J., 210
Stiris, 32
Stirling, James, 50, 52
Stirling University, 63
Stoa, 28
Stockholm, 62, 63, 130, 139
stone, 236-41, 248
Stone Age, 15-18, 77
Stourhead, 58

Strafor, 108
Street, George Edmund, 41
strength, 167-8
Strickland, William, 115, 270
structural elements, 171-82
structural systems, 182-93
structural theory, 166-71
Structuralism, 47
Strutt, William, 95, 201, 265
Stuart, James, 40
Stubbins (Hugh) and Associates, 218
stucco, 252
Studio BBPR, 130
Stupinigi, 90
Stussi, 259
Stuttgart, 48, 113
Sudan, 20
Sullivan, Louis H., 47, 107, 113, 247, 274
Sumer, 56
Sung Shan, 44
surveying instruments, 297-8
Susa, 22
suspension elements, 180
Swan, 107, 210
Sweden, 62, 104, 130, 139, 149, 229
Switzerland, 18-19, 34, 50, 51, 66, 92, 98, 103, 104, 122, 149, 231, 246
Sydney, 138, 178
Syracuse, 27, 156
Syria, 18, 21, 34, 42, 43, 84

Tabernacle, 162
Tacoma, 73, 74
Tatlin, Vladimir, 49
Taillibert, R., 143
Taliesin West Associates, 90-1
Tally, Thomas L., 138-9
Tampa, 74
Tange, Kenzo, 53
Taut, Bruno, 50, 283
Taylor, Augustine Deodat, 230
Team X, 52
techniques, 298-305
Tedesco, N. de, 257
Tefft, Thomas, 69
Tegea, 28
Telford, Thomas, 65, 67, 98, 170, 252, 255, 267-8
Tell Agrab, 77
Tell Brak, 21
tells, 21
temples, 21-8, 43-6, 88
temporary buildings, 161-2
tensile membranes, 180
tents, 162-3, 180
Teotihuacán, 25
terra-cotta, 246-7
thatching, 247-8
theaters, 28, 136-41, 203-4, 213-14, 270
Thebes, 24, 88, 181, 182-3, 245
Tiahuanaco, 26
Tiepolo, Giovanni Battista, 39
Tierra del Fuego, 15
Tijou, Jean, 264
Tikal, 25
tiles, 248-51
timber, 18-19, 66, 174, 229-36, 290-4, 306-7
Tiryns, 26, 155
Tite, Sir William, 115
tithe barns, 101
Tivoli, 57, 89, 101, 184
Tokyo, 52-3, 249
Toledo, 147
Toltecs, 25, 241
tombs, 23-4, 26, 27, 63
tools, 194, 290-8
Torcello, 32
Toronto, 74
Torrigiano, Pietro, 37
Torroja, Eduardo, 142, 170, 260
Toscanella, 32
Touraine, 57
towers, 183, 188-9; *see also* multistorey buildings
Town, Ithiel, 69, 232
trade fairs, 116

Tredgold, Thomas, 202, 265
Trelleborg, 17, 80, 156
Trinidad, 140
Trombe, F., 224-5
tropical climate, 12-14
Trucco, Matté, 98, 258
trusses, 66, 170, 178-9, 184-5, 230-2, 268
Tubbs, Ralph, 118
Tunisia, 93
Turin, 38, 82, 90, 98, 186, 259
Turkey, 13, 42-3, 85, 91, 192, 246, 268
Turner, C.A.P., 98, 103, 173, 258
Turner, Richard, 269, 271
turning, 303-4
Tyler, 115

Ukraine, 15
Ulm, 34
United States, airports, 71, 73, 74; apartments,
    87; banks, 115-16; bridges, 65-7, 231; building
    materials, 229-87 passim; circuses, 162; civic
    buildings, 120-2; exhibitions, 117, 118;
    fortifications, 159, 161; gardens and parks,
    59-62; hospitals, 148, 149, 205, 215-16; hotels,
    92-3; houses, 83, 84, 210-12, 223; industry, 63,
    98-100, 102-5; laboratories, 132-4; modern
    architecture, 52; movie theaters, 138-9, 140;
    museums, 129, 130; offices, 106-9, 206, 207,
    217-19; prefabricated buildings, 161, 162;
    prisons, 151-2; railroad stations, 68, 69-70,
    269; Romanticism, 41; schools, 125, 126, 208,
    209, 220-1; services, 201-15 passim; shops,
    112, 113-14, 270; sports buildings, 144;
    structural elements, 178; theaters, 137-8;
    universities, 62-3, 127-8
universities, 62-3, 126-9
Unwin, Sir Raymond, 83, 212
Ur, 21
Urbino, 63
Ure, Dr Andrew, 96
Uruk, 21
Utilitarianism, 52
Utrecht, 50
Utzon, Jørn, 138, 178, 260
Uxmal, 25

Valetta, 147
Van de Telde, Henry, 49
Van der Vlugt, L.C., 98, 258
Vanbrugh, Sir John, 37
Vancouver, 63
Vasquez, P.R., 143

Vauban, 159
vaults, 178, 180-2, 187, 189, 196; brick, 245;
    Gothic, 34, 185-6; Islamic, 42-3; Roman, 184
Vaux, Calvert, 121
Venice, 36, 89, 119, 154, 159
La Venta, 25
ventilation systems, 202-20 passim
Venturis, 53
verandas, 12
Versailles, 38, 57-8, 60, 83, 90, 120, 201, 280
Vicat, Louis-Joseph, 67, 255, 268
Vicenza, 36, 136
Vienna, 90, 115, 127, 128, 203
Vierzehnheiligne, 39
Vietnam, 161
Vignola, Giacomo Barozzi da, 36, 57
villas, 80, 81-2
Viollet-le-Duc, Eugène-Emanuel, 47, 256, 270
Virginia, 58
Virginia University, 62, 128, 245
Vitellozi, A., 143
Vitruvius Pollo, Marcus, 36, 37, 78, 136, 254
Vittone, Bernardo, 182
Voysey, Charles, 48

Wachsmann, Konrad, 100
Wagner, Otto, 115
Wagner, Richard, 137
Wales, 66, 157
Walker, Richard, 103, 161-2, 272
wall tiles, 250-1
walls, 173-4, 188-9
Walpole, Horace, 41, 82
Walter, Thomas U., 271
Wanamaker, John, 113
Waraka, 21
Ward, William E., 256
Ware, Samuel, 112
warehouses, 100-5
Washington, 60, 74, 130, 217-18, 238
Wasserburg Buchau, 19
waste disposal, 223-4
Watanabe, 53
water supply, 201, 208, 211
Waterhouse, Alfred, 47, 129, 247
watermills, 94-5, 101-2
Watt, James, 95, 96, 102, 201-2, 264, 265
wattle and daub, 242-3
Wayss, G.A., 257
Weeks, John, 134
welding, 275-6, 309-12
Welsbach, Carl von, 210
Welwyn Garden City, 61
West Indies, 13

Westinghouse, 107, 108
Whipple, 232
White, Canvas, 252, 255
White and Hazard, 268
Whitney, 259
Wilderspin, Samuel, 124
Wilkins, William, 127, 128
Wilkinson, John, 97, 264, 280
Wilkinson, W.B., 255
Wilkinson, William, 97
Williams, Charles, 256
Williams, Owen, 98, 117
Willis, George, 217
Wils, J., 143
Wilson (Josiah M.) and Brothers, 269
Wiltshire, 19
Winchester, 34, 124
Winckelmann, Johann Joachim, 39
Womersley, J.L., 88
wood, see timber
Wood, John, 60
Woodbury, 19
Woodhenge, 19
Woodward, Benjamin, 266
World War I, 160, 263
World War II, 52, 104, 114, 160-1, 162
Wotton, Sir Henry, 37
Wren, Sir Christopher, 37, 128, 170, 202, 238,
    244, 251, 264
Wright, Frank Lloyd, 52, 53, 59, 84, 106, 108,
    110, 129, 192, 207, 211-12, 223, 262, 278
Wright, Henry, 61
Würzburg, 39, 82, 90
Wyatt, Job and William, 230
Wyatt, M.D., 269
Wyatt, Samuel, 95

Yale, Linus, 115
Yemen, 84
Yorke, Rosenberg, and Mardall, 216, 219
Yorkshire, 19
Young, Charles D., 103, 266, 272
Ypres, 89
Yucatan, 25
Yugoslavia, 30

Zakharov, Adrian Dimitrievitch, 40
ziggurats, 21, 183, 243
zinc, 279
Zincirli, 155
Zoroastrianism, 23
Zores, C.F., 267
Zoser, King, 23
Zurich, 73, 115

Photo credits

Aerofilms Ltd, Air France; Amsterdam Parks
Department; Architectural Association;
Architectural Press; Arup Associates;
Automatic Building Components Ltd;
Automatic Pressings Ltd; A.J. Berman; British
Gypsum Ltd; British Industrial Plastics Ltd;
British Petroleum Company Ltd; British Steel
Corporation; Building Systems Development
(UK) Ltd; Cementation Chemicals Ltd; T.
Church; G. Coddington; James Collins
(Birmingham) Ltd; Colorific!; Crew & Sons
Ltd; I. Davis; J. Donat; J.V.P. Drury; R.
Evans; Finnish Plywood Development
Association; GKN Screws & Fasteners Ltd; P.
Guedes; Dr. D. Hawkes; K. Hudson; S.
Jellicoe; Lanarkshire Bolt Ltd; Dr. R.
Lewcock; London Brick Company Ltd;
London Building Centre; Lord Chamberlain's
Office; Dr. R.J. Mainstone; Marley Tile

Company Ltd; E. Martini & Associates; B.
Marx; Myford Ltd; Nylons & Alloys Ltd;
Piano & Rogers; Pilkington Brothers Ltd; Port ·
Authority of NY & NJ; Press-Bat Holdings
Ltd; Ransomes & Rapier Ltd; O.J. Roald; T.
Ronalds; Royal Academy of Arts; Ruberoid
Building Products Ltd; Sandell Perkins Ltd;
Science Museum, London; D. Sharp; Smithsonian
Institute; P. Stone; J. Tarlton; Thorsman &
Company (UK) Ltd; Timber Research and
Development Association; Tower
Manufacturing Ltd; Triborough Bridge and
Tunnel Authority; USIS, London; Dr. T.I.
Williams; Winget Ltd; Wolf Electric Tools Ltd.

The editors of this book have made every
attempt to give picture credit where it is due. If
any credits owed to copyright holders of
pictures used in this book have been omitted
we invite such copyright holders to contact us.